Lecture Notes in Artificial Intelligence 12854

Subseries of Lecture Notes in Computer Science

More information about this subseries at http://www.springer.com/series/1244

Leszek Rutkowski · Rafał Scherer ·
Marcin Korytkowski · Witold Pedrycz ·
Ryszard Tadeusiewicz ·
Jacek M. Zurada (Eds.)

Artificial Intelligence and Soft Computing

20th International Conference, ICAISC 2021 Virtual Event, June 21–23, 2021 Proceedings, Part I

Editors
Leszek Rutkowski
Czestochowa University of Technology
Częstochowa, Poland

Marcin Korytkowski
Częstochowa University of Technology
Częstochowa, Poland

Ryszard Tadeusiewicz
AGH University of Science and Technology
Krakow, Poland

Rafał Scherer
Częstochowa University of Technology
Częstochowa, Poland

Witold Pedrycz
Edmonton, AB, Canada

Jacek M. Zurada
Electrical and Computer Engineering
University of Louisville
Louisville, KY, USA

ISSN 0302-9743 ISSN 1611-3349 (electronic)
Lecture Notes in Artificial Intelligence
ISBN 978-3-030-87985-3 ISBN 978-3-030-87986-0 (eBook)
https://doi.org/10.1007/978-3-030-87986-0

LNCS Sublibrary: SL7 – Artificial Intelligence

This Springer imprint is published by the registered company Springer Nature Switzerland AG
The registered company address is: Gewerbestrasse 11, 6330 Cham, Switzerland

Preface

This volume constitutes the proceedings of 20th International Conference on Artificial Intelligence and Soft Computing (ICAISC 2021) held in Zakopane, Poland, during June 21–23, 2021, which took place virtually due to the COVID-19 pandemic. The conference was organized by the Polish Neural Network Society in cooperation with the University of Social Sciences in Łódż, the Department of Intelligent Computer Systems at the Częstochowa University of Technology, and the IEEE Computational Intelligence Society, Poland Chapter. Previous conferences took place in Kule (1994), Szczyrk (1996), Kule (1997), and Zakopane (1999, 2000, 2002, 2004, 2006, 2008, 2010, and 2012–2020) and attracted a large number of papers and internationally recognized speakers: Lotfi A. Zadeh, Hojjat Adeli, Rafal Angryk, Igor Aizenberg, Cesare Alippi, Shun-ichi Amari, Daniel Amit, Plamen Angelov, Albert Bifet, Piero P. Bonissone, Jim Bezdek, Zdzisław Bubnicki, Jan Chorowski, Andrzej Cichocki, Swagatam Das, Ewa Dudek-Dyduch, Włodzisław Duch, Adel S. Elmaghraby, Pablo A. Estévez, João Gama, Erol Gelenbe, Jerzy Grzymala-Busse, Martin Hagan, Yoichi Hayashi, Akira Hirose, Kaoru Hirota, Adrian Horzyk, Eyke Hüllermeier, Hisao Ishibuchi, Er Meng Joo, Janusz Kacprzyk, Jim Keller, Laszlo T. Koczy, Tomasz Kopacz, Jacek Koronacki, Zdzislaw Kowalczuk, Adam Krzyzak, Rudolf Kruse, James Tin-Yau Kwok, Soo-Young Lee, Derong Liu, Robert Marks, Ujjwal Maulik, Zbigniew Michalewicz, Evangelia Micheli-Tzanakou, Kaisa Miettinen, Krystian Mikołajczyk, Henning Müller, Ngoc Thanh Nguyen, Andrzej Obuchowicz, Erkki Oja, Nikhil R. Pal, Witold Pedrycz, Marios M. Polycarpou, José C. Príncipe, Jagath C. Rajapakse, Šarunas Raudys, Enrique Ruspini, Jörg Siekmann, Andrzej Skowron, Roman Słowiński, Igor Spiridonov, Boris Stilman, Ponnuthurai Nagaratnam Suganthan, Ryszard Tadeusiewicz, Ah-Hwee Tan, Dacheng Tao, Shiro Usui, Thomas Villmann, Fei-Yue Wang, Jun Wang, Bogdan M. Wilamowski, Ronald Y. Yager, Xin Yao, Syozo Yasui, Gary Yen, Ivan Zelinka, and Jacek Zurada. The aim of this conference is to build a bridge between traditional artificial intelligence techniques and so-called soft computing techniques. It was pointed out by Lotfi A. Zadeh that "soft computing (SC) is a coalition of methodologies which are oriented toward the conception and design of information/intelligent systems. The principal members of the coalition are: fuzzy logic (FL), neurocomputing (NC), evolutionary computing (EC), probabilistic computing (PC), chaotic computing (CC), and machine learning (ML). The constituent methodologies of SC are, for the most part, complementary and synergistic rather than competitive". These proceedings present both traditional artificial intelligence methods and soft computing techniques. Our goal is to bring together scientists representing both areas of research. This volume is divided into four parts:

- Neural Networks and Their Applications
- Fuzzy Systems and Their Applications
- Evolutionary Algorithms and Their Applications
- Artificial Intelligence in Modeling and Simulation

I would like to thank our participants, invited speakers, and reviewers of the papers for their scientific and personal contribution to the conference. Finally, I thank my co-workers, Łukasz Bartczuk, Piotr Dziwiński, Marcin Gabryel, Rafał, Grycuk, Marcin Korytkowski, and Rafał, Scherer, for their enormous efforts to make the conference a very successful event. Moreover, I appreciate the work of Marcin Korytkowski who was responsible for the Internet submission system.

June 2021 Leszek Rutkowski

Organization

ICAISC 2021 was organized by the Polish Neural Network Society in cooperation with the University of Social Sciences in Łódź and the Institute of Computational Intelligence at Częstochowa University of Technology.

Conference Chairs

General Chair

Leszek Rutkowski, Poland

Area Chairs

Fuzzy Systems

Witold Pedrycz, Canada

Evolutionary Algorithms

Zbigniew Michalewicz, Australia

Neural Networks

Jinde Cao, China

Computer Vision

Dacheng Tao, Australia

Machine Learning

Nikhil R. Pal, India

Artificial Intelligence with Applications

Janusz Kacprzyk, Poland

International Liaison

Jacek Żurada, USA

Program Committee

Hojjat Adeli, USA
Cesare Alippi, Italy
Shun-ichi Amari, Japan
Rafal A. Angryk, USA
Robert Babuska, The Netherlands
James C. Bezdek, Australia
Piero P. Bonissone, USA
Bernadette Bouchon-Meunier, France
Jinde Cao, China
Juan Luis Castro, Spain
Yen-Wei Chen, Japan
Andrzej Cichocki, Japan
Krzysztof Cios, USA
Ian Cloete, Germany
Oscar Cordón, Spain
Bernard De Baets, Belgium
Włodzisław Duch, Poland
Meng Joo Er, Singapore
Pablo Estevez, Chile
David B. Fogel, USA
Tom Gedeon, Australia
Erol Gelenbe, UK
Jerzy W. Grzymala-Busse, USA
Hani Hagras, UK
Saman Halgamuge, Australia
Yoichi Hayashi, Japan
Tim Hendtlass, Australia
Francisco Herrera, Spain
Kaoru Hirota, Japan
Tingwen Huang, USA
Hisao Ishibuchi, Japan
Mo Jamshidi, USA
Robert John, UK
Janusz Kacprzyk, Poland
Nikola Kasabov, New Zealand
Okyay Kaynak, Turkey
Vojislav Kecman, USA
James M. Keller, USA
Etienne Kerre, Belgium
Frank Klawonn, Germany
Robert Kozma, USA
László Kóczy, Hungary
Józef Korbicz, Poland
Rudolf Kruse, Germany
Adam Krzyzak, Canada
Věra Kůrková, Czech Republic
Soo-Young Lee, South Korea
Simon M. Lucas, UK
Luis Magdalena, Spain
Jerry M. Mendel, USA
Radko Mesiar, Slovakia
Zbigniew Michalewicz, Australia
Javier Montero, Spain
Eduard Montseny, Spain
Kazumi Nakamatsu, Japan
Detlef D. Nauck, Germany
Ngoc Thanh Nguyen, Poland
Erkki Oja, Finland
Nikhil R. Pal, India
Witold Pedrycz, Canada
Leonid Perlovsky, USA
Marios M. Polycarpou, Cyprus
Danil Prokhorov, USA
Vincenzo Piuri, Italy
Sarunas Raudys, Lithuania
Olga Rebrova, Russia
Vladimir Red'ko, Russia
Raúl Rojas, Germany
Imre J. Rudas, Hungary
Norihide Sano, Japan
Rudy Setiono, Singapore
Jennie Si, USA
Peter Sincak, Slovakia
Andrzej Skowron, Poland
Roman Słowiński, Poland
Pilar Sobrevilla, Spain
Janusz Starzyk, USA
Jerzy Stefanowski, Poland
Vitomir Štruc, Slovenia
Ron Sun, USA
Johan Suykens, Belgium
Ryszard Tadeusiewicz, Poland
Hideyuki Takagi, Japan
Dacheng Tao, Australia
Vicenç Torra, Spain
Burhan Turksen, Canada

Shiro Usui, Japan
Deliang Wang, USA
Jun Wang, Hong Kong
Lipo Wang, Singapore
Paul Werbos, USA
Bernard Widrow, USA
Kay C. Wiese, Canada
Bogdan M. Wilamowski, USA
Donald C. Wunsch, USA
Ronald R. Yager, USA
Xin-She Yang, UK
Gary Yen, USA
Sławomir Zadrożny, Poland
Jacek Zurada, USA

Organizing Committee

Rafał Scherer
Łukasz Bartczuk
Piotr Dziwiłski
Marcin Gabryel (Finance Chair)
Rafał Grycuk
Marcin Korytkowski (Databases and Internet Submissions)

Contents – Part I

Neural Networks and Their Applications

Contents – Part II

Data Mining

Various Problems of Ariticial Intelligence

Bioinformatics, Biometrics and Medical Applications

Neural Networks and Their Applications

Financial Portfolio Construction for Quantitative Trading Using Deep Learning Technique

Rasha Abdel Kawy[1(✉)], Walid M. Abdelmoez[2(✉)], and Amin Shoukry[3,4(✉)]

[1] Computer Science Department, College of Computing and Information Technology, Arab Academy for Science Technology and Maritime transport, Alexandria 1029, Egypt
Rasha.Shokry@student.aast.edu

[2] Software Engineering Department, College of Computing and Information Technology, Arab Academy for Science Technology and Maritime transport, Alexandria 1029, Egypt
walid.abdelmoez@aast.edu

[3] Computer Science and Engineering Department, Egypt-Japan University of Science and Technology, Alexandria 21934, Egypt
amin.shoukry@ejust.edu.eg

[4] Computer and Systems Engineering Department, Faculty of Engineering, Alexandria University, Alexandria 21544, Egypt

Abstract. Stock portfolio construction is a difficult task which involves the simultaneous consideration of dynamic financial data as well as investment criteria (e.g.: investors required return, risk tolerance, goals, and time frame). The objective of this research is to present a two phase deep learning module to csonstruct a financial stocks portfolio that can be used repeatedly to select the most promising stocks and adjust stocks allocations (namely quantitative trading system). A deep belief network is used to discover the complex regularities among the stocks while a long short-term memory network is used for time series financial data prediction. The proposed deep learning architecture has been tested on the american stock market and has outperformed other known machine learning techniques (support vector machine and random forests) in several prediction accuracy metrices. Furthermore, the results showed that our architecture as a portfolio construction model outperforms three benchmark models with several financial profitability and risk-adjusted metrics.

Keywords: Portfolio construction · Quantitative trading system · Deep learning · DBN · LSTM

1 Introduction

The Portfolio is a collection of financial investment assets such as stocks, bonds, commodities, etc. Portfolio management is the science of selecting a suitable

L. Rutkowski et al. (Eds.): ICAISC 2021, LNAI 12854, pp. 3–14, 2021.
https://doi.org/10.1007/978-3-030-87986-0_1

group of assets with the right proportions to meet the investors' strategic objectives. Managing the portfolio includes repeated evaluation of the portfolio value, making the necessary trading processes to generate a balance between the required return and the acceptable level of risk. Quantitative trading is defined as using mathematical and statistical methods to identify and build an automatic rule-based model to trade assets in the financial markets. Quantitative trading has the advantage of taking out emotions, rumors and fraud from the trading decisions which allow investors to invest with confidence and clarity. Machine learning techniques were widely used in quantitative trading. In [1], integrating support vector machines (SVM) with other classification methods has been proposed to forecast the weekly movement direction of NIKKEI 225 index. In [2], a forecasting model based on chaotic mapping firefly algorithm and SVR is proposed to predict stock market price, while [3] proposed a risk-adjusted profitable trading rule based on technical analysis and pattern recognition techniques. Moreover, [4,5] presented a comprehensive literature review on the application of evolutionary computation (EC), including genetic algorithms, genetic programing and multi-objective evolutionary algorithms, in stock trading and other financial applications, while [6] suggested the formation of a recursive clustering technique. Once the assets are hierarchically clustered, a risk- adjusted capital allocation is applied. In [7], the author explores the use of Gaussian processes and Bayesian optimization in modeling "the structure of interest rates", and building the trend-following optimization strategies. In [8], a comprehensive survey of particle swarm optimization (PSO) algorithm, in studying market behaviors is given, as well as, the potential future research directions for enhancing PSO-based stock market prediction.

The objective of this research is to present a deep learning quantitative trading architecture that can be used for stocks selection and stocks allocation to achieve maximum return with an accepted risk level. The proposed architecture has two modules. The first selects the efficient stocks for investment using deep learning while the second decides the budget to be invested in each selected stock. Our contributions can be summarized as:

1. Proposing a novel automatic quantitative trading model that relies on a Deep Belief Network (DBN) to extract the financial data regularities and reducing its dimensionality while using an LSTM (Long Short Term Memory Neural Network) to learn/model the dependency in the input financial data time series.
2. Conducting intensive experiments to evaluate and compare the performance, of the proposed deep learning architecture, to other conventional machine learning techniques in stocks' price prediction based on several prediction accuracy metrics.
3. Capturing the performance of the proposed architecture as a quantitative trading module and comparing it to other markets' strategies benchmarks using several risk-adjusted profitability metrics.

The remaining of this paper is organized as follows: Sect. 2 defines what is meant by a quantitative trading system and provides technical background on the deep learning networks used in this paper, in addition to presenting a review about the related work (using deep belief networks and long short-term memory networks (LSTM) in predicting stocks prices). Section 3 presents the proposed deep learning architecture. Section 4 describes the experiments performed on the proposed model and its performance evaluation. Finally, conclusions are summarized in Sect. 5.

2 Background and Literature Review

2.1 Quantitative Trading System

The quantitative trading system is based on forecasting the market movement. First, the proper mix of assets with the right proportions are chosen (portfolio construction) then - based on the market movement - trading rules are built that aim to achieve the maximum return given the risk tolerance set by the investor. A periodic evaluation is performed to determine whether the obtained performance is satisfactory or not. Necessary adjustments are made in the stocks to preserve the investment objectives.

2.2 Deep Learning Networks

2.2.1 Deep Belief Networks

Deep Belief Networks (DBN) [9] are probabilistic generative neural networks composed of multiple layers of restricted boltzmann machines (RBM) to capture higher-level representations of input features with the advantage of avoiding getting stuck at local optima.

2.2.2 Long Short-Term Memory (LSTM) Networks

Long Short-Term Memory (LSTM) [10] networks are special type of recurrent neural networks (RNN) capable of learning long-term dependence in time series data with the advantage of avoiding vanishing gradient problems existing in the training of RNN.

2.3 Related Work (LSTM and DBN in Financial Forecasting)

Numerous studies have shown that long short-term memory neural networks are very effective networks in the forecasting of financial times series data. Research in [11] introduced multivariate denoising wavelet transforms (WT), in order to eliminate the noise in the time-series data, then combined stacked autoencoders (SAEs) and long-short term memory (LSTM) networks to forecast six market indices. Fischer et al. [12], deployed LSTM networks for predicting out-of-sample directional movements for the S&P 500 stocks in the period from 1992 to 2009. Research in [13] proposed a long short-term memory (LSTM) network to predict stock movement in order to construct multiple portfolio optimization techniques using equal-weighted modeling (EQ), simulation modeling Monte Carlo simulation (MCS), and mean variant optimization (MVO). The work in [14] presented a DBN model with strong ability to generate high level features representation for accurate financial prediction that has been tested on a real dataset of French companies. In [15] a DBN has been used to forecast the currencies exchange rates. Conjugate gradient method was applied to accelerate the learning for DBN. The results showed that DBN outperforms other traditional methods. While in [16] an RBM is combined with SVM to detect trends in the Brazilian Stock Market prices. The obtained results were better compared to those obtained by SVMs only. Rasha et al. [17] proposed a novel multi-stock end-to-end trading model based on DBN and multi-agent deep reinforcement learning. Its efficiency, compared to existing techniques has been verified on datasets of different characteristics obtained from the American stock market.

3 Proposed Deep Learning Model Architecture

The proposed deep learning architecture is shown in Fig. 1 It is specialized in quantitative trading. Per each trading time period T, the model is fed with the data received from the financial market. Depending on this data, the model generates predictions on the stocks' returns during the next trading period. According to the errors between the predicted and the actual stocks' returns values, the back-propagation learning algorithm is applied to update the weights of the model's NNs. The model consists of two phases.

1. **Phase I (Deep Learning module):** in which financial raw data, for a set of M market stocks is received. For each stock, a DBN module extracts discriminant features from the high-dimensional raw financial data and reduces its dimensionality. Next, an LSTM module is fed with the stock DBN features and predicts the sum of the relative changes in the stock returns during the next trading period as given in Eq. (1) below.
2. **Phase II (Portfolio construction and assets allocation module):** the portfolio is dynamically constructed from the best N, as determined in Phase I, out of the available M stocks. The fund allocated to each selected stock depends on its predicted sum of relative changes in the stock returns, during the next period, as given in Eq. (2) below.

3.1 Phase I (Deep Learning Module)

3.1.1 Stocks' Features

The following types of data are used at the input:

1. **OHLCV data:** include the opening price, highest price, lowest price, closing price and the trading volume of each stock.

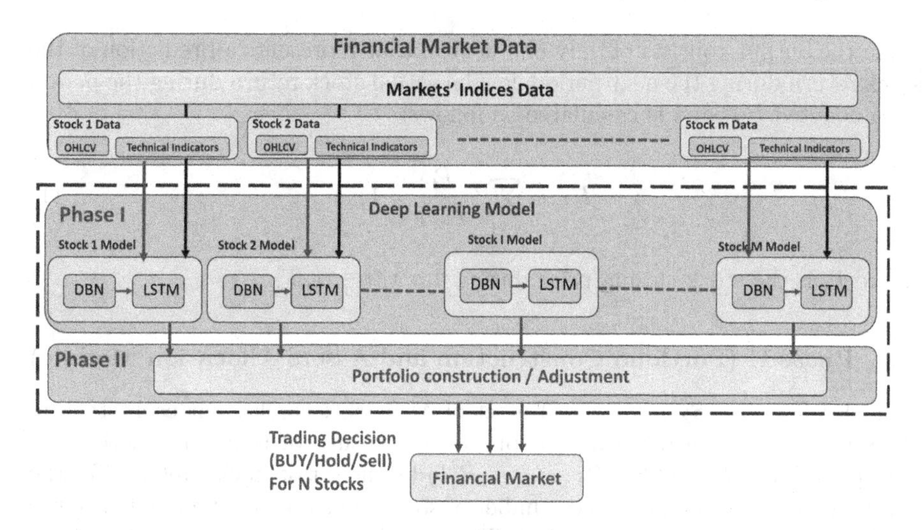

Fig. 1. Proposed deep learning model architecture.

2. **Technical indicators:** which correspond to equations applied on the OHLCV data for a stock to obtain indicators about its future position(price trend/oscillation/volatility/moving average) in the market: e.g. Directional Movement Indicator (DMI), Exponential Moving Average (EMA), Relative Strength Index (RSI), Stochastic Momentum Index (SMI), and Weighted Moving Average (WMA).
3. **Financial market indices:** which correspond to the averages of the OHLCV data for a group of stocks that is used to get an indication about the whole market direction. In our research, five indices are fed to the system, namely, (^GSPC, ^ DJI, ^IXIC, ^NYAC and ^XAX).

The data of the financial time series is fed to our proposed module in steps where each step T consists of 5 d. Each day includes 50 features corresponding to 5 features from the OHLCV data values of the stock, 25 features from the financial market indices and 20 features from the technical indicators of the stock. At each time step T, the features of each of the five days are arranged sequentially and in order to form a feature vector of length 250.

3.1.2 DBN Module

The deep DBN network consists of three hidden layers with 100-80-60 neurons and an output layer of 20 neurons (representing the compressed and uncorrelated features extracted by the DBN from the raw financial data relative to the higher dimensional input vector of length 250).

3.1.3 LSTM Module

The LSTM receives 20 input features from the DBN and has one hidden layer with 28 units (each unit consists of input, output, memory and forget gates) while the output consists of only one neuron that represents a prediction of the stock return during the next period T. The actual stock return during the period T, is denoted R_T, and is calculated as follows:

$$R_T = \sum_{t \in 1}^{T} (\frac{P_t}{P_{t-1}} - 1) \quad (1)$$

where P_t is the stock closing price during day t, $t \in [1, T]$.

3.2 Phase II (Portfolio Construction and Assets Allocation Module)

Consider $S = (S_1, S_2, \ldots\ldots, S_M)$ a set of M market stocks. For each stock in the set S, the model will run and predict the return of this stock during the next period. Let $R_T = (R_T^1, R_T^2, \ldots\ldots, R_T^M)$ be the set of stocks' returns for the period T. let N be the preferred number of stocks to trade in (the number of the stocks in the constructed portfolio). The set of largest N stocks' returns from the set R_T will be chosen $R_T^N = (R_T^1, R_T^2, \ldots\ldots, R_T^N)$ to construct the portfolio L, and the weight of each stock S in portfolio L will be estimated as :

$$W_T^S = \frac{R_T^S}{\sum_{s=1}^{N} R_T^S} \quad (2)$$

While other stocks will have zero weights. At time interval T = 0, the amount of fund F, is used to buy shares of the portfolio's stocks, using: $F_T^S = F * W_T^S$. However, in other time series steps the (Buy/Sell/Hold) signals are generated according to the following rules:

If $W_T^S > W_{T-1}^S$ then a Buy signal with $F * (W_T^S - W_{T-1}^S)$ fund is generated, If $W_T^S < W_{T-1}^S$ then a sell signal with $F * (W_{T-1}^S - W_T^S)$ fund is generated.

In case the fund required for Buying/Selling a stock is less than the minimum transaction cost allowed, a hold signal is generated and the fund will be reallocated to other stocks buying/selling transactions.

4 Experiments Design

4.1 Datasets Used

The thirty industrial stocks registered in the Dow Jones Industrial Average "DJI" index are used as the market's stocks to be traded in. All the financial data are available online at "Yahoo Finance". The datasets from January 2009 to December 2019 have been downloaded. Each input dataset is composed of a window of fixed length (Fifty weeks) moving along the time series (continuously shifted for 5 weeks periods) as shown in Fig. 2. Each input dataset has been divided into three parts, 80% of which (around 40 weeks) is used for training, i.e. to set the model parameters, while 10% (around 5 weeks), is used for validation, i.e. for hyper-parameters tuning, and the last 10% (around 5 weeks) is used for testing, i.e. to test the model predictions. The experiments were conducted on a personal computer (with Microsoft windows 10 Enterprise operating system) with Intel Core i5-3210M (2.50 GHz), 2 core(s), and 128 GB RAM. Python 3.7 with TensorFlow 2.0 backend have been used for implementation.

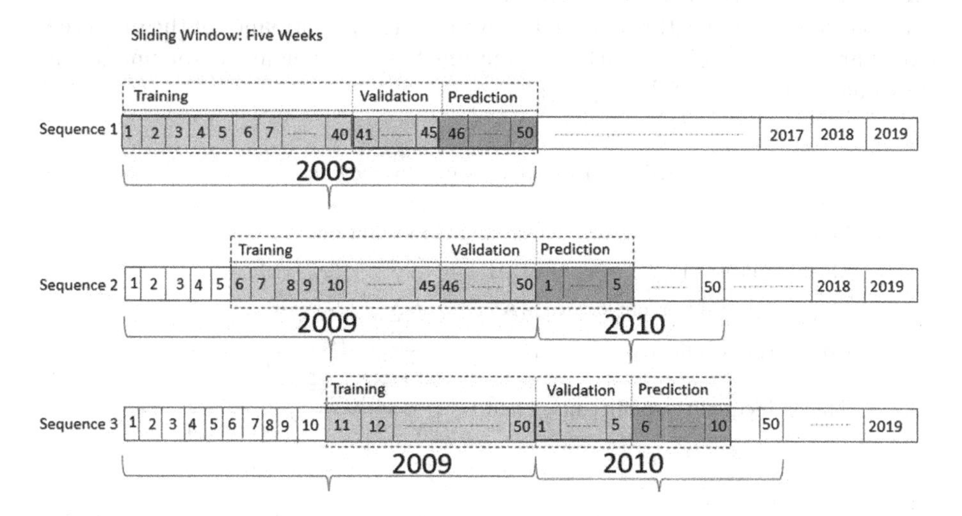

Fig. 2. Moving windows for training, validation and prediction (testing) datasets. The window is continuously shifted for 5 weeks periods.

4.2 Performance Evaluation

4.2.1 Benchmark Models

For benchmarking our proposed module Deep Learing (DBN with LSTM) the following conventional machine learning module are chosen:

1. **Random Forests (RAF):** introduced in [18] and developed in [19]. In this technique multiple decorrelated decision trees (ensemble of B trees each with a maximum depth J) are built on different samples of the training data. The prediction decision is based on the majority voting from the B committee trees. The number of trees B is set to be 100 while the maximum depth J is set to be 15.
2. **Support Vector Regression (SVR):** introduced in [20] and developed in [21]. The SVR uses the same principles as the SVM for classification. In this technique a linear regression function in a high dimensional feature space is computed to map the input data and minimize the generalization error bound to ensure that the distance between the input data points and the hyperplane generated by this function is not farther than epsilon. The hyper-parameters of SVR C, gamma and epsilon are set to 0.1, 0.01 and 0.1, respectively.

For financial bench-marking, the previous machine learning techniques are used to predict the stocks' prices, and the portfolio construction strategy along with the buy and hold strategy, described in Sect. 3.2, are used.

4.2.2 Predictive Accuracy Metrics

The conventional indicators adopted to evaluate the performance of the proposed model are given in Table 1. each row contains the metric name, acronym, how it is estimated and a brief description.

Table 1. Predictive accuracy metrices

Metric	Acronym	Estimation
Mean squared error	MSE	$\frac{1}{N}\sum_{t=1}^{N}(x_t - x_t^*)^2$
Mean absolute percent error	MAPE	$\frac{1}{N}\sum_{t=1}^{N}\lvert\frac{x_t - x_t^*}{x_t}\rvert$
Correlation coefficient	R	$\frac{\sum_{t=1}^{N}(x_t - x_t^-)(x_t^* - x_t^{*-})}{\sqrt{\sum_{t=1}^{N}(x_t - x_t^-)^2 (x_t^* - x_t^{*-})^2}}$
Theil's inequality coefficient	Theil U	$\frac{\sqrt{\frac{1}{N}\sum_{t=1}^{N}(x_t - x_t^*)^2}}{\sqrt{\frac{1}{N}\sum_{t=1}^{N}(x_t)^2} + \sqrt{\frac{1}{N}\sum_{t=1}^{N}(x_t^*)^2}}$

In Table 1, x_t and x_t^* stand for the actual and predicted values, respectively. N represents the number of test prediction periods. x_t^- is the mean value of the actual vector values $(x_1, x_2, \ldots, x_N)$, and x_t^{*-} is the mean value of the predicted vector values $(x_1^*, x_2^*, \ldots, x_N^*)$. The smaller the MSE/MSEP/Theil U values, the better the prediction results. On the other hand, the larger the R value the better the prediction results.

4.2.3 Profitability Metrics

The profitability test is implemented to find how the proposed model can earn real profits for the investors when implemented in real stock markets. Table 2 contains the profitability metrics used in our experiment. The higher the annualized return (AR) and the calmar ratio values, the better for the investors. While the lower the values of the standard deviation (SD), sharp ratio (SR) and maximum drawdown, the lower the risk taken in the investment and the better for the investors.

Table 2. Profitability metrics

Metric name	Short	Description
Volatility	SD	Annualized volatility, standard deviation of the profit and loss during the given time interval
Annualized Return	AR	The geometric average of the amounts of money earned by an investment each year over a given time period
Sharp ratio	SR	The ratio between (annualized return minus free-risk return) and annualized volatility
Maximum Drawdown	MDD	Maximum loss from a peak to a trough in the value of the trading portfolio during time interval T
Calmar Ratio	Calmar	The ratio between annualized return and max drawdown

4.3 Results

4.3.1 Predictive Accuracy Metrics Results

Figure 3 shows the average results obtained from the predicted data by the three models plus the average actual results. The results are given for the ten year periods from 2010 to 2019.

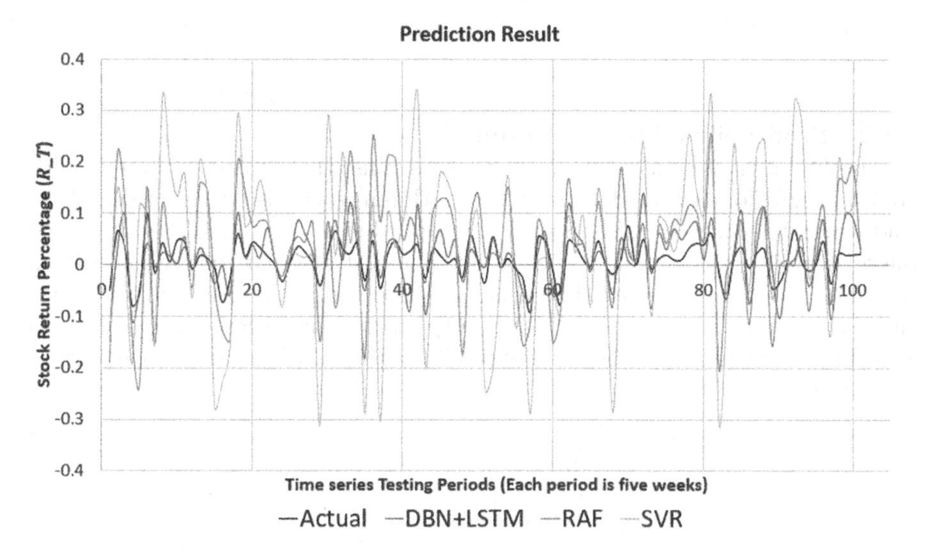

Fig. 3. Stocks' return percentage prediction data (R_T) versus actual data. Each interval of length 10 on the horizontal axis corresponds to one year.

Table 3 shows that the proposed Deep Learning model outperforms the other models since all metrics, except R, are required to be of low values. The SVR has the worst prediction result, as it has the largest deviations from the actual data.

Table 3. Prediction accuracy metrics of the different models calculated for the period from 2010 to 2019. For the MSE metric the shown results should be multiplied by 10^{-2}.

Deep Learning (DBN with LSTM)											
Metric	2010	2011	2012	2013	2014	2015	2016	2017	2018	2019	Average
MSE	0.189	0.087	0.076	0.142	0.294	0.058	0.16	0.199	0.241	0.23	0.168
MAPE	0.0879	0.0919	0.248	0.102	0.189	0.103	0.186	0.209	0.245	0.244	0.171
R	2.219	2.008	1.931	1.848	1.529	2.01	1.718	1.537	2.406	1.286	1.85
Theil U	0.356	0.362	0.392	0.379	0.588	0.278	0.419	0.471	0.442	0.509	0.42
Random Forests (RAF)											
Metric	2010	2011	2012	2013	2014	2015	2016	2017	2018	2019	Average
MSE	1.163	0.948	0.305	2.049	0.79	0.613	0.604	0.336	1.132	0.912	0.885
MAPE	0.276	0.4488	0.559	0.376	0.671	0.4.01	0.604	0.541	0.544	0.595	0.499
R	2.557	2.025	1.62	0.505	1.397	2.1	1.621	1.506	2.229	1.147	1.67
Theil U	0.515	0.594	0.547	0.721	0.654	0.564	0.549	0.54	0.634	0.675	0.6
Support Vector Regression (SVR)											
Metric	2010	2011	2012	2013	2014	2015	2016	2017	2018	2019	Average
MSE	1.997	2.431	1.68	1.967	2.27	1.949	1.014	1.252	2.992	1.979	1.953
MAPE	0.574	0.856	0.523	0.323	0.925	0.515	0.765	0.603	0.768	0.894	0.678
R	1.93	2.152	1.66	1.648	1.582	0.784	2.214	2.055	2.206	1.505	1.77
Theil U	0.626	0.695	0.742	0.679	0.755	0.712	0.627	0.677	0.745	0.725	0.699

4.3.2 Profitability Metrics Results

The profitability of each model is compared against the returns of the buy-and-hold strategy. For each model a portfolio, with ten stocks, is constructed and the trading strategy mentioned in Sect. 3.2 is applied to all models. Table 4 presents the profitability of each model. The results demonstrate that the proposed Deep learning model outperforms the other models in both the profitability return and the risk-return metrics. The RAF outperforms the SVR by a clear margin also. The Worst technique in the profitability return and the risk-return metrics is the buy and hold technique.

Table 4. Profitability metrics of the different models calculated for the period from 2010 to 2019.

Deep Learning (DBN with LSTM)											
Metric	2010	2011	2012	2013	2014	2015	2016	2017	2018	2019	Average
SD	0.069	0.044	0.041	0.054	0.066	0.046	0.052	0.046	0.076	0.063	0.0557
AR	19.5%	16.4%	20.6%	23.6%	18.6%	16.7%	18.6%	25.1%	16.8%	28.2%	20.4%
SR	2.1	2.59	3.8	3.4	2.06	2.54	2.61	4.37	1.55	3.689	2.77
MDD	0.226	0.159	0.126	0.172	0.214	0.144	0.162	0.154	0.203	0.274	0.184
Calmar	0.863	1.031	1.635	1.372	0.869	1.16	1.146	1.63	0.828	1.031	1.156
Random Forests (RAF)											
Metric	2010	2011	2012	2013	2014	2015	2016	2017	2018	2019	Average
SD	0.161	0.125	0.073	0.141	0.096	0.101	0.094	0.059	0.137	0.099	0.1086
AR	14.6%	12.2%	16.2%	18.8%	16.12%	16.65%	15.82%	19.9%	17.51%	15.3%	16.31%
SR	0.596	0.576	1.534	0.979	1.158	1.64	1.151	2.525	0.913	1.044	1.041
MDD	0.388	0.349	0.238	0.412	0.313	0.309	0.309	0.339	0.321	0.215	0.319
Calmar	0.376	0.35	0.681	0.456	0.515	0.539	0.512	0.587	0.545	0.713	0.524
Support Vector Regression (SVR)											
Metric	2010	2011	2012	2013	2014	2015	2016	2017	2018	2019	Average
SD	0.169	0.195	0.155	0.175	0.161	0.139	0.125	0.105	0.196	0.142	0.156
AR	10.8%	9.12%	10.35%	13.9%	12.68%	11.7%	13.4%	14.6%	8.88%	12.88%	11.83%
SR	0.343	0.211	0.345	0.509	0.477	0.482	0.672	0.914	0.198	0.555	0.437
MDD	0.452	0.388	0.289	0.427	0.429	0.462	0.463	0.434	0.352	0.453	0.415
Calmar	0.239	0.235	0.358	0.326	0.296	0.253	0.289	0.336	0.252	0.284	0.287
Buy and Hold											
Metric	2010	2011	2012	2013	2014	2015	2016	2017	2018	2019	Average
SD	0.058	0.039	0.026	0.035	0.026	0.045	0.043	0.019	0.04	0.029	0.036
AR	9.8%	5.53%	7.26%	11.2%	8.4%	2.3%	7.34%	9.3%	–5.63%	8.2%	6.37%
SR	0.828	0.136	0.869	1.771	1.308	–0.6	0.544	2.263	–2.66	1.103	0.381
MDD	0.67	0.56	0.594	0.593	0.522	0.544	0.62	0.52	0.64	0.622	0.589
Calmar	0.146	0.099	0.122	0.189	0.161	0.042	0.118	0.179	–0.088	0.132	0.11

5 Conclusion

This research proposes a novel framework to construct a financial stocks' portfolio, to be used as a quantitative trading system, that repeatedly adjust the number of stocks and their percentages based on a deep learning prediction module. The proposed module has the ability to extract useful knowledge from the input dynamic financial information using a DBN network, that is also used as a dimension reduction tool. A powerful prediction neural network (LSTM) is used to forecast the stocks' performance. The predicted stocks' performance affects the buy/sell/hold decisions taken by the model and the corresponding fund allocated to each stock.

References

1. Huang, W., Nakamori, Y., Wang, S.Y.: Forecasting stock market movement direction with support vector machine. Comput. Oper. Rese. **32**(10), 2513–2522 (2005)

2. Kazem, A., Sharifi, E., Hussian, F.K.: Support vector regression with chaos-based firefly algorithm for stock market price forecasting. Appl. Soft Comput. **13**(2), 947–958 (2013)
3. Cervelló-Royo, R., Guijarro, F., Michniuk, K.: Stock market trading rule based on pattern recognition and technical analysis: forecasting the DJIA index with intraday data. Expert Syst. Appl. **42**(14), 5963–5975 (2015)
4. Hu, Y., Liu, K., Zhang, X., Su, L., Ngai, E.W.T., Liu, M.: Application of evolutionary computation for rule discovery in stock algorithmic trading: a literature review. Appl. Soft Comput. **36**, 534–551 (2015)
5. Aguilar-Rivera, R., Valenzuela-Rend-on, M., Rodr-guez-Ortiz, J.: Genetic algorithms and Darwinian approaches in financial applications: a survey. Expert Syst. Appl. **42**, 7684–7697 (2015)
6. Raffinot, T.: Hierarchical clustering-based asset allocation. J. Portfolio Manage. Multi-Asset Special Issue **44**(2), 89–99 (2018)
7. Gonzalvez, J., Lezmi E., Roncalli, T., Xu J.: Financial Applications of Gaussian Processes and Bayesian Optimization. arXiv:1903.04841 (2019)
8. Thakkar, A., Chaudhari, K.: A comprehensive survey on portfolio optimization, stock price and trend prediction using particle swarm optimization. Arch. Comput. Meth. Eng. **28**(4), 2133–2164 (2020). https://doi.org/10.1007/s11831-020-09448-8
9. Hinton, G., Osindero, S., Teh, Y.W.: A fast learning algorithm for deep belief nets. Neural Comput. **18**(7), 527–554 (2006)
10. https://colah.github.io/posts/2015-08-Understanding-LSTMs/. Accessed on Nov 2020
11. Bao, W., Yue, J., Rao, Y.: A deep learning framework for financial time series using stacked autoencoders and long-short term memory. PLoS ONE **12**, e0180944 (2017)
12. Fischer, T., Krauss, C.: Deep learning with long short-term memory networks for financial market predictions. Euro. J. Oper. Res. **270**, 654–669 (2018)
13. Ta, V.-D., Liu, C.-M., Tadesse, D.A.: Portfolio optimization-based stock prediction using long-short term memory network in quantitative trading. Appl. Sci. **10**(2), 437 (2020)
14. Ribeiro, B., Lopes, N.: Deep Belief Networks for Financial Prediction. Lecture Notes in Computer Science, vol. 7064. Springer, Berlin, Heidelberg (2011). https://doi.org/10.1007/978-3-642-24965-5_86
15. Shen, F., Chao, J., Zhao, J.: Forecasting exchange rate using deep belief networks and conjugate gradient method. Neurocomputing **167**, 243–253 (2015)
16. Assis, C.A.S., Pereira, A.C.M., Carrano, E.G., Ramos, R., Dias, W.: Restricted boltzmann machines for the prediction of trends in financial time series. In: International Joint Conference on Neural Networks (IJCNN), pp. 1–18. Rio de Janeiro (2018)
17. AbdelKawy, R., Abdelmoez, W.M., Shoukry, A.: A synchronous deep reinforcement learning model for automated multi-stock trading. Progress Artif. Intell. **10**(1), 83–97 (2021). https://doi.org/10.1007/s13748-020-00225-z
18. Ho, T.K.: Random decision forests. In: Proceedings of the Third International Conference on Document Analysis and Recognition, pp. 278–282. IEEE (1995)
19. Breiman, L.: Random forests. Mach. Learn. **45**(1), 5–32 (2001)
20. Smola, A.J., Schölkopf, B.: A tutorial on support vector regression. Stat. Comput. **14**(1), 199–222 (2004)
21. Lu, C.J., Lee, T.S., Chiu, C.C.: Financial time series forecasting using independent component analysis and support vector regression. Decis. Support Syst. **47**(2), 115–125 (2009)

Factor Augmented Artificial Neural Network vs Deep Learning for Forecasting Global Liquidity Dynamics

David Alaminos(✉)

Department of Financial Management, Universidad Pontificia Comillas, Madrid, Spain
dalaminos@icade.comillas.edu

Abstract. This paper develops a global liquidity prediction model based on financial and macroeconomic information from different geographical areas. The methodology of the Factor Augmented Artificial Neural Network Model is applied to improve the predictive capacity of liquidity models compared to traditional econometric methodologies. This hybrid methodology based on dynamic factor models and neural networks is compared with Deep Learning methodologies such as Deep Recurrent Convolutional Neural Network and Deep Neural Decision Trees, which has recently shown great results. Our results show the superiority of the precision capacity of Factor Augmented Artificial Neural Network Model over the applied Deep Learning methodology, which demonstrates the importance of data treatment in International Macroeconomics and Finance with techniques from the Vector Autoregressive model. Our conclusions also show the importance of the impact of monetary policy, financial stability, and the real activity of the economy in the behavior of liquidity. This work may be useful for those interest groups in public and macroeconomic policy, showing the potential in the combination of conventional statistical methods with the envelope of Machine Learning techniques.

Keywords: Global liquidity · Capital flows · Factor Augmented Artificial Neural Network · Deep learning methods

JEL Codes: C5 · C63 · E44 · F3 · G01

MSC Codes: 91–08

1 Introduction

Global liquidity is defined as the availability of funds for the purchase of goods or assets provided by institutional and private channels [1]. Global liquidity is gaining importance because of the recent episodes of financial crises, becoming a vital concern in the design and analysis of public policy. In this context of the financial crisis, global liquidity is identified as a potentially important factor in the accumulation of financial imbalances before the crisis [2]. For example, the Asian crisis has been associated by different

L. Rutkowski et al. (Eds.): ICAISC 2021, LNAI 12854, pp. 15–28, 2021.
https://doi.org/10.1007/978-3-030-87986-0_2

researchers with the previous global liquidity conditions [1]. Also, the concept of global liquidity has been used in the debate on the effects of the propagation of accommodative monetary conditions from advanced economies to emerging economies. [3] showed that there is great regional heterogeneity in liquidity preferences, something that has been accentuated after the recent banking and sovereign debt crisis, which makes it difficult to synchronize monetary policies.

In the previous literature on the prediction of global liquidity, different models that have identified various liquidity indicators have been successfully developed. [4] applied a factor model and identified as predictors both indicators based on prices (market interest rates) and implied volatility indices of the stock market. [5, 6] pointed to the monetary base as a reflection of the initial condition of access to liquidity defined by the monetary authorities. [7] showed that foreign exchange reserves are another indicator of global liquidity that is very convenient for emerging countries.

Focusing this review of the literature on studies that have studied the factors that drive global liquidity, [8] focused their study of liquidity on international credit as a key factor. They showed using a linear regression that liquidity growth has been driven by the international issuance of debt securities, while the role of banks has declined, both as lenders and investors in debt securities. This has been more intense in advanced economies. [7] studied the factors influencing global liquidity movements and concluded after using panel data models that the uncertainty of economic policy and the yield differential of the US bond achieved adjustability of 0.40 setting. [2] studied the determinants of global li-quidity using data on cross-border bank flows for a logistic regression model. They showed that global liquidity is mainly driven by uncertainty (VIX), US monetary policy, and UK and euro area banking conditions. Their results ob-tained an adjustment of around 0.45. With the use of factor models, [4] analyzed global liquidity using a large set of financial and macroeconomic variables from advanced and emerging economies. Their results concluded that global liquidity conditions are driven by global monetary policy, global credit supply, and global credit demand. His level of adjustment was below 0.5. [6] studied the fluctuations in the estimated sensitivities to the US monetary policy through a logistic regres-sion showing a high degree of convergence between the monetary policies of the advanced economies. They also conclude that there was a drop after the 2007 crisis in the sensitivity of international bank loans to global risk. They obtained adjustability of 0.11. [1] conducted a review of recent work on global liquidity and its factors that determine it. This analysis highlighted the role of major inter-national financing currencies, such as the US dollar. The analysis also showed the need to take into account a set of global liquidity indicators based on the size and currency composition of balance sheets for future work, as well as to in-crease the search for empirical methods that allow accurate modeling of their future movements

Therefore, after the most recent studies on the determination of global liquidity dynamics, the need to explore the sensitivity of the factors that influence the trajectory of liquidity flows, as well as the lack of ineffective empirical tools for changes, is observed. in future scenarios of global liquidity [1, 6]. Even though advances in global liquidity prediction models have been important, the recent literature demands new research that resolves the differences between regional and global indicators [1–4], and on the properties of the global liquidity dynamics [1, 9]. To respond to these demands, the present study develops a new model based on the financial and macroeconomic information corresponding to a sample of 160 countries divided by geographic zones. The Factor Augmented Artificial Neural Network Model (FAANN) methodology, which is a novel hybrid technique that combines the properties of Factor-Augmented VAR (FAVAR) and artificial neural networks (ANN), has been applied to the data of our sample. Furthermore, it has been compared with two novel Deep Learning techniques, such as Deep Recurrent Convolutional Neural Networks (DRCNN) and Deep Neural Decision Trees (DNDT), to contribute to the methodological discussion in global liquidity dynamics. The results show as FAANN has allowed us to know which factors are the best predictors of global liquidity and achieve rates of accuracy greater than 90%, outperforming recent and popular Deep Learning methodologies. These results are important for the various interest groups in public policy, financial institutions, central banks, and other institutions with concern about macroeconomic forecasting and financial stability.

2 Methods

2.1 Factor Augmented Artificial Neural Network Model (FAANN)

The FAANN model is a hybrid technique of ANN and factor model, to obtain more accurate forecasts [10]. The ANN structure chosen for this methodology is the well-known Multilayer Perceptron (MLP). MLP is an advanced architecture of one entry, one or more hidden, and an exit layer [11]. The network structure provides a forward network connected to the neuron activation function. The input nodes are connected to the nodes in the hidden layer, and these nodes are joined to the single node in the output layer. The entries in this model serve as independent variables in the multiple regression model and are linked to the output node, which is similar to the dependent variable, through the hidden layer. Therefore, the following equations can be specified as follows:

$$n_{k,t} = \omega_0 + \sum\nolimits_{i=1}^{p} \omega_i y_{t-i} + \sum\nolimits_{j=1}^{J} \varnothing_j N_{t-i,j} \tag{1}$$

$$N_{k,t} = f\left(n_{k,t}\right) \tag{2}$$

$$y_t = \alpha_{i,0} + \sum\nolimits_{k=1}^{K} \alpha_{i,k} N_{k,t} + \sum\nolimits_{i=1}^{p} \beta_i y_{t-i} \tag{3}$$

where the inputs $y_{t\text{-}i}$ represent the lagged values of the independent variables and the output y_t is the result. ω_0 y $\alpha_{i,0}$ are the bias, and ω_i and $\alpha_{i,k}$ denote the weights that link the inputs to the hidden layer, and the hidden layer to the output layer, respectively. The j y i connect the entrance to the exit through the hidden layer [10]. The independent variables of p are linearly connected to form neurons K which then combine linearly to produce the output. Equations (1)–(3) define the entries of link $y_{t\text{-}i}$ to the output and through the hidden layer. The function f in a logistic function defined by $N_{k,t} = f(n_{k,t}) = 1/(1 + e^{-nk,t})$. In the FAANN model, the series represents a nonlinear function of several previous observations, and the factors are formed from a large data set that is related to the series analysed, as shown in (4).

$$y_t = f\left[\left(y_{t-1}, y_{t-2}, \ldots, y_{t-p}\right), \left(F_1, F_2, F_3, F_4, F_5\right)\right] \tag{4}$$

where f is the nonlinear functional form defined through ANN. In the first stage, the factor model is employed to obtain factors from an extensive dataset. In the second step, ANN is applied to model the nonlinear and linear relationships that exist between the factors and the data of the sample. Equation (5) defines the output layer of the FAANN model for our case study based on five factors.

$$y_t = \propto_0 + \sum_{i=1}^{q} \propto_j g\left(\beta_{0j} + \sum\nolimits_{i=1}^{p} \beta_{ij} y_{t-i} + \sum\nolimits_{i=p+1}^{p+5} \beta_{ij} F_{t,i}\right) + \varepsilon_t \quad (5)$$

As noted above, α_j ($j = 0, 1,\ldots, q$) and β_{ij} ($i = 0, 1,\ldots, p$; $j = 1, 2,\ldots, q$) are the parameters of the most known model as connection weights. In the same line, p and q are the number of input and hidden nodes, respectively, while ε_t is the error term.

2.2 Deep Recurrent Convolution Neural Network (DRCNN)

Recurrent neural networks (RNN) have been successfully used in many fields for time-series prediction due to their huge prediction performance. For a simple neural network, the inputs are assumed to be independent of each other. The structure of RNN is organized by the output of which is dependent on its previous computations [11–13]. Given an input sequence vector x, the hidden states of a recurrent layer s, and the output of a single hidden layer y, can be calculated as follows:

$$s_t = \sigma(W_{xs}x_t + W_{ss}s_{t-1} + b_s) \quad (6)$$

$$y_t = o\left(W_{so}s_t + b_y\right) \quad (7)$$

where W_{xs}, $W_{ss,}$ and W_{so} denote the weights from the input layer x to the hidden layer s, the hidden layer to itself, and the hidden layer to its output layer, respectively. b_y are the biases of the hidden layer and output layer. σ and o are the activation function

$$STFT\{z(t)\}(\tau, \omega) = \int_{-\infty}^{+\infty} z(t)\omega\left(t - \tau)e^{-j\omega t}dt\right) \quad (8)$$

where z (t) is the vibration signals, ω (t) is the Gaussian window function focused around 0. T (τ , ω) is a complex function that describes the vibration signals over time and frequency. When time-frequency features $\{T_i\}$ are used for global liquidity prediction with RNN, the convolutional operation is conducted in the state transition. To calculate the hidden layers with a convolutional operation the next equation is applied:

$$S_t = \sigma(W_{TS} * T_t + W_{SS} * S_{t-1} + B_s) \quad (9)$$

$$Y_t = o\left(W_{YS} * S_t + B_y\right) \quad (10)$$

where W term indicates the convolution kernels. The convolutional operation has been determined by local connections, weight sharing, and local grouping, which allow every unit to integrate time-frequency data in the current layer. Recurrent Convolutional

Neural Network (RCNN) can be heaped to establish a deep architecture, named deep recurrent convolutional neural networks [12]. When DRCNN for liquidity prediction, the last part of the model is a supervised learning layer for liquidity, which is determined as:

$$\hat{r} = \sigma(W_h * h + b_h) \tag{11}$$

$$L(r, \hat{r}) = \frac{1}{2}\|r - \hat{r}\|_2^2 \tag{12}$$

Stochastic gradient descent is applied for optimization to learn the parameters. The gradient of loss function regarding parameters W_h and b_h are determined as follows:

$$\frac{\partial L}{\partial w_h} = -(r - \hat{r})\sigma'(.)h \tag{13}$$

$$\frac{\partial L}{\partial b_h} = -(r - \hat{r})\sigma'(.) \tag{14}$$

2.3 Deep Neural Decision Trees (DNDT)

DNDT are decision tree models executed by deep-learning neural networks, where a configuration of DNDT weightings corresponds to a specific decision tree and is thus interpretable [11, 13]. The algorithm begins by implementing a soft binning function to calculate the error rate for each node, making it possible to make decisions divided into DNDT [11, 14]. In general, the input of a binning function is a real scalar x, which generates an index of the containers to which x belongs. Assuming x is a continuous variable, group it into n + 1 intervals. The cut-off points are denoted as [β1, β2,. . ., βn] and are strictly ascending such that β1 < β2 <...< βn.

The activation function of the DNDT algorithm is implemented based on the NN defined in Eq. (15).

$$\pi = fw, b, \tau(x) = soft\max((wx + b)/\tau \tag{15}$$

where w is a constant with value w = [1, 2,. . . , n + 1], τ > 0 is a temperature factor and b is defined in Eq. (16).

$$b = [0, -\beta 1, -\beta 1 - \beta 2, \ldots, -\beta 1 - \beta 2 \cdots\cdots - \beta n] \tag{16}$$

The NN defined in Eq. (1) gives a coding of the binning function x. Additionally, if τ tends to *0*, (often the most common case), the vector sampling is implemented using the Straight-Through (ST) Gumbel–Softmax method [11, 14]. Given the binning function described above, the key idea is to build the DT using the Kronecker product. Assuming we have an input instance $x \in R^D$ with D characteristics. Associating each characteristic x_d with its NN $f_d(x_d)$, we can determine all the final nodes of the DT, in line with Eq. (17).

$$z = f1(x1) \otimes f2(x2) \otimes \cdots\cdots \otimes fD(xD) \tag{17}$$

where z is now also a vector that indicates the index of the leaf node reached by instance x. Finally, we assume that a linear classifier on each leaf z classifies the instances that reach it.

2.4 Sensitivity Analysis

To improve the interpretation of the results offered by the computational methods, a sensitivity analysis has been applied in the three methodologies used to detect the most significant variables and their impact on the objective studied. For this, the most common sensitivity analysis in neural network techniques has been chosen [15]. This analysis consists of taking 100% of the data and dividing them into groups, and each group of data is processed in the network constructed as many times as there are variables of the model. Each time the value of one of the variables is modified, placing it with zero value. The answers of the network are evaluated about the objective values or classification values already known, using expression (18).

$$S_{x_i} = \sum\nolimits_{j=1}^{n} \left(\Phi x_{ij}(0) - \Phi x_{ij}\right)^2 \quad (18)$$

3 Sample and Factors

The sample used in this study is composed of 160 developed and emerging countries, which have been grouped into seven regions: Advanced Economies (United States, Europe, and Japan), Emerging Europe, Latin America, Africa & Middle East, Asia-Pacific, Emerging Economies, and Global. The selection of countries is mainly guided by the availability of data and covers the main regions to meet the needs of the literature regarding the differentiation of factors between regions [1]. From the sample of countries, financial and macroeconomic information has been available for the period 1995q1-2018q4. This information has been extracted from the databases of IMF International Financial Statistics, World Bank Development Indicators, World Economic Outlook, the World Bank Global Financial Database, Bank for International Settlements (BIS) Statistics, and Federal Reserve Economic Data (FRED). The sample is divided randomly into two sets: training data set (70%) and testing data set (30%). In this process, we used the cross-validation method with 10-fold and 500 iterations to estimate RMSE ratios [16]. The first data set is used for model training, that is, for parameter estimation. Finally, the second data set is used to evaluate the prediction accuracy of the model. Two four-core Intel Core I7-6500U are used as computing resources and the code is made from MATLAB package (R2016b version).

With this information, and based on the previous literature, a set of 23 predictors has been constructed, of quarterly frequency, which has subsequently been transformed into 5 factors through the analysis of main components using the method of Principal Component Analysis. These predictors have been chosen from the literature review made in this study [1, 4, 6–8]. Besides, and as a proxy for global liquidity, cross-border credit has been used [4, 6, 8]. Table 1 shows the factors and indicators used for the construction of liquidity models.

4 Results

Tables 2, 3, and 4 show the results obtained by the FAANN, DRCNN, and DNDT models for each region and the global sample of countries. Prediction models have been built

Table 1. Factors and indicators

Factors	Code	Indicators
Factor exchange rate	F^{exch}	Bilateral Exchange Rate Against USD
Factor real activity	F^{act}	Consumption (% of GDP)
		Gross Fixed Capital Formation (% of GDP)
		Government Spending (% of GDP)
		Industrial Production (% of GDP)
		Credit Growth National Savings (% of GDP)
		Real Interest Rate
		Real GDP
		Unemployment Rate
		Imports and Exports (% of GDP)
Factor inflation	F^{cpi}	Consumption Price Index
Factor financial stability	F^{fstab}	House Price Index
		3-month Treasury Bond Rate
		10-year Government Bond Rate
		Stock Exchange Index
		VIX[a]
Factor monetary aggregates	F^{magg}	Global Indicator based on Monetary Base
		Global Indicator of Foreign Reserves
		Global Indicator of Central Bank Interest Rate
		Global Indicator of Narrow Monetary Aggregate (M1)
		Global Indicator of Broad Monetary Aggregate (M2)
		Global Indicator of Credit to Private Sector

[a] Chicago Board Options Exchange Market Volatility Index

3 months ahead (t-3), 6 months (t-6), and 12 months (t-12) concerning the current time (t).

The results show that the factor of aggregate monetary (F^{magg}) is the most significant factor in the estimated models, showing an impact higher than 60%. This factor reflects the influence of monetary policy decisions of central banks on the interest rate, the money supply, and currency reserves, among others. This result is in line with some previous works [2], but shows a consistent impact throughout the applied time horizon, something that co ntradicts the results previously thrown [4]. This factor has been more important for emerging regions such as Asia-Pacific and Latin America, and with a lower impact in advanced areas such as Europe and Japan. For its part, the financial stability factor (F^{fstab}) is shown as another factor with high significance, and with an impact of more than 20%. This factor indicates the level of volatility existing in the markets and access to them, in addition to the monetary policy implemented. The trend of the results of this factor increases throughout the study horizon, contradicting what was evidenced by the previous literature [5]. Financial stability has shown a greater impact for emerging regions such as Asia-Pacific and Latin America, and with a lower impact in advanced areas such as Europe and Japan. Finally, the factor of real activity (F^{act}) has been shown to have a significant impact, like the F^{fstab} factor, and also higher than 20%. This factor demonstrates the importance of the real economy in the level of liquidity and summarizes indicators such as the behavior of credit, economic growth, and the external sector. Also, its results have been of greater importance in advanced regions such as the United States and Europe, and less so in emerging regions such as Africa & Middle East. Several previous works have not shown such significant results on this factor [4, 5, 8].

On the other hand, the precision of our FAANN models shows a level higher than 90%, both in the training sample and in the testing, and throughout the entire time horizon used. Similarly, the average errors obtained show reduced levels. These levels of precision obtained improve those shown by the previous works [2, 6, 8]. They also exceed the precision shown by the models built from the applied Deep Learning techniques, even though in the most recent previous literature the superiority of these mentioned techniques over the rest has been common [5, 6]. Finally, Tables 1, 2, 3 and 4 show the time-lapses required for each technique used to estimate all the proposed models and for each of the time horizons considered. It is observed that the FAANN method needs a slightly shorter period than the rest of the methodologies, where it stands out that DNDT is the one that obtains the highest levels of time. Finally, the overall of time lapses required for each technique used to estimate all the proposed models and for each of the time horizons have been 20.12 s for FAANN, 24.65 s for DRCNN, and 27.29 s for DNDT.

Table 2. Results of forecasting evaluation for FAANN Model

Horizon time	Regions	Classification matrix (%)		RMSE		Impact of factors (%)				
		Training	Testing	Training	Testing	F^{exch}	F^{act}	F^{cpi}	F^{fstab}	F^{magg}
t										
	United States	91.74	89.47	0.24	0.29	8.86	21.52	9.86	18.65	71.46
	Europe	93.07	89.79	0.22	0.26	5.72	21.22	10.51	19.99	67.82
	Japan	92.28	90.01	0.21	0.23	7.43	19.08	8.78	18.21	65.74
	Emerging Europe	92.61	90.33	0.17	0.22	11.24	22.14	10.97	21.62	62.85
	Latin America	92.82	90.53	0.16	0.21	15.07	18.55	9.11	21.2	69.62
	Africa & Middle East	92.97	90.73	0.22	0.24	13.44	16.64	9.53	20.74	74.83
	Asia-Pacific	93.19	90.89	0.16	0.18	18.65	17.73	10.08	22.51	81.04
	Global	93.39	91.09	0.14	0.18	12.82	18.96	9.15	20.95	68.89
t + 3										
	United States	93.51	91.26	0.16	0.19	11.81	31.83	14.79	23.31	76.96
	Europe	93.56	91.37	0.15	0.18	7.63	28.62	15.77	24.99	73.04
	Japan	93.69	91.38	0.14	0.16	9.91	33.21	13.17	22.76	70.80
	Emerging Europe	93.99	91.69	0.11	0.15	14.99	27.83	16.46	27.03	67.68
	Latin America	94.06	91.81	0.15	0.18	20.09	24.96	13.67	26.53	74.98
	Africa & Middle East	94.19	92.01	0.18	0.21	17.92	26.6	14.31	25.93	80.59
	Asia-Pacific	94.43	92.17	0.14	0.16	24.87	32.28	15.12	28.14	87.27
	Global	94.64	92.37	0.13	0.16	17.09	28.44	13.73	26.19	74.19
t + 6										
	United States	94.74	92.56	0.16	0.17	13.52	35.81	16.64	25.92	69.96
	Europe	94.78	92.59	0.12	0.15	8.72	32.27	17.74	27.76	66.40
	Japan	94.94	92.67	0.11	0.13	11.32	37.36	14.82	25.29	64.36
	Emerging Europe	95.25	92.98	0.09	0.12	17.13	31.35	18.51	30.03	61.53
	Latin America	95.3	93.11	0.12	0.16	22.96	28.08	15.37	29.44	68.16
	Africa & Middle East	95.67	93.48	0.11	0.15	20.48	29.92	16.08	28.81	73.26
	Asia-Pacific	95.76	93.24	0.16	0.19	28.42	36.32	17.01	31.26	79.34
	Global	96.1	93.57	0.14	0.18	19.54	32.04	15.44	29.13	67.44
t + 12										
	United States	96.19	93.98	0.07	0.11	12.67	38.56	15.53	28.26	63.62

(*continued*)

Table 2. (*continued*)

Horizon time	Regions	Classification matrix (%)		RMSE		Impact of factors (%)				
		Training	Testing	Training	Testing	F^{exch}	F^{act}	F^{cpi}	F^{fstab}	F^{magg}
	Europe	96.32	93.79	0.11	0.14	8.14	34.67	16.55	30.29	60.36
	Japan	96.36	93.83	0.11	0.15	10.57	40.24	13.83	27.59	58.51
	Emerging Europe	96.67	94.13	0.10	0.13	15.99	33.71	17.28	32.76	55.94
	Latin America	96.89	94.37	0.09	0.14	21.43	30.24	14.35	32.12	61.96
	Africa & Middle East	97.27	94.73	0.08	0.12	19.11	32.22	15.01	31.42	66.63
	Asia-Pacific	97.48	94.93	0.07	0.11	26.52	39.11	15.88	34.11	72.13
	Global	96.11	94.56	0.12	0.15	18.23	34.46	14.41	31.74	61.31

Table 3. Results of forecasting evaluation for DRCNN Model

Horizon time	Regions	Classification matrix (%)		RMSE		Impact of factors (%)				
		Training	Testing	Training	Testing	F^{exch}	F^{act}	F^{cpi}	F^{fstab}	F^{magg}
t										
	United States	85.40	83.23	0.41	0.44	8.76	24.54	8.54	19.65	70.89
	Europe	86.61	83.62	0.33	0.42	5.03	21.66	9.44	20.56	67.65
	Japan	85.89	83.81	0.33	0.39	6.73	22.88	8.35	20.32	60.88
	Emerging Europe	86.19	84.11	0.35	0.40	10.86	24.17	10.26	24.22	61.14
	Latin America	86.38	84.30	0.33	0.38	14.58	18.65	7.64	22.77	68.68
	Africa & Middle East	86.52	84.48	0.33	0.43	13.33	18.05	9.23	23.57	71.51
	Asia-Pacific	86.72	84.62	0.21	0.27	18.23	18.93	9.86	23.01	80.36
	Global	86.90	84.81	0.36	0.43	12.62	19.97	7.84	21.05	67.56
$t+3$										
	United States	87.01	84.96	0.28	0.33	11.27	33.39	14.25	24.12	75.39
	Europe	87.05	85.00	0.33	0.36	6.82	30.24	15.55	26.14	71.99
	Japan	87.17	85.07	0.32	0.37	9.30	35.79	12.16	24.98	66.78
	Emerging Europe	87.45	85.35	0.35	0.36	14.29	31.01	15.99	29.36	66.82
	Latin America	87.51	85.46	0.31	0.37	19.53	28.61	13.24	27.30	70.71
	Africa & Middle East	87.63	85.64	0.38	0.35	17.71	30.10	13.66	28.81	77.65
	Asia-Pacific	87.85	85.79	0.28	0.28	24.65	33.57	14.37	29.63	87.11
	Global	88.04	85.97	0.17	0.29	16.99	29.68	13.56	26.45	71.44
$t+6$										
	United States	88.13	86.14	0.31	0.35	12.63	36.53	15.73	28.25	66.64
	Europe	88.17	86.17	0.28	0.30	8.53	35.16	16.29	29.00	63.98

(*continued*)

Table 3. *(continued)*

Horizon time	Regions	Classification matrix (%)		RMSE		Impact of factors (%)				
		Training	Testing	Training	Testing	F^{exch}	F^{act}	F^{cpi}	F^{fstab}	F^{magg}
	Japan	88.31	86.24	0.24	0.28	11.11	40.53	14.01	26.29	64.27
	Emerging Europe	88.59	86.53	0.28	0.30	16.46	33.26	18.08	30.76	58.09
	Latin America	88.64	86.64	0.27	0.31	22.29	28.79	14.71	30.72	63.63
	Africa & Middle East	88.98	86.98	0.37	0.41	19.67	33.05	15.70	29.39	69.17
	Asia-Pacific	89.06	86.76	0.26	0.37	28.00	37.95	16.89	33.94	78.34
	Global	89.37	87.06	0.31	0.38	19.34	34.42	14.39	31.18	65.41
$t + 12$										
	United States	89.45	87.44	0.26	0.37	12.10	38.72	14.42	30.84	59.10
	Europe	89.57	87.26	0.33	0.45	8.12	38.60	15.82	32.70	57.01
	Japan	89.60	87.30	0.33	0.41	10.30	43.41	12.41	29.98	57.55
	Emerging Europe	89.89	87.57	0.35	0.39	15.51	35.56	17.25	33.18	53.02
	Latin America	90.09	87.77	0.30	0.38	21.12	30.37	13.86	34.86	61.94
	Africa & Middle East	90.43	88.12	0.23	0.31	18.29	36.21	13.81	33.28	65.82
	Asia-Pacific	90.62	88.30	0.22	0.26	25.88	42.54	15.37	36.40	68.02
	Global	89.38	87.96	0.26	0.35	17.52	34.62	13.63	31.85	57.60

Table 4. Results of forecasting evaluation for DNDT Model

Horizon time	Regions	Classification matrix (%)		RMSE		Impact of factors (%)				
		Training	Testing	Training	Testing	F^{exch}	F^{act}	F^{cpi}	F^{fstab}	F^{magg}
t										
	United States	86.91	84.11	0.34	0.38	9.09	20.93	10.58	18.06	70.28
	Europe	89.12	89.19	0.32	0.37	6.16	20.16	10.95	19.34	66.85
	Japan	88.36	89.42	0.24	0.33	7.56	18.95	8.82	17.33	64.35
	Emerging Europe	88.68	89.73	0.23	0.31	11.35	21.93	11.39	20.87	60.98
	Latin America	88.88	89.93	0.22	0.31	15.57	18.23	9.95	20.65	69.21
	Africa & Middle East	89.02	90.13	0.26	0.28	13.8	14.59	9.87	19.94	74.45
	Asia-Pacific	89.24	90.29	0.29	0.26	18.88	16.89	10.57	21.83	80.06
	Global	89.43	90.49	0.19	0.23	13.04	18.78	9.74	20.7	67.19
$t + 3$										
	United States	93.06	90.66	0.16	0.21	11.95	30.11	15.16	22.71	75.41

(continued)

Table 4. *(continued)*

Horizon time	Regions	Classification matrix (%)		RMSE		Impact of factors (%)				
		Training	Testing	Training	Testing	F^{exch}	F^{act}	F^{cpi}	F^{fstab}	F^{magg}
	Europe	93.13	90.75	0.18	0.27	8.2	27.11	15.95	24.57	71.59
	Japan	93.24	90.78	0.17	0.20	10.4	32.61	14.05	22.58	69.76
	Emerging Europe	93.54	91.09	0.15	0.18	15.5	26.87	16.58	26.92	67.54
	Latin America	93.61	91.21	0.20	0.22	20.5	24.64	14.54	25.98	74.51
	Africa & Middle East	93.74	91.41	0.27	0.31	18.18	25.88	14.77	25.32	79.81
	Asia-Pacific	93.98	91.57	0.21	0.26	25.39	31.52	15.22	28.14	85.65
	Global	94.19	91.77	0.22	0.24	17.16	27.61	13.95	25.86	73.89
$t+6$										
	United States	94.29	91.96	0.2	0.27	13.53	33.44	16.69	25.19	68.52
	Europe	94.33	91.99	0.22	0.27	8.91	31.32	18.08	27.53	66.29
	Japan	94.49	92.07	0.19	0.22	11.39	37.08	15.45	24.83	63.46
	Emerging Europe	94.82	92.38	0.21	0.24	17.66	30.68	18.85	29.9	60.47
	Latin America	94.85	92.51	0.24	0.26	23.41	27.82	15.37	29.42	67.08
	Africa & Middle East	95.22	92.88	0.21	0.26	20.77	29.04	16.90	28.05	71.77
	Asia-Pacific	95.31	92.64	0.19	0.21	28.73	35.86	17.15	31.05	78.81
	Global	95.65	92.97	0.18	0.20	20.04	31.28	16.04	28.83	66.24
$t+12$										
	United States	95.74	93.38	0.11	0.18	12.88	37.3	16.24	27.87	62.88
	Europe	95.87	93.19	0.11	0.14	8.65	34.59	16.61	29.69	59.93
	Japan	95.91	93.23	0.11	0.19	10.61	38.66	14.43	27.12	56.81
	Emerging Europe	96.22	93.53	0.14	0.18	16.46	32.31	17.84	32.22	55.57
	Latin America	96.44	93.82	0.16	0.22	21.96	27.9	14.37	31.23	60.51
	Africa & Middle East	96.82	93.23	0.15	0.21	19.11	29.82	15.66	31.34	66.18
	Asia-Pacific	97.03	94.03	0.11	0.16	26.87	37.05	15.96	33.79	70.71
	Global	95.66	93.59	0.12	0.19	18.54	32.74	14.91	30.92	60.98

5 Conclusions

The results obtained in the present study show that monetary aggregates, financial stability, and real activity are the main factors to predict the dynamism of global liquidity. Aggregates represent the most important factor in determining liquidity flows, in many cases exceeding 60% of the impact, regardless of the area and the time of the projection. On the other hand, the levels of impact of the factors of stability and real activity show some differences between regions, indicating that financial stability is more important in emerging economies, while real activity is more significant in advanced economies. Therefore, the present study offers new models of the main regions of the world, trying to cover the need to show the regional differences in the prediction of capital flows that the previous literature requested.

The FAANN methodology applied in this work has shown high levels of accuracy, exceeding 90% in all treated cases. Besides, together with the application of the sensitivity analysis that measures the impact of the factors, this methodology shows an improvement in the interpretability of the results and in the extension of predictions to different future time horizons, showing itself as a promising alternative to other methodologies widely used in this line of study such as the dynamic factor model or Factor-Augmented Vector Autoregressive (FAVAR), not only to predict liquidity flows but also other financial stability and macroeconomic policy concerns. On the other hand, Deep Learning methodologies have shown a high and robust level of success, but lower than the FAANN model. Despite this, our results show that these techniques can also be used efficiently for the prediction of liquidity flows for those professionals and academics interested in the development of Deep Learning methodology.

These conclusions can be extremely useful for institutions, both public and private, that need to make forecasts about public policy financial and macroeconomic analysis. Finally, as a future line of research, it could deal with the possible influence of political decisions and the management of their management on the differences in liquidity flows between regions of the world. It would also be interesting to analyze the shocks created by the unconventional programs of the world's major central banks and their long-term impact on liquidity patterns between different world regions, both in advanced and emerging economies.

References

1. Cohen, B.H., Domanski, D., Fender, I., Shin, H.S.: Global liquidity: a selective review. Ann. Rev. Econ. **9**, 587–612 (2017)
2. Cerutti, E.M., Claessens, S., Ratnovski, L.: Global liquidity and cross-border bank flows. Econ. Policy **32**(89), 81–125 (2017)
3. El-Shagi, M., Kelly, L.: What can we learn from country-level liquidity in the EMU? J. Financ. Stab. **42**, 75–83 (2019)
4. Eickmeier, S., Gambacorta, L., Hofmann, B.: Understanding global liquidity. Eur. Econ. Rev. **68**, 1–18 (2014)
5. Angrick, S.: Global liquidity and monetary policy autonomy: an examination of open-economy policy constraints. Camb. J. Econ. **42**, 117–135 (2018)
6. Avdjiev, S. Gambacorta, L. Goldberg, L.S., Schiaff, S.: The shifting drivers of global liquidity. J. Int. Econ. **125**, 103324 (2020)

7. Belke, A., Volz, U.: Capital flows to emerging market and developing economies. Global liquidity and uncertainty versus country-specific pull factors. German Development Institute, Bonn. Discussion Paper No. 23/2018 (2018)
8. Aldasoro, I. Ehlers, T.: Global liquidity: changing instrument and currency patterns. BIS Quarterly Review, September 2018 (2018)
9. Chen, S.F., Liu, P., Maechler, A.M., Marsh, C., Saksonovs, S., Shin, H.S.: Exploring the Dynamics of Global Liquidity. IMF Working Paper, No. 12/246 (2012)
10. Babikir, A., Mwambi, H.: Factor augmented artificial neural network model. Neural Process. Lett. **45**(2), 507–521 (2017)
11. Lamothe-Fernández, P., Alaminos, D., Lamothe-López, P., Fernández-Gámez, M.A.: Deep learning methods for modeling bitcoin price. Mathematics **8**, 1245 (2020)
12. Mishra, D., Naik, B., Sahoo, R.M., Nayak, J.: Deep Recurrent Neural Network (Deep-RNN) for classification of nonlinear data. In: Das, A.K., Nayak, J., Naik, B., Dutta, S., Pelusi, D. (eds.) Computational Intelligence in Pattern Recognition. AISC, vol. 1120, pp. 207–215. Springer, Singapore (2020). https://doi.org/10.1007/978-981-15-2449-3_17
13. Taud, H., Mas, J.F.: Multilayer Perceptron (MLP). In: Camacho Olmedo, M.T., Paegelow, M., Mas, J.-F., Escobar, F. (eds.) Geomatic Approaches for Modeling Land Change Scenarios. LNGC, pp. 451–455. Springer, Cham (2018). https://doi.org/10.1007/978-3-319-60801-3_27
14. Yang, Y.; Garcia-Morillo, I.; Hospedales, T.M.: Deep neural decision trees. In: Proceedings of the 2018 ICML Workshop on Human Interpretability in Machine Learning (WHI 2018), Stockholm, Sweden, 14 July 2018
15. Samarasinghe, S.: Neural Networks for Applied Sciences and Engineering: From Fundamentals to Complex Pattern Recognition. Auerbach Publications, Taylor & Francis Group. New York (2016)
16. Tsamardinos, I.G., E., Borboudakis, G. : Bootstrapping the out-of-sample predictions for efficient and accurate cross-validation. Mach. Learn. **107**(12), 1895–1922 (2018)

Integrate-and-Fire Neurons for Low-Powered Pattern Recognition

Florian Bacho(✉) and Dominique Chu

CEMS, School of Computing, University of Kent, Canterbury CT2 7NF, UK
fb320@kent.ac.uk

Abstract. Embedded systems acquire information about the real world from sensors and process it to make decisions and/or for transmission. In some situations, the relationship between the data and the decision is complex and/or the amount of data to transmit is large (e.g. in biologgers). Artificial Neural Networks (ANNs) can efficiently detect patterns in the input data which makes them suitable for decision making or compression of information for data transmission. However, ANNs require a substantial amount of energy which reduces the lifetime of battery-powered devices. Therefore, the use of Spiking Neural Networks can improve such systems by providing a way to efficiently process sensory data without being too energy-consuming. In this work, we introduce a low-powered neuron model called Integrate-and-Fire which exploits the charge and discharge properties of the capacitor. Using parallel and series RC circuits, we developed a trainable neuron model that can be expressed in a recurrent form. Finally, we trained its simulation with an artificially generated dataset of dog postures and implemented it as hardware that showed promising energetic properties.

Keywords: Remote system · Spiking Neural Networks · Integrate-and-fire · Neuromorphic hardware

1 Introduction

Embedded systems acquire physical measurements of the real world from sensors before performing simple computations [13]. From signal acquisition, these systems often require a transformation of the data to make decisions or compress the information for transmission. Pattern recognition is an important area in the emergence of intelligent systems the classification of patterns from sensory information into categories is necessary to achieve a goal [13]. For example, recent years have seen the development of new animal-attached devices called *Biologgers* which are used to monitor the environment, track locations and quantify the behaviour of certain species [1]. These devices sometimes use transmission technologies such as Very High Frequency (VHF), acoustic telemetry or, more recently, orbiting satellites to monitor certain species over a long period. However, data transmission has a high cost not only financially, but also in terms of

L. Rutkowski et al. (Eds.): ICAISC 2021, LNAI 12854, pp. 29–40, 2021.
https://doi.org/10.1007/978-3-030-87986-0_3

energy. This can be problematic on battery-powered devices. Thus, to optimise the lifetime of remote devices, the number of transmissions must be minimized. As some sensors often run at a high sampling frequency – typically 10 Hz and 1000 hz for inertial sensors – the amount of collected data becomes so large that transmission becomes difficult without any compression or processing. To reduce this amount of data, embedded classifiers can directly process the sensor values, which significantly reduces the information to transmit. For example, some methods have been used on biologgers to classify animal activities from inertial data using machine learning approaches, especially Artificial Neural Networks [7,11,17]

Artificial Neural Networks (ANNs) are one of the most powerful methods to solve classification problems. ANNs try to mimic the behaviour of biological neurons to find complex relationships between input signals and desired outputs. However, the computation of these artificial neurons is computationally expensive due to complex operations that require substantial amounts of energy or sometimes the use of Graphics Processing Units (GPUs) which makes them unsuitable for battery-powered devices [15]. Contrary to the abstracted models used in Deep Learning, Spiking Neural Networks (SNN) are biologically plausible artificial neuron models [8] that transmit information through discrete electrical signals called spike trains [8,15]. Spiking neurons integrate synaptic events only when they occur and fire action potentials when the membrane potential reaches a defined threshold [8]. This event integration property makes them relatively easy to simulate and can also be implemented as energy-efficient dedicated hardware (called neuromorphic chips) [2,3,5,6]. To the best of our knowledge, there is no hardware implementation of SNNs embedded in small remote devices such as biologgers – mainly because of the size of the current neuromorphic hardware. Therefore, it is necessary to bring new simple and non-energy-consuming solutions for embedded pattern recognition in remote systems.

In this paper, we present a simple neuron circuit that can be used for basic pattern recognition in remote systems. This model developed is the *Integrate and Fire* (IF) which is easily implementable as energy-efficient hardware with low-cost components. It integrates successive currents during different amounts of time – according to the inputs – and exploits the charge and discharge capabilities of capacitors to create a trainable and electronically implementable neuron. The model has both excitatory and inhibitory synapses and we introduce it as a recurrent form which makes it suitable for gradient descent optimisations. This model has been chosen for the simplicity of its hardware implementation and its simulation. To validate it, we trained three neurons to classify dog postures using inclination vectors (calculated from inertial data) and implemented them with electronic components to compare the hardware and its simulation.

2 Results

The capacitor is an electronic passive component that creates a potential difference between two conductive plates, analogous to the difference of electric potential of the biological neuron membrane created by ions that flow in and out of the cell. Therefore, the capacitor is often used in computational models of spiking neural networks to reproduce membrane potentials of the biological neurons. Connected in series or parallel with a resistance, the capacitor forms two circuits with distinct charge and discharge properties respectively called series and parallel Resistor-Capacitor circuits (RC). Thus, the IF neuron is mainly composed of passive components: a capacitor that reproduces the membrane potential and resistors that charge (excite) or discharge (inhibit) the neuron.

2.1 Series RC Circuit for Excitatory Stimulations

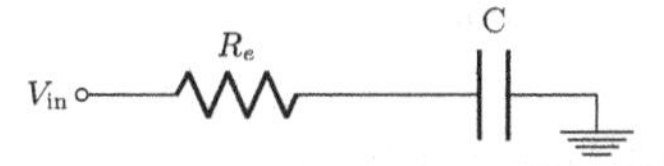

Fig. 1. Electric diagram of the series Resistor-Capacitor (RC) circuit. The circuit is composed of a voltage supplier V_{in}, a resistor R_e that is analog to the excitatory synapses of the neuron and a capacitor C that reproduces the membrane potential.

The series RC circuit is defined by a successive resistor R_e which represents the excitatory synapses of the biological neuron and a capacitor C which reproduces the membrane potential – see Fig. 1. Taking into consideration Ohm's law ($I = \frac{V}{R}$), the fact that the current I_{R_e} flowing through the resistor R_e is equal to the current I_C flowing through the capacitor C ($I_{R_e} = I_C$) and that the capacitor component theoretically does not produce any resistance, the current flowing through the circuit depends only on the input voltage V_{in} and the excitatory resistor R_e. Thus, the resistor can be seen as a weight defined as $w = \frac{1}{R_e}$ which scales the input value V_{in} such as $I_{R_e} = wV_{in}$. Consequently, the higher the value of the resistor, the lower the current will flow through the capacitor and vice versa. Knowing that the total voltage V_{in} of the circuit is defined as the sum of the voltages V_R and V_C respectively across the resistor and the capacitor ($V_{in} = V_R + V_C$), and the Ohm's law, we can define the following equation:

$$\begin{aligned} V_{in} &= R_e I_{R_e} + V_C \\ \Leftrightarrow I_{R_e} &= \frac{V_{in} - V_C}{R_e} \end{aligned} \tag{1}$$

Equation 1 shows that the current flows through the excitatory resistor does not only depend on the input voltage and the resistance but also depends on the voltage across the capacitor. Therefore, the higher the voltage across the capacitor, the lower the current flowing in the circuit will be. To describe the dynamic of the capacitor, the instantaneous rate of voltage change $\frac{dV}{dt}$ of the capacitor is introduced as the current I flowing through the capacitor divided by the capacitance C ($\frac{dV}{dt} = \frac{I}{C}$). Equation 1 can be reformulated as:

$$\tau_e \frac{dV_C}{dt} = V_{in} - V_C \tag{2}$$

where $\tau_e = R_e C$ is the time constant of the series RC circuit which represents the number of seconds needed to reach approximately 63.2% of the input voltage V_{in} – this value is explained below. For a constant input voltage and a given initial voltage $V_C(t)$ at time t, the capacitor voltage $V_C(t+\Delta t)$ after an amount of time Δt can be found by integrating Eq. 2:

$$V_C(t + \Delta t) = V_{in} - (V_{in} - V_C(t))e^{-\frac{\Delta t}{\tau_e}} \tag{3}$$

The fact that the time constant τ_e represents the amount of time to reach a voltage of approximately 63.2% of the input voltage is due to of the exponential property of Eq. 3. Indeed, with an initial voltage of 0, the voltage $V_C(\tau_e)$ reached by the capacitor after a stimulation of τ_e seconds with an input voltage V_{in} is $V_{in}(1 - e^{-1})$ where $1 - e^{-1} \approx 0.632$.

2.2 Parallel RC Circuit for Inhibition

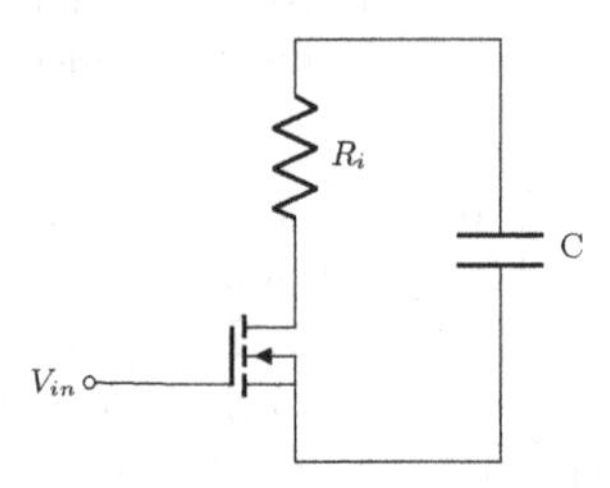

Fig. 2. Electric diagram of the parallel Resistor-Capacitor (RC) circuit controlled by a N-Channel MOSFET transistor.

Inhibitory neurons represent 10%–20% of brain population and their activity plays a major role in cognition [16]. By producing stop signals of excitation and therefore decreasing the membrane potentials of neurons receiving inhibitory stimulus, inhibitory neurons can be seen as regulators of firing rates by maintaining neurons to sub-threshold regimes. In the IF neuron, an inhibitory connection is implementable with a controlled leakage – similar to the leak of the leaky-integrate-and-fire neuron (LIF). As Fig. 2 shows, the parallel Resistor-Capacitor (RC) of the LIF neuron circuit can be improved with an N-Channel MOSFET transistor to control the current flowing out of the capacitor. In the parallel RC circuit, the current I_C flowing through the capacitor is equal to the current I_{R_i}:

$$I_C = I_{R_i} = \frac{V_C}{R_i} \tag{4}$$

Kirchhoff's voltage law states that the voltage of the capacitor is equal to the voltage drop across the resistor R_i – and the transistor – is equivalent to the voltage V_C of the capacitor:

$$\begin{aligned} &V_{R_i} + V_C = 0 \\ \Leftrightarrow\ &I_{R_i} R_i = -V_C \\ \Leftrightarrow\ &I_C R_i = -V_C \end{aligned} \tag{5}$$

Finally, as mentioned in Sect. 2.1, the instantaneous rate of voltage change $\frac{dV_C}{dt}$ of the capacitor can replace the capacitor's current term in the previous equation:

$$\tau_i \frac{dV_C}{dt} = -V_C \tag{6}$$

As for the series RC circuit, the time constant $\tau_i = R_i C$ is introduced which also represents the time required by the discharged capacitor to lose approximately 63.2% of its voltage. Thus, the previous equation can be integrated to obtain the capacitor's voltage $V_C(t + \Delta t)$ after a stimulation time Δt:

$$V_C(t + \Delta t) = V_C(t) e^{-\frac{\Delta t}{\tau_i}} \tag{7}$$

2.3 Integrate-and-Fire Neuron

Biological neurons receive several stimuli (excitatory and inhibitory) at their dendrites and having multiple inputs is a necessary condition to allow the IF model to compute separations of multi-dimensional spaces. Both excitatory and inhibitory can be combined to obtain several inputs – see an example in Fig. 3. In some specific situations, no excitation is provided by inputs and, for this reason, a bias connection – i.e. a connection always set to 1, as in rate-based models – is introduced to provide a constant stimulation. This allows a permanent charge of the capacitor and the neuron can become excited even if no pattern is provided. In such configuration, the total resistance of parallel resistors is not a simple sum of all the resistances but the inverse of the total resistance is the sum of all inverted resistances ($\frac{1}{R_{total}} = \sum_i^n \frac{1}{R_i}$). For this reason, computation of the IF model can become complex due to the differences of input stimulation times. For a lack of simplicity, inputs are stimulated one by one and as the capacitor must be charged to allow inhibition of the membrane potential inhibitory stimulations must follow excitatory ones. Therefore, the inference of the IF model becomes sequential and can be represented under a recurrent form where synapses are stimulated independently.

2.4 Integrate-and-Fire Neuron as a Recurrent Model

Sequential data are sequences with chronological order. In the deep learning field, this type of data is processed using recurrent units which are feedforward neural networks augmented with the inclusion of internal states of units, introducing a time dimension to the model [10]. At each step t of the inference of a recurrent neural network, the states at $t-1$ of the neurons are integrated into the computation. Intrinsically, the integration of stimulus in the IF model depends on the capacitor voltage – see Eqs. 3 and 7 – and can be expressed as a recurrent

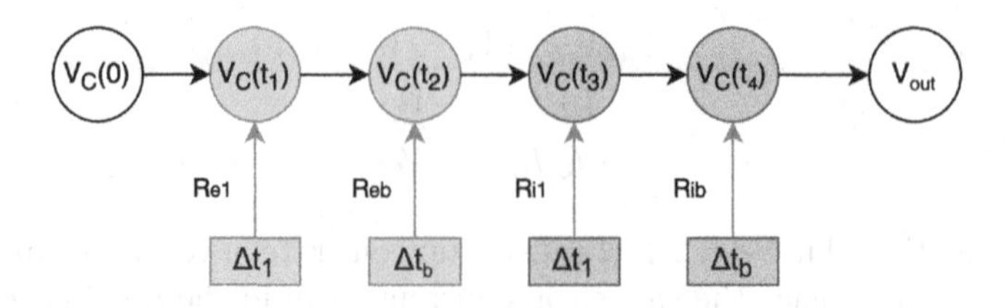

Fig. 3. Recurrent representation of the inference of the integrate and fire model. Green states represent excitatory stimulations and red states represent inhibition. (Color figure online)

form where each step is a precise synapse stimulation. As presented in Fig. 3, each step represents the stimulation of a synapse (excitatory ones first) and the hidden state is the potential of the neuron at time $t-1$ with an initial voltage of 0 – i.e. fully discharged capacitor. Thus, the final hidden state represents the membrane potential of the inferred neuron that can be compared with the voltage threshold to determine if the unit must release a spike or not – this step is achieved by the micro-controller controlling the circuit. The IF model defined as a recurrent form is a continuous and differentiable function which makes it suitable for the gradient descent algorithm. Therefore, some particular set of resistance values makes the IF neuron reach sub-threshold or super-threshold regimes for specific input and this behavior is exploited to achieve classification of patterns. To find the right combinations of resistances, optimisation algorithms can be used such as the well known gradient descent [9]. The loss of the model can be defined as the Mean Squared Error (MSE) between output membrane potentials and target potentials and the gradient used in the algorithm is computed with respect to resistance values.

2.5 Dataset, Network Architecture and Training

To demonstrate our model, we generated an artificial dataset of dog postures and trained a network of three IF neuron on it. It has been generated by using the average inclination vector for each class – i.e. we determined the average tilt of the device for each class – and created many samples by augmenting these vectors with random noise. The tilt of the device can be computed using both accelerometer and gyroscope data from inertial sensors [14] which gives a three-dimension vector (pitch, roll and yaw). In this work, three distinct classes of dog postures have been used: stand, sit and lay on the side. The average tilt vectors of each class can be determined by only the pitch and roll axis as following: $\begin{pmatrix}0 & 0\end{pmatrix}$ for stand, $\begin{pmatrix}0 & 0.25\end{pmatrix}$ for sit and $\begin{pmatrix}0.5 & 0\end{pmatrix}$ for lying – the maximum value for each axis is 1. The yaw axis is ignored because it corresponds to the horizontal angle of the device and is irrelevant in this case. From these average tilt vectors, we can generate new input samples with a normally distributed random noise $\epsilon \sim \mathcal{N}(0, 0.04^2)$.

The model has been implemented as a 3 neuron network – one per class – with both excitatory and inhibitory connections for every input to allow the model to have both types of connection and be flexible enough to achieve correct separations of the input space. The chosen capacitance value for the capacitors is $1e{-}6$ which is small enough to have a low charging time, but high enough to have fine control of the charge, to limit noise and voltage dissipation when implemented as hardware. The maximum stimulation time per input is defined as 50 ms – e.g. an input of value 1.0 will stimulate the corresponding synapse during 50 ms and an input of 0.5 during 25 ms. Finally, the output class is determined by the unit with the highest membrane potential using an argmax operator and the model has been trained using the gradient descent algorithm with a learning rate $\alpha = 5e^{-4}$.

During the training, the algorithm did not converge properly due to the scale of resistance values (between 10^3 and 10^6) which produces exceedingly large gradients. As the resistances are large and computed into gradients, the scale of gradients also becomes large. This very well known problem is known as *exploding gradient* in machine learning [12]. Many solutions exist to solve the exploding gradient problem such as gradient clipping [12]. However, the gradient clipping method makes gradients too small to converge in an acceptable amount of time – again due to the large scale of resistance values in the model. Another solution has been found to solve the issue: reduce the scale of resistances (between 1^{-3} and 1) and compensate with the capacitance value C of the unit. As the charge and discharge are driven by RC time constants $\tau = RC$, decreasing the resistance R can be balanced by increasing the capacitance C. Thus, by scaling down the resistance value, calculated gradients become small enough to obtain stable learning.

2.6 Weights Selection and Hardware Validation

Table 1. Resistance values (weights) of stand, lie and sit units. *Excit.* is for *Excitatory* and *Inhib.* is for *Inhibitory*. All values are given in kilohms ($k\Omega$).

Output neuron	Excit. x	Excit. y	Excit. bias	Inhib. x	Inhib. y	Inhib. bias
Stand	20.33	101.47	1.53	9.77	6.65	1000.00
Lie	7.61	1000.00	1000.00	1000.00	22.44	1000.00
Sit	1000.00	5.42	1000.00	19.57	1000.00	1000.00

Table 1 presents the weights of the model after training. In the IF neuron, a low resistance gives high weight to the input because it lets more current flow in or out of the capacitor and thus has a high contribution in its charge or discharge. Therefore, the contribution of very high resistances is insignificant and can be ignored. For this reason, all resistance values that converged to the maximum resistance (1000 kΩ) can be ignored in the trained model presented in Table 1 and consequently only 9 synapses remain out of the 18.

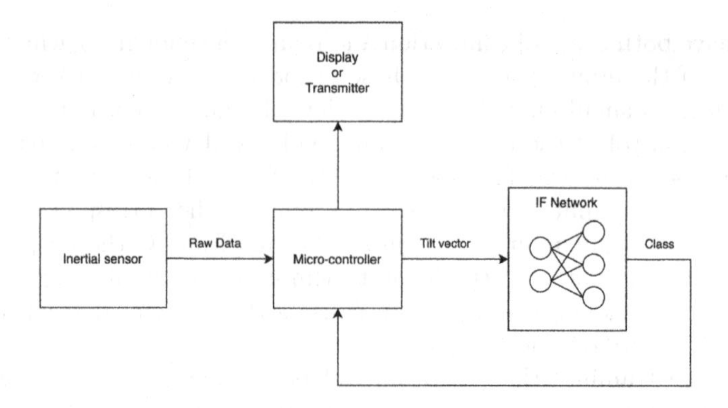

Fig. 4. Diagram of the experimental setup. The micro-controller collects the raw accelerometry and gyroscopic data from the inertial sensor, pre-process it to obtain tilt vectors that are sent to a network of IF neuron implemented as hardware. The class inferred by the network can be read by the micro-controller before being sent to the serial display (or a transmitter in real situations of remote systems).

After a weight selection (i.e. removing weights that converged to the maximum value), the three IF units for dog posture classification have been implemented as hardware to validate the training. The microcontroller used in this work is an ATmega328P on an Arduino Uno to ease its programming. An inertial unit (MPU-6050) is used to obtain accelerometry and gyroscopic data. Therefore the accelerometry data is used by the microcontroller to determine the gravity vector and the gyroscope data is integrated and combined with the previously computed vector to obtain the precise orientation of the device. Then, the microcontroller stimulates the synapses one by one during variable times depending on the pitch and roll of the device. The synapses charge (excitatory) then discharge (inhibitory) the capacitors using the digital pins. Finally, the microcontroller can read the membrane potential of each neuron by reading the voltage of the capacitors. See Fig. 4 for a diagram of the setup.

The model has been validated by sending all the possible inputs to the simulation and the hardware and comparing their responses. To achieve this, the hardware has not been tested using the inertial sensor but stimulated with the same tilt vectors as used in the simulation. Therefore, a mapping of the units' responses for both the simulation and the hardware has been generated – see Fig. 5. It appears that the behaviour of the electronic implementation is close to the simulation and the slight variations in voltage are due to noise and rounding of resistance values – e.g. a resistance of 3230 Ω in the simulation is rounded to 3000 Ω in the electronic implementation. Once the hardware is implemented and the model accuracy is validated, the power consumption of the device can be measured and compared to the use of simulated artificial neural networks.

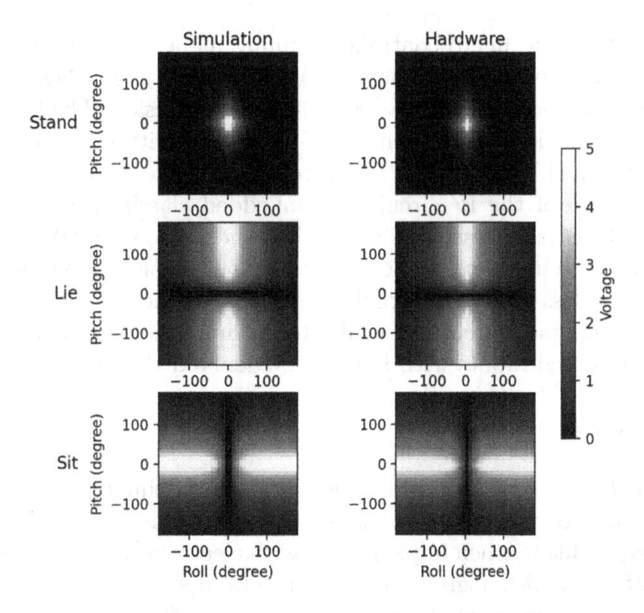

Fig. 5. Comparison of neural responses between simulated and electronic IF neurons. Each simulated unit (i.e. Stand, Sit and Lie) is compared with its corresponding electronic implementation by measuring the capacitor voltage for all possible inputs (pitch and roll). The hardware implementation of the model is very close to the simulation behavior and only varies due to electric noise and rounded resistance values.

Table 2. Comparison of average power consumptions of the micro-controller only, a Logistic Regression model running on the micro-controller and the micro-controller with the designed IF neurons. The values are given with and without the micro-controller power consumption to ease understanding.

Setup	Average power consumption with the micro-controller (in Watt)	Average power consumption without the micro-controller (in Watt)
Micro-controller only	0.2155	–
Logistic Regression on the micro-controller	0.2265	0.011
Micro-controller + IF circuits	**0.218**	**0.0025**

2.7 Power Consumption Analysis

To measure the power efficiency of the hardware, the current consumed by the device (i.e. the micro-controller, inertial sensors and IF circuits together) has been recorded while performing real-time classification of dog postures. The measures were done using a power analyzer and power supply (Otii ARC) and performed on the micro-controller running alone, on the micro-controller classifying postures with the IF neurons implemented as hardware and on the micro-controller classifying postures with logistic regression. The following average power consumptions have been determined and written in Table 2. Most of the power consumed by the device is due to the micro-controller, but the results show that the use of the IF circuit for embedded classification consumes less than a simple logistic regression performed on the micro-controller. If the power consumption of the micro-controller is ignored, the implemented hardware consumes 4.4 times less than the logistic regression method, which is significant. Therefore, the designed circuit is faithful to its simulation and able to recognize patterns in presented inputs with less energy demand than simulated ANNs.

3 Discussion

In this work, the Integrate-and-Fire model has been simulated and trained to achieve dog posture classification and showed promising results with relatively low energy expenditure when implemented with electronic components compared to the use of embedded logistic regression. The designed model implemented as hardware can be integrated into remote systems for embedded and energy-efficient pattern recognition, reducing the amount of data to transmit and thus reducing the number of transmissions, leading to low energy consumption. The simulation of the IF neuron is faithful to the electronic implementation which makes it possible to train using the gradient descent algorithm. Once trained, the resistances that do not contribute to the pattern detection – i.e. those that converge to the highest value – are removed from the final circuit and the remaining are implemented with the final hardware. This hardware implementation has been done with only a few passive components (resistors, diodes and capacitors) and one active component (N-MOSFET transistors) which all have low costs. It has been implemented using prototyping boards but can be miniaturised on Printed Circuit Boards (PCBs) with Surface Mount Technology (SMT) that provides miniature components to produce a version of the hardware small enough to be integrated into small devices.

In terms of power, the measured consumptions are almost identical due to the power demand of the micro-controller. However, the lifetime of battery-powered devices is very important and no aspect of the entire device should be overlooked, including the power usage of data processing. Therefore, by disregarding consumption of the micro-controller, the IF model consumes four times less when it is electronically implemented than a trained logistic regression running on the micro-controller. With this setup, the battery life-time is improved by 3.75%, but it can be enhanced even more by using a low-powered micro-controller. Moreover, an implementation of the model with spike trains should significantly reduce energy consumption. Therefore, it would be wise to rethink the way of communicating features given to the model using spike trains to further reduce the power consumption of the circuit.

One main issue of the IF approach is the time dependence of the inference. As the stimulation time of synapses varies according to the input values, the inference time is also variable. Thus, the higher the inputs, the longer the inference time will be. This maximum inference time can be calculated by summing the maximum stimulation time of inputs or can be compensated by varying the capacitance value of units. Another issue of the IF model is that the leak channel, specific to the LIF neuron, has been removed and the time dimension disappeared. This model is thus no longer able to process animal dynamics to infer its activity and only the posture – i.e. static patterns – can be classified. To achieve this task, the LIF model should be used which involves transforming features into spike trains. However, due to the non-continuity of spike trains in spiking neural networks, algorithms based on differentiation – such as the gradient descent algorithm used in this work – cannot be applied for training.

4 Future Work

In future works, the time capabilities of the Leaky Integrate-and-Fire model must be exploited to classify time-series patterns using spike trains. As the gradient descent algorithm is not suitable to train such models, other training algorithms must be explored to find new ways to classify patterns or compress sensory data into a spike code generated by a spiking neural network. Recent advances in neurosciences permitted the development of unsupervised learning algorithms such as Spike Time Dependent Plasticity (STDP) which is a biologically plausible Hebbian learning rule that adjusts the strength of connections between neurons in the brain [2,4,8]. Based on the timing of pre and post-synaptic spikes, STDP allows neurons to learn time-dependent correlations in spike trains and thus a relevant representation of input features [2,4,8]. Therefore, such algorithms may be able to find correlations between some sensory inputs and achieve a compression of recorded data.

References

1. Arkwright, A.C., et al.: Behavioral biomarkers for animal health: a case study using animal-attached technology on loggerhead turtles. Front. Ecol. Evol. **7**, 504 (2020)
2. Babacan, Y., Kaçar, F.: Memristor emulator with spike-timing-dependent-plasticity. AEU-Int. J. Electron. C. **73**, 16–22 (2017)
3. Davies, M., et al.: Loihi: a neuromorphic manycore processor with on-chip learning. IEEE Micro **38**, 1–1 (2018)
4. Feldman, D.: The spike-timing dependence of plasticity. Neuron **75**(4), 556–571 (2012)
5. Frenkel, C., Lefebvre, M.C., Legat, J.D., Bol, D.: A 0.086-mm^2 12.7-pj/sop 64k-synapse 256-neuron online-learning digital spiking neuromorphic processor in 28-nm CMOS. IEEE Trans. Biomed. Circ. Syst. **13**, 145–158 (2019)
6. Furber, S.B., Galluppi, F., Temple, S., Plana, L.A.: The spinnaker project. Proc. IEEE **102**(5), 652–665 (2014)
7. Gerencsér, L., Vásárhelyi, G., Nagy, M., Vicsek, T., Miklósi, A.: Identification of behaviour in freely moving dogs (canis familiaris) using inertial sensors. PLOS ONE **8** (2013)

8. Gerstner, W., Kistler, W.M.: Spiking Neuron Models: Single Neurons, Populations, Plasticity. Cambridge University Press, Cambridge (2002)
9. Jiawei, Z.: Gradient descent based optimization algorithms for deep learning models training (2019)
10. Lipton, Z.C.: A critical review of recurrent neural networks for sequence learning (2015)
11. Nathan, R., Spiegel, O., Fortmann-Roe, S., Harel, R., Wikelski, M., Getz, W.M.: Using tri-axial acceleration data to identify behavioral modes of free-ranging animals: general concepts and tools illustrated for griffon vultures. J. Exp. Biol. **215**(6), 986–996 (2012)
12. Pascanu, R., Mikolov, T., Bengio, Y.: On the difficulty of training recurrent neural networks (2012)
13. Perez-Cortes, J.C.: Pattern recognition in embedded systems: An overview. In: Chen, Q., Hameurlain, A., Toumani, F., Wagner, R., Decker, H. (eds.) Database and Expert Systems Applications, pp. 3–10. Springer International Publishing, Cham (2015). https://doi.org/10.1007/978-3-319-22849-5_1
14. Shahdloo, M., Khalkhali Sharifi, S.S., Vossoughi, G.: Precise tilt angle detection using gyro and accelerometer sensor fusion. In: The Bi-Annual International Conference on Experimental Solid Mechanics (2014)
15. Sorbaro, M., Liu, Q., Bortone, M., Sheik, S.: Optimizing the energy consumption of spiking neural networks for neuromorphic applications. Front. Neurosci. **14**, 662 (2020)
16. Swanson, O.K., Maffei, A.: From hiring to firing: activation of inhibitory neurons and their recruitment in behavior. Front. Mol. Neurosci. **12**, 168 (2019)
17. Williams, H., et al.: Identification of animal movement patterns using tri-axial magnetometry. Move. Ecol. **5** (2017)

A New Variant of the GQR Algorithm for Feedforward Neural Networks Training

Jarosław Bilski(✉) and Bartosz Kowalczyk

Department of Computational Intelligence, Częstochowa University of Technology, Częstochowa, Poland
{Jaroslaw.Bilski,Bartosz.Kowalczyk}@pcz.pl

Abstract. This paper presents an application of the scaled Givens rotations in the process of feedforward artificial neural networks training. This method bases on the QR decomposition. The paper describes mathematical background that needs to be considered during the application of the scaled Givens rotations in neural networks training. The paper concludes with sample simulation results.

Keywords: Neural network training algorithm · QR decomposition · Scaled givens rotations · Approximation · Classification

1 Introduction

Artificial intelligence is one of the most popular research areas worldwide. Together with the evolution of the industry towards digitalization and big data, a growing number of companies are starting to benefit from AI. In recent years many research projects have been conducted in multiple areas including security and surveillance [14,17,30], environment protection and air quality [37,41], vehicle and automation industries [12,13,27,28], classification, image and speech recognition [20,21,32–34] and others [29,36,39]. This produces an everlasting need for developing high performance algorithms for neural networks as attempted in [3,5,11,16,17,23,24,31,35].

Many modern training algorithms such as Adam and its derivatives [15,26,40] originate from the classic Backpropagation method initially presented in [38]. Some of the well performing algorithms such as the Levenberg-Marquardt [22] are burdened with a very high computational complexity, which makes them impractical in use for larger data sets [11]. Despite that, such algorithms are being developed due to continuous improvements of the processing devices such as CPUs and GPUs.

In this paper a novel approach to neural networks training is presented. The idea originates from the scaled Givens rotations and the GQR algorithm [4]. As a result of reducing the computational complexity of the classic GQR algorithm,

This work has been supported by the Polish National Science Center under Grant 2017/27/B/ST6/02852.

L. Rutkowski et al. (Eds.): ICAISC 2021, LNAI 12854, pp. 41–53, 2021.
https://doi.org/10.1007/978-3-030-87986-0_4

the scaled Givens rotations (SGQR) is expected to achieve better results in terms of the performance when compared to the classic GQR training method. This comparison is shown in the last section of this paper.

2 Givens Rotations Basics

The Givens rotation originates from elementary orthogonal transformations. The most commonly used rotation is limited to a single plain which is stretched between two vectors $span\{e_p, e_q\}(1 \le p < q \le n)$. The rotation itself is represented by an orthogonal matrix of the following structure [19,25]:

$$\mathbf{G}_{pq} = \begin{bmatrix} 1 & & & \cdots & & & 0 \\ & \ddots & & & & & \\ & & c & \cdots & s & & \\ \vdots & & \vdots & \ddots & \vdots & & \vdots \\ & & -s & \cdots & c & & \\ & & & & & \ddots & \\ 0 & & & \cdots & & & 1 \end{bmatrix} \begin{matrix} \\ \\ p \\ \\ q \\ \\ \\ \end{matrix} \quad (1)$$
$$\qquad\qquad p \qquad q$$

To keep it simple, matrix $\mathbf{G}_{pq}$ is referred to as the rotation matrix or rotation. By definition, this matrix differs from the Identity matrix only in terms of four elements $g_{pp} = g_{qq} = c$ and $g_{pq} = -g_{qp} = s$, where

$$c^2 + s^2 = 1 \quad (2)$$

From (2), it is known that $\mathbf{G}_{pq}^T \mathbf{G}_{pq} = \mathbf{I}$, which proves that matrix $\mathbf{G}_{pq}$ is an orthogonal matrix. Let $\mathbf{a} \in \mathbb{R}^n$. The rotation is performed by the orthogonal transformation given as

$$\mathbf{a} \rightarrow \bar{\mathbf{a}} = \mathbf{G}_{pq}\mathbf{a} \quad (3)$$

Due to that, only two elements of vector $\mathbf{a}$ are directly affected by the rotation. Based on that property, we are able to find values of c and s, so the a_q element equals 0 after being rotated. Let us consider

$$\bar{a}_q = -sa_p + ca_q = 0 \quad (4)$$

Parameters c and s of rotation matrix $\mathbf{G}_{pq}$ are calculated as follows

$$c = \frac{a_p}{\rho}, \quad s = \frac{a_q}{\rho}, \quad \text{where} \quad \rho = \sqrt{a_p^2 + a_q^2} \quad (5)$$

3 The Scaled Givens Rotation

For vector $\mathbf{a} \in \mathbb{R}^n$, consider transformation given by (3) [18,25]. Matrix $\mathbf{G}_{pq}$ has to meet the condition (4). The scaled Givens rotation is obtained by using scaled multipliers $\mathbf{K}^2$ and $\bar{\mathbf{K}}^2$:

$$\begin{aligned} \mathbf{a} &= \mathbf{K}\mathbf{d}, \quad \text{where} \quad \mathbf{K} = diag\left(\sqrt{\chi_l}\right) \\ \bar{\mathbf{a}} &= \bar{\mathbf{K}}\bar{\mathbf{d}}, \quad \text{where} \quad \bar{\mathbf{K}} = diag\left(\sqrt{\bar{\chi}_l}\right) \end{aligned} \tag{6}$$

where $\chi_l,\, \bar{\chi}_l > 0\,(l = 1, \dots, n)$. Also matrix $\mathbf{G}_{pq}$ will be presented in a scalable form

$$\mathbf{G}_{pq} = \mathbf{K}\mathbf{F}_{pq}\mathbf{K}^{-1} \tag{7}$$

where $\mathbf{F}_{pq}$ is:

$$\mathbf{F}_{pq} = \begin{bmatrix} 1 & & \cdots & & 0 \\ & \ddots & & & \\ & & \alpha \;\cdots\; \beta & & \\ \vdots & & \vdots \;\; \ddots \;\; \vdots & & \vdots \\ & & -\gamma \;\cdots\; \delta & & \\ & & & \ddots & \\ 0 & & \cdots & & 1 \end{bmatrix} \begin{matrix} \\ \\ p \\ \\ q \\ \\ \\ \end{matrix} \tag{8}$$
$$\qquad\qquad p \quad q$$

Equation (3) takes the form

$$\begin{aligned} &\mathbf{K}^2 \to \bar{\mathbf{K}}^2 \\ &\mathbf{d} \to \bar{\mathbf{d}} = \mathbf{F}_{pq}\mathbf{d} \end{aligned} \tag{9}$$

and Eq. (4) becomes the following

$$\bar{d_q} = -\gamma d_p + \delta d_q = 0 \tag{10}$$

From (7) the following is obtained

$$\bar{\chi}_l = \chi_l \quad \text{for} \quad (l \neq p, q; l = 1, ..., n) \tag{11}$$

$$c = \alpha\sqrt{\frac{\bar{\chi}_p}{\chi_p}} = \delta\sqrt{\frac{\bar{\chi}_q}{\chi_q}}, \quad s = \beta\sqrt{\frac{\bar{\chi}_p}{\chi_q}} = \gamma\sqrt{\frac{\bar{\chi}_q}{\chi_p}} \tag{12}$$

Equation (2) must also be satisfied.

Because there are six variables $\alpha, \beta, \delta, \gamma, \bar{\chi}_p, \bar{\chi}_q$ and only four Eqs. (10), (12) and (2), two cases have to be treated as parameters. Two variants are possible, see the important parts of the $\mathbf{F}_{pq}$ matrix

$$\begin{bmatrix} 1 & \beta \\ -\gamma & 1 \end{bmatrix} \quad \text{and} \quad \begin{bmatrix} \alpha & 1 \\ -1 & \delta \end{bmatrix} \tag{13}$$

From (5) and (6) the following is obtained

$$c^2 = \frac{a_p^2}{a_p^2 + a_q^2} = \frac{\chi_p d_p^2}{\chi_p d_p^2 + \chi_q d_q^2}, \quad s^2 = \frac{a_q^2}{a_p^2 + a_q^2} = \frac{\chi_q d_q^2}{\chi_p d_p^2 + \chi_q d_q^2} \tag{14}$$

There are two computational cases:

Case 1: $c \neq 0$ i.e. $d_p \neq 0$. The two parameters are set as follows

$$\alpha = \delta = 1 \tag{15}$$

from (10), (12) and (14) the following is obtained

$$\gamma = \frac{d_q}{d_p}, \quad \beta = \frac{\gamma \chi_q}{\chi_p} = \frac{\gamma \bar{\chi}_q}{\bar{\chi}_p}. \tag{16}$$

From (12) is $\bar{\chi}_i = \chi_i c^2$ for $i = p, q$. Taking into account equation

$$\frac{1}{c^2} = 1 + \beta\gamma \overset{def}{=} \tau \tag{17}$$

and (12), the following values are obtained

$$\bar{\chi}_p = \frac{\chi_p}{\tau}, \quad \bar{\chi}_q = \frac{\chi_q}{\tau}, \quad \bar{d}_p = d_p \tau. \tag{18}$$

Case 2: $s \neq 0$ i.e. $d_q \neq 0$. The two parameters are set as follows

$$\beta = \gamma = 1 \tag{19}$$

from (10) we obtain

$$\delta = \frac{d_p}{d_q}, \quad \alpha = \frac{\delta \chi_p}{\chi_q} = \frac{\delta \bar{\chi}_q}{\bar{\chi}_p}. \tag{20}$$

From (12) is $\bar{\chi}_p = \chi_q s^2$ and $\bar{\chi}_q = \chi_p s^2$. Taking into account equation

$$\frac{1}{s^2} = 1 + \alpha\delta \overset{def}{=} \tau \tag{21}$$

and (12) the obtained values are

$$\bar{\chi}_p = \frac{\chi_q}{\tau}, \quad \bar{\chi}_q = \frac{\chi_p}{\tau}, \quad \bar{d}_p = d_q \tau. \tag{22}$$

Equations (11, 15–22) allow to determine parameters $\alpha, \beta, \gamma, \delta$ of matrix $\mathbf{F}_{pq}$ and scaling multipliers $\bar{\chi}_i$.

4 The Scaled Givens Rotation in the QR Decomposition

The QR decomposition method assumes that any non-singular matrix regular by columns can be depicted by the product of the upper triangle and orthogonal matrices.

$$\mathbf{A} = \mathbf{QR}, \tag{23}$$

where

$$\mathbf{Q}^T\mathbf{Q} = \mathbf{I}, \tag{24}$$

$$\mathbf{Q}^T = \mathbf{Q}^{-1}, \tag{25}$$

$$r_{ij} = 0 \quad for\ i > j. \tag{26}$$

The presented process of the QR decomposition is called the Givens orthogonalization [25]. According to the previous equations, for any vector $\mathbf{a} \in \mathbb{R}^n$ and matrix $\mathbf{A} \in \mathbb{R}^{n,n}$, there exists a sequence of the scaled Givens rotations of $\mathbf{a} = \mathbf{Kd}$, and also $\mathbf{A} = \mathbf{KE}$, where $\mathbf{K} = diag(\sqrt{\chi_l})$, which leads to $\bar{\mathbf{a}} = \bar{\mathbf{K}}\bar{\mathbf{d}}$, $\bar{\mathbf{A}} = \bar{\mathbf{K}}\bar{\mathbf{E}}$, where $\bar{\mathbf{K}} = diag(\sqrt{\bar{\chi}_l})$

$$\left.\begin{array}{ll} \mathbf{K}_{11}^2 = \mathbf{K}^2, & \mathbf{K}_{1,i-1}^2 \rightarrow \mathbf{K}_{1,i}^2 \\ \mathbf{d}_1 = \mathbf{d}, & \mathbf{d}_{i-1} \rightarrow \mathbf{d}_i = \mathbf{F}_{1i}\mathbf{d} \\ \mathbf{E}_{11} = \mathbf{E}, & \mathbf{E}_{1,i-1} \rightarrow \mathbf{E}_{1,i} = \mathbf{F}_{1i}\mathbf{E}_{1,i-1} \end{array}\right\} (i = 2, \cdots, n), \begin{array}{l} \bar{\mathbf{K}}^2 = \mathbf{K}_1^2 = \mathbf{K}_{1n}^2 \\ \bar{\mathbf{d}} = \mathbf{d}_n = \prod_{i=2}^n \mathbf{F}_{1i}\mathbf{d} \\ \bar{\mathbf{E}} = \mathbf{E}_{1,n} = \prod_{i=2}^n \mathbf{F}_{1i}\mathbf{E}. \end{array} \tag{27}$$

Parameters $\alpha, \beta, \gamma, \delta$ of matrix $\mathbf{F}_{pq}$ and scaling multipliers $\bar{\chi}_i$ can be applied to matrix $\mathbf{A} = \mathbf{KE}$ to obtain matrix $\bar{\mathbf{A}} = \bar{\mathbf{K}}\bar{\mathbf{E}} = \bar{\mathbf{G}}_{pq}\bar{\mathbf{E}}$, where $\bar{\mathbf{E}}$ has the following values

$$\begin{array}{l} \bar{e}_{i,j} = e_{i,j} \quad \text{for} \quad (j = 1, ..., r; i \neq p, q; i = 1, ..., n) \\ \bar{e}_{p,j} = e_{p,j} + \beta e_{q,j}, \quad \bar{e}_{q,j} = -\gamma e_{p,j} + e_{q,j} \quad \text{for} \quad (j = 1, ..., r), \alpha = \delta = 1 \\ \bar{e}_{p,j} = \alpha e_{p,j} + e_{q,j}, \quad \bar{e}_{q,j} = -e_{p,j} + \delta e_{q,j} \quad \text{for} \quad (j = 1, ..., r), \beta = \gamma = 1 \end{array} \tag{28}$$

Matrix $\mathbf{F}_1$ is able to perform multiple rotations at once and transform vector $\mathbf{a}$ to the pattern given by the following form

$$\bar{\mathbf{a}} = \bar{\mathbf{K}}\bar{\mathbf{d}} = \mathbf{K}_1\mathbf{F}_1\mathbf{d} = \mathbf{K}_1 e_1 \rho = \mathbf{K}_1[\rho, 0, \ldots, 0]^T, \rho = \pm\|\mathbf{a}\|_2 \tag{29}$$

where

$$\mathbf{F}_1 = \prod_{i=2}^{n} \mathbf{F}_{1i} = \mathbf{F}_{12}\mathbf{F}_{13}\cdots\mathbf{F}_{1n} \tag{30}$$

In the scaled rotation, Eq. (27) are also able to transform the whole matrix. Let $\mathbf{A}$ be a non-singular matrix regular by columns and let $\mathbf{A} \in \mathbb{R}^{m,n}$. The left-sided multiplication of matrix

$$\mathbf{A} = \mathbf{A}_1 = \mathbf{M}_1 = \left[\mathbf{a}_1\ \mathbf{B}_1\right] \tag{31}$$

by matrices $\mathbf{K}_1$ and $\mathbf{F}_1$ results in a pattern shown in Eq. (32)

$$\mathbf{A}_2 = \mathbf{K}_1\mathbf{F}_1\mathbf{M}_1 = \mathbf{K}_1\left[\bar{\mathbf{a}}_1\ \bar{\mathbf{B}}_1\right] = \mathbf{K}_1\left[\begin{array}{c|c} \rho_1 & \bar{\mathbf{B}}_1 \\ \mathbf{0} & \end{array}\right] = \mathbf{K}_1\left[\begin{array}{c|c} r_{11} & r_{12}\cdots r_{1n} \\ \hline \mathbf{0} & \mathbf{M}_2 \end{array}\right] \tag{32}$$

At this point, the very left column vector of matrix $\mathbf{A}$ equals as shown in Eq. (29). The top row of matrix $\mathbf{A}$ is also already rotated as desired in the final upper-triangle form. In the next steps, new sequences of rotations need to be performed

$$\mathbf{K}_k\mathbf{F}_k = \mathbf{K}_k\prod_{i=k+1}^{n} \mathbf{F}_{ki} = \mathbf{K}_k\mathbf{F}_{k,k+1}\mathbf{F}_{k,k+2}\cdots\mathbf{F}_{kn} \tag{33}$$

By performing similar transformations of matrix $\mathbf{M}_k$, each time the input matrix is one step closer to the desired upper-triangle form

$$\mathbf{A}_{k+1} = \mathbf{K}_k \mathbf{F}_k \mathbf{M}_k = \mathbf{K}_k \left[\bar{\mathbf{a}}_k \ \bar{\mathbf{B}}_k \right] = \mathbf{K}_k \left[\begin{array}{c|c} \rho_k & \bar{\mathbf{B}}_k \\ \mathbf{0} & \end{array} \right] = \mathbf{K}_k \left[\begin{array}{c|c} r_{kk} & r_{k,k+1} \cdots r_{k,n} \\ \hline \mathbf{0} & \mathbf{M}_{k+1} \end{array} \right] \tag{34}$$

The algorithm is operational until it reaches $n-1$ steps. Then, the input matrix is fully transformed into the upper-triangle form

$$\mathbf{R} = \mathbf{K}_n \mathbf{F}_n \ldots \mathbf{F}_1 \mathbf{A}_1 = \mathbf{Q}^T \mathbf{A} \tag{35}$$

The full QR decomposition has been accomplished by the scaled Givens rotations as given in Eq. (23).

5 Weights Update in the SGQR Algorithm

The SGQR algorithm is designed for any multi-layered neural network with any differentiable activation function. The weight update is computed based on the error measure given as

$$J(n) = \sum_{t=1}^{n} \lambda^{n-t} \sum_{j=1}^{N_L} \varepsilon_j^{(L)2}(t) = \sum_{t=1}^{n} \lambda^{n-t} \sum_{j=1}^{N_L} \left[d_j^{(L)}(t) - f\left(\mathbf{x}^{(L)T}(t)\, \mathbf{w}_j^{(L)}(n) \right) \right]^2 \tag{36}$$

Finding the minimum of function (36) is a primary target for the SGQR algorithm. It starts with the classic error back propagation phase followed by linearisation of the activation function, which yields

$$\sum_{t=1}^{n} \lambda^{n-t} f'^2 \left(s_i^{(l)}(t) \right) \left[b_i^{(l)}(t) - \mathbf{x}^{(l)T}(t)\, \mathbf{w}_i^{(l)}(n) \right] \mathbf{x}^{(l)T}(t) = \mathbf{0} \tag{37}$$

The SGQR algorithm is using rotation matrices, hence Eq. (37) needs to be presented in the matrix notation as follows

$$\mathbf{A}_i^{(l)}(n)\, \mathbf{w}_i^{(l)}(n) = \mathbf{h}_i^{(l)}(n) \tag{38}$$

where

$$\mathbf{A}_i^{(l)}(n) = \sum_{t=1}^{n} \lambda^{n-t} \mathbf{z}_i^{(l)}(t)\, \mathbf{z}_i^{(l)T}(t) \tag{39}$$

$$\mathbf{h}_i^{(l)}(n) = \sum_{t=1}^{n} \lambda^{n-t} f' \left(s_i^{(l)}(t) \right) b_i^{(l)}(t)\, \mathbf{z}_i^{(l)}(t) \tag{40}$$

and

$$\mathbf{z}_i^{(l)}(t) = f' \left(s_i^{(l)}(t) \right) \mathbf{x}^{(l)}(t) \tag{41}$$

$$b_i^{(l)}(n) = \begin{cases} f^{-1}\left(d_i^{(l)}(n)\right) & \text{for} \quad l = L \\ s_i^{(l)}(n) + e_i^{(l)}(n) & \text{for} \quad l = 1 \dots L-1 \end{cases} \tag{42}$$

$$e_i^{(k)}(n) = \sum_{j=1}^{N_{k+1}} f'\left(s_i^{(k)}(n)\right) w_{ji}^{(k+1)}(n)\, e_j^{(k+1)}(n) \quad \text{for} \quad k = 1 \dots L-1 \tag{43}$$

All neurons of the network compute their own linear response ($s_i^{(l)}$). Due to that, Eq. (38) needs to be solved for each neuron. Equation (38) is solved by the Givens QR decomposition as described in the previous section. During the process, orthogonal matrix $\mathbf{Q}^T$ is implicitly calculated, but it is not stored in the memory. Only the rotations described by the c and s parameters are applied

$$\mathbf{Q}_i^{(l)T}(n)\,\mathbf{A}_i^{(l)}(n)\,\mathbf{w}_i^{(l)}(n) = \mathbf{Q}_i^{(l)T}(n)\,\mathbf{h}_i^{(l)}(n) \tag{44}$$

$$\mathbf{R}_i^{(l)}(n)\,\mathbf{w}_i^{(l)}(n) = \mathbf{Q}_i^{(l)T}(n)\,\mathbf{h}_i^{(l)}(n) \tag{45}$$

Equation (45) yields fully transformed matrix $\mathbf{A}_i^{(l)}(n)$ to its upper-triangle given as $\mathbf{R}_i^{(l)}(n)$. Since $\mathbf{R}_i^{(l)}(n)$ is an upper-triangle matrix, its inversion is no longer so expensive. The weight update formula of the i-th neuron takes the following form

$$\hat{\mathbf{w}}_i^{(l)}(n) = \mathbf{R}_i^{(l)-1}(n)\;\mathbf{Q}_i^{(l)T}(n)\;\mathbf{h}_i^{(l)}(n) \tag{46}$$

$$\mathbf{w}_i^{(l)}(n) = (1-\eta)\,\mathbf{w}_i^{(l)}(n-1) + \eta\,\hat{\mathbf{w}}_i^{(l)}(n) \tag{47}$$

6 Experimental Results

The SGQR algorithm has been tested against the classic GQR variant. The scope of the experiment includes three types of feedforward neural networks, i.e. MLP—the Multi-Layered Perceptron, FCMLP—the Fully Connected Multi-Layered Perceptron, and FCC—the Fully Connected Cascade. The performance of the presented SGQR algorithm has been measured in two areas: SR—Success Ratio and T—average training time in milliseconds. The presented results have been gathered according to the best combination of SGQR's hyperparameters η and λ. The initial phase of the experiment assumed a search for the highest value of the algorithm's performance factor given as (48). Each training has been attempted 100 times to gather valuable statistics data.

$$\xi = \frac{\mathrm{SR}}{\mathrm{Ep} \cdot \mathrm{T}} \tag{48}$$

where Ep is the average epoch count.

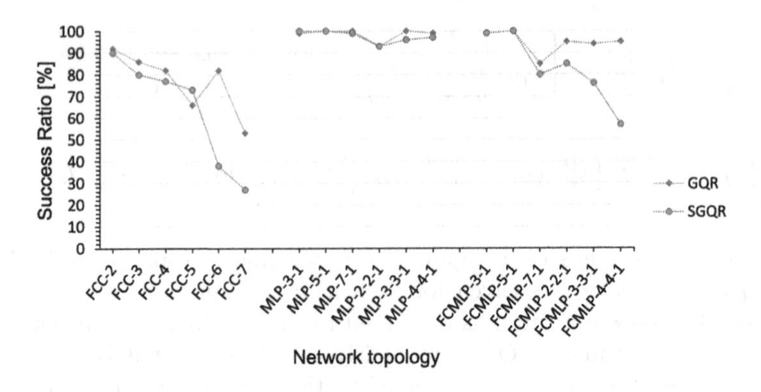

Fig. 1. The logistic function success ratio.

6.1 Logistic Function Approximation

The logistic function performs a single argument mapping according to the following equation

$$y = f(x) = 4x(1-x) \tag{49}$$

The teaching sequence contains 11 samples where $x \in [0,1]$ with the average accepted error threshold set to 0.001.

In Fig. 1 the success ratio across all tested networks is given. The overall success ratio of the SGQR algorithm is similar as for the classic GQR variant. One can observe lower values of the SR for the bigger FCC and FCMLP networks.

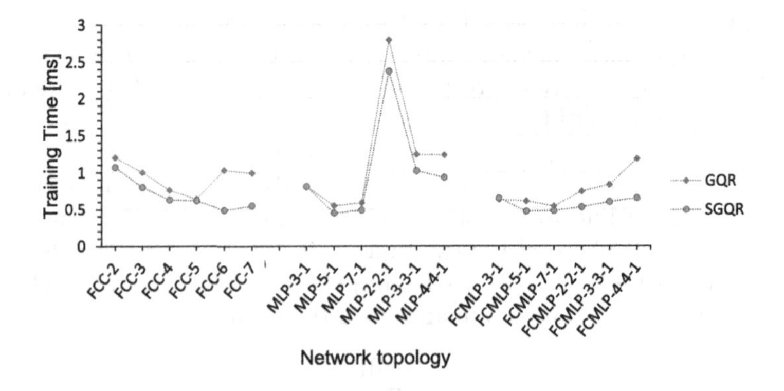

Fig. 2. The logistic function average time.

In terms of the average training time (Fig. 2), the SGQR convergence time is on average 22% shorter when compared with GQR. The time difference grows together with the network's size.

6.2 Hang Function Approximation

The Hang function performs a non-linear mapping of two arguments x_1 and x_2 with respect to the following formula

$$y = f(x_1, x_2) = \left(1 + x_1^{-2} + \sqrt{x_2^{-3}}\right)^2 \tag{50}$$

The Hang teaching sequence contains 50 samples which cover arguments in the range of $x_1, x_2 \in [1, 5]$. The target error threshold was set to 0.001 as the epoch average.

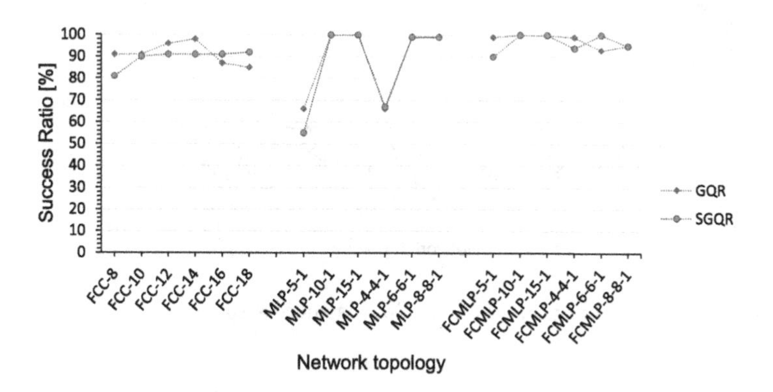

Fig. 3. The Hang success ratio.

The success ratio for the Hang benchmark is shown in Fig. 3. Both tested algorithms showed a similar performance in this scope.

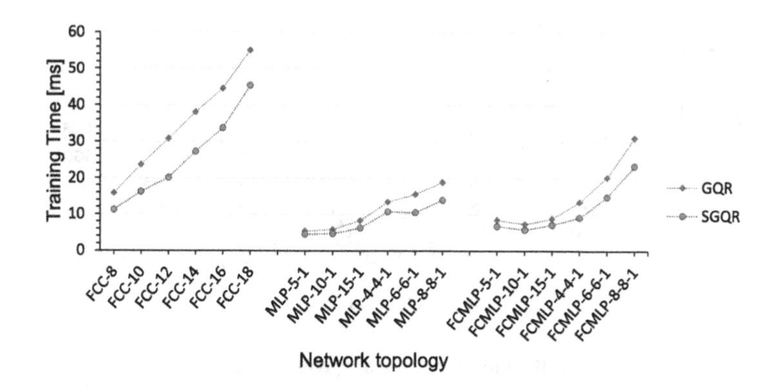

Fig. 4. The Hang average time.

In Fig. 4 the average training time comparison is shown. The SGQR convergence time is on average 26% shorter than for the classic GQR algorithm. The time gap grows together with the network's size.

6.3 The Two Spirals Classification

The Two Spirals is a well known classification problem, where a neural network needs to group incoming two-dimensional samples into one of two spirals. The training sequence in this benchmark contains 96 samples. The trial is assumed to be successful if the average epoch error goes below the 0.05 threshold.

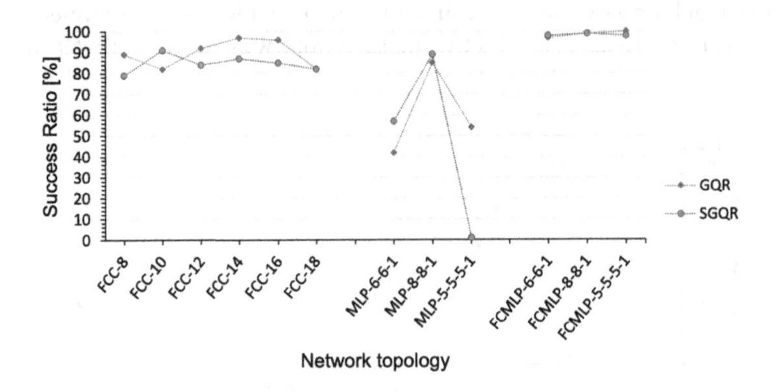

Fig. 5. The Two Spirals success ratio.

The Two Spirals success ratio is shown in Fig. 5. One can observe that the SGQR performance drops low for the MLP-5-5-5-1 network. For the FCC and FCMLP networks the success ratio for both algorithms is similar.

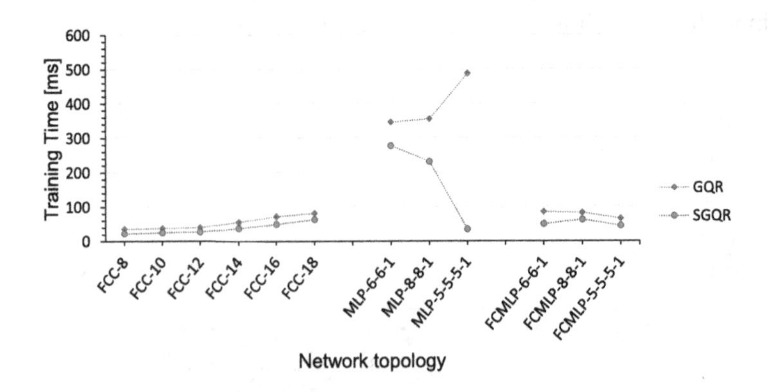

Fig. 6. The Two Spirals average time.

In terms of the average training time (Fig. 6), one can observe that the SGQR training converges on average 31% faster than the classic GQR algorithm (excluding the MLP networks).

7 Conclusion

In this paper the novel approach to the GQR algorithm has been presented. The proposed modification utilizes the scaled Givens rotations, hence its name—SGQR. The main section of the paper presents a comprehensive discussion of the mathematical background of the algorithm. The experiments concluded in Sect. 6 show that the proposed method is superior for the classic GQR variant in terms of the average training time. The SGQR algorithm maintains a very high success ratio. Due to the close relation of the SGQR to the GQR, it can be applied to any feedforward neural network. The flexibility of this method opens a lot of opportunities for further research projects, such as momentum [1] or parallel variants as shown in [2,6–10].

References

1. Bilski, J.: Momentum modification of the RLS algorithms. In: Rutkowski, L., Siekmann, J.H., Tadeusiewicz, R., Zadeh, L.A. (eds.) ICAISC 2004. LNCS (LNAI), vol. 3070, pp. 151–157. Springer, Heidelberg (2004). https://doi.org/10.1007/978-3-540-24844-6_18
2. Bilski, J.: Parallel structures for feedforward and dynamic neural networks. (In Polish) Akademicka Oficyna Wydawnicza EXIT (2013)
3. Bilski, J., Kowalczyk, B., Grzanek, K.: The parallel modification to the Levenberg-Marquardt algorithm. In: Rutkowski, L., Scherer, R., Korytkowski, M., Pedrycz, W., Tadeusiewicz, R., Zurada, J.M. (eds.) ICAISC 2018. LNCS (LNAI), vol. 10841, pp. 15–24. Springer, Cham (2018). https://doi.org/10.1007/978-3-319-91253-0_2
4. Bilski,, J., Kowalczyk, B., Żurada, J.M.: Application of the Givens rotations in the neural network learning algorithm. In: Artificial Intelligence and Soft Computing, volume 9602 of Lecture Notes in Artificial Intelligence, pp. 46–56. Springer-Verlag, Berlin, Heidelberg (2016). https://doi.org/10.1007/978-3-319-39378-0_5
5. Bilski, J., Kowalczyk, B., Żurada, J.M.: Parallel implementation of the givens rotations in the neural network learning algorithm. In: Rutkowski, L., Korytkowski, M., Scherer, R., Tadeusiewicz, R., Zadeh, L.A., Zurada, J.M. (eds.) ICAISC 2017. LNCS (LNAI), vol. 10245, pp. 14–24. Springer, Cham (2017). https://doi.org/10.1007/978-3-319-59063-9_2
6. Bilski, J., Smoląg, J.: Parallel realisation of the recurrent multi layer perceptron learning. Artificial Intelligence and Soft Computing, pp. 12–20. Springer-Verlag, Berlin, Heidelberg, (LNAI 7267) (2012). https://doi.org/10.1007/978-3-642-13232-2_3
7. Bilski, J., Smoląg, J.: Parallel approach to learning of the recurrent Jordan neural network. Artificial Intelligence and Soft Computing, pp. 32–40. Springer-Verlag, Berlin, Heidelberg (LNAI 7895) (2013)
8. Bilski, J., Smoląg, J.: Parallel architectures for learning the RTRN and Elman dynamic neural network. IEEE Trans. Parallel Distrib. Syst. **26**(9), 2561–2570 (2015)
9. Bilski, J., Smoląg, J., Galushkin, A.I.: The parallel approach to the conjugate gradient learning algorithm for the feedforward neural networks. In: Artificial Intelligence and Soft Computing, volume 8467 of Lecture Notes in Computer Science, pp. 12–21. Springer-Verlag, Berlin, Heidelberg (2014). https://doi.org/10.1007/978-3-319-07173-2_2

10. Bilski, J., Smoląg, J., Żurada, J.M.: Parallel approach to the Levenberg-Marquardt learning algorithm for feedforward neural networks. In: Rutkowski, L., Korytkowski, M., Scherer, R., Tadeusiewicz, R., Zadeh, L.A., Zurada, J.M. (eds.) ICAISC 2015. LNCS (LNAI), vol. 9119, pp. 3–14. Springer, Cham (2015). https://doi.org/10.1007/978-3-319-19324-3_1
11. Bilski, J., Kowalczyk, B., Marchlewska, A., Zurada, J.M.: Local levenberg-marquardt algorithm for learning feedforwad neural networks. J. Artif. Intell. Soft Comput. Res. **10**(4), 299–316 (2020)
12. Cao, Y., Samidurai, R., Sriraman, R.: Stability and dissipativity analysis for neutral type stochastic markovian jump static neural networks with time delays. J. Artif. Intell. Soft Comput. Res. **9**(3), 189–204 (2019)
13. Costa, M., Oliveira, D., Pinto, S., Tavares, A.: Detecting driver's fatigue, distraction and activity using a non-intrusive AI-based monitoring system. J. Artif. Intell. Soft Comput. Rese. **9**(4), 247–266 (2019)
14. de Souza, G.B., da Silva Santos, D.F., Pires, R.G., Marananil, A.N., Papa, J.P.: Deep features extraction for robust fingerprint spoofing attack detection. J. Artif. Intell. Soft Comput. Res. **9**(1), 41–49 (2019)
15. Duchi, J., Hazan, E., Singer, Y.: Adaptive subgradient methods for online learning and stochastic optimization. J. Mach. Learn. Res. **12**, 2121–2159 (2011)
16. Duda, P., Jaworski, M., Cader, A., Wang, L.: On training deep neural networks using a streaming approach. J. Artif. Intell. Soft Computi. Res. **10**(1), 15–26 (2020)
17. Gabryel, M., Grzanek, K., Hayashi, Y.: Browser fingerprint coding methods increasing the effectiveness of user identification in the web traffic. J. Artif. Intell. Soft Computi. Res. **10**(4), 243–253 (2020)
18. Gentleman, W.M.: Least squares computations by givens transformations without square roots. IMA J. Appl. Math. **12**(3), 329–336 (1973)
19. Givens, W.: Computation of plain unitary rotations transforming a general matrix to triangular form. J. Soc. Indust. Appl. Math. **6**, 26–50 (1958)
20. Grycuk, R., Najgebauer, P., Kordos, M., Scherer, M.M., Marchlewska, A.: Fast image index for database management engines. J. Artif. Intell. Soft Computi. Res. **10**(2), 113–123 (2020)
21. Grycuk, R., Wojciechowski, A., Wei, W., Siwocha, A.: Detecting visual objects by edge crawling. J. Artif. Intell. Soft Computi. Res. **10**(3), 223–237 (2020)
22. Hagan, M.T., Menhaj, M.B.: Training feedforward networks with the marquardt algorithm. IEEE Trans. Neuralnetw. **5**, 989–993 (1994)
23. Hou, Y., Holder, L.B.: On graph mining with deep learning: introducing model r for link weight prediction. J. Artif. Intell. Soft Computi. Res. **9**(1), 21–40 (2019)
24. Kamimura, R.: Supposed maximum mutual information for improving generalization and interpretation of multi-layered neural networks. J. Artif. Intell. Soft Computi. Res. **9**(2), 123–147 (2019)
25. Kiełbasiński, A., Schwetlick, H.: Numeryczna Algebra Liniowa: Wprowadzenie do Obliczeń Zautomatyzowanych. Wydawnictwa Naukowo-Techniczne, Warszawa (1992)
26. Kingma, D.P., Ba, J.: Adam: A method for stochastic optimization (2014)
27. Krell, E., Sheta, A., Balasubramanian, A.P.R., King, S.A.: Collision-free autonomous robot navigation in unknown environments utilizing pso for path planning. J. Artif. Intell. Soft Comput. Res. **9**(4), 267–282 (2019)
28. Kumarratneshk, R., Weilleweill, E., Aghdasi, F., Sriram, P.: A strong and efficient baseline for vehicle re-identification using deep triplet embedding. J. Artif. Intell. Soft Comput. Res. **10**(1), 27–45 (2020)

29. Łapa, K., Cpałka, K., Wang, L.: New method for design of fuzzy systems for nonlinear modelling using different criteria of interpretability. In: Rutkowski, L., Korytkowski, M., Rafał Scherer, R., Tadeusiewicz, L.A.Z., Zurada, J.M. (eds.) Artificial Intelligence and Soft Computing, pp. 217–232. Springer International Publishing, Cham (2014). https://doi.org/10.1007/978-3-319-07173-2_20
30. Ludwig, S.A.: Applying a neural network ensemble to intrusion detection. J. Artif. Intell. Soft Comput. Res. **9**(3), 177–188 (2019)
31. Abbas, M. Javaid, M., Jia-Bao, L., Teh, W.C., Jinde, C.: Topological properties of four-layered neural networks. J. Artif. Intell. Soft Comput. Res. **9**(2), 111–122 (2019)
32. Nobukawa, S., Nishimura, H., Yamanishi, T.: Pattern classification by spiking neural networks combining self-organized and reward-related spike-timing-dependent plasticity. J. Artif. Intell. Soft Comput. Res. **9**(4), 283–291 (2019)
33. Nowicki, R.K., Grzanek, K., Hayashi, Y.: Rough support vector machine for classification with interval and incomplete data. J. Artif. Intell. Soft Comput. Res. **10**(1), 47–56 (2020)
34. Shewalkar, A., Nyavanandi, D., Ludwig, S.A.: Performance evaluation of deep neural networks applied to speech recognition: RNN, LSTM and GRU. J. Artif. Intell. Soft Comput. Res. **9**(4), 235–245 (2019)
35. Simões, D., Lau, N., Reis, L.P.: Multi agent deep learning with cooperative communication. J. Artif. Intell. Soft Comput. Res. **10**(3), 189–207 (2020)
36. Szczypta, J., Przybył, A., Cpałka, K.: Some aspects of evolutionary designing optimal controllers. In: Rutkowski, L., Korytkowski, M., Scherer, R., Tadeusiewicz, R., Zadeh, L.A., Zurada, J.M. (eds.) ICAISC 2013. LNCS (LNAI), vol. 7895, pp. 91–100. Springer, Heidelberg (2013). https://doi.org/10.1007/978-3-642-38610-7_9
37. Wei, Y., Ying, Yu., Lifeng, X., Huang, W., Guo, J., Wan, Y., Cao, J.: Vehicle emission computation through microscopic traffic simulation calibrated using genetic algorithm. J. Artif. Intell. Soft Comput. Res. **9**(1), 67–80 (2019)
38. Werbos, J.: Beyond Regression: New Tools for Prediction and Analysis in the Behavioral Sciences. Harvard University (1974)
39. Zalasiński, M., Cpałka, K., Er, M.J.: New method for dynamic signature verification using hybrid partitioning. In: Rutkowski, L., Korytkowski, M., Rafał Scherer, R., Tadeusiewicz, L.A.Z., Zurada, J.M. (eds.) Artificial Intelligence and Soft Computing, pp. 216–230. Springer International Publishing, Cham (2014). https://doi.org/10.1007/978-3-319-07176-3_20
40. Zeiler, M.D.: An adaptive learning rate method, Adadelta (2012)
41. Zhao, X., Song, M., Liu, A., Wang, Y., Wang, T., Cao, J.: Data-driven temporal-spatial model for the prediction of AQI in nanjing. J. Artif. Intell. Soft Comput. Res. **10**(4), 255–270 (2020)

Modification of Learning Feedforward Neural Networks with the BP Method

Jarosław Bilski(✉), Jacek Smoląg, and Patryk Najgebauer

Department of Computer Engineering, Częstochowa University of Technology, al. Armii Krajowej 36, 42-200 Częstochowa, Poland
{Jaroslaw.Bilski,Jacek.Smolag,Patryk.Najgebauer}@pcz.pl

Abstract. The backpropagation (BP) algorithm is a worldwide used method for learning neural networks. The BP has a low computational load. Unfortunately, this method converges relatively slowly. In this paper a new approach to the backpropagation algorithm is presented. The proposed solution speeds up the BP method by using vector calculations. This modification of the BP algorithm was tested on a few standard examples. The obtained performance of both methods was compared.

Keywords: Feedforward neural network · Neural network learning algorithm · Backpropagation algorithm · Parallel computation

1 Introduction

Artificial intelligence methods have become an increasingly important and interesting field for science and industry. They are studied by many authors, e.g. [1–11]. Feedforward neural networks (FNN) are often used in numerous research projects and applications. They are used in many fields, e.g.: approximation, classification, prediction, pattern recognition, or signal processing [14–18]. Many researchers are working on the development of the FNN, e.g. [19–23]. This results in a lot of works on neural networks application [24,25] and parallel processing in neural networks [26–33]. Learning of artificial neural networks is very important. To train FNNs, gradient methods [34–36], the conjugate gradient (CG) algorithm [37–42], the Levenberg-Marquardt [43] and others [44], are usually used.

The bacpropagation algorithm is a gradient algorithm based on the steepest descent method for each iteration of the learning process. Generally, the neural networks learning algorithms are implemented on a serial computer. The learning algorithms require a high computational load. Fortunately, many new processors have many cores and vector instructions, which can be used in parallelization computations.

This paper presents a new approach to learning algorithms based on parallel computations using vector instructions. The proposed solution executes a few

This work has been supported by the Polish National Science Center under Grant 2017/27/B/ST6/02852.

L. Rutkowski et al. (Eds.): ICAISC 2021, LNAI 12854, pp. 54–65, 2021.
https://doi.org/10.1007/978-3-030-87986-0_5

learning steps with a different learning coefficient, simultaneously. Then, based on the best result suitable, new steps are chosen for the next calculation step. This is achieved by using vector calculations. This significantly reduces the computation time. The results part of this paper shows a very promising performance of the proposed approach.

2 Background

In this paper, the backpropagation algorithm is used for FNNs. FNNs have various structures. The data are delivered to the input of the neural network, they are processed by neurons and the result is obtained at the output of the network. In this paper three structures of FNNs are used. The first structure is the multilayer perceptron (MLP), see Fig. 1. This network is divided into a few layers, and each layer is built from neurons. The first layer is connected to the input of the network. The next layer is connected to the output of the previous layer, and so on. The last layer outputs are at the same time the FNN outputs. All layers except the last (output) layer are called hidden layers.

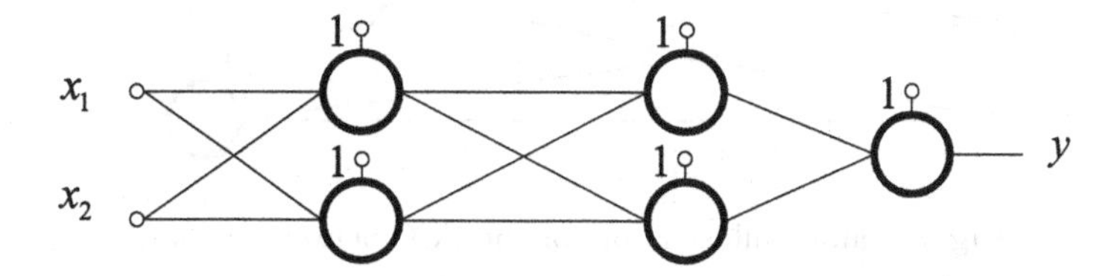

Fig. 1. Sample illustration for the MLP neural network.

The second FNN structure is the fully connected multilayer perceptron (FCMLP), see Fig. 2. This type of network is similar to the classic MLP, but its layers are connected to the outputs of all the previous layers and to the network inputs.

The third considered structure is the fully connected cascade network (FCC), see Fig. 3. In this network each layer contains only one neuron and the connections are identical to the FCMLP network.

Consider the FNN with L layers, N_l neurons in each $l-th$ layer and N_L outputs. The input vector has N_0 input values. The FNN recall phase is defined by formulas

$$\begin{aligned} s_i^{(l)} &= \sum_{j=0}^{N_{l-1}} w_{ij}^{(l)} x_i^{(l)} \\ y_i^{(l)}(t) &= f(s_i^{(l)}(t)) \end{aligned} \tag{1}$$

The BP algorithm [36] is used to train FNNs. The following target criterion is minimized

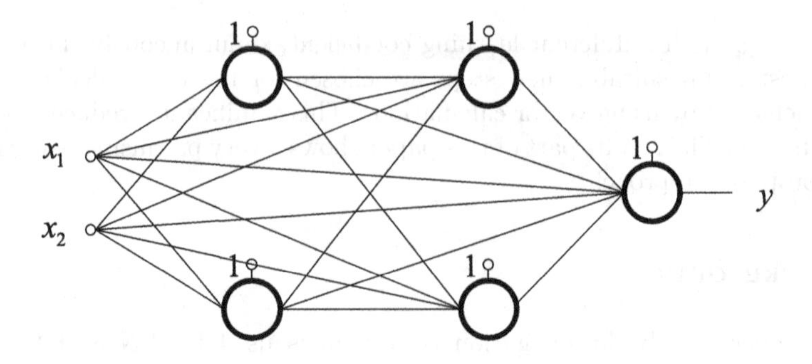

Fig. 2. Sample illustration for the FCMLP neural network.

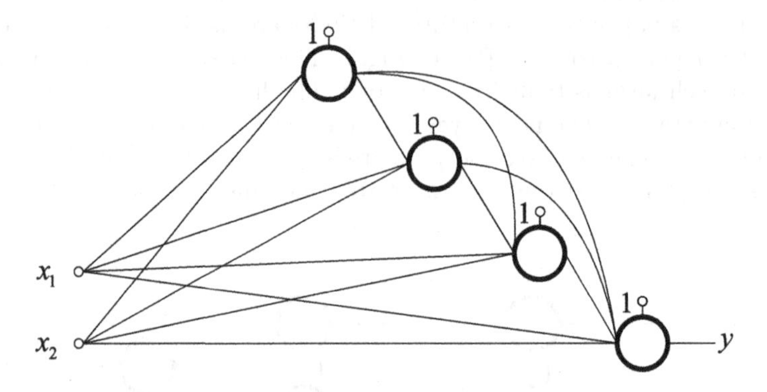

Fig. 3. Sample illustration for the FCC neural network.

$$J(t) = \frac{1}{2} \sum\nolimits_{i=1}^{N_L} \varepsilon_i^{(L)^2}(t) = \frac{1}{2} \sum\nolimits_{i=1}^{N_L} \left(d_i^{(L)}(t) - y_i^{(L)}(t) \right)^2 \tag{2}$$

where $\varepsilon_i^{(L)}$ is defined as

$$\varepsilon_i^{(L)}(t) = d_i^{(L)}(t) - y_i^{(L)}(t) \tag{3}$$

and $y_i^{(L)}(t)$ is the $i-th$ network output and $d_i^{(L)}(t)$ is the $i-th$ desired output. Using the steepest descent method, the new weight $w_{ij}^{(l)}(n+1)$ can be calculated by

$$w_{ij}^{(l)}(n+1) = w_{ij}^{(l)}(n) + \eta \left(-\nabla J_{ij}^{(l)}(n) \right) \tag{4}$$

where $\nabla J_{ij}^{(l)}(n)$ is the gradient computed by

$$\nabla J_{ij}^{(l)}(n) = \frac{\partial J(n)}{\partial w_{ij}^{(l)}(n)} = \frac{\partial J(n)}{\partial s_i^{(l)}(n)} x_j^{(l)} = -\delta_i^{(l)} x_j^{(l)} \tag{5}$$

and where $\delta_i^{(l)}(n)$ is defined by $\delta_i^{(l)}(n) \triangleq -\frac{\partial J(n)}{\partial s_i^{(l)}(n)}$. Errors $\varepsilon_i^{(l)}$ and deltas $\delta_i^{(l)}(n)$ in the hidden layers are calculated as follows

$$\varepsilon_i^{(l)}(t) \triangleq \sum_{m=1}^{N_{l+1}} \delta_i^{(l+1)}(t)\, w_{mi}^{(l+1)}(t), \tag{6}$$

$$\delta_i^{(l)}(t) = \varepsilon_i^{(l)}(t)\, f'\left(s_i^{(l)}(t)\right). \tag{7}$$

Finally, weights $w_{ij}^{(l)}(n+1)$ are determined by

$$w_{ij}^{(l)}(n+1) = w_{ij}^{(l)}(n) + \eta \delta_i^{(l)} x_j^{(l)}. \tag{8}$$

3 Vector Approach to Learning Neural Networks

In the proposed solution, the possibilities offered by modern processors are used. They have vector registers containing a few values (2^k) and they can compute simultaneously using all vector values in one instruction. In the vector approach, the 2^k starting points are used. The recall phase for all points is computed by formulas

$$s_{iv}^{(l)} = \sum_{j=0}^{N_{l-1}} w_{ijv}^{(l)} x_i^{(l)}, \quad y_{iv}^{(l)}(t) = f(s_{iv}^{(l)}(t)) \tag{9}$$

where $v = 0, \ldots, 2^k - 1$. Index v means the v-th point number. All values with different v indexes are placed in the vector register and $2^k - 1$ computations are carried out simultaneously. Next, the target criteria are computed

$$J_v(t) = \frac{1}{2}\sum_{i=1}^{N_L} \varepsilon_{iv}^{(L)^2}(t) = \frac{1}{2}\sum_{i=1}^{N_L} \left(d_i^{(L)}(t) - y_{iv}^{(L)}(t)\right)^2 \tag{10}$$

After that, the lowest criterion value from all points needs to be found

$$\exists v_{\min} \forall v \left(J_{v_{\min}}(t) \le J_v(t)\right) \text{ for } v = 0, \ldots, 2^k - 1 \tag{11}$$

and the best point is chosen (the smallest point)

$$w_{ij}^{(l)}(n+1) = w_{ijv_{min}}^{(l)}(n). \tag{12}$$

For this point errors $\varepsilon_i^{(l)}$ and deltas $\delta_i^{(l)}(n)$ are designated by (6, 7). Now, the new 2^k points are calculated

$$w_{ijv}^{(l)}(n+1) = w_{ij}^{(l)}(n) + \eta_v \delta_i^{(l)} x_j^{(l)} \quad \text{where} \quad v = 0, \ldots, 2^k - 1. \tag{13}$$

Generally, the η learning rate is a constant (8). In the proposed approach, the 2^k coefficients η_v are used. They are used to calculate the 2^k new points. Note that by using vector instructions, the time needed to compute the 2^k new points is identical to the computation time of one point. All new points are calculated

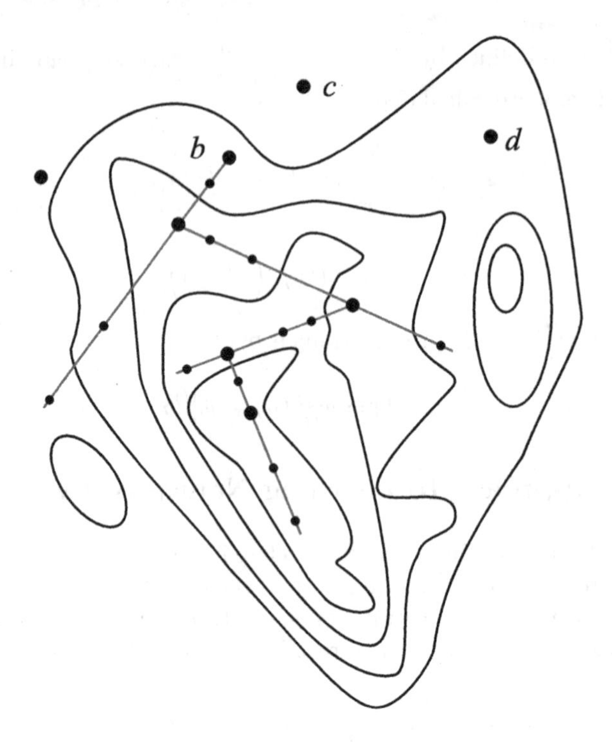

Fig. 4. Sample illustration for directional minimization.

according to the steepest descent rule and are placed all in one direction. The η_v coefficients are selected in such a way so as to increase the scope of the search for new points, from which the one with the lowest value of the minimizing criterion is chosen, see Fig. 4. This causes that in each iteration a lower minimization criterion value is obtained compared to the value obtained with the constant learning rate. This leads to faster algorithm convergence to the minimum point reducing computation time without using more than one processor core.

4 Experimental Results

In the paper a new approach to the error backpropagation algorithm is presented. The new approach was compared with the classic BP algorithm. All the simulations were carried out on a computer with the Intel i9-9980X processor and 128 GiB memory. As part of the tests, three selected problems were simulated; i.e., the logistic function, the SINC function and the circle problem. All problems were simulated for various networks including the MLP, FCMLP and FCC with a select few architectures. Each problem was tested using nine architectures. The simulations results are presented in the tables. In each case, 100 experiments were performed with a maximum of 1000 epochs and at a specified error and learning rate. In all cases, the weights were set at random values in the range $[-0.5 - 0.5]$. The first column of each table presents network architectures with the number of hidden neurons and the number of outputs for the MLP and FCMLP networks or with the number of neurons for the FCC network. The next four columns present the used learning rate, the success ratio (SR), the average number of epochs and the average time for the classic BP method. The next four columns show these values for the new approach. The last two columns show how the number of epochs (ED) and the time (TD) of the new approach decrease compared to the classical BP algorithm.

4.1 The Logistic Function

The logistic function is given by equation:

$$f(x) = 4x(1 - x) \quad x \in [0, 1]. \tag{14}$$

The training set has 11 samples from the range of $x \in [0, 1]$. The target error is 0.005. All simulation results are presented in Table 1. It should be noted that the success ratio is similar in both cases, but the number of epochs and the time for the new approach to the BP algorithm is significantly shorter. The average acceleration for epochs is 50.3% and for time is 41.6%. A sample graph for the FCMLP 13-1 network is depicted in Fig. 5.

Table 1. Training results for the logistic function.

	Classic BP				New version					
Network	η	SR [%]	Ep	T [ms]	η base	SR [%]	Ep	T [ms]	ED	TD
FCC10	0.015	94	568.1	3.0	0.015	91	361.0	1.9	0.635	0.641
FCC12	0.015	99	490.5	3.3	0.015	97	285.3	2.0	0.582	0.595
FCC14	0.015	98	411.5	3.7	0.015	95	251.2	2.3	0.610	0.621
FCMLP12-1	0.03	97	656.1	2.0	0.03	92	273.6	1.2	0.417	0.595
FCMLP13-1	0.03	99	637	2.3	0.03	86	275.8	1.2	0.433	0.506
FCMLP15-1	0.03	98	608.2	2.5	0.03	97	259.3	1.4	0.426	0.559
MLP12-1	0.03	91	690.3	2.2	0.03	96	336.2	1.5	0.487	0.687
MLP13-1	0.4	100	173.2	0.5	0.4	99	73.7	0.2	0.425	0.516
MLP15-1	0.03	97	658.5	2.8	0.03	91	302.8	1.5	0.460	0.539

4.2 The SINC 2D Function

The SINC two-dimensional function is defined by equation

$$y = f(x_1, x_2) = \begin{cases} 1 & \text{for } x_1 = x_2 = 0 \\ \frac{\sin(x_2)}{x_2} & \text{for } x_1 = 0 \wedge x_2 \neq 0 \\ \frac{\sin(x_1)}{x_1} & \text{for } x_1 \neq 0 \wedge x_2 = 0 \\ \frac{\sin(x_1)}{x_1}\frac{\sin(x_2)}{x_2} & \text{for } x_1 \neq 0 \wedge x_2 \neq 0 \end{cases} \tag{15}$$

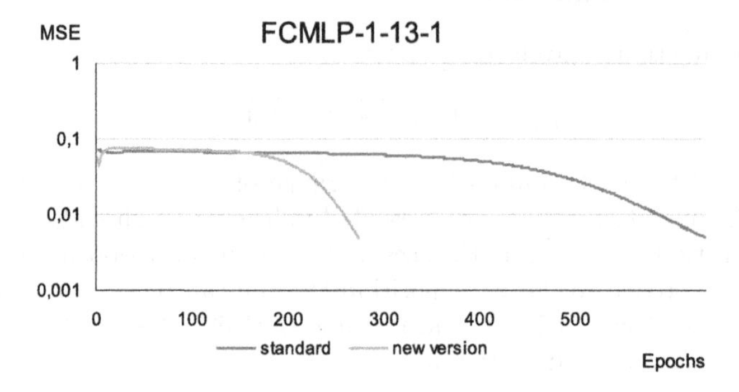

Fig. 5. Exemplary training process of the logistic function using the FCMLP 13-1 network.

Table 2. Training results for the SINC function.

	Classic BP				New version					
Network	η	SR [%]	Ep	T [ms]	η base	SR [%]	Ep	T [ms]	ED	TD
FCC14	0.009	98	122.8	11.8	0.009	98	83.2	8.3	0.678	0.706
FCC16	0.009	99	99.7	12.0	0.009	100	78.8	9.68	0.790	0.800
FCC18	0.009	100	86.1	12.7	0.009	98	72.8	10.3	0.839	0.812
FCMLP4-4-1	0.01	96	469.5	16.3	0.01	98	292.1	12.5	0.622	0.769
FCMLP6-6-1	0.009	100	252.9	14.7	0.009	100	155.0	9.9	0.613	0.674
FCMLP8-8-1	0.005	98	166.8	14.3	0.005	99	111.5	10.3	0.669	0.723
MLP4-4-1	0.009	95	518.4	14.4	0.009	99	150.7	5.6	0.291	0.388
MLP6-6-1	0.009	100	242.1	11.6	0.009	100	186.4	10.9	0.770	0.939
MLP8-8-1	0.009	100	139.2	10.3	0.009	100	39.5	3.3	0.284	0.314

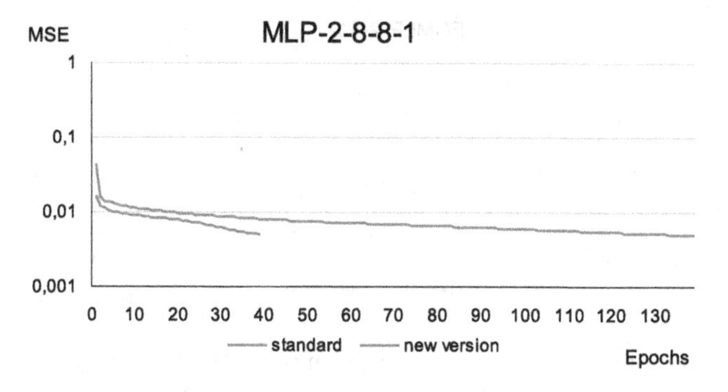

Fig. 6. Exemplary training process of the SINC function using the MLP 8-8-1 network.

The training set consists of 50 points from the range of $x_1 \in (1, 5)$ and $x_2 \in (1, 5)$. The target error is 0.005. Table 2 shows the simulation results. The success ratio is similar in both cases, but the number of epochs and the time for the new approach are significantly shorter. The average speedup is 38.3% for a the epoch and 31.9% for time. A sample graph for the FCMLP8-8-1 network is shown in Fig. 6.

4.3 The Circle Problem

The circle is a classification problem that says if a given point is inside or outside the circle. The training set has 100 samples from the range of $x_1, x_2 \in [-5, 5]$. The target error is 0.09. Table 3 presents the simulation results. As previously, the success ratio is similar in both approaches, but the number of epochs and the time for the new approach is significantly shorter. The average acceleration is 56.1% and 50.0%. A sample graph for the FCC 18 network is shown in Fig. 7.

Table 3. Training results for the circle problem.

	Classic BP				New version					
Network	η	SR [%]	Ep	T [ms]	η base	SR [%]	Ep	T [ms]	ED	TD
FCC10	0.0016	100	3.36	0.22	0.0016	100	1.74	0.11	0.518	0.5
FCC12	0.0016	100	4.2	0.34	0.0016	100	1.47	0.1	0.35	0.294
FCC14	0.0016	100	3.31	0.33	0.0016	100	1.55	0.15	0.468	0.455
FCMLP11-1	0.0016	100	4.23	0.16	0.0016	100	1.54	0.07	0.364	0.438
FCMLP13-1	0.0016	100	4.67	0.23	0.0016	100	1.71	0.1	0.366	0.435
FCMLP15-1	0.0016	100	3.76	0.21	0.0016	100	1.8	0.13	0.479	0.619
MLP11-1	0.0016	100	3.29	0.13	0.0016	100	1.33	0.07	0.404	0.538
MLP13-1	0.0016	100	2.79	0.13	0.0016	100	1.58	0.09	0.566	0.692
MLP15-1	0.0016	100	2.95	0.17	0.0016	100	1.28	0.09	0.434	0.529

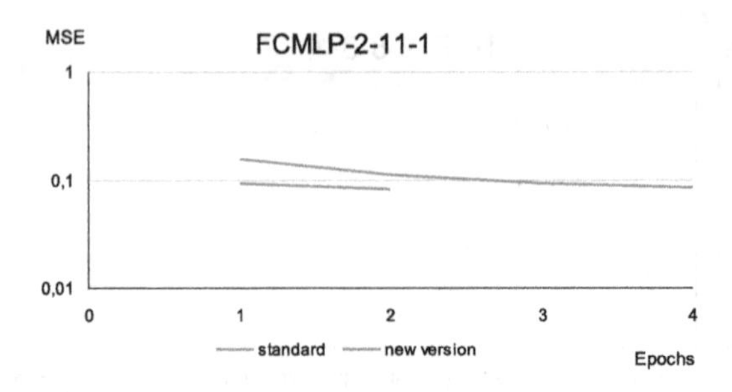

Fig. 7. Training process of the circle problem using the FCMLP-2-11-1 network.

5 Conclusion

In this paper a new approach to learning neural networks is presented. As an example, the error backpropagation algorithm was used. Our modification is based on using the $2^k = 8$ learning rates simultaneously, which results in an acceleration of the learning process. The experiments show that our approach to the learning algorithm significantly decreases the neural network's learning time (about 40% on average). The obtained results are very satisfying and testify to a high efficiency of the proposed solution.

In our future work we plan to test our approach with other algorithms and with different neural networks. Our approach can be implemented to solve several industrial problems, see e.g. [45–47].

References

1. Wang, Z., Cao, J., Cai, Z., Rutkowski, L.: Anti-synchronization in fixed time for discontinuous reaction–diffusion neural networks with time-varying coefficients and time delay. IEEE Trans. Cybern. **50**(6), 2758–2769 (2020). https://doi.org/10.1109/TCYB.2019.2913200
2. Duda, P., Rutkowski, L., Jaworski, M., Rutkowska, D.: On the parzen kernel-based probability density function learning procedures over time-varying streaming data with applications to pattern classification. IEEE Trans. Cybern. **50**(4), 1683–1696 (2020). https://doi.org/10.1109/TCYB.2018.2877611
3. Lin, L., Cao, J., Rutkowski, L.: Robust event-triggered control invariance of probabilistic boolean control networks. IEEE Trans. Neural Netw. Learn. Syst. **31**(3), 1060–1065 (2020). https://doi.org/10.1109/TNNLS.2019.2917753
4. Liu, Y., Zheng, Y., Lu, J., Cao, J., Rutkowski, L.: Constrained quaternion-variable convex optimization: a quaternion-valued recurrent neural network approach. IEEE Trans. Neural Netw. Learn. Syst. **31**(3), 1022–1035 (2020). https://doi.org/10.1109/TNNLS.2019.2916597
5. Gabryel, M., Przybyszewski, K.: The dynamically modified BoW algorithm used in assessing clicks in online ads. In: Rutkowski, L., Scherer, R., Korytkowski, M., Pedrycz, W., Tadeusiewicz, R., Zurada, J.M. (eds.) ICAISC 2019, Part II. LNCS (LNAI), vol. 11509, pp. 350–360. Springer, Cham (2019). https://doi.org/10.1007/978-3-030-20915-5_32

6. Gabryel, M.: The bag-of-words method with different types of image features and dictionary analysis. J. Univ. Comput. Sci. **24**(4), 357–371 (2018)
7. Starczewski, A.: A new validity index for crisp clusters. Pattern Anal. Appl. **20**(3), 687–700 (2015). https://doi.org/10.1007/s10044-015-0525-8
8. Starczewski, A., Goetzen, P., Er, M.J.: A new method for automatic determining of the DBSCAN parameters. J. Artif. Intell. Soft Comput. Res. **10**(3), 209–221 (2019)
9. Łapa, K., Cpałka, K., Wang, L.: New method for design of fuzzy systems for nonlinear modelling using different criteria of interpretability. In: Rutkowski, L., Korytkowski, M., Scherer, R., Tadeusiewicz, R., Zadeh, L.A., Zurada, J.M. (eds.) ICAISC 2014, Part I. LNCS (LNAI), vol. 8467, pp. 217–232. Springer, Cham (2014). https://doi.org/10.1007/978-3-319-07173-2_20
10. Zalasiński, M., Cpałka, K., Er, M.J.: New method for dynamic signature verification using hybrid partitioning. In: Rutkowski, L., Korytkowski, M., Scherer, R., Tadeusiewicz, R., Zadeh, L.A., Zurada, J.M. (eds.) ICAISC 2014, Part II. LNCS (LNAI), vol. 8468, pp. 216–230. Springer, Cham (2014). https://doi.org/10.1007/978-3-319-07176-3_20
11. Szczypta, J., Przybył, A., Cpałka, K.: Some aspects of evolutionary designing optimal controllers. In: Rutkowski, L., Korytkowski, M., Scherer, R., Tadeusiewicz, R., Zadeh, L.A., Zurada, J.M. (eds.) ICAISC 2013, Part II. LNCS (LNAI), vol. 7895, pp. 91–100. Springer, Heidelberg (2013). https://doi.org/10.1007/978-3-642-38610-7_9
12. Rutkowski, T., Romanowski, J., Woldan, P., Staszewski, P., Nielek, R., Rutkowski, L.: A content-based recommendation system using neuro-fuzzy approach. In: 2018 IEEE International Conference on Fuzzy Systems (FUZZ-IEEE), pp. 1–8 (2018)
13. Rutkowski, T., Łapa, K., Nowicki, R., Nielek, R., Grzanek, K.: On explainable recommender systems based on fuzzy rule generation techniques. In: Rutkowski, L., Scherer, R., Korytkowski, M., Pedrycz, W., Tadeusiewicz, R., Zurada, J.M. (eds.) ICAISC 2019, Part I. LNCS (LNAI), vol. 11508, pp. 358–372. Springer, Cham (2019). https://doi.org/10.1007/978-3-030-20912-4_34
14. Rutkowski, L.: Computational Intelligence. Methods and Techniques. Springer, Heidelberg (2008). https://doi.org/10.1007/978-3-540-76288-1
15. Liao, J., Liu, T., Liu, M., Wang, J., Wang, Y., Sun, H.: Multi-context integrated deep neural network model for next location prediction. IEEE Access **6**, 21980–21990 (2018)
16. Zalasiński, M., Łapa, K., Cpałka, K., Przybyszewski, K., Yen, G.G.: On-line signature partitioning using a population based algorithm. J. Artif. Intell. Soft Comput. Res. **10**(1), 5–13 (2020)
17. Nobukawa, S., Nishimura, H., Yamanishi, T.: Pattern classification by spiking neural networks combining self-organized and reward-related spike-timing-dependent plasticity. J. Artif. Intell. Soft Comput. Res. **9**(4), 283–291 (2019)
18. de Souza, G.B., da Silva Santos, D.F., Gonçalves Pires, R., Nilceu Marananil, A., Paulo Papa, J.: Deep features extraction for robust fingerprint spoofing attack detection. J. Artif. Intell. Soft Comput. Res. **9**(1), 41–49 (2019)
19. Bilski, J.: The UD RLS algorithm for training the feedforward neural networks. Int. J. Appl. Math. Comput. Sci. **15**(1), 101–109 (2005)
20. Rumelhart, D.E., Hinton, G.E., Williams, R.J.: Learning Internal Representations by Error Propagation, Parallel Distributed Processing, vol. 1. The MIT Press, Cambridge (1986).Ch. 8, Rumelhart dE. and McCelland J. (red.)
21. Wilamowski, B.M., Yo, H.: Neural network learning without backpropagation. IEEE Trans. Neural Netw. **21**(11), 1793–1803 (2010)

22. Żurada, J.: Introduction to Artificial Neural Systems. West Publishing Co., St. Paul (1992)
23. Grycuk, R., Najgebauer, P., Kordos, M., Scherer, M.M., Marchlewska, A.: On the topological properties of the certain neural networks. J. Artif. Intell. Soft Comput. Res. **10**(2), 257–268 (2018)
24. Shewalkar, A., Nyavanandi, D., Ludwig, S.A.: Fast image index for database management engines. J. Artif. Intell. Soft Comput. Res. **9**(4), 95–111 (2020)
25. Scherer, R.: Computer Vision Methods for Fast Image Classification and Retrieval. Springer, Cham (2020). https://doi.org/10.1007/978-3-030-12195-2
26. Bilski, J., Litwiński, S., Smoląg, J.: Parallel realisation of QR algorithm for neural networks learning. In: Rutkowski, L., Siekmann, J.H., Tadeusiewicz, R., Zadeh, L.A. (eds.) ICAISC 2004. LNCS (LNAI), vol. 3070, pp. 158–165. Springer, Heidelberg (2004). https://doi.org/10.1007/978-3-540-24844-6_19
27. Bilski, J., Smoląg, J.: Parallel realisation of the recurrent RTRN neural network learning. In: Rutkowski, L., Tadeusiewicz, R., Zadeh, L.A., Zurada, J.M. (eds.) ICAISC 2008. LNCS (LNAI), vol. 5097, pp. 11–16. Springer, Heidelberg (2008). https://doi.org/10.1007/978-3-540-69731-2_2
28. Bilski, J., Smoląg, J.: Parallel realisation of the recurrent elman neural network learning. In: Rutkowski, L., Scherer, R., Tadeusiewicz, R., Zadeh, L.A., Zurada, J.M. (eds.) ICAISC 2010, Part II. LNCS (LNAI), vol. 6114, pp. 19–25. Springer, Heidelberg (2010). https://doi.org/10.1007/978-3-642-13232-2_3
29. Bilski, J., Smoląg, J.: Parallel realisation of the recurrent multi layer perceptron learning. In: Rutkowski, L., Korytkowski, M., Scherer, R., Tadeusiewicz, R., Zadeh, L.A., Zurada, J.M. (eds.) ICAISC 2012, Part I. LNCS (LNAI), vol. 7267, pp. 12–20. Springer, Heidelberg (2012). https://doi.org/10.1007/978-3-642-29347-4_2
30. Bilski, J., Smoląg, J., Żurada, J.M.: Parallel approach to the levenberg-marquardt learning algorithm for feedforward neural networks. In: Rutkowski, L., Korytkowski, M., Scherer, R., Tadeusiewicz, R., Zadeh, L.A., Zurada, J.M. (eds.) ICAISC 2015, Part I. LNCS (LNAI), vol. 9119, pp. 3–14. Springer, Cham (2015). https://doi.org/10.1007/978-3-319-19324-3_1
31. Bilski, J., Wilamowski, B.M.: Parallel levenberg-marquardt algorithm without error backpropagation. In: Rutkowski, L., Korytkowski, M., Scherer, R., Tadeusiewicz, R., Zadeh, L.A., Zurada, J.M. (eds.) ICAISC 2017, Part I. LNCS (LNAI), vol. 10245, pp. 25–39. Springer, Cham (2017). https://doi.org/10.1007/978-3-319-59063-9_3
32. Smoląg, J., Bilski, J.: A systolic array for fast learning of neural networks. In: Proceedings of V Conference Neural Networks and Soft Computing, Zakopane, pp. 754–758 (2000)
33. Smoląg, J., Rutkowski, L., Bilski, J.: Systolic array for neural networks. In: Proceedings of IV Conference Neural Networks and Their Applications, Zakopane, pp. 487–497 (1999)
34. Fahlman, S.: Faster learning variations on backpropagation: an empirical study. In: Proceedings of Connectionist Models Summer School, Los Atos (1988)
35. Riedmiller, M., Braun, H.: A direct method for faster backpropagation learning: the RPROP algorithm. In: IEEE International Conference on Neural Networks, San Francisco (1993)
36. Werbos, J.: Backpropagation through time: what it does and how to do it. Proc. IEEE **78**(10), 1550–1560 (1990)
37. Charalambous, C.: Conjugate gradient algorithm for efficient training of artificial neural networks. IEE Proc. G - Circ. Devices Syst. **139**(3), 301–310 (1992)

38. Fletcher, R., Powell, M.J.D.: A rapidly convergent descent method for minimization. Comput. J. **6**, 163–168 (1963)
39. Fletcher, R., Reeves, C.M.: Function minimization by conjugate gradients. Comput. J. **7**, 149–154 (1964)
40. Nocedal, J., Wright, S.J.: Conjugate Gradient Methods in Numerical Optimization, pp. 497–528. Springer, New York (2006)
41. Polak, E.: Computational Methods in Optimization: A Unified Approach. Academic Press, New York (1971)
42. Jin, X.-B., Zhang, X.-Y., Huang, K., Geng, G.-G.: Stochastic conjugate gradient algorithm with variance reduction. IEEE Trans. Neural Netw. Learn. Syst. **30**(5), 1360–1369 (2019)
43. Hagan, M.T., Menhaj, M.B.: Training feedforward networks with the Marquardt algorithm. IEEE Trans. Neural Netw. **5**(6), 989–993 (1994)
44. Bilski, J., Kowalczyk, B., Marchlewska, A., Zurada, J.M.: Local levenberg-marquardt algorithm for learning feedforwad neural networks. J. Artif. Intell. Soft Comput. Res. **10**(4), 299–316 (2020). https://doi.org/10.2478/jaiscr-2020-0020
45. Rafajłowicz, E., Rafajłowicz, W.: Iterative learning in optimal control of linear dynamic processes. Int. J. Control **91**(7), 1522–1540 (2018)
46. Rafajłowicz, E., Rafajłowicz, W.: Iterative learning in repetitive optimal control of linear dynamic processes. In: Rutkowski, L., Korytkowski, M., Scherer, R., Tadeusiewicz, R., Zadeh, L.A., Zurada, J.M. (eds.) ICAISC 2016, Part I. LNCS (LNAI), vol. 9692, pp. 705–717. Springer, Cham (2016). https://doi.org/10.1007/978-3-319-39378-0_60
47. Jurewicz, P., Rafajłowicz, W., Reiner, J., Rafajłowicz, E.: Simulations for tuning a laser power control system of the cladding process. In: Saeed, K., Homenda, W. (eds.) CISIM 2016. LNCS, vol. 9842, pp. 218–229. Springer, Cham (2016). https://doi.org/10.1007/978-3-319-45378-1_20

Data-Driven Learning of Feedforward Neural Networks with Different Activation Functions

Grzegorz Dudek(✉)

Częstochowa University of Technology, Częstochowa, Poland
grzegorz.dudek@pcz.pl

Abstract. This work contributes to the development of a new data-driven method (D-DM) of feedforward neural networks (FNNs) learning. This method was proposed recently as a way of improving randomized learning of FNNs by adjusting the network parameters to the target function fluctuations. The method employs logistic sigmoid activation functions for hidden nodes. In this study, we introduce other activation functions, such as bipolar sigmoid, sine function, saturating linear functions, reLU, and softplus. We derive formulas for their parameters, i.e. weights and biases. In the simulation study, we evaluate the performance of FNN data-driven learning with different activation functions. The results indicate that the sigmoid activation functions perform much better than others in the approximation of complex, fluctuated target functions.

Keywords: Data-driven learning · Feedforward neural networks · Randomized learning algorithms

1 Introduction

FNNs are widely used as predictive models to fit data distribution. They learn using gradient descent methods and ensure a universal approximation property. However, gradient-based algorithms suffer from many drawbacks which make the learning process ineffective and time-consuming. This is because gradient learning is sensitive to local minima, flat regions, and saddle points of the loss function. Moreover, its application is time-consuming for complex target functions (TFs), big data, and large FNN architectures. Randomized learning was proposed as an alternative to gradient-based learning. In this approach, the parameters of the hidden nodes are selected randomly from any interval, and stay fixed. Only the output weights are learned. The optimization problem in randomized learning becomes convex and can be solved by a standard linear least-squares method [1]. This leads to very fast training. The the universal approximation property is kept when the random parameters are selected from a symmetric interval according

Supported by Grant 2017/27/B/ST6/01804 from the National Science Centre, Poland.

L. Rutkowski et al. (Eds.): ICAISC 2021, LNAI 12854, pp. 66–77, 2021.
https://doi.org/10.1007/978-3-030-87986-0_6

to any continuous sampling distribution [2]. The main problems in randomized learning are [3,4]: how to select the interval and distribution for the random parameters, and whether the weights and biases should be chosen from the same interval and distribution.

It was shown in [5] and [6] that the weights and biases of hidden nodes have different functions and should be selected separately. The weights decide about the activation function (AF) slopes and should reflect the TF complexity, while the biases decide about the AF shift and should ensure the placement of the most nonlinear fragments of AFs into the input hypercube. These fragments are most useful for modeling TF fluctuations. The method proposed in [5] selects the proper interval for the weights based on AF features and TF properties. The biases are calculated based on the weights and data scope. This approach introduces the AFs into the input hypercube and adjusts the interval for weights to TF complexity. In [6], instead of generating the weights, the slope angles of AFs were randomly selected. This changed the distribution of weights, which typically is a uniform one. This new distribution ensured that the slope angles of AFs were uniformly distributed, which improved results by preventing overfitting, especially for highly nonlinear TFs.

To improve further FNN randomized learning, in [7], a D-DM was proposed. This method introduces the AFs into randomly selected regions of the input space and adjusts the AF slopes to the TF slopes in these regions. As a result, the AFs mimic the TF locally, and their linear combination approximates smoothly the entire TF. This work contributes to the development of data-driven FNN learning by introducing different AFs, i.e. bipolar sigmoid, sine function, saturating linear functions, reLU, and softplus. For each AF, the formulas for weights and biases are derived.

The remainder of this paper is structured as follows. In Sect. 2, the framework of D-DM is presented. The formulas for hidden nodes parameters for different AFs are derived in Sect. 3. The performance of FNN data-driven learning with different AFs is evaluated in Sect. 4. Finally, Sect. 5 concludes the work.

2 Framework of the Data-Driven FNN Learning

Let us consider a shallow FNN architecture with n inputs, a single-hidden layer, and a single output. AFs of hidden nodes, $h(\mathbf{x})$, map nonlinearly input vectors $\mathbf{x} = [x_1, x_2, ..., x_n]^T \in \mathbb{R}^n$ into an m-dimensional feature space. An output node combines linearly m nonlinear transformations of the inputs. The function expressed by this FNN has the form:

$$\varphi(\mathbf{x}) = \sum_{i=1}^{m} \beta_i h_i(\mathbf{x}) \tag{1}$$

where β_i is the output weight linking the i-th hidden node with the output node.

Such FNN architecture has a universal approximation property, even when the hidden layer parameters are not trained but generated randomly from the proper distribution [2,8].

The output weights $\boldsymbol{\beta} = [\beta_1, \beta_2, ..., \beta_m]^T$ can be determined by solving the following linear problem: $\mathbf{H}\boldsymbol{\beta} = \mathbf{Y}$, where $\mathbf{H} = [\mathbf{h}(\mathbf{x}_1), \mathbf{h}(\mathbf{x}_2), ..., \mathbf{h}(\mathbf{x}_N)]^T \in \mathbb{R}^{N \times m}$ is the hidden layer output matrix, and $\mathbf{Y} = [y_1, y_2, ..., y_N]^T$ is a vector of target outputs. The optimal solution for $\boldsymbol{\beta}$ is given by:

$$\boldsymbol{\beta} = \mathbf{H}^{+}\mathbf{Y} \tag{2}$$

where $\mathbf{H}^{+}$ denotes the Moore–Penrose generalized inverse of matrix $\mathbf{H}$.

The hidden node parameters, i.e. weights $\mathbf{a} = [a_1, a_2, ..., a_n]^T$ and a bias b, control slopes and position of AF in the input space. For a sigmoid AF given by the formula:

$$h(\mathbf{x}) = \frac{1}{1 + \exp\left(-\left(\mathbf{a}^T\mathbf{x} + b\right)\right)} \tag{3}$$

weight a_j decides about the sigmoid slope in the j-th direction and bias b decides about the sigmoid shift along a hyperplane containing all x-axes. The appropriate selection of the slopes and shifts of all sigmoids determine the fitting accuracy of FNN to the TF. To adjust the sigmoids to the local features of the TF, in [7], a D-DM for FNN learning was proposed. This method selects an input space region by randomly choosing one of the training points for each sigmoid. Then, it places the sigmoid in this region and adjusts the sigmoid slopes to the TF slopes in the neighborhood of the chosen point. By combining linearly all the sigmoids randomly placed in the input space, we obtain a fitted surface which reflects the TF shape in different regions.

The D-DM algorithm, in the first step, selects randomly training point $\mathbf{x}^*$. Then, sigmoid S is placed in the input space in such a way that one of its inflection points, P, is in $\mathbf{x}^*$. The sigmoid value at the inflection point is 0.5:

$$h(\mathbf{x}^*) = \frac{1}{1 + \exp\left(-\left(\mathbf{a}^T\mathbf{x}^* + b\right)\right)} = 0.5 \tag{4}$$

From this equation we obtain the sigmoid bias as:

$$b = -\mathbf{a}^T\mathbf{x}^* \tag{5}$$

The slopes of sigmoid S are adjusted to the TF slopes in $\mathbf{x}^*$. The TF slopes in $\mathbf{x}^*$ are estimated by fitting hyperplane T to the neighborhood of $\mathbf{x}^*$. The neighborhood, $\Psi(\mathbf{x}^*)$, contains point $\mathbf{x}^*$ and k training points nearest to it. Hyperplane T has the form:

$$y = a_1'x_1 + a_2'x_2 + ... + a_n'x_n + b' \tag{6}$$

where coefficient a_j' expresses a slope of T in the j-th direction.

We assume that sigmoid S is tangent to hyperplane T in point $\mathbf{x}^*$. This means that the partial derivatives of S and T in $\mathbf{x}^*$ are the same. Comparing the formulas for partial derivatives of both functions, we obtain an equation for the sigmoid weights (see [7] for details):

$$a_j = 4a_j', \quad j = 1, 2, ..., n \tag{7}$$

To generate all the FNN hidden nodes, the D-DM algorithm repeats the procedure described above m times. So, for each node it randomly selects training point $\mathbf{x}^*$, fits hyperplane T to its neighborhood $\Psi(\mathbf{x}^*)$, calculates weights a_j according to (7), and calculates biases b according to (5). Finally, it calculates hidden layer output matrix $\mathbf{H}$, and output weights from (2). The resulting function, $\varphi(\mathbf{x})$, constructed in line with such data-driven learning, reflects TF fluctuations.

The D-DM has two hyperparameters: the number of hidden nodes m and neighbourhood size k. They control the fitting performance of the model and its bias-variance tradeoff. Their optimal values for a given TF should be tuned during cross-validation.

3 Data-Driven FNN Learning with Different Activation Functions

When we employ other AFs instead of logistic sigmoids, the projection matrix $\mathbf{H}$ changes in a way which can entail changes in the approximation properties of the model. Using other AFs requires the derivation of new formulas for the hidden node parameters in the following ways.

Bipolar sigmoid SIGMOID_B. Usually the bipolar sigmoid is defined as a hyperbolic tangent function. In this study, we define it slightly differently:

$$h_{sigb}(\mathbf{x}) = \frac{2}{1+\exp\left(-\left(\mathbf{a}^T\mathbf{x}+b\right)\right)} - 1 \tag{8}$$

D-DM places SIGMOID_B in the input space in such a way that one of its inflection points is in the randomly selected training point, $\mathbf{x}^*$. The SIGMOID_B value at the inflection points is 0, so, $h_{sigb}(\mathbf{x}^*) = 0$. From this equation we obtain the formula for the bias, which is the same as for the unipolar sigmoid (SIGMOID_U), (5).
To find weights a_j, we equate the partial derivatives of SIGMOID_B in $\mathbf{x}^*$ to the partial derivatives of hyperplane T, (6):

$$\frac{\partial h_{sigb}(\mathbf{x}^*)}{\partial x_j} = \frac{1}{2}a_j(1+h_{sigb}(\mathbf{x}^*))(1-h_{sigb}(\mathbf{x}^*)) = a'_j \tag{9}$$

From this equation, taking into account that $h_{sigb}(\mathbf{x}^*) = 0$, we obtain:

$$a_j = 2a'_j, \quad j = 1, 2, ..., n \tag{10}$$

Sine function SINE. Let us place the SINE AF, $h_{sin}(\mathbf{x}) = \sin(\mathbf{a}^T\mathbf{x}+b)$, in the input space in such a way that it has one of its inflection point in randomly selected training point $\mathbf{x}^*$. The SINE value in the inflection points is 0, so, $h_{sin}(\mathbf{x}^*) = 0$. From this equation we obtain the formula for bias, which is the same as for both sigmoid AFs, (5).

To determine equations for the weights for SINE, we equate the partial derivatives of SINE in $\mathbf{x}^*$ to the partial derivatives of hyperplane T, (6):

$$\frac{\partial h_{sin}(\mathbf{x}^*)}{\partial x_j} = a_j \cos(\mathbf{a}^T\mathbf{x}^* + b) = a'_j \tag{11}$$

Taking into account that $\sin(\mathbf{a}^T\mathbf{x}^* + b) = 0$ implies $\cos(\mathbf{a}^T\mathbf{x}^* + b) = 1$, from (11) we obtain:

$$a_j = a'_j, \quad j = 1, 2, ..., n \tag{12}$$

Saturating linear unipolar function SATLIN_U**.** This is a linearized version of SIGMOID_U defined as follows:

$$h_{satu}(\mathbf{x}) = \begin{cases} 0 \text{ if } z \leq 0 \\ z \text{ if } 0 < z < 1 \\ 1 \text{ if } z \geq 1 \end{cases} \tag{13}$$

where $z = \mathbf{a}^T\mathbf{x} + b$.
SATLIN_U is placed in the input space in such a way that it has a value of 0.5 in $\mathbf{x}^*$. This is analogous to SIGMOID_U to which SATLIN_U has a similar shape. Thus, $\mathbf{a}^T\mathbf{x}^* + b = 0.5$. From this equation we obtain:

$$b = 0.5 - \mathbf{a}^T\mathbf{x}^* \tag{14}$$

We assume that the middle segment of $h_{satu}(\mathbf{x})$, $\mathbf{a}^T\mathbf{x} + b$, has the same slopes as hyperplane T, thus:

$$a_j = a'_j, \quad j = 1, 2, ..., n \tag{15}$$

Saturating linear bipolar function SATLIN_B**.** This AF is a linearized version of bipolar sigmoid SIGMOID_B:

$$h_{satb}(\mathbf{x}) = \begin{cases} -1 & \text{if } z \leq -1 \\ z & \text{if } -1 < z < 1 \\ 1 & \text{if } z \geq 1 \end{cases} \tag{16}$$

where $z = \mathbf{a}^T\mathbf{x} + b$.
SATLIN_B is placed in the input space in such a way that it has a value of 0 in $\mathbf{x}^*$. Thus, $\mathbf{a}^T\mathbf{x}^* + b = 0$. From this equation we obtain the same formula for a bias as for sigmoid AFs, (5).
As with SATLIN_U, we assume that the middle segment of SATLIN_B has the same slopes as hyperplane T. Thus, weights a_j are the same as the T coefficients, (15).

Rectified linear unit RELU**.** This is an AF commonly used in deep learning. It is expressed by:

$$h_{reLU}(\mathbf{x}) = \begin{cases} 0 \text{ if } z \leq 0 \\ z \text{ if } z > 0 \end{cases} \tag{17}$$

where $z = \mathbf{a}^T\mathbf{x} + b$.

RELU is composed of two half-hyperplanes: the first being $y = 0$ and the second $y = \mathbf{a}^T\mathbf{x}+b$. D-DM places the RELU AF in the input space so that the second half-hyperplane coincides with hyperplane T. Thus, their coefficients are the same:

$$b = b', \quad a_j = a'_j, \quad j = 1, 2, ..., n \tag{18}$$

Softplus SOFTPLUS**.** This is a smooth approximation of the RELU. It is expressed by:

$$h_{soft}(\mathbf{x}) = \ln\left(1 + \exp\left(\mathbf{a}^T\mathbf{x} + b\right)\right) \tag{19}$$

For $\mathbf{x} = [0, 0, ..., 0]$ and $b = 0$, the value of $h_{soft}(\mathbf{x}) = \ln(2)$. Let us shift this function in such a way that it has the value of $\ln(2)$ in $\mathbf{x}^*$. In such a case $\ln(1 + \exp(\mathbf{a}^T\mathbf{x}^* + b)) = \ln(2)$. From this equation we obtain a formula for b, which is the same as for the sigmoids (5).

Now, let us assume that the slopes of SOFTPLUS in $\mathbf{x}^*$ are the same as the slopes of T. Equating the partial derivative of both functions we obtain:

$$\frac{\partial h_{soft}(\mathbf{x}^*)}{\partial x_j} = \frac{a_j}{1 + \exp(-(\mathbf{a}^T\mathbf{x}^* + b))} = a'_j \tag{20}$$

From $\ln(1+\exp(\mathbf{a}^T\mathbf{x}^* + b)) = \ln(2)$ we obtain $1+\exp(\mathbf{a}^T\mathbf{x}^* + b) = 2$. Substituting this into (20), we obtain the weights of hidden nodes with SOFTPLUS AFs:

$$a_j = 2a'_j, \quad j = 1, 2, ..., n \tag{21}$$

Table 1 details the hidden nodes parameters determined by D-DM for different AFs. Note that in all cases, weights a_j reflect hyperplane T coefficients a'_j. Biases for all AFs, excluding RELU, are expressed using a dot product of the weight vector and $\mathbf{x}^*$ vector.

Table 1. Hidden nodes parameters for different activation functions.

	Activation function	Weights a_j	Bias b
SIGMOID_U:	$h_{sigu}(\mathbf{x}) = \frac{1}{1+\exp(-(\mathbf{a}^T\mathbf{x}+b))}$	$4a'_j$	$-\mathbf{a}^T\mathbf{x}^*$
SIGMOID_B:	$h_{sigb}(\mathbf{x}) = \frac{2}{1+\exp(-(\mathbf{a}^T\mathbf{x}+b))} - 1$	$2a'_j$	$-\mathbf{a}^T\mathbf{x}^*$
SINE:	$h_{sin}(\mathbf{x}) = \sin(\mathbf{a}^T\mathbf{x} + b)$	a'_j	$-\mathbf{a}^T\mathbf{x}^*$
SATLIN_U:	$h_{satu}(\mathbf{x}) = \begin{cases} 0 & \text{if } z \le 0 \\ z & \text{if } 0 < z < 1 \\ 1 & \text{if } z \ge 1 \end{cases}$	a'_j	$0.5 - \mathbf{a}^T\mathbf{x}^*$
SATLIN_B:	$h_{satb}(\mathbf{x}) = \begin{cases} -1 & \text{if } z \le -1 \\ z & \text{if } -1 < z < 1 \\ 1 & \text{if } z \ge 1 \end{cases}$	a'_j	$-\mathbf{a}^T\mathbf{x}^*$
RELU:	$h_{reLU}(\mathbf{x}) = \begin{cases} 0 & \text{if } z \le 0 \\ z & \text{if } z > 0 \end{cases}$	a'_j	b'
SOFTPLUS:	$h_{soft}(\mathbf{x}) = \ln\left(1 + \exp\left(\mathbf{a}^T\mathbf{x} + b\right)\right)$	$2a'_j$	$-\mathbf{a}^T\mathbf{x}^*$

where a'_j and b' are coefficients of hyperplane T, $y = a'_1x_1 + a'_2x_2 + ... + a'_nx_n + b'$, adjusted to the TF in the neighborhood $\Psi(\mathbf{x}^*)$ of randomly selected training point $\mathbf{x}^*$; $z = \mathbf{a}^T\mathbf{x} + b$.

Figure 1 shows AFs of different types introduced into the input space by D-DM. The training points belonging to the neighborhood of $\mathbf{x}^*$, $\Psi(\mathbf{x}^*)$, are shown as red dots. Note that the AFs in all cases have the same slopes in $\mathbf{x}^*$ as the slope of line T, which estimates the TF slope in $\mathbf{x}^*$. D-DM introduces m AFs in different regions of the input space.

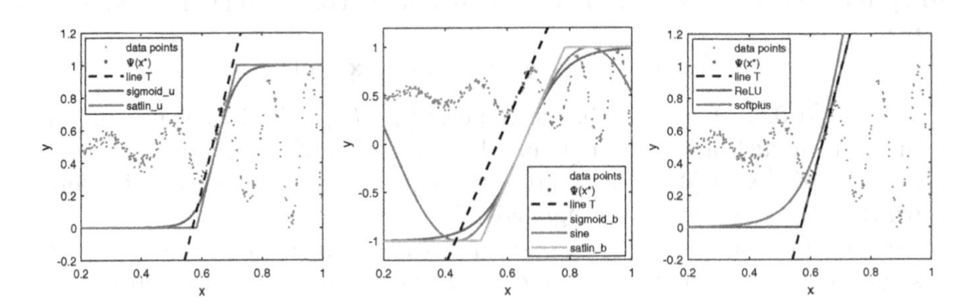

Fig. 1. AFs of different types introduced into the input space in $\mathbf{x}^*$ by D-DM. (Color figure online)

4 Simulation Study

In this section, we report the experimental results over several regression problems in order to compare the fitting properties of D-DM with different AFs. They include an approximation of extremely nonlinear TFs:

TF1 $g(x) = \sin(20 \cdot \exp x) \cdot x^2$, $x \in [0, 1]$
TF2 $g(x) = 0.2e^{-(10x-4)^2} + 0.5e^{-(80x-40)^2} + 0.3e^{-(80x-20)^2}$, $x \in [0, 1]$.
TF3 $g(\mathbf{x}) = \sum_{j=1}^{n} \sin(20 \cdot \exp x_j) \cdot x_j^2$, $x_i \in [0, 1]$
TF4 $g(\mathbf{x}) = -\sum_{i=1}^{n} \sin(x_i) \sin^{20}\left(\frac{ix_i^2}{\pi}\right)$, $x_i \in [0, \pi]$
TF5 $g(\mathbf{x}) = 418.9829n - \sum_{i=1}^{n} x_i \sin(\sqrt{|x_i|})$, $x_i \in [-500, 500]$

Both the training and test sets for TF1 and TF2 included 5000 points. For the training set, argument x was generated randomly from $U(0, 1)$, and for the test set, it was evenly distributed in $[0, 1]$. The function values were normalized in the range $[0, 1]$. Note that TF1 starts flat, near $x = 0$, then has increasing fluctuations (see Fig. 3). TF2 has two spikes that could be difficult to model with FNN (see Fig. 5).

TF3–TF5 are multivariate functions. We considered these functions with $n = 2, 5$ and 10 arguments. The sizes of the training and test sets depended on the number of arguments. They were 5000 for $n = 2$, 20,000 for $n = 5$, and

50,000 for $n = 10$. All arguments for TF3–TF5 were normalized to $[0, 1]$, and the function values were normalized to $[-1, 1]$. Two-argument functions TF3–TF5 are shown in Fig. 2. Note that TF3 is a multivariate variant of TF1. It combines flat regions with strongly fluctuated regions. TF4 expresses flat regions with perpendicular grooves. TF5 fluctuates strongly, showing the greatest amplitude at the borders.

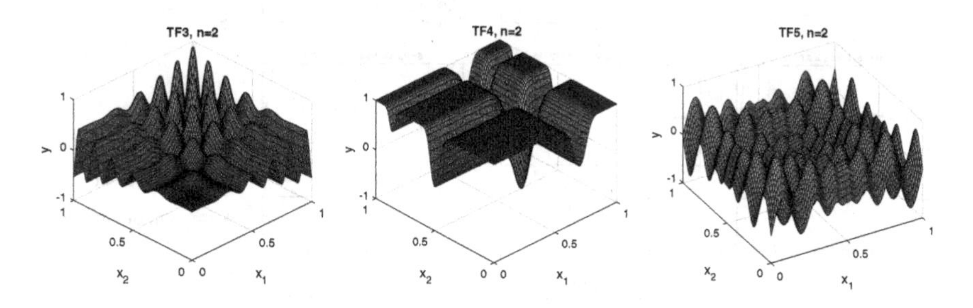

Fig. 2. Target functions TF3–TF5.

Figure 3 shows the results of TF1 fitting. The fitted lines are composed of AFs of different shapes. The AFs distributed by D-DM in the input interval (shown by the gray field) are shown in the lower panels. FNN included 30 hidden nodes. The neighborhood size was 2 ($k = 1$). As you can see from Fig. 3, the slopes of the AFs reflect the TF slopes. D-DM introduces the steepest fragments of the AFs into the input interval. These fragments are the most useful for modeling the TF fluctuations. The saturated AF fragments in the input interval are avoided. The best fitting results were achieved for both sigmoid AFs. SINE cannot cope with a TF with variable intensity of fluctuations. Neither RELU, which yielded the highest fitting error, nor the saturating linear functions are not able to fit smoothly to TF1. The smooth counterpart of RELU, SOFTPLUS, improves significantly on the RELU fitting results by offering a smooth approximation of TF1. Obviously, the results are dependent on the number of hidden nodes. The left panel of Fig. 4 shows the TF1 fitting error for different numbers of hidden nodes. As can be seen from this figure, the sigmoid AFs outperformed all the others. Slightly worse results were achieved for SOFTPLUS, while the highest error was observed for RELU. Detailed results for each AF, i.e. RMSE for the maximal number of hidden nodes shown in the figures, are presented in Table 2. The lowest errors, i.e. those that are at least 5% lower than the others, are marked in bold in this table.

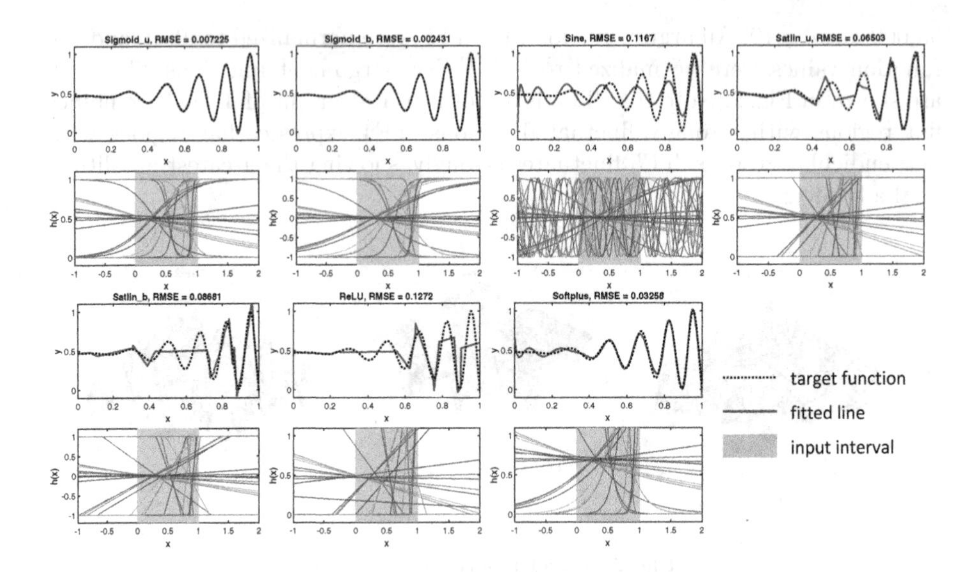

Fig. 3. TF1: Results of D-DM fitting for different AFs.

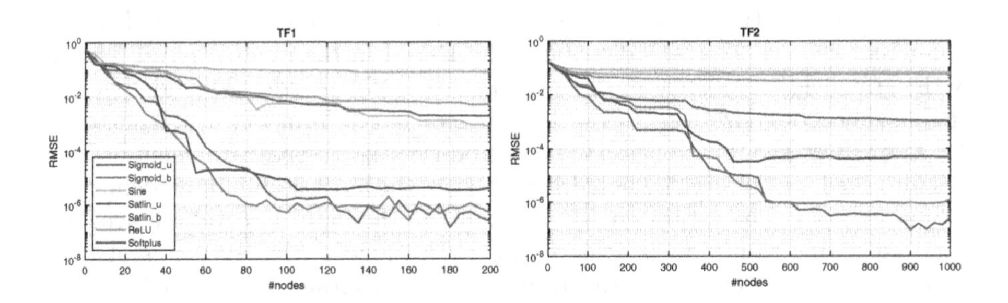

Fig. 4. Convergence of FNN for TF1 and TF2.

Table 2. Fitting errors (RMSE).

		SIGMOID_U	SIGMOID_B	SINE	SATLIN_U	SATLIN_B	RELU	SOFTPLUS
TF1		**2.39E−7**	4.74E−7	7.44E−4	1.86E−3	4.78E−3	7.84E−2	4.00E−6
TF2		**2.63E−7**	1.23E−6	6.65E−2	9.93E−4	2.93E−2	5.46E−2	4.89E−5
TF3	$n = 2$	2.19E−5	**2.26E−6**	1.64E−3	5.81E−3	9.01E−3	1.87E−2	–
TF3	$n = 5$	0.2214	0.2215	0.2213	0.2214	0.2215	0.2212	–
TF3	$n = 10$	0.2329	0.2328	0.2331	0.2329	0.2328	0.2329	–
TF4	$n = 2$	**6.69E−7**	4.87E−6	3.95E−2	2.65E−3	9.05E−3	5.18E−2	–
TF4	$n = 5$	0.2419	0.2412	0.2411	0.2381	0.2433	0.2418	–
TF4	$n = 10$	0.2611	0.2723	0.3095	0.2618	0.2738	0.2571	–
TF5	$n = 2$	**0.0083**	0.0116	0.0426	0.0257	0.0258	0.0319	–
TF5	$n = 5$	0.2385	0.2380	0.2404	0.2390	0.2381	0.2405	–
TF5	$n = 10$	0.2246	0.2243	0.2260	0.2247	0.2243	0.2238	–

Figure 5 shows fitting results for TF2 (120 hidden nodes and $k = 1$). In this case, SIGMOID_U and SIGMOID_B provided the best fitting, while SATLIN_U and SOFTPLUS provided a slightly worse fitting. Other AFs could not cope with this TF. For them, increasing the number of hidden nodes did not improve results and RMSE remained outside the acceptable level of 0.01 (see right panel of Fig. 4 and Table 2).

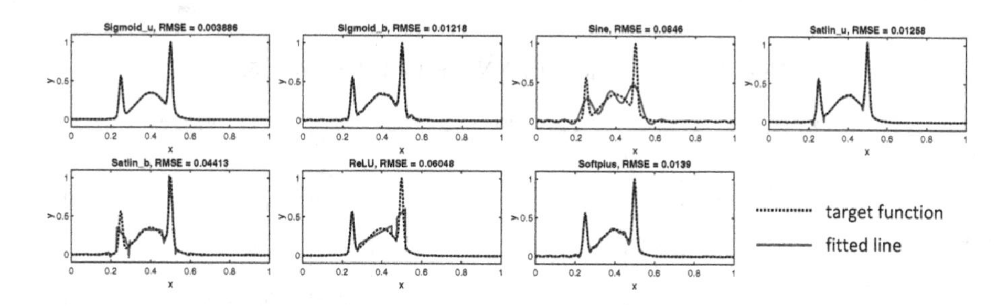

Fig. 5. TF2: Results of D-DM fitting for different AFs.

Figure 6 shows the convergence curves of FNN trained using D-DM for two-argument TF3-TF5 ($k = n$). In all these cases, the sigmoid AFs yielded the best results, while RELU, SINE and both saturating linear functions yielded the worst results. SOFTPLUS suffered from numerical problems related to the rapid growth of this function and exceeding the limit for double precision numbers. So, in Table 2, no results for SOFTPLUS are given.

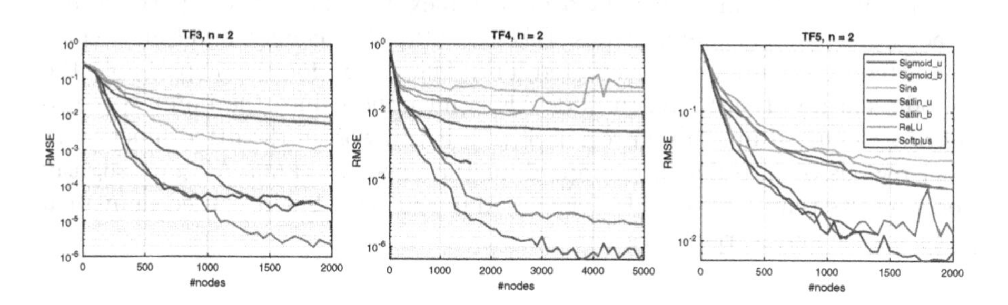

Fig. 6. Convergence of FNN for TF3–TF5, $n = 2$.

In the case of multidimensional modeling ($n = 5$ and 10), results for all AFs were comparable (see Figs. 7 and 8; $k = n$). This could be explained by the change in the TF landscape, which flattens with an increasing number of dimensions. When modeling flat TF, the AF shape turned out not to be as important as in the case of TF with strong fluctuations.

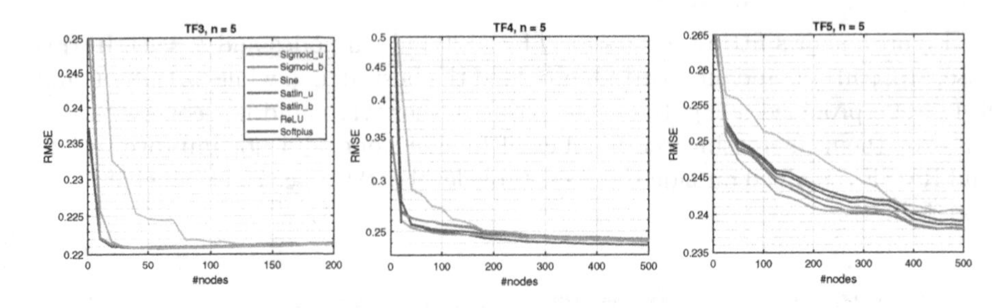

Fig. 7. Convergence of FNN for TF3–TF5, $n = 5$.

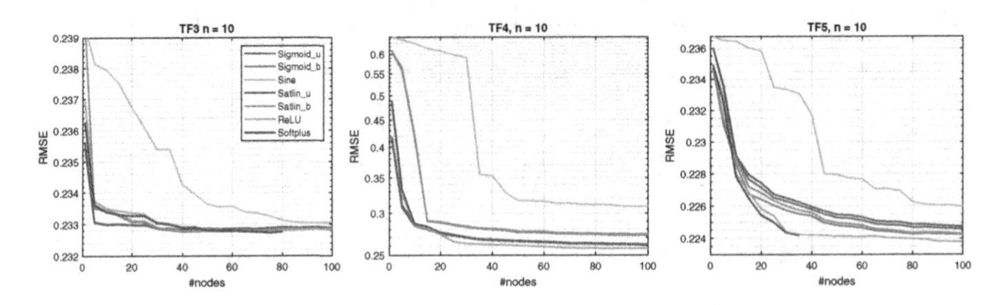

Fig. 8. Convergence of FNN for TF3–TF5, $n = 10$.

It is obvious from the performed simulations that the approximation properties of FNN trained using D-DM strongly depend on the AF type. The most useful for smoothing highly nonlinear TFs with fluctuations turned out to be the sigmoid AFs. The piecewise linear functions, i.e. RELU, SATLIN_U, and SATLIN_B, have problems with modeling smoothly complex TFs. Their linear parts do not fit accurately to TF nonlinearities. Likewise SINE AFs cannot build an acceptable fitted function for the fluctuated TFs. The reason for this is probably the periodic nature of SINE. When SINE AF is introduced into the input space to improve the fitted function in region $\Psi(\mathbf{x}^*)$, it can worsen the fitted function in other regions by introducing unwanted fluctuations. SOFTPLUS AF gave slightly worse results than sigmoid AFs for one-argument TFs, but it caused numerical problems for multivariate TFs.

5 Conclusion

The data-driven FNN learning described in this study is an alternative to both standard gradient-based learning and randomized learning. It allows us to bypass the tedious iterative process of tuning weights based on gradients. In the proposed approach, the parameters of hidden nodes are calculated based on the local properties of the TF. The AFs, which compose the fitted function, are introduced into the input space in randomly selected regions and their slopes are adjusted to the TF slopes in these regions. Consequently, the set of AFs reflects the TF fluctuations in different regions, which leads to accurate approximation.

Our approach is completely different from typical randomized learning, where the AF parameters are chosen randomly and do not reflect the TF landscape. D-DM finds the network parameters quickly, without repeatedly presenting the training set.

FNN performance strongly depends on AF shape. In this work, using a data-driven approach, we derived equations for the hidden node parameters for different AFs. As our experimental study has shown, the best FNN performance in smoothing highly nonlinear TFs was achieved by the sigmoid AFs. They were able to fit to the TF fluctuations. RELU AF, which is very popular in deep learning, fared very poorly in fluctuation modeling due to its piecewise linear nature. Its smooth counterpart, SOFTPLUS, produced much better results but suffered from numerical problems related to rapid growth.

References

1. Principe, J., Chen, B.: Universal approximation with convex optimization: gimmick or reality? IEEE Comput. Intell. Mag. **10**(2), 68–77 (2015)
2. Husmeier, D.: Random vector functional link (RVFL) networks. In: Husmeier, D. (ed.) Neural Networks for Conditional Probability Estimation. PERSPECT.NEURAL, pp. 87–97. Springer, London (1999). https://doi.org/10.1007/978-1-4471-0847-4_6
3. Cao, W., Wang, X., Ming, Z., Gao, J.: A review on neural networks with random weights. Neurocomputing **275**, 278–287 (2018)
4. Zhang, L., Suganthan, P.: A survey of randomized algorithms for training neural networks. Inf. Sci. **364–365**, 146–155 (2016)
5. Dudek, G.: Generating random weights and biases in feedforward neural networks with random hidden nodes. Inf. Sci. **481**, 33–56 (2019)
6. Dudek, G.: Generating random parameters in feedforward neural networks with random hidden nodes: drawbacks of the standard method and how to improve it. In: Yang, H., Pasupa, K., Leung, A.C.-S., Kwok, J.T., Chan, J.H., King, I. (eds.) ICONIP 2020. CCIS, vol. 1333, pp. 598–606. Springer, Cham (2020). https://doi.org/10.1007/978-3-030-63823-8_68
7. Dudek, G.: Data-driven randomized learning of feedforward neural networks. In: 2020 International Joint Conference on Neural Networks (IJCNN), Glasgow, United Kingdom, pp. 1–8 (2020). https://doi.org/10.1109/IJCNN48605.2020.9207353
8. Igelnik, B., Pao, Y.H.: Stochastic choice of basis functions in adaptive function approximation and the functional-link net. IEEE Trans. Neural Netw. **6**(6), 1320–1329 (1995)

Time-Domain Signal Synthesis with Style-Based Generative Adversarial Networks Applied to Guided Waves

Mateusz Heesch(✉), Krzysztof Mendrok, and Ziemowit Dworakowski

AGH, University of Science and Science and Technology,
Al. A. Mickiewicza 30, 30-059 Krakow, Poland
heesch@agh.edu.pl

Abstract. Data scarcity is a significant problem when it comes to designing machine learning systems for structural health monitoring applications, especially those based around data-hungry algorithms and methods, such as deep learning. Synthetic data generation could potentially alleviate this problem, lowering the number of measurements that need to be acquired in slow and often expensive conventional experiments. Such synthesis can be done by Generative Adversarial Networks, potentially creating unlimited synthetic samples recreating the original data distribution. While most of the research about these networks is centered around using them on image data, they have also been applied to audio waves - going as far as successfully synthesizing human speech. This suggests that these networks should also apply to synthesizing time-domain signals in various fields of structural health monitoring, guided waves in particular, as they are in many ways similar to audio wave signals. This work proposes an adaptation of style-based GAN architecture to time-domain signal generation, and presents its viability for guided waves synthesis, utilizing a database of signals collected in series of pitch-catch experiments on a composite plate.

Keywords: Neural network applications · Neural network theory and architectures · Supervised and unsupervised learning

1 Introduction

Guided wave based monitoring is one of many methods utilized for structural health monitoring. They're based on analyzing how high-frequency vibration propagates on a multitude of paths between transducers placed on the structure. Though the concept is fairly simple, the analysis of the output signals is complex and may overshadow the advantages offered by the method - like their sensitivity to small damage, being able to monitor large structures with a low amount of transducers, or their comparably low mass and cost [8]. This prompts attempts to apply various advances in machine learning to the analysis of these signals, however, these are generally very data-hungry algorithms. While this is often not

L. Rutkowski et al. (Eds.): ICAISC 2021, LNAI 12854, pp. 78–88, 2021.
https://doi.org/10.1007/978-3-030-87986-0_7

a problem for tasks like image recognition with plentiful rich publicly available datasets, the same is not true for guided waves. In this case, data is at best scarce and time-consuming to acquire - and at worst expensive or borderline impossible, as measurements for damaged scenarios require introducing damage to the part we are collecting data for, destroying it in the process.

Similar problems have been found in some more niche areas of application for deep learning, where the availability of publicly accessible data is lower due to various reasons - e.g. medicine with relatively difficult data collection as well as privacy concerns. Due to that, some methods for combating data scarcity have been researched and established as viable. The simplest one is data augmentation, based on passing the available samples through various transformations which despite altering it, still present viable (and possible to encounter) data. For image recognition examples would include random cropping, blurring, or adding noise to the image - because while it does alter the image, it doesn't change the nature of what the image represents (e.g. a blurry cat is still a cat).

Among newer methods, are Generative Adversarial Networks (GANs) [4] that create synthetic data. They are a pair of deep neural networks trained to imitate the distribution of the data they were supplied with for the training process. During training, these two networks are pitted against each other, with the Discriminator network attempting to distinguish synthetic data provided by Generator from real examples. Both of these networks are trained to progressively get better at their respective tasks so that the Discriminator can continuously provide Generator with information on how to create more "believable" data.

There has been a significant amount of research and improvements on how to utilize GANs, as well as what problems they can be applied to [5]. Their use has also since expanded from image data to various other types of data - such as audio wave synthesis [2]. Using the synthetic data for training in addition to real samples has also been shown to improve the capabilities of models across multiple domains, including kidney CT scan segmentation [10], machine fault detection [11], card fraud detection [3] and many others, underlining their potential as a tool for creating viable training data.

This work proposes a GW-GAN (Guided Wave - Generative Adversarial Network) model based on a state-of-the-art style based architecture for image synthesis StyleGAN2 [6,7], which besides generating high-quality images allows for controlling the output via style vectors. The motivation behind this work was the success of GANs in multiple fields as a data generation tool - in particular when it comes to audio signals which in principle are similar to guided waves. Additionally, to the best of authors' knowledge, style-based GANs (besides style transfer) have not been used to work with other time-domain signals, and while the presented model was created for guided waves, it can very well be used with any time-domain data - though it may need minor adjustments depending on the nature of the signals in question. For the guided waves proof of concept, the model was trained using OpenGuidedWaves dataset [9] and used to generate new signals that match the original distribution.

2 Style-Based Generative Adversarial Networks

2.1 Generative Adversarial Networks

Generative Adversarial Networks are generative models defined by their adversarial training procedure involving 2 networks, referred to as the Generator and the Discriminator. The Generator network is usually fed random noise as input, and its task is to create some data out of this random noise. The training gradients for making its output similar to the training data are given by the Discriminator, which is tasked with distinguishing the output produced by Generator from the training examples. These two nets are trained in turns, with Discriminator progressively getting better at distinguishing these fake examples from original data, and giving the Generator information on how to better fool it.

2.2 StyleGAN

Many of the newer GAN architectures differ from the original in various ways. StyleGAN [6,7] has re-imagined the way in which the random controlling (also called latent) vector is applied to the Generator. While normally it is the input to the first layer, in the case of Stylegan that input has been replaced with a constant vector which is modified during training to provide optimal initialization for the data generation. The random vector is instead fed into a multilayer perceptron mapping network - which is also trained together with the generator - whose goal is to provide a style vector. The goal of that is to allow the Generator to extract arbitrary learned features from the random vector. These styles are then applied to the generator at multiple scales (between upscales), influencing the output both in terms of large-scale information (e.g. background) as well as fine details. After the training phase, this mapping network can be disconnected, and the styles can be provided directly by the user, resulting in a large level of control over the output of the generator.

The original StyleGAN2 Generator is constructed out of an arbitrary amount of convolutional blocks (depending on the desired output size). Each of these wraps upsampling with gaussian blur, followed by 2 convolutional layers together with weight demodulation, which is used in place of instance normalization. The signals at each scale are then appropriately upsampled to much final output size and added together using skip-connections, with the output of the Generator being a sum of the signals from each scale. The goal of such architecture is to simplify the training process in comparison to classic progressive-growing [4], and avoid some of its drawbacks [7].

2.3 Finding Styles for a Given Image

After training a StyleGAN, besides generating random images and experimenting with manually making changes to the style vector, it is also possible to find the style vectors for specific examples. This comes in handy when a database of labeled examples is available - this way identifying style vectors for them may allow for finding some connections between the style values and contents of the images.

Finding the style vector for a set image is done using just the Generator, which has its weights frozen throughout this process. An arbitrary set of styles is supplied to the network as a starting point (ideally one that is somewhat close to the result, if it is possible to create an accurate "style-guessing" network). This style vector then undergoes adjustment via backpropagation, with the loss function being defined as the distance between the desired image, and the image produced by the Generator. This distance can be a simple mean square error, however, it often leads to the process getting stuck in local minima. A better distance metric is using the differences between the output of the last convolutional layer of a robust image classifier as perceptual distance [1].

3 GW-GAN

3.1 Generator

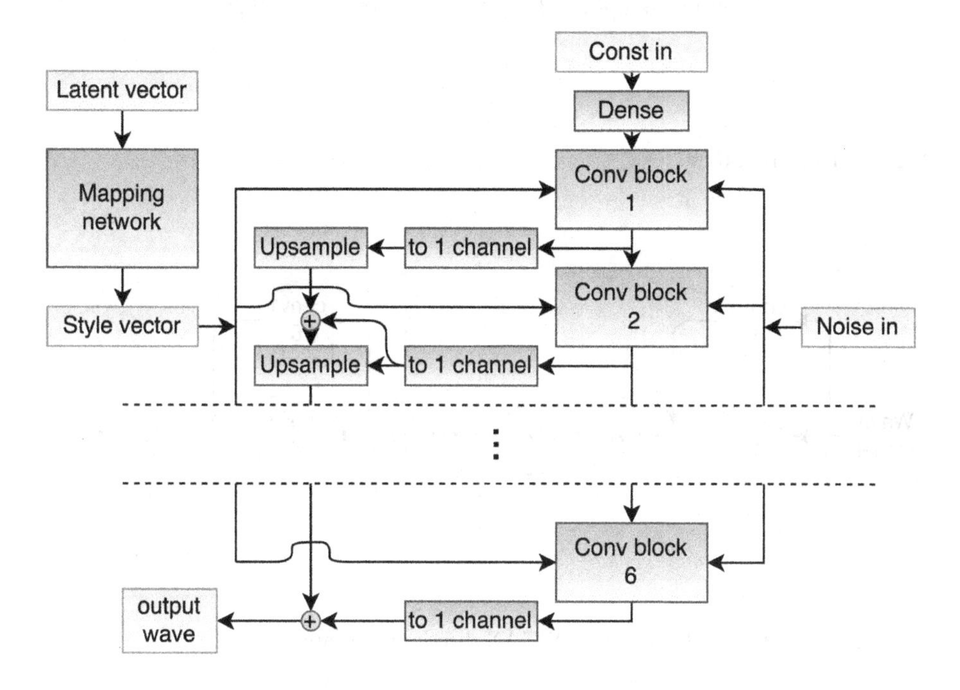

Fig. 1. Proposed GW-GAN generator architecture

The GW-GAN Generator is composed of 6 convolutional blocks with the initialization constant vector that has a size of 32×384 for output signal length of 2ˆ13. Due to the various scales reflecting frequency bands in the output signal, as well as the fact that the output signals are oscillating, a non-uniform upscaling approach was used to increase the number of high-frequency components at the cost of low-frequency ones. The first block does not have an upsampling operation, following 3 upsample the signal by the factor of 4, and last 2 by the factor of 2. Since the values of signals can have both positive and negative values, tanh activations were used in place of more common relu. Besides styles, each block is supplied with random noise like in StyleGAN2 [7]. The schematic of generator architecture can be seen in Fig. 1, and the number of filters in each convolutional block, as well as the output samples in Table 1. All upsampling operations are done using bilinear interpolation.

The mapping network is a 4-layer multilayer perceptron with 128 nodes per layer, and leaky relu activations (alpha = 0.2).

Table 1. Filter count and output samples for convolutional blocks in the Generator

Block id	1	2	3	4	5	6
Filter count	384	192	144	96	48	24
Output samples	32	128	512	2048	4096	8192

3.2 Discriminator

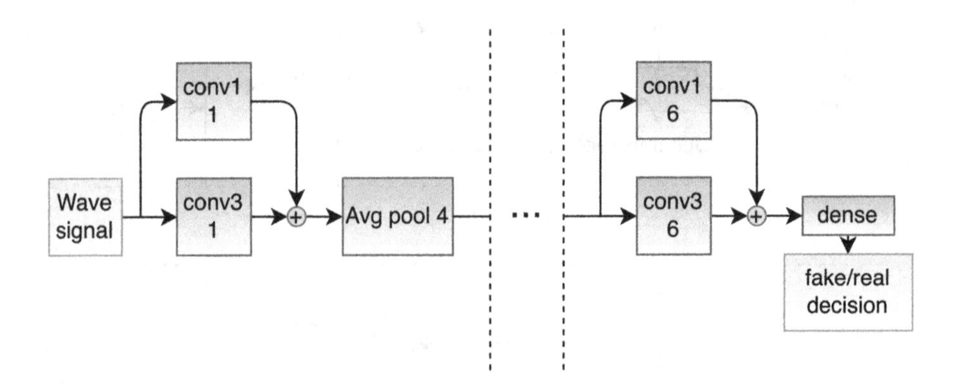

Fig. 2. Proposed GW-GAN discriminator architecture

The Discriminator is a simple 6-block residual network using the same filter counts as Generator (Table 1) however in reverse order. The blocks consist of splitting the signal into 2 copies, one of which goes through a convolution layer with a set filter count and kernel size of 3, while the other goes through one with kernel size of 1. Lastly, the outputs of these 2 operations are added, and (besides the last block) go through average pooling at a size of 4 to downsample them. The Discriminator ends with a dense layer for the fake/real decision-making. The schematic of the discriminator can be seen in Fig. 2.

3.3 Training

The proposed model was trained on data from OpenGuidedWaves dataset [9]. The dataset is composed of a series of measurements of guided waves between 12 transducers spread along 2 opposite edges of a composite plate. A total of 28 damage states are simulated by attaching magnets to various fragments of the plate. The measurements are also performed across a wide spectrum of excitation frequencies, ranging from 60 kHz to 240 kHz. For the training procedure, a total of 160 signals have been hand-picked from damaged measurements for damage positions 1–24 at an excitation frequency of 60 kHz, selecting instances where the damage was either directly on the measured path or close to it. The selected signals have been cropped to 8192 samples and passed through band-pass filtration with the lower band at 20 kHz and upper at 110 kHz. The filtration was performed due to a significant amount of measurement noise present in the signals (as can be seen in Fig. 3), to prevent the Generator from attempting to mimic that noise and instead focus on the actual signal. The final model (mapping network + discriminator + generator) had around 1.8 million parameters in total and followed the same training procedures as StyleGAN2 [7]. It was trained for a total of 326 epochs, which took 120 h on an Nvidia GeForce 1060ti graphical processing unit, using a batch size of 8.

3.4 Style-Finding for a Given Signal

Finding the style vector for a desired signal has proven to be more difficult than initially anticipated. The attempts at training a residual convolutional network to perform style-guessing and give the style backpropagation an approximate starting position have failed to produce satisfying results. Due to random initialization not being reliable enough, in an effort to increase that reliability, a series of random style-signal pairs are generated and compared against the desired signal, resulting in the closest one (in terms of mean square error) being selected as the starting position.

Due to the lack of a robust feature extractor for guided wave signals, as well as a very prominent local minimum at $y(x) = 0$ due to the oscillating nature of these signals, the backpropagation was broken down into 2 steps:

1. loss = perceptual difference + raw signal mean square error
2. loss = raw signal mean square error

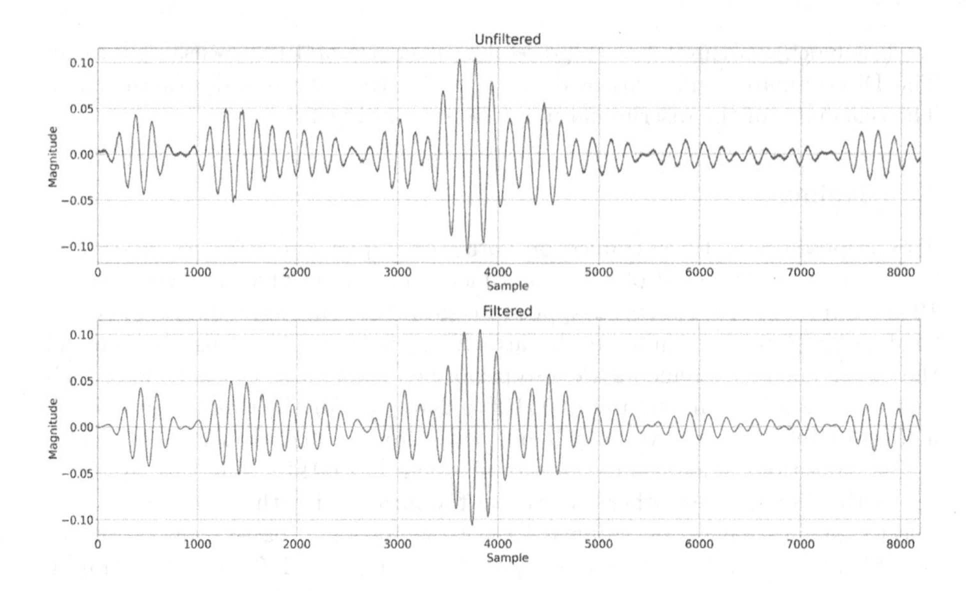

Fig. 3. Effect of bandpass filtering the training data

The network used for the perceptual difference was the aforementioned style-guesser. While it wasn't particularly good at the task it was trained to perform, the features from the last convolution layer have proven to work well for the style-finding process. The reasoning behind breaking it down into these 2 steps is the fact that starting with raw MSE loss practically always resulted in getting stuck in the local minimum of $y(x) = 0$, and relying on the combined perceptual + raw loss produced close, however not satisfying results. Instead, the result is initially approached using the combined loss and later fine-tuned with raw MSE loss.

4 Results

4.1 Generation from Random Noise

As can be seen in Fig. 4, the GW-GAN training was successful despite a very small amount of training data. Random generation produces a variety of signals, showing that mode collapse did not occur. It is a situation in which the network "collapses" into only ever producing the same signal, as it finds it to be the easiest way of fooling the discriminator. The generated signals also resemble the measured ones (presented in Fig. 3): they contain clearly visible wave packets that tend to merge in later portions of the signal. At this point, an experienced operator likely wouldn't be able to distinguish between real and fake data.

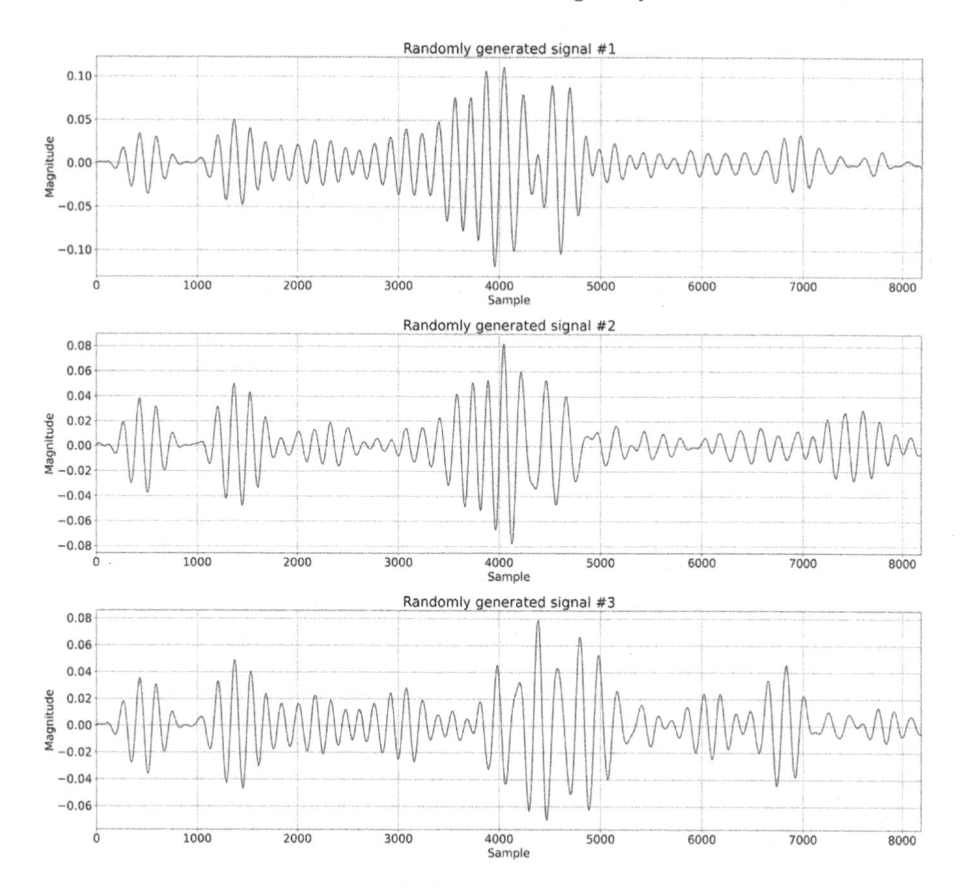

Fig. 4. Example of several signals generated from random noise

4.2 Finding Style Vector for a Given Signal

The plots in Fig. 5 demonstrate that the process of finding style vectors for a given signal works and that manipulating the style vectors can change the output of the network in various ways.

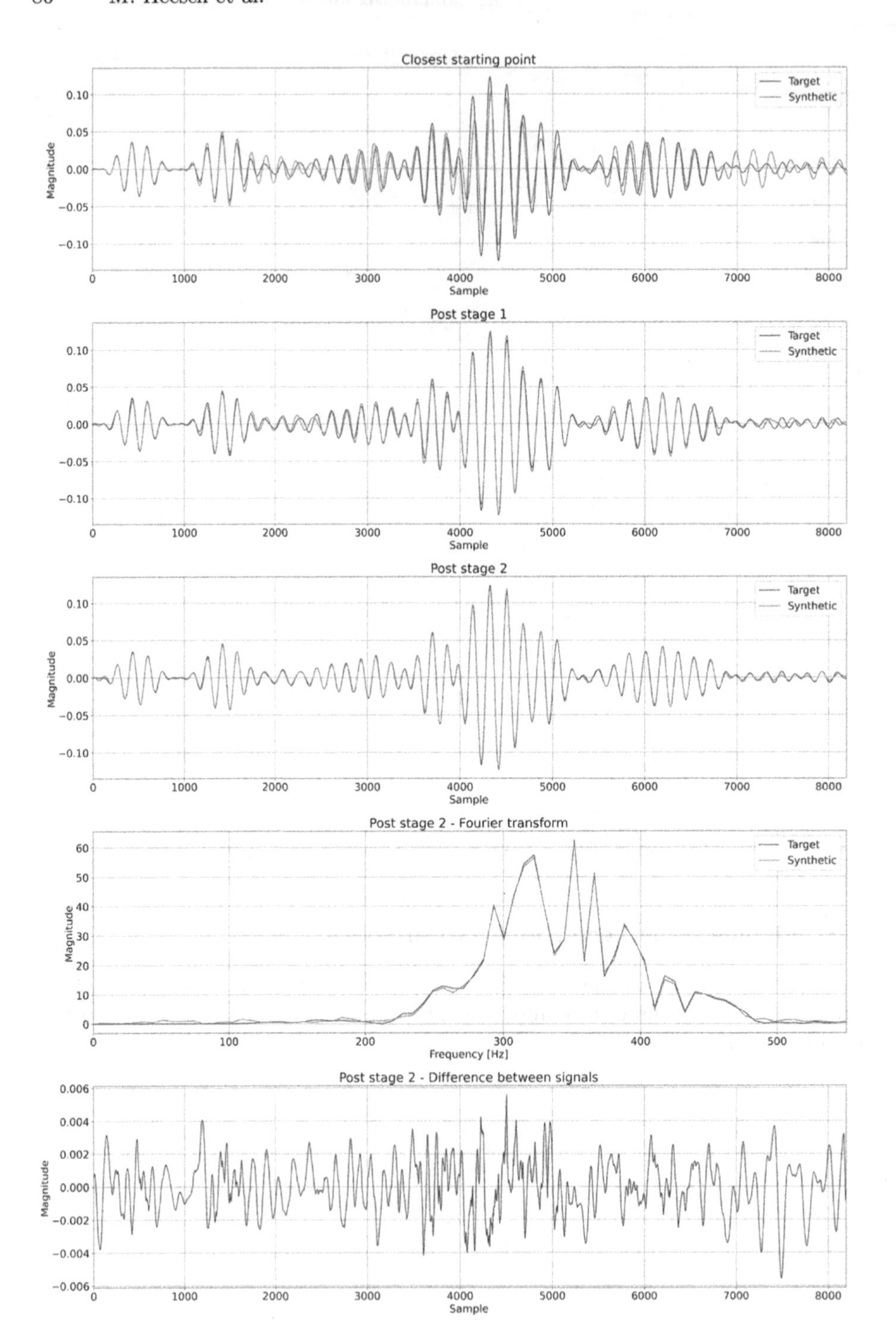

Fig. 5. Example of a process of finding style vector for a specific signal

5 Discussion

As presented in the previous section, the generator can successfully synthesize various guided wave signals - be it via random generation or manipulating the style vector to achieve a specific result. This is a promising stepping stone on the road to creating a comprehensive tool for guided waves database augmentation and synthesis - as it suggests that it is indeed possible in the first place.

Since adjusting the input values in style vector allowed the generator to synthesize signals of a desired shape, establishing how to work directly on the style vectors by investigating the connections between labeled signals and the style vectors required to generate them should lead to formulating a control scheme that would allow for the synthesis of various desired signals - ideally ones that were not present in the original training distribution - e.g. synthesizing signals at new sensor locations.

This work did not explore the effects that adding synthetic data to training would have on classifiers trained using it, as it still preliminary research, at this moment focused on the controlled synthesis of high-quality data. Though using GAN generated data for dataset expansion has proven to result in more accurate or robust classifiers in other domains [3,10,11], the exact impact will have to be investigated once proper synthesis control is formulated.

Lastly, while GW-GAN was designed with guided waves in mind, it should be usable for any time-series signals. This may require minor changes in the architecture, such as altering activations (which will certainly be required if the signal in question should not produce negative values), scaling the output signal, or adding/removing convolution blocks, and altering scale distribution.

6 Conclusions

The proposed GW-GAN was successfully trained to generate guided wave signals. The initial attempts of controlling the result via adjusting style vectors have worked, and prompt further research to achieve better control over the generation outputs. Once such control is achieved, the effects of adding various amounts of synthetic data on the efficacy of classifiers trained on it should be investigated. The applicability of the architecture to other time-series problems is yet to be verified.

Acknowledgement. The work presented in this paper was supported by the National Center for Research and Development in Poland, under project number LIDER/3/0005/L-9/17/NCBR/2018.

References

1. Abdal, R., Qin, Y., Wonka, P.: Image2styleGAN: how to embed images into the styleGAN latent space? (2019)

2. Donahue, C., McAuley, J.J., Puckette, M.S.: Synthesizing audio with generative adversarial networks. CoRR abs/1802.04208 (2018). http://arxiv.org/abs/1802.04208
3. Fiore, U., Santis, A., Perla, F., Zanetti, P., Palmieri, F.: Using generative adversarial networks for improving classification effectiveness in credit card fraud detection. Inf. Sci. **479**, 448–455 (2017). https://doi.org/10.1016/j.ins.2017.12.030
4. Goodfellow, I.J., et al.: Generative adversarial nets. In: Proceedings of the 27th International Conference on Neural Information Processing Systems - Volume 2, NIPS 2014, pp. 2672–2680. MIT Press, Cambridge (2014)
5. Gui, J., Sun, Z., Wen, Y., Tao, D., Ye, J.: A review on generative adversarial networks: algorithms, theory, and applications (2020)
6. Karras, T., Laine, S., Aila, T.: A style-based generator architecture for generative adversarial networks (2018)
7. Karras, T., Laine, S., Aittala, M., Hellsten, J., Lehtinen, J., Aila, T.: Analyzing and improving the image quality of styleGAN (2019)
8. Mitra, M., Gopalakrishnan, S.: Guided wave based structural health monitoring: a review. Smart Mater. Struct. **25**, 053001 (2016). https://doi.org/10.1088/0964-1726/25/5/053001
9. Moll, J., et al.: Open guided waves: online platform for ultrasonic guided wave measurements. Struct. Health Monit. **18**(5–6), 1903–1914 (2019)
10. Sandfort, V., Yan, K., Pickhardt, P.J., Summers, R.M.: Data augmentation using generative adversarial networks (cycleGAN) to improve generalizability in CT segmentation tasks. Sci. Rep. **9**(1), 1–9 (2019)
11. Shao, S., Wang, P., Yan, R.: Generative adversarial networks for data augmentation in machine fault diagnosis. Comput. Ind. **106**, 85–93 (2019). https://doi.org/10.1016/j.compind.2019.01.001

A Comparison of Trend Estimators Under Heteroscedasticity

Jan Kalina, Petra Vidnerová(✉), and Jan Tichavský

The Czech Academy of Sciences, Institute of Computer Science, Pod Vodárenskou věží 2, 18207 Prague 8, Czech Republic
{kalina,petra,tichavsky}@cs.cas.cz

Abstract. Trend estimation, i.e. estimating or smoothing a nonlinear function without any independent variables, belongs to important tasks in various applications within signal and image processing, engineering, biomedicine, analysis of economic time series, etc. We are interested in estimating trend under the presence of heteroscedastic errors in the model. So far, there seem no available studies of the performance of robust neural networks or the taut string (stretched string) algorithm under heteroscedasticity. We consider here the Aitken-type model, analogous to known models for linear regression, which take heteroscedasticity into account. Numerical studies with heteroscedastic data possibly contaminated by outliers yield improved results, if the Aitken model is used. The results of robust neural networks turn out to be especially favorable in our examples. On the other hand, the taut string (and especially its robust L_1-version) inclines to overfitting and suffers from heteroscedasticity.

Keywords: Nonlinear regression · Robust neural networks · Taut string · Outliers · Heteroscedasticity

1 Introduction

Estimating and predicting a nonlinear trend of an observed continuous variable represents an important task with various applications in signal and image processing, engineering, biomedicine, economics, etc. A broad spectrum of trend estimation methods has been proposed and investigated in machine learning as well as statistics. If we consider (only) one-dimensional regressors $X_1, \ldots, X_n \in \mathbb{R}$, the nonlinear regression model can be expressed as

$$Y_i = f(X_i) + e_i \quad \text{for} \quad i = 1, \ldots, n, \tag{1}$$

where $Y_1, \ldots, Y_n$ are observed values of an unknown continuous nonlinear function f under the presence of random errors $e_1, \ldots, e_n$. The errors must be independent and identically distributed (i.i.d.) but we do not want to assume any particular probabilistic distribution of the errors. The aim is to estimate (predict) the response variable based on given regressors.

L. Rutkowski et al. (Eds.): ICAISC 2021, LNAI 12854, pp. 89–98, 2021.
https://doi.org/10.1007/978-3-030-87986-0_8

We consider however a special case of (1) in the form

$$Y_i = f(i) + e_i \quad \text{for} \quad i = 1, \ldots, n, \tag{2}$$

with equidistant observations and denote the task to estimate the unknown nonlinear function f as trend estimation (in this context, also commonly denoted as smoothing, function approximation, pattern estimation, etc.). The values of the response in (2) are not modeled by means of any independent variables (regressors, predictors). In our regression point of view, we take the horizontal axis, i.e. the vector $(1, \ldots, n)^T$, as a regressor, which is a common approach in time series [2], but our approach here is not limited only to time series. The estimated values will be denoted as $\hat{f}_1, \ldots, \hat{f}_n$, and an estimate of f considered as a function will be denoted by $\hat{f}$. We are interested in neural networks and the taut string algorithm, while our aim is to discuss and investigate the performance of both approaches (as well as their robust versions) under heteroscedasticity.

Heteroscedasticity is well known to complicate regression modeling. In linear regression, numerous hypothesis tests of heteroscedasticity are available; if a suitable heteroscedasticity test is significant, the least squares estimator b_{LS} of the parameters β is not efficient and the standard estimates of $\mathsf{var}\, b_{LS}$ are biased. Therefore, the least squares estimator in the linear model is commonly replaced by a generalized least squares approach (also known as weighted least squares, Aitken estimator) [7]. Also, robust statistical estimation in linear regression, i.e. robust with respect to the presence of outliers in the data [9], has to be adapted for heteroscedastic data accordingly [1,25]; in fact, there have been heteroscedasticity tests proposed as tailor-made tools for robust regression [11].

Also in nonlinear regression (1), standard estimation methods such as neural networks lose their efficiency under heteroscedasticity. Nevertheless, the literature on modeling heteroscedastic data by neural networks remains rather rare. Heteroscedasticity was revealed in numerical experiments to have effect on neural networks (however without attempts for its modeling) in [18]. A specific likelihood-based approach assuming gamma distribution of the data was used in [16]. A heteroscedastic model combined with clustering (based however on the non-robust maximum likelihood principle) of the data was exploited in [19]. While there exist several robust approaches to training neural networks, which have dealt with robustness to outlying values (outliers) in the data [12], their performance under heteroscedasticity has not been investigated. In fact, some robust neural networks (e.g. those of [23]) are clearly not much suitable for heteroscedasticity, because they may consider a large portion of observations in such parts of the data, where the variability of the errors is the largest, to be outlying. In this paper, we propose a transformation of the basic model and to exploit robust neural networks in a heteroscedastic regression model.

The taut string (stretched string) algorithm [3,4] for (2) represents one of available nonlinear (and nonparametric) estimators for (2). It estimates a nonlinear trend by a piecewise monotone function, which is desirable in various applications e.g. in signal and image processing. However, we are not aware of any study of the performance of the taut string under heteroscedastic errors, which at any case appear quite commonly in real data. Section 2 of this paper recalls the taut string algorithm and some available robust neural networks. It also proposes a heteroscedastic-adapted (Aitken) model, which is used in numerical experiments with heteroscedastic data in Sect. 3. Section 4 concludes the paper.

2 Methodology

In this section, we give a list of regression methods, which will be used in the subsequent computations in Sect. 3. Here, we also specify our particular selection of parameters for each method. If necessary, we also describe other details necessary in order to uniquely describe our computational approaches.

2.1 (Robust) Neural Networks for Nonlinear Regression

In our computations, we use the following versions of neural networks. While we do not need to recall definitions of their standard types (see e.g. [8]), we comment their recently proposed robust versions. field of numerical mathematics, i.e. in mathematics by means of deterministic tools without specifying any uncertainty or randomness (e.g. splines, Chebyshev polynomials,
to the free space.

- MLP: Multilayer perceptron with 2 hidden layers, which contain 16 and 2 neurons, respectively. We use a sigmoid activation function in the hidden layers and a linear output layer.
- RBF network: Radial basis function network with a Gaussian kernel and 10 radial units.
- LWS-MLP, LWS-RBF: LWS-based versions of MLPs or RBF networks, replacing the common loss (i.e. minimal sum of squares residuals) by the loss of the least weighted squares (LWS) estimator [10,25].
- LTS-MLP, LTS-RBF: LTS-based versions of MLPs or RBF networks, exploiting the loss function of the least trimmed squares (LTS) estimator [21,22,24].
- Back-MLP, Back-RBF: Robust approaches to training MLPs and RBF networks using the backward subsample selection of [12], i.e. a backward search of the least outlying subsamples.

Robust versions of MLPs (LWS-MLP, LTS-MLP, Back-MLP) have 2 hidden layers, which contain 16 and 2 neurons, respectively. They use a sigmoid activation function in the hidden layers and a linear output layer. Robust versions of RBF networks (LWS-RBF, LTS-RBF, Back-RBF) use a Gaussian kernel and 10 radial units. For LWS-MLP and LWS-RBF, we use the particular choice of the weight function in the form

$$\psi_{LWS}(t) = \psi_0\left(\frac{t}{\tau}\right) \cdot \mathbb{1}[t < \tau], \quad t \in [0,1], \tag{3}$$

with $\tau = 3/4$, where

$$\psi_0(t) = \exp\left\{-\frac{t^2}{2\sigma^2}\right\}, \quad t \in [0,1]. \tag{4}$$

With again $\tau = 3/4$, LTS-MLP and LWS-RBF use the weight function

$$\psi_{LTS}(t) = \mathbb{1}[t < \tau], \quad t \in [0,1]. \tag{5}$$

2.2 Taut String

The taut string algorithm [4], also denoted as stretched string or piecewise monotone regression with taut strings, has received great attention in signal and image processing, especially in image deblurring and restoration, or in real-time communication systems; see [15] or [17] and the references cited therein. It is formally defined for the model (2) for equidistant data as a regularized smoother penalizing the total variation. It has not been however discussed whether the method is suitable under heteroscedasticity or not.

Appealing properties of the taut string include its ability to keep the number of local extremes of the fitted curve under control. In other words, the method does not have a tendency to include redundant local extremes in the estimated trend [4]. The taut string was proven in [17] to be equivalent to the Rudin-Osher-Fatemi model, which is a popular tool for image denoising based on total variation. The taut string method was also exploited in [13] within proofs of asymptotic properties of other more complex approaches to solving statistical inverse problems; it belongs to a broader class of non-parametric procedures with an adaptive choice of the regulation parameter [14]. Other properties of the taut string may be rather limiting; particularly, the taut string is defined only equidistant observations in (2) and its trajectory is non-smooth. In addition, the method does not perform a prediction for a new (independent) observation, but only provides a data-driven approximation. Thus, it is not possible to evaluate its performance within a cross validation. Neither can be cross validation used for finding a suitable value of the regularization parameter; therefore, an approach based on a multiresolution analysis of the residuals was proposed in [4].

In addition to the plain taut string, there exists an alternative robust approach of [6] estimating the conditional median of Y_i (rather than its expectation) in (2). We are however not aware of applications of such L_1-version of the taut string. The computation of the plain as well as robust taut string may exploit the combinatoric algorithm of [4]. This minimizes the trajectory of $\hat{f}$ under constraints, which require $\hat{f}$ to lie between the largest convex minorant and smallest concave majorant computed from $Y_1, \dots, Y_n$ (and depending on the chosen regularization parameter). Thus, the estimator corresponds to an intuitive idea of a taut string with the shortest possible trajectory.

2.3 Neural Networks and Heteroscedasticity: The Aitken Model

Let us now assume the model (1) with $\mathsf{var}\, e_i = \sigma^2 k_i$ for $i = 1, \dots, n$, where the positive constants $k_1, \dots, k_n$ are known. We propose now an alternative model, which is more suitable for the modeling under heteroscedasticity. It is appropriate to replace (1) by an alternative model,

$$\frac{Y_i}{\sqrt{k_i}} = f\left(\frac{X_i}{\sqrt{k_i}}\right) + \frac{e_i}{\sqrt{k_i}}, \quad i = 1, \dots, n, \tag{6}$$

which takes heteroscedasticity into account. We call (6) the Aitken model, as it is usual to denote an analogous approach for linear regression (as discussed in Sect. 1). The errors in (6) are homoscedastic, as it holds

$$\mathsf{var}\,\frac{e_i}{\sqrt{k_i}} = \frac{1}{k_i}\mathsf{var}\,e_i = \sigma^2. \tag{7}$$

None of available studies of robust neural networks considered heteroscedastic data. As our novelty, we propose to use also robust neural networks of Sect. 2.1 in the transformed model (6).

3 Numerical Experiments

3.1 Data Description

The aim of the numerical experiments is to compare various methods on two different datasets, which are considered in different versions with a different severity of heteroscedasticity. We use the notation of the model (2) here.

- Dataset A: Simulated heteroscedastic data ($n = 500$) using

$$Y_i = \frac{(i-100)^2}{200} - 50 + e_i, \quad i = 1, \ldots, n, \tag{8}$$

  where $e_1, \ldots, e_n$ are generated as i.i.d. random variables with $e_i \sim \mathsf{N}(0, \sigma_i^2)$ and $\sigma_i^2 = 0.3i$ for $i = 1, \ldots, n$. in the data (however $c = 0$ would correspond to exact fit
- Dataset B: Simulated heteroscedastic data ($n = 500$) with

$$Y_i = \frac{i}{5} + e_i, \tag{9}$$

  where $e_1, \ldots, e_n$ are generated as i.i.d. random variables from the normal distribution with $e_i \sim \mathsf{N}(0, \sigma_i^2)$ and $\sigma_i^2 = 0.6i$ for $i = 1, \ldots, n$.

In addition, we also consider the datasets after an artificial contamination.

- Contaminated version of Dataset A. We replace $\lfloor n/6 \rfloor = 83$ observations by (deterministic) values $Y_{6k} = -40 + k$ for $k = 1, \ldots, \lfloor n/6 \rfloor$, where $\lfloor x \rfloor$ denotes the integer part of $x \in \mathbb{R}$.
- Contaminated version of Dataset B, we replace (again) $\lfloor n/6 \rfloor = 83$ observations by values $Y_{6k} = 80 - 0.9k$ for $k = 1, \ldots, \lfloor n/6 \rfloor$.

In all examples, the construction of the model (6) considers $k_i = i$ for $i = 1, \ldots, n$. We use either Python or R software [20] for the computations, as indicated in Table 1 for each of the trend estimation methods.

3.2 Evaluating the Prediction Ability

Because the taut string does not allow to be computed within a cross validation (see Sect. 2.2), we evaluate the approximation ability of all methods by means of the mean square error (MSE) in an autovalidation (autoverification). Only for all the methods without the taut string, we perform a standard 10-fold cross validation; this divides a particular dataset to 10 groups, repeatedly performs the trend estimation jointly over 9 groups, and applies the prediction to the omitted group. Denoting the cross validation prediction error of the i-th observation for $i = 1, \ldots, n$, as $\hat{Y}_i$, MSE is finally aggregated as $\mathsf{MSE} = (1/n)\sum_{i=1}^{n}(Y_i - \hat{Y}_i)^2$.

Because MSE is not suitable for evaluating the prediction performance under contamination, we consider its robust version known as the trimmed mean square error (TMSE). Let us denote by $r_1, \ldots, r_n$ prediction errors of individual observations and by $r_{(1)}^2 \leq \cdots \leq r_{(n)}^2$ their ordered squared values. TMSE is defined as $\mathsf{TMSE} = (1/h)\sum_{i=1}^{h} r_{(i)}^2$, where we use $h = \lfloor 0.8n \rfloor$.

Specifically, if we perform the prediction of the response in (6), MSE there is however not directly comparable with values of MSE in (1). Nevertheless, we are able to re-transform the predictions from (6) back to the original model (1). If the fitted value of the response of the i-th observation in (6) is denoted as $\hat{Y}_i^*$, then the corresponding fitted value in (1) is $\hat{Y}_i = \sqrt{k_i}\hat{Y}_i^*$ and

MSE transformed to (1) is thus obtained as

$$\mathsf{MSE} = \frac{1}{n}\sum_{i=1}^{n}\left(\sqrt{k_i}\hat{Y}_i^* - Y_i\right)^2. \tag{10}$$

TMSE in (6) is evaluated in an analogous way.

3.3 Results

Results for Dataset A are presented in Table 1, which contains values of the mean square error (MSE) within autovalidation, and Table 2 with values obtained in the 10-fold cross validation. Results for Dataset B are presented in Table 3 for autovalidation and Table 4 for 10-fold cross validation. Let us now discuss the results.

For contaminated data, robust neural networks improve the prediction error compared to standard ones. If the data are non-contaminated, the performance of robust neural networks stays only slightly behind that of standard ones. This is analogous to results of [12] on data without heteroscedasticity. The Aitken model brings benefits compared to estimating in the plain nonlinear model. This is true basically for each (heteroscedastic) situation.

The best results in both examples are obtained in the Aitken model, if robust neural networks are used. Particularly, the best results in the two datasets are obtained with LWS-RBF and LTS-RBF. Their robustness is beneficial much more in the Aitken model compared to the plain nonlinear model. The taut string and especially its L_1-version overfits the data. This is clearly revealed by graphical visualizations (not shown here). Moreover, no version of the taut string method turns out to be suitable for data, which are heteroscedastic.

Table 1. Autovalidation results of model (2) in Dataset A, evaluated as MSE and TMSE for raw or contaminated data. For each method, we also present the software used in the computations (and in parentheses, particular packages or functions are given, or a statement that we used our code).

Method	Software	Raw data		Contam. data	
		MSE	TMSE	MSE	TMSE
Taut string	R (ftnonpar [5])	76.0	18.6	211.4	46.4
L_1-taut string	R (ftnonpar [5])	56.9	16.8	164.9	45.7
MLP	Python (Keras)	77.2	23.7	225.1	72.3
LWS-MLP	Python (own)	78.5	23.0	253.6	51.6
LTS-MLP	Python (own)	80.3	22.9	262.1	54.2
Back-MLP	R (own)	82.4	23.4	270.4	61.8
RBF network	Python (Keras)	73.6	21.1	203.9	60.8
LWS-RBF	Python (own)	75.4	20.5	247.0	47.7
LTS-RBF	Python (own)	79.8	19.4	252.8	50.3
Back-RBF	R (own)	80.1	20.8	264.3	52.6

Table 2. 10-fold cross validation results in Dataset A, evaluated as MSE and TMSE for raw or contaminated data.

Method	Raw data		Contam. data	
	MSE	TMSE	MSE	TMSE
	Model (2)			
MLP	79.3	27.8	243.7	83.4
LWS-MLP	81.4	26.1	271.8	56.0
LTS-MLP	83.5	25.8	279.3	57.6
Back-MLP	85.0	26.9	286.4	63.1
RBF network	76.7	25.8	232.2	76.3
LWS-RBF	80.6	25.2	251.8	52.5
LTS-RBF	82.2	24.8	256.1	54.9
Back-RBF	84.9	25.3	263.2	59.4
	Model (6)			
MLP	75.9	26.7	241.4	74.3
LWS-MLP	79.4	25.2	275.2	31.9
LTS-MLP	80.3	25.0	286.5	33.0
Back-MLP	83.6	27.6	303.6	37.4
RBF network	72.2	24.3	237.0	69.8
LWS-RBF	74.9	25.7	259.3	30.7
LTS-RBF	75.1	24.7	262.1	31.6
Back-RBF	78.5	26.8	280.7	34.7

Table 3. Autovalidation results of model (2) in Dataset B, evaluated as MSE and TMSE for raw or contaminated data.

Method	Raw data		Contam. data	
	MSE	TMSE	MSE	TMSE
Taut string	142.8	34.2	500.2	62.6
L_1-taut string	98.0	27.1	204.2	53.0
MLP	134.4	37.8	480.8	98.7
LWS-MLP	145.1	37.5	494.3	64.1
LTS-MLP	148.9	37.2	491.6	66.4
Back-MLP	156.3	38.1	503.7	70.3
RBF network	125.3	36.1	463.7	97.0
LWS-RBF	136.6	35.7	482.5	61.2
LTS-RBF	138.2	35.3	481.1	63.9
Back-RBF	147.0	36.8	487.8	65.4

Table 4. 10-fold cross validation results in Dataset B, evaluated as MSE and TMSE for raw or contaminated data.

Method	Raw data		Contam. data	
	MSE	TMSE	MSE	TMSE
	Model (2)			
MLP	145.1	40.4	516.7	115.6
LWS-MLP	148.2	38.5	541.9	70.5
LTS-MLP	147.6	39.9	538.4	72.6
Back-MLP	153.7	38.8	568.1	78.2
RBF network	136.8	38.8	483.2	109.4
LWS-RBF	139.4	37.0	509.7	65.3
LTS-RBF	138.3	36.2	506.3	67.8
Back-RBF	145.7	38.1	511.0	73.9
	Model (6)			
MLP	133.2	37.0	452.3	104.6
LWS-MLP	139.1	35.3	490.7	46.1
LTS-MLP	137.6	34.9	494.8	45.3
Back-MLP	145.7	36.1	503.4	49.4
RBF network	128.2	33.5	437.9	99.2
LWS-RBF	132.1	32.2	474.7	42.0
LTS-RBF	130.5	32.8	478.0	44.8
Back-RBF	141.8	33.1	488.1	47.5

4 Conclusions

Estimating trend in the model (2), i.e. smoothing a continuous variable, represents an important (and only seemingly simple) task in the analysis of real data in various fields. We consider heteroscedastic data possibly contaminated by outliers and compare standard and robust versions of neural networks and the taut string method. As a novelty, we consider very recently proposed robust neural networks of [12] in the Aitken model (6), which takes heteroscedasticity into account.

While standard as well as robust methods for training neural networks suffer from heteroscedasticity, the Aitken model turns out to be more suitable in our examples and we recommend to perform the estimation in it; a suitable model for the heteroscedasticity can be (at least in the situation with a single regressor) easily formulated based on a visual inspection of the data. In the examples, we achieved the best performance with LWS- and LTS-based robust versions of RBF networks if the Aitken model (6) was used. As future research, we plan to perform additional computations for robust neural networks, especially for models with more-dimensional regressors.

The taut string algorithm is appealing for very specific applications of signal and image processing. The taut string, which does not allow a cross validation, is clearly overcome by recently proposed robust neural networks in our heteroscedastic examples. Indeed, it is the idea of the multiresolution analysis of the residuals, which implicitly assumes homoscedastic errors within the taut string method (although not explicitly stated in the literature); the same is true for the robust L_1-version. Under heteroscedasticity, the taut string (and especially its L_1-version) turns out to overfit the data heavily. In addition, we are not aware of any reliable extension of the taut string algorithm to regressors with more than one dimension.

Acknowledgements. The research is supported by the projects GA19-05704S and GA18-23827S of the Czech Science Foundation. Jiří Tumpach and Patrik Janáček provided technical support.

References

1. Atkinson, A.C., Riani, M., Torti, F.: Robust methods for heteroskedastic regression. Comput. Stat. Data Anal. **104**, 209–222 (2016)
2. Brockwell, P.J., Davis, R.A.: Introduction to Time Series and Forecasting. STS, 2nd edn. Springer, Cham (2016). https://doi.org/10.1007/978-3-319-29854-2
3. Davies, P.L.: Data Analysis and Approximate Models. CRC Press, Boca Raton (2014)
4. Davies, P.L., Kovac, A.: Local extremes, runs, strings and multiresolution. Ann. Statist. **29**, 1–65 (2001)
5. Davies, L., Kovac, A.: Ftnonpar: features and strings for nonparametric regression. R package version 0.1-88 (2019). https://CRAN.R-project.org/package=ftnonpar
6. Dümgen, L., Kovac, A.: Extensions of smoothing via taut strings. Electron. J. Stat. **3**, 41–75 (2009)

7. Greene, W.H.: Econometric Analysis, 8th edn. Pearson, London (2017)
8. Haykin, S.O.: Neural Networks and Learning Machines: A Comprehensive Foundation, 2nd edn. Prentice Hall, Upper Saddle River (2009)
9. Jurečková, J., Picek, J., Schindler, M.: Robust Statistical Methods with R, 2nd edn. CRC Press, Boca Raton (2019)
10. Kalina, J., Schlenker, A.: A robust supervised variable selection for noisy high-dimensional data. BioMed Res. Int. **2015**, Article 320385 (2015)
11. Kalina, J., Tichavský, J.: On robust estimation of error variance in (highly) robust regression. Meas. Sci. Rev. **20**, 6–14 (2020)
12. Kalina, J., Vidnerová, P.: Robust training of radial basis function neural networks. In: Rutkowski, L., Scherer, R., Korytkowski, M., Pedrycz, W., Tadeusiewicz, R., Zurada, J.M. (eds.) ICAISC 2019. LNCS (LNAI), vol. 11508, pp. 113–124. Springer, Cham (2019). https://doi.org/10.1007/978-3-030-20912-4_11
13. Kim, S., Pokojovy, M., Wan, X.: The taut string approach to statistical inverse problems: theory and applications. J. Comput. Appl. Math. **382**, Article 113098 (2021)
14. Koenker, R., Mizera, I.: The alter egos of the regularized maximum likelihood density estimators: Deregularized maximum-entropy, Shannon, Rényi, Simpson, Gini, and stretched strings. In: Hušková, M., Janžura, M. (eds.) Prague Stochastics, pp. 145–157. Matfyzpress, Prague (2006)
15. Makovetskii, A., Voronin, S., Kober, V., Voronin, A.: Tube-based taut string algorithms for total variation regularization. Mathematics **8**, Article 1141 (2020)
16. Ng, N.H., Gabriel, R.A., McAuley, J., Elkan, C., Lipton, Z.C.: Predicting surgery duration with neural heteroscecastic regression. Proc. Mach. Learn. Res. **68**(26), 100–111 (2017)
17. Overgaard, N.C.: On the taut string interpretation and other properties of the Rudin-Osher-Fatemi model in one dimension. J. Math. Imaging Vis. **61**, 1276–1300 (2019)
18. Paliwal, M., Kumar, U.A.: The predictive accuracy of feed forward neural networks and multiple regression in the case of heteroscedastic data. Appl. Soft Comput. **11**, 3859–3869 (2011)
19. Paul, C., Vishwakarma, G.K.: Back propagation neural networks and multiple regressions in the case of heteroscedasticity. Comm. Stat. Simul. Comput. **46**, 6772–6789 (2017)
20. R Core Team: R: A language and environment for statistical computing. R Foundation for Statistical Computing, Vienna (2019). https://www.R-project.org/
21. Rousseeuw, P.J., Leroy, A.M.: Robust Regression and Outlier Detection. Wiley, New York (1987)
22. Rousseeuw, P.J., Van Driessen, K.: Computing LTS regression for large data sets. Data Min. Knowl. Disc. **12**, 29–45 (2006)
23. Rusiecki, A.: Robust LTS backpropagation learning algorithm. In: Sandoval, F., Prieto, A., Cabestany, J., Graña, M. (eds.) IWANN 2007. LNCS, vol. 4507, pp. 102–109. Springer, Heidelberg (2007). https://doi.org/10.1007/978-3-540-73007-1_13
24. Víšek, J.Á.: The least trimmed squares. Part I: consistency. ybernetika **42**, 1–36 (2006)
25. Víšek, J.Á.: Consistency of the least weighted squares under heteroscedasticity. Kybernetika **47**, 179–206 (2011)

Canine Behavior Interpretation Framework Using Deep Graph Model

Jongmin Lim(✉), Donghee Kim, and Kwangsu Kim

Sungkyunkwan University, Suwon, Korea
jm.lim@g.skku.edu, {ym.dhkim,kim.kwangsu}@skku.edu

Abstract. Humans have long aspired to understand dog behavior. While research on the Calming signal has achieved substantial progress in understanding dog behavior, it remains an unfamiliar concept to non-expertise. Therefore, in this paper, we introduce a framework for analyzing dog behavior, which defines the interrelationship between dog postures through a graph model without any additional devices but a camera. First of all, our framework classifies the dog posture in frame units, using a machine learning model based on the position information of the dog's body part in the video captured by the camera. We then analyze dog behavior using graph models that define interrelationships among classified dog postures. We expect that our approach will help non-expertise to understand dog behavior.

Keywords: Canine behavior analysis · Object detection · Graph model

1 Introduction

For centuries, dogs have socially interacted with humans by playing various roles such as hunters, security guards, and friends. As such, dogs have come to be thought of as spiritual companions rather than mere possessions for one's pleasure. Nevertheless, humans and dogs have many differences. In particular, unlike a human, dogs communicate non-verbally using body language. To interpret their behavior is more difficult for humans because of these differences. Therefore, keeping dogs would become handy if it is possible to understand dog's behavior.

Norwegian dog trainer Turid Rugaas defines at least 30 types of "Calming signals [6]" to understand dog behavior. The calming signal is a communication method with dogs each other, but also known to use these signals with humans

This research was partly supported by the MSIT(Ministry of Science and ICT), Korea, under the Grand Information Technology Research Center support program(IITP-2021-2015-0-00742) and Institute of Information & communications Technology Planning & Evaluation (IITP) (No. 2020-0-00990, Platform Development and Proof of High Trust & Low Latency Processing for Heterogeneous·Atypical·Large Scaled Data in 5G-IoT Environment).

L. Rutkowski et al. (Eds.): ICAISC 2021, LNAI 12854, pp. 99–110, 2021.
https://doi.org/10.1007/978-3-030-87986-0_9

too. Research on the calming signal has helped many dog trainers and behaviorists and has led to the development of dog ethology. However, it is still arduous for non-expertise to understand dog ethology without any knowledge of biology, genetics, evolution, etc.

Moreover, various studies have been done using high-quality equipment, yet attaching the equipment might cause stress for dogs, and it is too price for commercializing.

Also, Previous studies have been defined dog behaviors by analyzing a single posture. But it neglects the continuity of dog postures. For example, a dog's tail-raising posture is usually considered a friendly signal. Nevertheless, the tail-raising posture could sign fear, anxiety, danger, or warning depending on the previous postures.

Therefore, in this paper, we will introduce a framework for dog behavior analysis by defining the interrelationship between dog postures, using a graph model without any additional devices but a camera. The graph model uses various dog postures as nodes and defines the continuous interrelationships between nodes in reference to the calming signal. First, Our framework uses object detection to detect dogs and their body parts from the video captured by the camera. Then, the detected dog area is set as absolute coordinate space to utilize the location information of each body part detected regardless of the size and position of the dog in the image. And, a dog's posture is classified in frame units through a machine learning model based on the absolute coordinate values of the body parts. Finally, the graph model determines the dogs' behavior using the postures classified on frame units. Our framework is extensible to adding a new dog posture as a node and redefining the continuous interrelationships between nodes.

Organization. In Sect. 2, We discuss about basic knowledge on dog behaviors. In Sect. 3, we explain our posture classification method and graph model. In Sect. 4, we evaluate the training process and performance of dog posture classification models. Lastly, we conclude the study in Sect. 5.

2 Dog Behaviors

A dog's body language is a sophisticated non-verbal system that non-understand in a single posture. A single posture is only part of the package that displays its mood. Therefore, skilled dog trainers and behaviorists do not analyze a dog's behavior with just a single posture but observe every posture expressed in succession. For example, a beginner trainer may believe that when a dog lifts one paw, it is emotional such as anxiety, fear, or stress, or describe as being hurt. However, experienced trainers take into account the previous postures the dog has shown. If a dog raises its paw after tilting its face, it can express curiosity and expectation. Also, if the dog raises its paws after raising its tail, this may be an action to get someone's attention. Therefore, it is imperative to consider all the postures they have shown when decoding dog behavior. In the next chapter, we explain how a dog's behavior can be analyzed through dog posture continuity (Fig. 1).

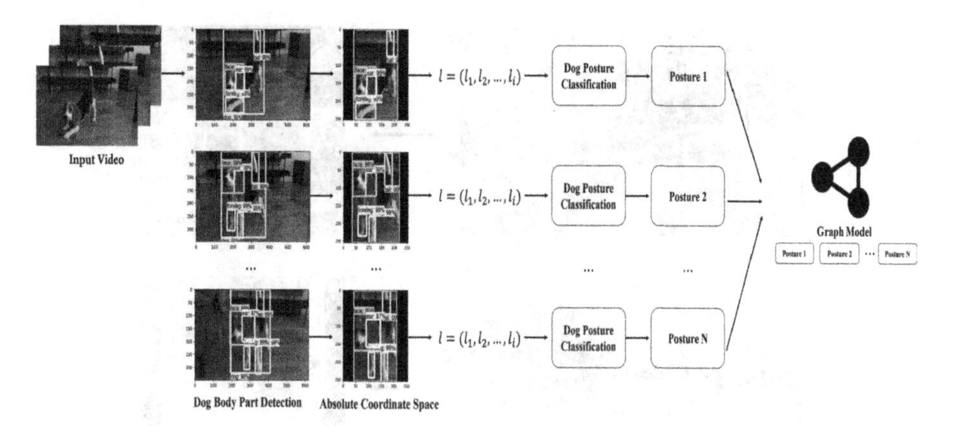

Fig. 1. Overall process of Canine Behavior Interpretation Framework. Given a video, detects the dog's body part in a frame unit. Each body part's location information is extracted after setting the detected dog area as an absolute coordinate space. Then, the dog posture is classified through a machine learning algorithm based on the extracted body part location information. Later, dog behavior is analyzed through a graph model that defines postures' continuous interrelationships classified in frame units.

3 Method

We introduce a graph model that sequentially represents an interrelationship of a dog's postures. This graph model puts the nodes defined in various dog postures and explores the successive interrelationships between each node to understand dog behavior. To implement this model, we first detect dogs and their body parts using the object detection method [4,5] in the video. We then project the detected dog area as absolute coordinate space and classify the dog's posture in frame units using each body part's coordinates. After that, the graph model determines the dogs' behavior using the posture classified in frame units.

3.1 Dog Body Parts Detection Using Object Detection

It is vital to observe the dog's body parts' location to understand its body language. In previous studies, researchers attached sensors to dog body parts for movement analysis. However, dogs were reluctant to have the sensors attached.

Therefore, we established our goal to detect the dog's body part through a deep learning-based object detection method without any additional equipment but a camera to extract the dog's body part's location information. For this goal, we collected an image dataset of beagles among dog breeds through YouTube. Yet, considering that the deep learning-based object detection method requires a vast dataset, we encountered a problem that it is too expensive to collect and label additional datasets.

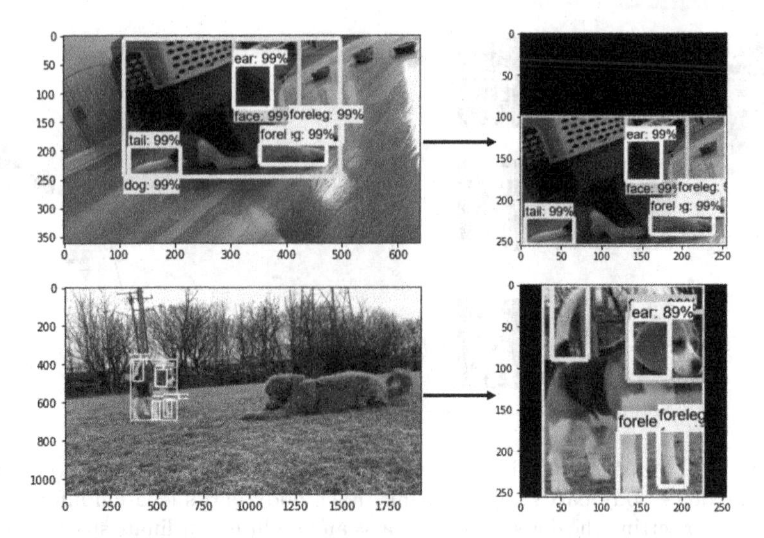

Fig. 2. We set the absolute coordinate space to utilize the dog's body part's position information without space constraints. Black is padding to maintain spatial information of the dog's body part.

To address this problem, we decided to use transfer learning [7]. Transfer learning is a training method that uses pre-trained models in similar domains to the corresponding model when data is deficient. The advantage of transfer learning use is previous learning experiences are adaptable for related tasks. Thus, a model is trained to detect dog body parts by applying transfer learning with a few samples, and as a result, it is possible to get the location information of dog and body parts from the image without using expensive equipment.

3.2 Dog Posture Classification Using Machine Learning

Once the dog and its body parts are detected from the video, the next step is to classify the dog's postures using the detected body parts' coordinates. Setting an absolute coordinate space is necessary to accomplish the exact coordinate. As shown in Fig. 2, the absolute coordinate space allows us to obtain fixed position information of body parts regardless of the size and position of the dog in the image. Our absolute coordinate space is established in the following process. First, To maintain the spatial information of the detected dog area, padding is added its area according to the ratio of width and height. We then resize the dog area where the padding was added in the same size and set it as absolute coordinate space. Once the absolute coordinate space is set, the detected dog's body parts center coordinate values are extracted based on the absolute coordinate space. And the dog's postures are classified in frame units through a machine learning model based on the absolute coordinate values. Our dog posture classifier yields an expected percentile value.

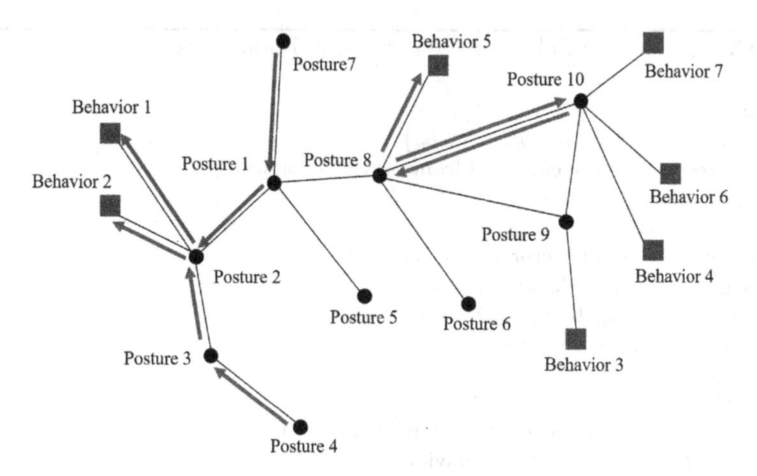

Fig. 3. We analyze dog behavior by considering posture change based on a graph model. For example, posture 2, which is finally classified, can be analyzed into two behaviors according to the postures classified in the previous frame. (Blue vs. purple). (Color figure online)

In this paper, following notation for the extracted absolute coordinate values. We define the *body part vector* as $\mathbf{l} = (\mathbf{l_1}, \mathbf{l_2}, .., \mathbf{l_i}), i \in \{1, 2..., k\}$, where $\mathbf{l_i}$ is the (x, y) absolute coordinate value of the kth dog body parts.

3.3 Dog Behavior Graph Model Generation

Deep learning based object detection methods and machine learning are applied to classify dog postures in frame units. However, a single posture is merely part of the package that expresses a dog's mood. Therefore, we generated a graph model to analyze the behavior using the posture patterns that show continuously. The process of generating a graph model is as follows. First, combine successive postures and define them as behavior concerning the calming signal. After that, generate a graph that expresses the posture as nodes and the posture order as an edge. Also, the defined behavior is added as the end node of the graph, and the behavior node contains the information of the dog posture sequence. Finally, compose various graphs that define dog behavior through posture sequences into one graph model as shown in Fig. 3.

We denote for dog behavior set as $B = \{b_1, b_2, b_3..., b_n\}$, where b_n is a representation of variable length sequences data and o_i indicates one element of the dog posture set $P = \{p_1, p_2, .., p_n\}$.

$$b_n = (o_1 \rightarrow o_2 \rightarrow o_3 \rightarrow ... \rightarrow o_T)^T \tag{1}$$

Algorithm 1. Canine Behavior Interpretation Framework

```
Input: Video frames f
Output: Behavior
  Dog Behavior set B = {b1, b2, b3, ..., bn}
  prevposture ← Posture classified from previous frame
  route ← graph travel route
  for f = 1 to n do
     Obtain body parts vector l = {l1, l2..., li}
     posture ← PostureClassifier(l)
     if posture ∈ neighbors(prevposture) then
        route ←Insert(posture)
     end if
  end for
  if route.endnode is connected to behavior nodes then
     for behavior in connected behavior nodes do
        Choose higher-similarity behavior to the route
     end for
  else
     for bj ∈ B do
        Chooses highest-similarity behavior to the route
     end for
  end if
  return behavior
```

3.4 Dog Behavior Analysis Using Graph Model

Once the graph model is built, our framework analyzes the dog's behavior as in Algorithm 1. First, approaches from the first frame to the n^{th}frame in the input video. And, the posture classifier expects the dog posture for each frame based on the *body part vector* **l** by setting a threshold with a probability of 0.8. With the first classified posture as the starting node, the graph model is explored by sequentially traveling to the next frame's classified posture node. However, incorrect posture classification can lead to an error in analyzing dog behavior. Therefore, if the classified posture from the current frame is not a neighbor node of the previous posture node in the graph model that defines the relationship between postures, it is considered misclassified and does not travel.

After exploring the graph up to the nth frame of the input video with this method, our framework determined the dog behavior by dividing the information into 2 cases. If the end node is connected to the behavior node, measure the similarity between the travel route and posture sequence information of each connected behavior node and determine the behavior node with a higher similarity as the final behavior. And, if the similarity under the threshold is gained, or the end node is unconnected to the behavior node, the similarity is computed between the travel route and posture sequence information of all behavior nodes. Then, the dog behavior of the highest similarity value is considered as an output.

The similarity between the travel route and the posture sequence of the behavior node is measured through the graph editing distance [3] with the following Eq. 2.

$$d(g1, g2) = 1 - \frac{|mcs(g1, g2)|}{max(|g1|, |g2|)} \tag{2}$$

where, $d(g1, g2)$ means graph edit distance and $mcs(g1, g2)$ indicate the maximum common sub graph [1]. Also, $|g|$ indicate the size of graph. If the graph edit distance [3] is close, it means that the two graphs have high similarity.

4 Experiment

4.1 Evaluation for Dog Body Parts Detection

This section evaluates the object detection method's training process and performance to detect dog body parts.

Dataset and Traning. One of the most critical tasks is to prepare quality datasets when training the deep learning-based object detection method. To prepare a high-quality dataset of dog's body parts, we collected a total of 9,282 images of beagles with various backgrounds from YouTube. All images are annotated by drawing a bounding box on the dog and body part utilizing LabelImg [8] Then the images are converted to XML files in PASCAL VOC format [2]. Although high-quality datasets have been collected adequately through this process, an enormous amount of data is required to train the object detection method. However, acquiring a larger dataset is cost-prohibitive.

To overcome these barriers, we applied transfer learning [7]. A pre-trained Faster RCNN and SSD Mobilenet model is taken that uses a COCO dataset containing a large amount of animal data provided by the TensorFlow Model Zoo [9].

Result. We compared the performance of Faster R-CNN and SSD Mobilenet trained through transfer learning. We first set the IOU (Intersection over Union) threshold to compare the two models' mean average precision (mAP). The mAP is the result of precision and recalls "precision-recall" calculations on determining bounding boxes. Also, we observed the FPS(Frame per second), an important index in real-time detection. Table 1 shows mAP and FPS of Faster R-CNN and SSD Mobilenet. Faster R-CNN was an accurate model with high mAP but showed a poor FPS to detect dog's body parts in real-time. In contrast, SSD Mobilenet was effective in detecting dog body parts in real-time by showing good mAP and FPS.

Table 1. Dog body part detection comparison

Method	mAP@0.50	mAP@0.75	FPS
Faster R-CNN	0.97	0.91	2
SSD Mobilenet	0.89	0.63	32

4.2 Evaluation for Dog Posture Classification

This section evaluates the machine learning model's training process and performance classified into six postures(stand, playbow, lie, sit, stand on two paws, tail raising) as shown in Fig. 4.

Dataset and Traning. Our dog's body part detection model emits the coordinate values of each body part. We constructed a dataset of dog posture classifiers by converting the emitted coordinate values into absolute coordinate values. Then, trained the Decision Tree, Neural Network, and Support Vector Machine among machine learning algorithms with the Configured dataset.

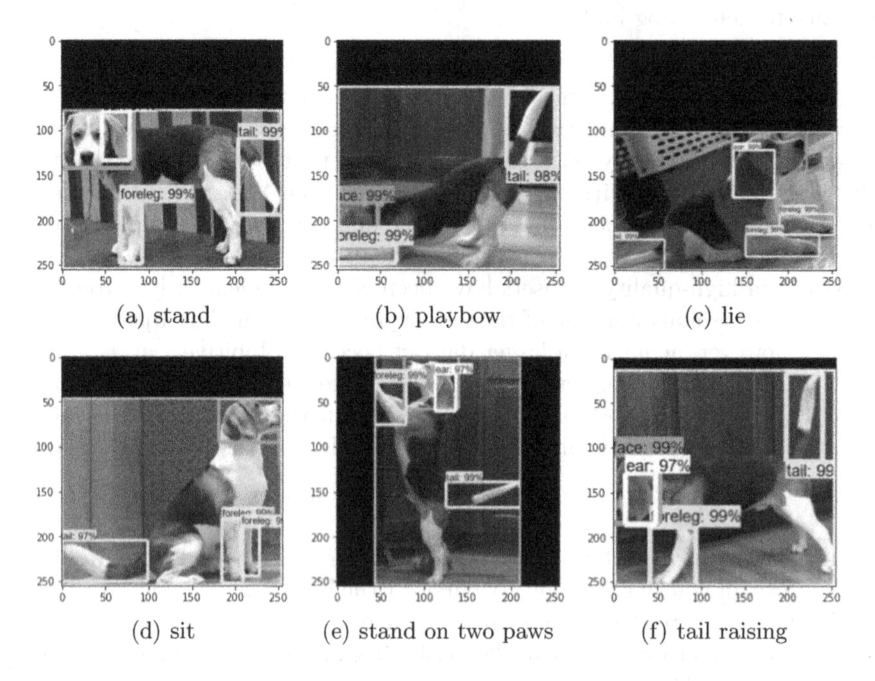

(a) stand (b) playbow (c) lie

(d) sit (e) stand on two paws (f) tail raising

Fig. 4. Example of classifying dog postures through machine learning based on the absolute coordinates of each body part.

Result. We first observed the accuracy of machine learning algorithms. As shown in Table 2, the Support Vector Machine's accuracy was the highest at 99.9%.

Also, we evaluated the Support Vector Machine's performance, the most accurate algorithm through the Confusion Matrix as shown in Fig. 5. A small misclassification occurred mainly between Tail raising and Playbow. This happens because the absolute coordinate values are similar between the two classes. As shown in Figs. 4(f) and 4(b), the difference between the two postures is in the position of the dog's upper body. However, the dog's upper body is in

Table 2. Dog posture classification compare

Algorithms	Accuracy
Decision Tree	85.6%
Neural Network	92.4%
Support Vector Machine	**99.9%**

an ambiguous position when analyzing the misclassified images. Although the postures classifier caused some misclassification, it is not a big problem in our framework for analyzing dog behavior by exploring a graph model that defines the interrelationships between postures.

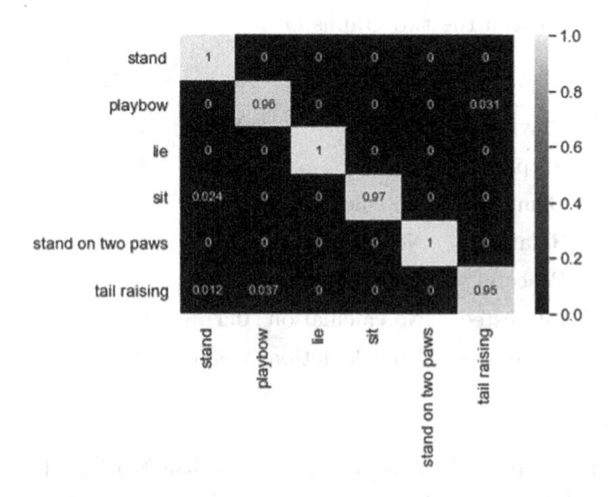

Fig. 5. Confusion matrix.

4.3 Experiment for Dog Behavior Analysis Using Graph Model

In this section, we describe the dog behavior defined according to the calming signal [6] and the experiment that analyzes the behavior for various videos based on the graph model.

Definition. As shown in Table 3, we defined the dog's behavior into seven types (Playful, Suspicious, Demanding, Relaxed, Peaceful, Interested, Joyful) according to the dog's posture change, and then generated a dog behavior graph model. However, according to dog trainers and behaviorists' needs, the specific behaviors are open to addition or modification.

Table 3. Defined dog behavior according to posture sequence

Posture sequence	Behavior
playbow→tail raising	Playful
tail raising→tail raising	Suspicious
tail raising→sit	Demanding
sit→sit	Relaxed
lie→lie	Peaceful
stand→tail raising→stand on two paws	Interested
sit→stand on two paws	Joyful

Table 4. Graph Edit Distance. The closer the graph edit distance is to 0, the higher the similarity between the two graphs. Conversely, as the graph edit distance is closer to 1, the similarity between the two graphs low.

	Video 1	Video 2	Video 3
Playful	**0.0**	**0.833**	1.0
Suspicious	0.666	**0.833**	1.0
Demanding	No calculation	0.833	0.75
Relaxed	No calculation	1.0	0.75
Peaceful	No calculation	1.0	1.0
Interested	No calculation	**0.166**	0.8
Joyful	No calculation	0.833	**0.25**

When the End Node is Connected to Behavior Nodes. Figure 6(a) shows the travel route(playbow→tail raising) created from Sample Video 1. The end node of the created travel route is connected to the behavior node Playful and Suspicious when according to Table 3. Our framework evaluates the similarity between each posture sequence information of connected behavior nodes(Playful and Suspicious) and the travel route to determine the final behavior between Playful and Suspicious. As a result, the behavior is regarded as Playful with higher similarity, as shown in Table 4.

However, as shown in Fig. 6(b), the end node of the travel route(stand→tail raising→stand on tow paws→tail raising) created from Sample video 2 is also a Tail raising, but when referring to Table 4, the similarity with the connected behavior node (Playful, Suspicious) is very low. Therefore, the similarity between the travel route and posture sequence of all the behavior nodes is measured, and the final behavior is determined as Interested with the highest similarity as shown in Table 4.

When the End Node is Not Connected to Behavior Nodes. As shown in Fig. 6(c), the end node Stand of the travel route(sit→stand on two paws→stand)

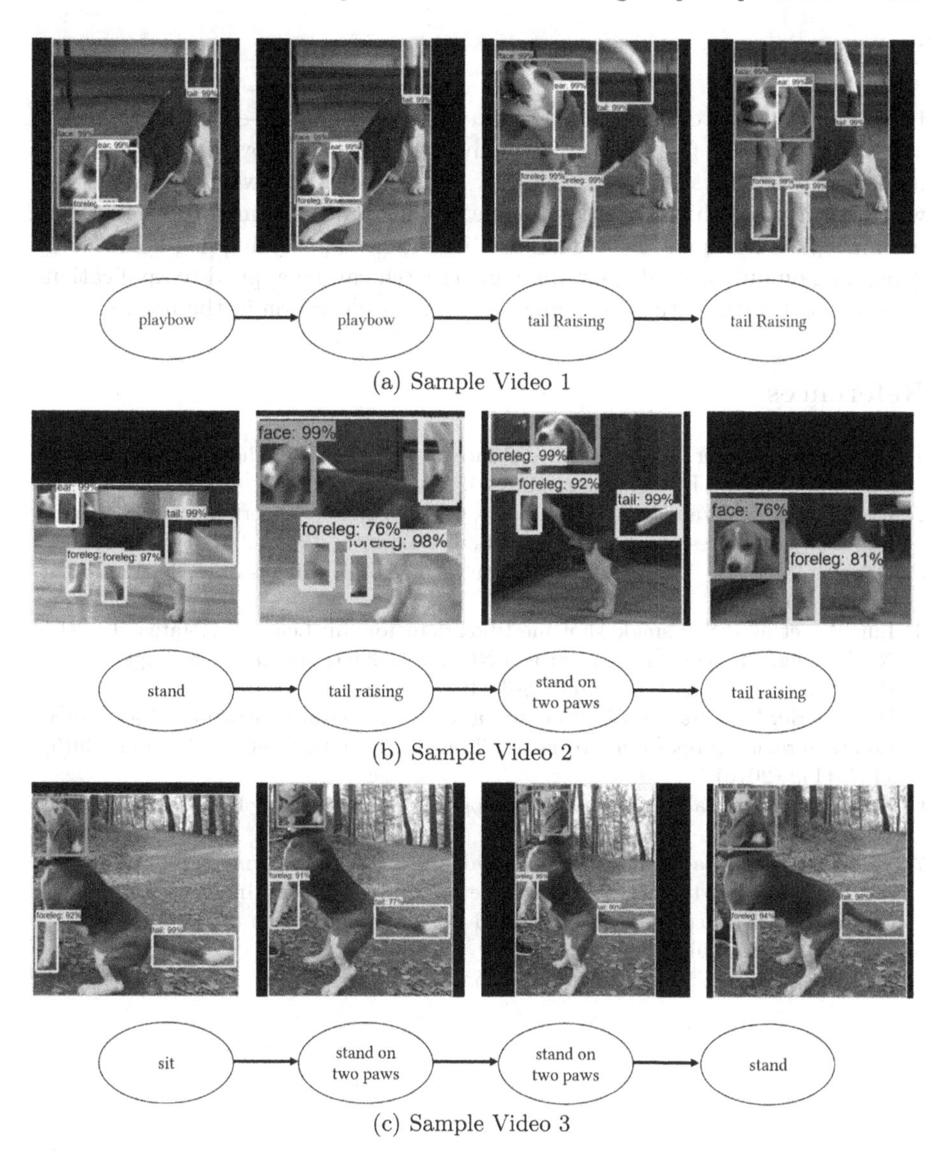

(a) Sample Video 1

(b) Sample Video 2

(c) Sample Video 3

Fig. 6. Example of Behavior Graph travel route created from Input Video

created in sample video 3 is not connected to any behavior node. In order to analyze the behavior of video 3, the similarity between the travel route and posture sequence of all the behavior nodes is measured, and the final behavior is determined as Joyful with the highest similarity as shown in Table 4.

5 Conclusion

In this paper, we proposed a noble framework for help non-expertise to understand dog behavior. Our framework analyzes dog behavior by exploring a graph model that defines successive postures' interrelationships with only a camera without additional high-quality equipment. This process eliminates the need to use multiple sensors and overcomes the limitation of analyzing dog behavior by depending on only a single posture. Our research can be applied to applications such as dog health status and abnormal behavior detection in the future.

References

1. Bunke, H., Shearer, K.: A graph distance metric based on the maximal common subgraph. Pattern Recogn. Lett. **19**(3–4), 255–259 (1998)
2. Everingham, M., Van Gool, L., Williams, C.K., Winn, J., Zisserman, A.: The pascal visual object classes (voc) challenge. Int. J. Comput. Vis. **88**(2), 303–338 (2010)
3. Gao, X., Xiao, B., Tao, D., Li, X.: A survey of graph edit distance. Pattern Anal. Appl. **13**(1), 113–129 (2010)
4. Liu, W., et al.: SSD: single shot multibox detector. In: Leibe, B., Matas, J., Sebe, N., Welling, M. (eds.) ECCV 2016. LNCS, vol. 9905, pp. 21–37. Springer, Cham (2016). https://doi.org/10.1007/978-3-319-46448-0_2
5. Ren, S., He, K., Girshick, R., Sun, J.: Faster r-cnn: towards real-time object detection with region proposal networks. IEEE Trans. Pattern Anal. Mach. Intell. **39**(6), 1137–1149 (2016)
6. Rugaas, T.: On Talking Terms with Dogs: Calming Signals. Dogwise publishing, Wenatchee (2005)
7. Talukdar, J., Gupta, S., Rajpura, P., Hegde, R.S.: Transfer learning for object detection using state-of-the-art deep neural networks. In: 2018 5th International Conference on Signal Processing and Integrated Networks (SPIN), pp. 78–83. IEEE (2018)
8. Tzutalin, D.: Labelimg. git code (2015)
9. Wu, N., Rathod, V.: Tensorflow detection model zoo (2017)

Artificial Neural Network Based Empty Container Fleet Forecast

Joan Meseguer Llopis[1](✉), Salvador Furio Prunonosa[1], Miguel Llop Chabrera[1], and Francisco J. Cubas Rodriguez[2]

[1] R&D Projects, Fundacion Valenciaport, Valencia, Spain
{jmeseguer,sfurio,mllop}@fundacion.valenciaport.com
[2] Cosco Shipping Lines Spain, SA, Valencia, Spain
javier.cubas@coscospain.com

Abstract. Global trade imbalances and poor, partial and unreliable information about available equipment make the coordination of empty containers a very challenging issue for shipping lines. The cancellation of transport operations once started or the extraordinary repositioning of containers are some of the problems faced by the local shipping agencies. In this paper, we selected the Artificial Neural Networks technique to predict the reception and withdrawal of empty containers in depots to forecast their future stock. To train the predictive models we used the different messages generated along the containers' shipment journey together with the temporal data related to these events. The evaluation of the models with the test dataset confirmed the possibility of using ANN to predict the number of empty containers in depots.

Keywords: Artificial Neural Networks · Empty Container Depots · Container Flow Prediction · Forecasting

1 Introduction

The management of the fleet of empty containers carried out by the local agencies of large shipping companies is an activity difficult to optimize. Among the different reasons, the level of uncertainty in the operations of delivery and reception of containers and the difficulty to predict them stands out [1]. This uncertainty leads to repetition of problems that cause extra costs and discontent of customers. Repeated lack of equipment in certain places (i.e. container depots) and times when attending export operations, and the extraordinary movements of repositioned equipment due to bad decisions are some of these problems. According to a study by the Boston Consulting Group the repositioning of empty containers costs the shipping industry between $15 and $20 billion a year. The same study estimates that 33% of these costs are derived from company inefficiencies and poor management decisions [2].

Nowadays, shipping agencies do not have systems to forecast the availability of empty containers. They keep track of the movements of containers and the ordered transport operations while maintaining an updated stock, which is contrasted with the

L. Rutkowski et al. (Eds.): ICAISC 2021, LNAI 12854, pp. 111–123, 2021.
https://doi.org/10.1007/978-3-030-87986-0_10

information received from empty container depots. Therefore, shipping agencies make decisions based on the current stock of containers and the experience of the equipment control department. However, they would make better decisions with the support of good predictions of the stock of containers using the information available.

With the help of artificial intelligence, this work aims to improve the effectiveness in short-term and mid-term planning operations of the Spanish agency of COSCO Shipping Lines, the world's third-largest shipping company in terms of container traffic. It is expected an improvement in the quality of service level to clients by avoiding situations of lack of availability of empty containers thanks to the quality of the forecasts.

The real-time prediction of the empty container delivery and demand in certain inland depots is a complex task. Therefore, previous work has been concentrated in optimizing the container transport operations by reducing the empty container movements [3] and in optimizing the repositioning of empty containers using various techniques such as evolutionary algorithms [4], policy-based decision making [5], and Multi-Agent Reinforcement Learning techniques [6].

In this paper, we selected Artificial Neural Networks (ANN) as it is used to cope with complex forecasting problems in applications with a high level of uncertainty that need to take into account several variables at a time. We wanted to verify if ANN models can be effectively used to make short/mid-term predictions of the delivery and withdrawal of empty containers from the warehouses. To train our ANN models, we used the date and time related parameters, and the different events generated along the containers' shipment journey as input variables. In this paper, we show the results for the most frequently used container type (i.e. 20 feet containers) by the local agencies. The same approach can be used for the other types.

The paper is structured as follows: in Sect. 2, the empty container management process is described and the different data sources are identified. In Sect. 3, we provide the Exploratory Data Analysis of the variables used to fit our models. The data preprocessing steps are presented in Sect. 4. In Sect. 5, we describe the ANN architecture of our solution and the method used to build our models. Results of the evaluation and its further discussion are presented in Sect. 6. Section 7 provides the main conclusions resulting from this work and proposes some guidelines for future work.

2 Empty Container Management Process

Generally, a container can have 3 states: empty, full and in maintenance. A depot is a warehouse used by the shipping and logistic companies to keep their empty containers until it is time for reloading of goods. In the process of export of goods, the empty container (EC) leaves a warehouse to be loaded at a client, transported to the port of origin, loaded on a vessel and transported to the destination country. In an import process, the loaded container arrives at the port of destination, is unloaded at the port terminal and is delivered to the carrier that takes it to the goods' destination. The emptied container is finally taken to the EC depot selected by the shipping agent.

The flow of ECs in a depot is generally determined by the delivery of ECs by the carriers to the depot (i.e. confirmed entries) and the delivery of ECs by the depot to the carriers (i.e. confirmed exits). These operations are also known as Acceptance and

Release of ECs. The number of confirmed entries and the number of confirmed exits per day are the target variables that we want to predict. These two variables will help the shipping agents to know the actual stock of ECs in the near future.

Two different transport orders lead to the entries and exits of ECs in a depot:

Empty Acceptance Order (EAO): it is the order generated for the delivery of the EC by the carrier to the depot in an import of goods (i.e. a new entry)
Empty Release Order (ERO): it is the order generated for the delivery of the EC by the depot to the truck driver in the export of goods. (i.e. a new exit)

Each of the above orders can have two states: generation and confirmed. The order generation serves as a statement of intent to inform the various agents involved in shipping operations that such EC acceptance or release will occur soon. The confirmation message is sent once the operation has been carried out. These events are:

Empty Acceptance Order Generation (EAOG): it is the event triggered when a new EAO is generated.
Empty Release Order Generation (EROG): it is the event triggered when a new ERO is generated.
Full Release Confirmation (FRC): it is the event generated when the port terminal delivers the full container to the carrier (i.e. a new EC about to enter at the depot).
Empty Acceptance Confirmation (EAC): it is the message generated in the moment when the carrier delivers the EC to the depot.
Empty Release Confirmation (ERC): it is the message generated in the moment when the EC is withdrawn from the warehouse by the truck driver.

A history of such events has been the main data source for our training and validation datasets. More precisely, EAOG and FRC events have been the source of data for the input variables to predict the quantity of EAC while the number of ERC was forecasted using the EROG events.

In addition to the above events, the temporal information associated with these events has also been taken into account as input variables for the models. We use the month, the day of the week and the day of the month as additional predictors.

All the above order and confirmation messages are exchanged between the shipping agent and the different transport actors (i.e. port terminals, carriers and depots) through its internal fleet management system. These messages are also stored in an internal database. The dataset used for the training and validation of our models is built from three years-long history of messages extracted from the database of Cosco's Spanish agency. A total of 163.441 messages generated over 1154 days has been used.

3 Exploratory Data Analysis

This section describes the Exploratory Data Analysis (EDA) process realized before the model training phase. The EDA refers to the important process of conducting initial research on the data to discover patterns, detect outliers, check correlations between variables, and test assumptions. In this paper, we show the most relevant results due to lack of space.

Our EDA has been performed to the numeric variables: EAOG, EROG, FRC, EAC and ERC. These variables refer to the number of messages registered in a given day. The parameters EAOG, EROG, FRC are the predictors for our predictive models, whereas EAC and ERC are the target variables to predict.

The tools used in the EDA phase are: a) Pandas [7] Python library for the management and analysis of our different data structures; b) Seaborn [8] Python library for the statistical data visualization of our dataset; and c) Jupyter Notebook [9] as our web-based development environment to run the above libraries and to build and train our models in the following phase.

A summary of the descriptive statistics for all the numeric variables in the dataset is shown in Table 1.

Table 1. Input and Output variables statistics.

Statistics	EAOG	EROG	FRC	EAC	ERC
Max	87.0	71.0	53.0	53.0	83.0
Mean	8.01	7.32	7.99	7.99	7.31
Median	4.0	4.0	5.00	5.0	5.0
Std. dev	10.53	9.31	9.31	9.3	8.86
Skewness	1.98	1.99	1.43	1.42	2.05
Kurtosis	5.92	5.80	2.12	2.18	7.55

As we can observe from the above table, all variables have similar statistical values which indicate that they all follow a similar behavior. All variables present relatively high maximum values which may show the presence of outliers if we look at the values of their variance. These outliers can also be seen in Fig. 1. As can be seen in the boxplot on the left, all variables have a zero value as the minimum. This is normal as there are many days (especially the weekends) where no transport orders nor confirmations occur. The quantity per day of EAOG, EROG and ERC have a high value of kurtosis which indicates a higher degree of concentration around the mean and a sharper distribution of their observations [10]. A lower kurtosis in FRC and EAC variables shows a more normal distribution of their recorded quantities. As for the bias (i.e. skewness), the values between 1.5 and 2 indicate a slight bias towards the upper side of the mean, also visible in the boxplots' upper quartile of Fig. 1.

As it can be observed from the above boxplot, all variables have their whisker (i.e. 1.5 * [Q3 − Q1]) between 30 and 35. The number of EAOG per day has the biggest outliers. This could be also noted from its higher deviation from the rest.

Last but not least, it is worth analyzing the time lapse between the container's EAOG/EROG and its final EAC/ERC at the depot. For this purpose, the boxplot on the right-hand side of Fig. 1 is shown. As it can be noted, the orders usually take from 0

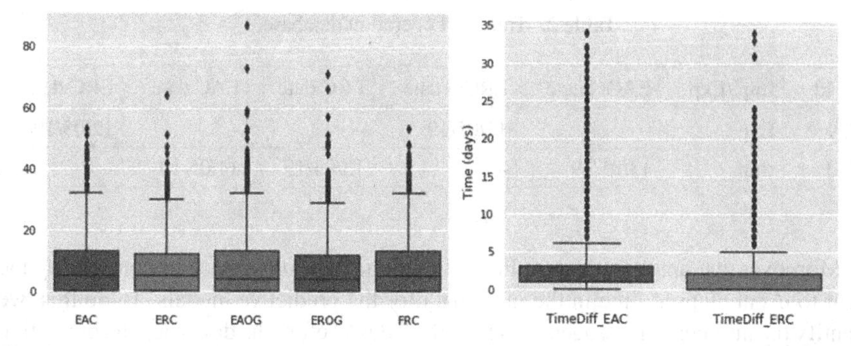

Fig. 1. Number of events per day (left) and time lapse between order and confirmation (right)

to 7 days to be completed. As it will be shown in Sect. 4, this is relevant when representing the input data for the prediction models.

Now we analyze the existing correlations between the different numeric variables considered. We run the *heatmap()* method of the Python Seaborn tool to obtain the correlation matrix for the entire dataset. As for the strength of the relationship, a value of ± 1 indicates a perfect degree of association between the two variables. And a value of 0 means a weak correlation between variables [11].

The matrix showed a correlation 0.63 between EAC and the EAOG. This correlation is between moderate and strong with a positive linear relationship as expected (the more orders the more acceptances). A correlation of 0.86 between confirmed acceptances and the number of full container releases (i.e. FRC) is also strong. Both correlations can be confirmed as their Pearson's p-values obtained are 1.75E-127 and 0.0 respectively. A value of 0.71 between ERC and EROG reflects a strong correlation between these two variables as well. This is also confirmed with a p-value of 2.0E-175. The p-values have been calculated using Python's Scipy library [12].

4 Data Preprocessing

To train the models with data of good quality a historical data preprocessing of the three years' messages from the database is required. As mentioned in Sect. 2.1, before the EC is delivered to the depot a EAO is issued first followed by a FRC. Similarly, before an EC withdrawal from the depot takes place, a ERO is generated. Each row in our database corresponds to a registry of timestamps for the events generated within each container import or export operation that took place along these 3 years of data. An extract example of this database is shown in Table 2.

Table 2. Transport operation database.

Op. Id	Imp./Exp	EAOG date	EROG date	FRC date	EAC date	ERC date
7739	Exp	–	11/05/19	–	–	12/05/19
7740	Imp	13/05/19	–	13/05/19	14/05/19	–

Moreover, the analysis done in Fig. 1 highlights the importance of considering the event to event elapsed time in the input data for the predictive models. To do this, we quantify the number of messages in each of the 7 days before the day when the prediction is made (from now on the PD day). These are the ECs that have not yet been delivered or withdrawn.

For the prediction of the quantity of EAC, we build two arrays as input data to the model. In Function 1, the steps to build these arrays are shown. The first one (EAOGs) is an array of length 7 where the first value corresponds to the number of EAOG issued the day before the prediction day (i.e. PD-1), the second value is the number of EAOG two days before the PD, and so forth until the last value. The last value of this array refers to the number of EAOG (and not confirmed yet) equal or older than 7 days before the PD (i.e. d $\leq$ PD-7). The second array (FRCs) of length 5, corresponds to the number of FRC messages for the days PD-1, PD-2, PD-3, PD-4 and d $\leq$ PD-5. As for the forecast of ERC, an array of length 7 (EROGs), with the number of EROG in the past days before the PD, is also computed. A new array is built for every day from the start date (StartDate) until the end date (EndDate) of the whole history of registries. At the end, it is obtained a matrix where each row represents a new observation in the training dataset. The arrays EAOGs, FRCs and EROGs are appended as new rows in the matrixes [EAOG], [FRC] and [EROG] respectively.

Function 1 Build Prediction Input and Output Arrays

```
Input: StartDate, EndDate
Output: [EAOG], [FRC], [EROG], [EAC], [ERC]
1:  let [EAOG], [FRC], [EROG], EAOGs, FRCs, EROGs = new Array []
2:  let [EAC], [ERC], EACs, ERCs = new Array []
3:  Let currentDay = StartDate
4:  while currentDay ≤ EndDate do
5:      for d in range(0,6):
6:          dayBack = currentDay - timedelta(days= d)
7:          let N_EAOG_orders = COUNT registries WHERE EAOG_Date = dayBack
            and EAC_Date > currentDay
8:          let N_EROG_orders = COUNT registries WHERE EROG_Date = dayBack
            and ERC_Date > currentDay
9:          EAOGs.append(N_EAOG_orders); EROGs.append(N_EROG_orders)
10:     for d in range(0,4):
11:         dayBack = currentDay - timedelta(days= d)
12:         let N_FRC_orders = COUNT registries WHERE FRC_Date = dayBack and
            EAC_Date > currentDay
13:         FRCs.append(N_FRC_orders)
14:     for d in range(0,6):
15:         dayForec = currentDay + timedelta(days= d)
16:         let N_conf_EAC = COUNT registries WHERE EAC_Date = dayForec
17:         let N_conf_ERC = COUNT registries WHERE ERC_Date = dayForec
18:         EACs.append(N_conf_EAC); ERCs.append(N_conf_ERC)
19:     [EAOG].append(EAOGs); [FRC].append(FRCs); [EROG].append(EROGs)
20:     [EAC].append(EACs); [ERC].append(ERCs)
21:     currentDay = currentDate + timedelta(days=1)
22: end while
23: return [EAOG], [FRC], [EROG], [EAC], [ERC]
```

In regards to the output data for the model's training dataset, a similar approach is followed. In this case, the third *for* loop of Function 1 collects the number of deliveries and withdrawals confirmed in the following 7 days from the prediction day (i.e. EACs and ERCs arrays). In each *while* loop iteration, each EACs and ERCs array is added to the matrixes [EAC] and [ERC] respectively.

Besides the above matrixes, another one ([Temp]) with the temporal parameters is also created. Each row in this temporal matrix contains an array with 3 elements: day of the month, day of the week and the month. From the above functions, these values are taken as follows: [*currentDay.Day, currentDay.DayWeek, currentDay.Month*].

Before proceeding to the models' training step, we first need to merge all above matrixes into a single dataset using Python's Numpy [13] and Pandas libraries (lines 2–4 in Function 2), split the whole dataset into a train set and a test set, and finally, normalize these data partitions. The code implemented to perform these steps is described in Function 2. This function is a reduced version of the one used which only shows the steps for the dataset to train the model that forecast the number of EACs. In lines 5–7 we remove the rows with outliers greater than the 99% percentile of the dataset's values.

To find out the best configuration parameters (a.k.a hyperparameters) that control the learning process of our models, the whole set of observations is split into a train set and a test set. The former is used to initially fit the model while the latter is used to evaluate the predictions done by the fitted model with the true values from this partition. A good rule of thumb is to divide the whole dataset into 80% train set and 20% test set [14]. In our case, we made a random split of our observations (see line 8 in Function 2) using the Panda's *sample*() function of our dataset *class*.

The next step is to normalize the train and test input datasets for the model fitting process. The data normalization process is used to train models with homogeneous data and without outliers. This is a well-known procedure that considerably improves the performance of predictive models [15]. In our case, the min-max technique has been used (lines 12–13 in Function 2), which leverages the Pandas' DataFrame *describe()* function [7] that returns the *min* and *max* values for each column in our dataset.

Function 2 Get Train and Test Data

```
Input: [EAC], [EAOG], [FRC], [Temp]
Output: x_train, x_test, y_train, y_test
1:  let dataset = [EAC]
2:  dataset = column_stack((dataset, [EAOG]))
3:  dataset = column_stack((dataset, [FRC].))
4:  dataset = column_stack((dataset, [Temp]))
5:  q = dataset.quantile(0.99)
6:  dataset = dataset[dataset < q]
7:  dataset = dataset.dropna()
8:  train_dataset = dataset.sample(frac=0.80,ra dom_state=0)
9:  test_dataset = dataset.drop(train_dataset.index)
10: x_train, x_test, y_train, y_test =   split_to_x_and_y(train_dataset, test_da-
    taset)
11: train_stats = dataset.describe()
12: x_train = (x_train - train_stats['min']) / (train_stats['max']-train_stats['min'])
13: x_test = (x_test - train_stats['min']) / (train_stats['max']- train_stats['min'])
14: return x_train, x_test, y_train, y_test
```

5 Methodology: ANN for the EC Fleet Forecast

ANN basically performs like a human brain. A neural network is a dense parallel-distributed processor composed of simple computing nodes, also known as neurons [16]. This network is made of groups or layers of interconnected nodes. Each node is an artificial neuron and contains a function (i.e. activation function) that aggregates and correlates the weighted input parameters by weights (w_{ij}) into an output parameter that is transmitted to the next node. ANNs are trained by an optimization process (e.g. Gradient Descent Algorithm [17]) which is an iterative task aiming to find the value of the network weights that minimizes a loss function. This loss function is used to calculate the model's error. In regression problems, like the one described in this work, it is common to use the mean squared error as the loss function (1).

$$MSE = \frac{1}{n}\sum_{i=1}^{n}\left(\tilde{Y}_i - Y_i\right)^2 \quad MAE = \frac{1}{n}\sum_{i=1}^{n}\left|\tilde{Y}_i - Y_i\right| \tag{1}$$

In our approach, we use a multi-layer neural network composed of an input layer (i.e. input data), N hidden layers and an output layer. In this work, we had to find the value of N and the number of neurons in each layer that provided a higher performance in the predictions. The output layer consists of 7 nodes; each one returns the number of delivered/withdrawn ECs for each of the following 7 days' prediction. As we aim to forecast numeric values greater than the unit, the Relu activation function is used [16]. This architecture is visually described in Fig. 2.

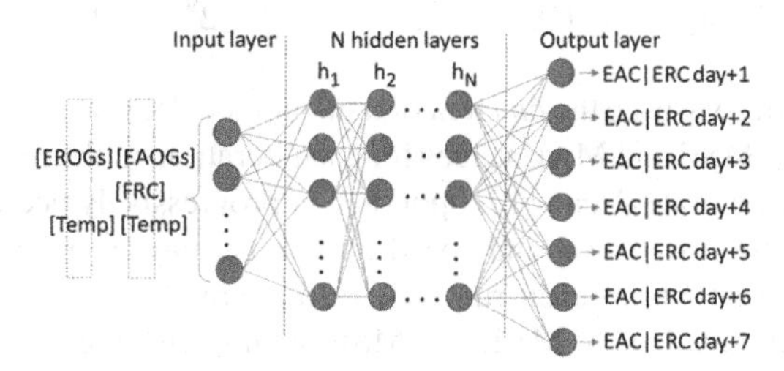

Fig. 2. ANN architecture for ECF forecast

The input layer is a concatenation of the normalized arrays [EAOG], [FRC] and [Temp] for the 7-day prediction, and [EROG] with [Temp] for the ERC forecast.

The architecture of Fig. 2, is built using Python's Tensorflow [17] and Keras [18] libraries. We made an iterative function (see Function 3) that automatically builds a NN following the architecture of Fig. 2. The input parameter "*layers*" is an array with N (i.e. the number of layers) values, each one referring to the number of nodes in each layer. The function *Sequential()* instantiates a model structure with a plain stack of layers. In steps 2 and 4, the *add()* function of the model class is used to add the layers with the number of nodes given by the *layers* array. Finally, the function *compile()* configures the model for training, which in this case we have indicated the MSE as loss function and the Mean Absolute Error (MAE) with the MSE as metrics for evaluating the model in the training and testing phases.

Function 3 Build NN model

Input: layers
Output: x_train, x_test, y_train, y_test

```
1:  model = keras.Sequential()
2:  model.add(layers.Dense(architecture[0],activation='relu',input_shape=[layers(0)]))
3:  for d in range(1, len[layers]):
4:      model.add(keras.layers.Dense(architecture[i],activation = 'relu'))
5:  model.compile(loss='mse', metrics=['mse','mae'])
6:  return model
```

6 Results: EC Fleet Forecast Evaluation

Now, the two neural networks to predict the number of EAC and ERC are trained with the datasets obtained from Function 2. As mention before, we had to train several models with a different number of layers and neurons in each layer that approximates us to a NN architecture that performs better in terms of MAE and MSE metrics. To do so, Function 3 was run for all the possible combinations of the possible values in the 3 hidden layers in Eq. (2). A total of 294 NN architectures was tried. As a heuristic widely adopted by the community, we also used powers of 2 number of nodes in each hidden layer.

$$\begin{aligned} layers &= [l_1, l_2, l_3] \\ l_1 &= \left(2^4, 2^5, 2^6, 2^7, 2^8, 2^9, 2^{10}\right) \\ l_2, l_3 &= \left(0, 2^4, 2^5, 2^6, 2^7, 2^8, 2^9, 2^{10}\right) \end{aligned} \tag{2}$$

To train the model, we used the *fit()* function of our model instance created in Function 3, which returns the MSE and MAE values for each iteration on the test data set. In each run, we used a learning rate decay [19] approach to progressively decrease the learning rate to avoid weights' oscillation and speed up the learning of our model. We fit our model using 500 iterations in each run. The results are shown in Fig. 3. As we can observe, the minimum values of MAE and MSE obtained for the EAC forecast are 2.2 and 16.6 respectively. These values correspond to the [1024,518,128,7] architecture of the NN. As for the ERC forecast (see Fig. 4), the found architecture is [1024,256,32,7] whose MSE and MAE are 3.04 and 38.8 respectively.

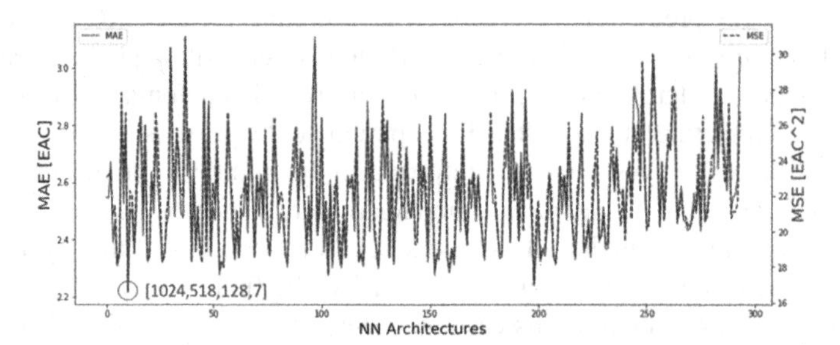

Fig. 3. EAC NN architecture selection

We have used the *predict()* function of our model to compute the predictions for the test dataset. The scatter plots of the true and predicted values for the EAC's model are shown in Fig. 5. In this figure, we can see that our model predicts reasonably well.

Also in Fig. 5 the scatter plots of the true and predicted values for the ERC's model are presented.

In this case, the plotted points on the first and second day are slightly farther from the true line than in the EAC forecast. One reason for this could be the lack of an event previous to the actual withdrawal of the container from the depot similar to the FRC event in the import operations.

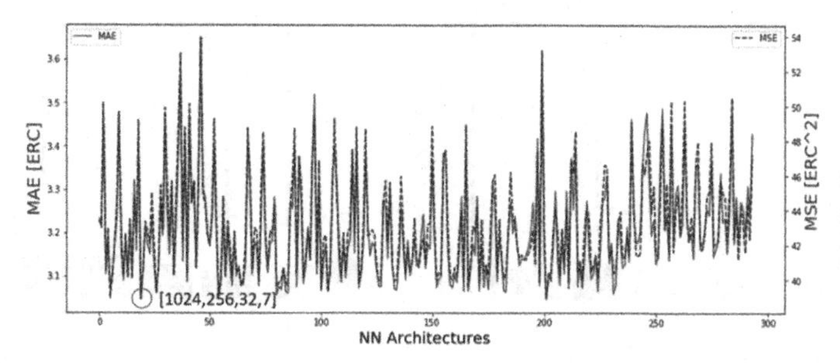

Fig. 4. ERC NN architecture selection

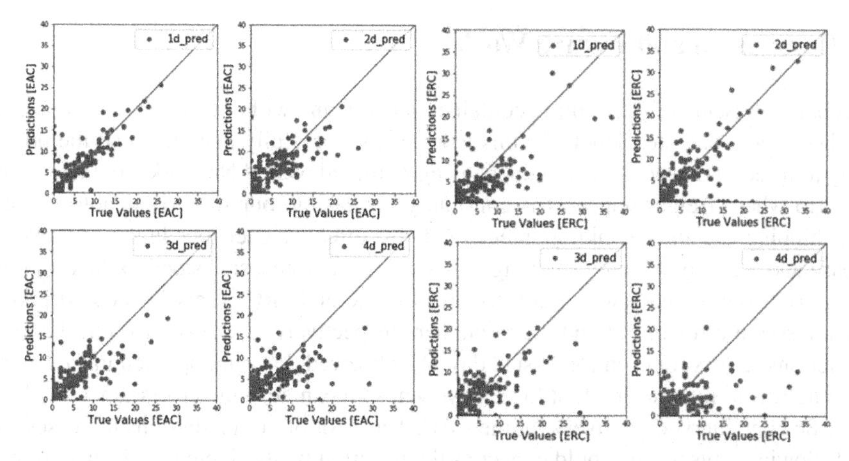

Fig. 5. Scatter plot for EAC and ERC 4-day predictions evaluation

To see the performance of our model for the 7 days of forecast we use the boxplot of the MAE for each forecast sample (see Fig. 6 Left and Right). As it can be observed, from days 3 to 6, the forecasted error is noticeably higher than the first 3 days. In Sect. 3.1, we saw that most of the transport orders (75% approx.) are carried out within 0 to 4 days. The rest of the orders introduces more uncertainty for longer-term predictions. Nevertheless, shipping agents still can make decisions based on the predictions of days 3–6 by aggregating the forecasted values.

The above results show the possibility to use our method to forecast the flow of ECs in depots. The future stock would be calculated by adding and subtracting these forecasted values to the actual amounts in each depot on the day when the forecast is computed. These stock predictions can then be effectively used to support shipping agents to make decisions in selecting the depot from which a EC should be withdrawn to attend an export operation as well as the depot to which the EC should be delivered to correct in advance a predicted lack of equipment. A direct consequence of this would be a reduction of the extraordinary displacements of trucks between depots. This, in turn, contributes to a CO2 reduction too.

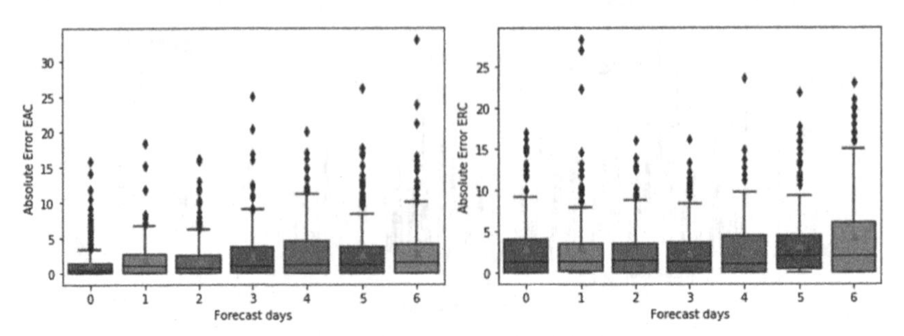

Fig. 6. Absolute 7-day EAC forecast error (left) and Absolute ERC forecast error (right)

7 Conclusion and Future Work

This paper presents a 7-day empty container delivery and withdrawal forecast approach for the empty container depots. In this work, we successfully demonstrated the possibility to forecast the stock of containers using Artificial Neural Networks. In the current scenario where freight transport is constantly increasing, our solution is able to support shipping agencies in taking greener and cost-effective decisions in the containers' release and pick-up operations, taking into account the forecasted stock in the different locations. In our test data, we are able to forecast EAC and ERC events for the next 7 days in an absolute error of 2.2 and 3.2 containers respectively. We have also checked the predictions' errors for each forecasted day. We observed that our approach makes more accurate predictions for the first three days whose mean absolute errors are 1.24, 1.79 and 1.86 containers respectively for the EAC. The error of our predictions increases for the following days as one could expect as the uncertainty also increases. Hence, future work could involve the usage of additional parameters such as customer class, weather variables or location of clients. These variables may affect the midterm forecast of the number of EAC and ERC. If these parameters will also be taken into account, the result of forecasting can be further improved. Other further work includes the usage of other techniques such as Recurrent Neural Networks [16] which is also used for time series prediction.

References

1. Sanchez-Rodrigues, V., Potter, A., Naim, M.M.: The impact of logistics uncertainty on sustainable transport operations. Int. J. Logist. Manage. (2010)
2. Sanders, U., Roeloffs, C., Schlingmeier, J., Riedl, J.: Think Outside Your Boxes: Solving the Global Container-Repositioning Puzzle. bgc.perspectives, The Boston Consulting Group, Inc. (2015)
3. Furio, S., Andres, C., Adenso-Diaz, B., Lozano, S.: Optimization of empty container movements using street-turn: application to Valencia hinterland Comput. Ind. Eng. **66**(4), 909–917 (2013)
4. Wong, E., Yeung, H., Lau, H.: Immunity-based hybrid evolutionary algorithm for multi-objective optimization in global container repositioning. Eng. Appl. Artif. Intell. **22**(6), 842–854 (2009)

5. Won Young, Y., Yu Mi, L., Yong Seok, C.: Optimal inventory control of empty containers in inland transportation system. Int. J. Prod. Econ. **133**(1), 451–457 (2011)
6. Luo, Q., Huang, X.: Multi-agent reinforcement learning for empty container repositioning. In: IEEE 9th International Conference on Software Engineering and Service Science. IEEE, Beijing (2018)
7. Python Data Analysis Library. https://pandas.pydata.org/docs/ (2020)
8. Seaborn v0.11.0. Statistical data visualization. https://seaborn.pydata.org (2020)
9. Jupyter Notebook. https://jupyter.org/documentation (2020)
10. Jain, V.K.: Data Science and Analytics (with Python, R and SPSS Programming). Khanna Publishing House (2018)
11. Phillips, D., Romano, F., Vo T. H., P., Czygan, M., Layton, R., Raschka, S.: Python: Real-World Data Science. Packt Publishing Ltd. (2016)
12. SciPy v1.5.3. Python-based ecosystem of open-source software for mathematics, science, and engineering. https://www.scipy.org/ (2020)
13. NumPy v1.19.0. The fundamental package for scientific computing with Python. https://numpy.org/doc/stable/ (2020)
14. EMC: Data Science & Big Data Analytics: Discovering, Analyzing, Visualizing and Presenting Data. EMC Education Services (2015)
15. Nawi, N. M., Atomi, W. H., Rehman, M. Z.: The effect of data pre-processing on optimized training of artificial neural networks. In: 4th ICEEI Conference 2013. Selangor (2013)
16. Aggarwal, C.C.: Neural Networks and Deep Learning: A Textbook. Springer, Cham (2018). https://doi.org/10.1007/978-3-319-94463-0
17. TensorFlow v2.1.0. An end-to-end open source machine learning platform. Google Brain Team. https://www.tensorflow.org/versions/r2.1/api_docs/python/tf (2020)
18. Keras v2.3.1. Keras: The Python deep learning API. Google Brain Team. https://keras.io/getting_started/intro_to_keras_for_researchers (2020)
19. Smith, L.N.: A disciplined approach to neural network hyper-parameters: Part 1 – learning rate, batch size, momentum, and weight decay. US Naval Research Laboratory Technical Report, Washington DC (2018)

Efficient Recurrent Neural Network Architecture for Musical Style Transfer

Mateusz Modrzejewski(✉), Konrad Bereda, and Przemysław Rokita

Division of Computer Graphics, Institute of Computer Science,
The Faculty of Electronics and Information Technology,
Warsaw University of Technology, Nowowiejska 15/19, 00-665 Warsaw, Poland
{M.Modrzejewski,P.Rokita}@ii.pw.edu.pl

Abstract. In this paper we present an original method for style transfer between music tracks. We have used a recurrent model consisting of LSTM layers enclosed within an encoder-decoder architecture. In addition, a method for programmatic synthesis of sufficient, paired training datasets using MIDI data was presented. The representation of the data in the form of a real and an imaginary part of short-time Fourier transformation allowed for independent modeling of the music components. The proposed architecture allowed us to improve upon the state of the art solutions in terms of efficiency and range of applications while achieving high precision of the network.

Keywords: Artificial intelligence · Neural networks · LSTM · Style transfer · Timbre · MIDI · STFT · Synthesizer · AI in music

1 Introduction

Recent years have brought a great number of innovations in machine learning and artificial intelligence. Some of the most prominent applications of such technologies are associated with image and language processing. Generative-adversarial [1] has promised significant improvements to image generation, while attention-based transformer networks [2] have been a breakthrough in language processing.

Some of the applications of machine learning algorithms to music include generating new musical content [3–6], music genre classification [7–9] and various forms of automated music information retrieval [10,11]. Another interesting issue, originating from image processing [12], is called *style transfer*. The general idea is to combine two images, i.e. a photograph and a painting, in order to produce a "version" of the photograph as if it has been painted by a given painter. A similar idea may be applied to music.

In this paper, we propose an original recurrent neural network architecture suitable for style transfer between pairs of instruments, along with a method for generating sufficient, synthesized training datasets. Our solution, given a piece played on a particular instrument, is able to produce a piece with the same content but played on a different instrument. The proposed method achieves the

L. Rutkowski et al. (Eds.): ICAISC 2021, LNAI 12854, pp. 124–132, 2021.
https://doi.org/10.1007/978-3-030-87986-0_11

state of the art precision, while greatly decreasing the computational cost when compared to existing approaches. We have chosen piano to guitar style transfer as the main focus and benchmark of our experiments, but the presented architecture is versatile and allows for style transfer between any two instruments.

2 Previous Work

2.1 Musical Style Transfer

In musical style transfer [13] we consider the composition (harmony, rhythm, arrangement) as the *content* of the piece of music, while the particular sound (timbre, sonic qualities, particular instrument) as the general *style*.

Previous work on the subject of musical style transfer includes applications of VAEs (*variational autoencoders*) i.e. MoVE [14] and convolutional neural networks (i.e. WaveNet [15]). MoVE consists of convolutional, dense and *FiLM* [16] layers in an encoder-decoder architecture. The Non-Stationary Gabor Transform [17] is used to obtain the time-frequency representation of the audio signal. However, the Transform introduces some loss of information, making it difficult to reconstruct the original signal and lowering the quality of the resulting sound files.

[18] proposes the Universal Music Translate Network a network capable of style transfer between instruments and composers. The authors use raw audio signals as the representation of music. The usage of raw audio has also been described in [19] and [15] - WaveNet has shown significant improvements over previous works, especially in speech generation. The usage of raw audio signal allows to maintain full information about the signal but requires massive amounts of computational power to process.

Another approach to the style transfer is the usage of relativistic-average generative adversarial networks [20] with a complex, compound representation of music using Mel-spectrograms, MFCCs, spectral difference, and spectral envelope. In [21], the authors use this representation to perform multi-modal one-to-many style transfer. [22] proposes the use of symbolic MIDI data representation and a variational autoencoder, while [23] uses an AlexNet model and spectrogram representation.

2.2 Approach to the Usage of MIDI

The domain of music continues to prove a challenge for the development of AI algorithms. There has been no single, agreed-upon representation of music suitable for machine learning applications [13] and the availability of datasets is much sparser when compared to image and text processing. Many of the perceivable qualities, like the harmony, rhythm, instrumentation, and composition qualities of music are abstract and difficult to capture simultaneously. Also, the psychoacoustic side of music [24,25] and the non-linearities of the human hearing apparatus introduce the need for additional arbitral representations [26].

The plethora of approaches has also led to some criticism. The available datasets are often either sparse and vague [27,28] or highly-specific [29], while many are undisclosed. [30] and [31] also highlight the need for the practical impact of ML algorithms in music and state that no music data is independent of its context and function. Having a background in music production and performance, we also include this idea in our approach by proposing a streamlined method of generating training datasets by synthesizing MIDI data into sound. This is very close to the context of how MIDI data would be used by a musician in professional music production. Another advantage of this approach is that sufficient pure MIDI datasets already exist (i.e. [32,33]).

3 Data Processing

We have used data from the *LMD-matched* subset of the *Lakh MIDI Dataset v0.1* [34] which consists of 45129 samples and from the *130,000 MIDI File Collection* [35]. We have compared all of the files using their MD5 hash and eliminated direct duplicates. However, the same piece may be present more than once, since there may be different versions of its way (i.e. different key, different arrangements). The presence of different arrangements of the same piece is beneficial to the network's ability to generalize. We have selected a subgroup of songs performed in at least 95% on a single instrument.

The collected dataset consists of 2000 piano songs - networks with LSTM cells have previously been shown to perform well on moderate datasets [36,37]. On top of that, the amount of songs we have synthesized translates to roughly 130 h of music, which is sufficient enough for our needs. The dataset was divided into training and test subsets. The test dataset contains 200 tracks and represents about 10% of the volume of the training dataset.

The MIDI files were synthesized using a state of the art programmatic synthesizer *FluidSynth*, which allowed us to obtain high-quality .WAV files with music encoded with a pulse-code modulation. *FluidSynth* is very similar to VST synths applied in professional music production, which further brings musical context to our approach.

For the purpose of this work, we have decided to represent the model input in terms of the Short-Time Fourier Transform (see Eq. 1) of the audio signal for a time-frequency representation, as shown in Fig. 1. In order to keep complete information about the provided signal, both the real part and the imaginary part of the STFT are passed to the model in the form of two independent inputs.

$$\mathbf{STFT}\{x(t)\}(\tau, \omega) \equiv X(\tau, \omega) = \int_{-\infty}^{\infty} x(t)w(t-\tau)e^{-i\omega t}\, dt \tag{1}$$

The STFT is invertible - the signal can be recreated from the transform by the inverse STFT. We use this to recover playable .WAV files from the outputs of our network.

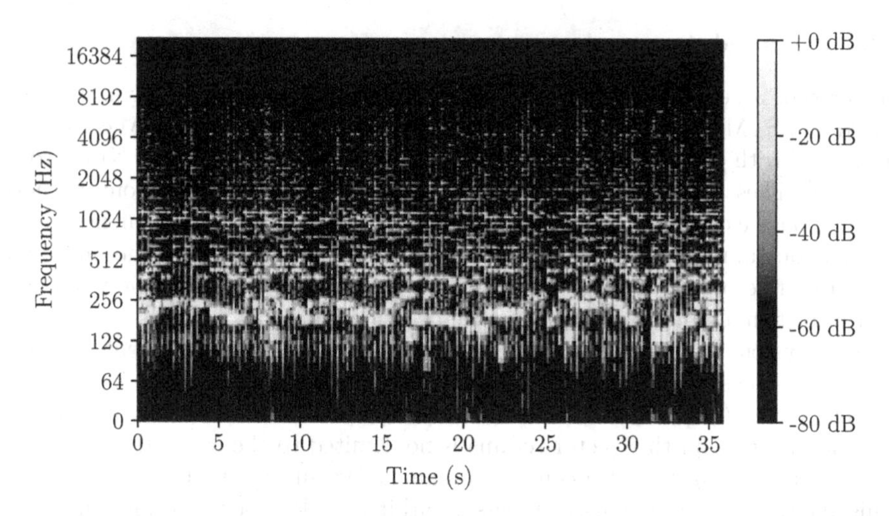

Fig. 1. Example of input data passed to the model. This spectrogram contains the full range of frequencies present in the song and must be processed serially due to its length.

4 Method

4.1 Baseline Encoder

For the purpose of this task, we assume the input X and the output Y samples come from two different distributions that share some high-level characteristics. We humans naturally perceive this common part as the content of the music. However, it is challenging to precisely indicate the parameters of an audio signal that separate the content from the timbre. Therefore, we propose an encoder-decoder model that reduces all the attributes of a song to the latent space and encodes the only common part of X and Y—the content, ignoring the instrument-specific representation (timbre). Later, we use the decoder network to restore the desired style of an output instrument from the encoded content.

The baseline model we prepared is a symmetrical encoder-decoder consisting of fully connected layers with a fixed-size bottleneck in the middle. This bottleneck layer separates the encoder from the decoder and is responsible for storing information needed to transform samples between domains. We use this model for comparison purposes in succeeding experiments.

The main disadvantage of the described baseline benchmark model is its limited context. Music is sequential with respect to time and should be processed in a manner that accounts for temporal dependencies. This is not the case in the baseline model, which forces long tracks to be processed in independent chunks. This results in poor overall performance but is especially noticeable between the boundaries of consecutive blocks, where context continuity is not preserved.

4.2 Proposed Approach

We present a new artificial neural network architecture for music style transfer based on LSTM layers that addresses the issues outlined above. LSTM cells have been used with great success in a similar encoder-decoder fashion for NLP problems [38]. These tasks share many similarities that we can leverage in our method to build on the conclusions from previous research, as music has a sequential representation similar to speech and written text. Although we have chosen piano to guitar style transfer for our particular experiments, style transfer between any pair of instruments is possible using the described method.

Our system has two main parts, as shown in Fig. 2. The recurrent encoder takes one time step of the STFT signal x_t as an input and combines it with the context vector h_{t-1}^{enc} to produce compressed representation of the content vector c_t. The information this vector retains is not limited to the currently processed section of the song but also contains a cumulative summary of previous parts. This allows us to continuously process an arbitrarily long song without the need to explicitly segment it into artificial chunks.

The last state vector h_T^{enc} produced by the encoder summarizes the whole sequence. Then it populates the initial context vector h_0^{dec} of the decoder. The decoder processes the data in a manner analogous to the encoder. It uses the encoded portions of the current signal and its hidden state, to step-by-step produce a consistent next sequence prediction containing the song's features in the domain of the target instrument.

Each sequence is passed to the second stage of the decoder where nonlinear transformations are applied and the data is projected onto the final output shape. We have selected the ReLU as an activation function between the final layers.

The Mean Squared Error was used as a loss function. As our model processes real and imaginary parts separately, the sum of both components is included in the final optimization criterion. We also use this metric to evaluate the quality of the models. The optimizer we have chosen in our network was Adam [39].

While the current trend for the state of the art solutions is to create extremely large models with dozens of repeating blocks - each with multiple layers, our approach consists only of a few simple components. This allows us to train the network on a single GPU card within a reasonable amount of time. The model is fast to converge, as it takes roughly ~4–5 h of training time on a standard Nvidia Tesla T4 card.

4.3 Hyperparameters Tuning

The appropriate choice of hyperparameters can have a significant impact on model results. During our experiments, we have considered a wide range of settings in order to select the setup that was used in the final version of the proposed approach.

Among the main parameters are the size and location of the network layers. In general, increasing the state vector in the recurrent layers shows a tendency to improve performance and to enhance the ability to remember a wider context.

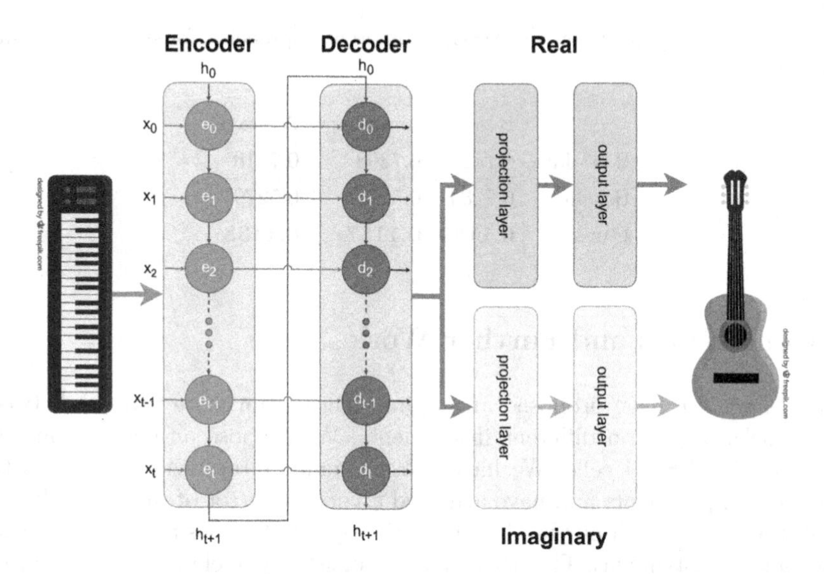

Fig. 2. Proposed architecture of our model. The input audio of piano songs is processed by the network to create samples corresponding to the guitar timbre.

On the other hand, this results in increased usage of the graphics card memory, so a trade-off is needed. In the end, we opted for recurrent connections with a state vector of length 2048, followed by dense layers matching the output size.

We also noted the positive effect of using the Hann window in signal processing on the model's ability to produce samples with finer sound qualities. Overall, of the cost functions we tested, MSE was a superior choice over Mean Squared Logarithmic Error.

5 Results

To establish a reference point for further comparisons, the loss value of the test dataset was calculated. The difference between the input given to the model X (piano) and the expected output Y (guitar) is used as a relative benchmark of models' performance. Every result below this score should be treated as a contribution of the model.

As shown in Table 1, the final architecture has achieved the loss of 0.1138 on average. Our method is several times better than the conventional autoencoder and introduces very low noise values. The content of the music (i.e. the notes, rhythm, etc.) has not been shifted or modified by the network. The samples produced by the network have clear characteristics of the output instrument, which in this case was a guitar. As the guitar is a plucked string instrument, we expected to hear a brighter, slightly more "metallic" sound quality with more attack and pronounced middle range when compared to a piano, which was exactly the case with our output samples. No additional, strange sonic artifacts or unwanted sound qualities were audible in the output samples.

Table 1. The mean squared error (MSE) of different approaches for style transfer task on test dataset.

	Real	Imaginary	Average
Raw data	0.7328	0.7308	0.7318
Baseline	0.5284	0.5273	0.5279
Ours	**0.1139**	**0.1137**	**0.1138**

6 Conclusions and Further Work

In this paper, we have proposed an original method for style transfer between music tracks played on different instruments. We propose an encoder-decoder structure with LSTM cells. We have chosen piano-guitar style transfer as the base of our experiments and have achieved a very low error of the network. The samples produced with our method have clearly audible, desired sonic qualities of the output instrument. The network also persists the melody and structure of the original sample with no distortion, additional noise, or sonic artifacts. The method is also suitable for style transfer between any pair of instruments.

We have also proposed a method for generating paired datasets by using music written in the MIDI format, which allowed us to train the network in a supervised manner. We programmatically synthesize MIDI data into audible files and compute their STFT for a time-frequency representation. The MIDI data format has been a standard in the music industry since its introduction and is the base of music production with electronic instruments, which is most of today's music. MIDI also allows us to incorporate dynamics, rhythm, pitch (including bends), pressure, control change, and other musical information in our style transfer. Our usage of MIDI and synthesizers ensures our approach is close to the context and subject of style transfer in music.

With a parallel dataset, our network is also much faster to converge when compared to existing approaches and can be used without massive computational resources. To the best of our knowledge, we have proposed an original, efficient, and highly applicable solution for musical style transfer.

In the future, we plan to conduct further experiments with the usage of LSTM encoder-decoder structures (including VAE architectures) and use our approach to build creative, usable tools for musicians.

References

1. Goodfellow, I., et al. Generative adversarial nets. In: Ghahramani, Z., Welling, M., Cortes, C., Lawrence, N.D., Weinberger, K.Q. (eds.) Advances in Neural Information Processing Systems 27, pp. 2672–2680. Curran Associates Inc (2014)
2. Vaswani, A., et al.: Attention is all you need (2017)
3. Liang, F.T., Gotham, M., Johnson, M., Shotton, J.: Automatic stylistic composition of Bach Chorales with deep LSTM. In: ISMIR (2017)

4. Roberts, A., Engel, J., Raffel, C., Hawthorne, C., Eck, D.: A hierarchical latent vector model for learning long-term structure in music (2019)
5. Johnson, D.D., Keller, R.M., Weintraut, N.: Learning to create jazz melodies using a product of experts. In: ICCC (2017)
6. Wu, J., Hu, C., Wang, Y., Hu, X., Zhu, J.: A hierarchical recurrent neural network for symbolic melody generation (2018)
7. Choi, K., Fazekas, G., Sandler, M., Cho, K.: Convolutional recurrent neural networks for music classification. In: 2017 IEEE International Conference on Acoustics, Speech and Signal Processing (ICASSP), pp. 2392–2396. IEEE (2017)
8. Feng, L., Liu, S., Yao, J.: Music genre classification with paralleling recurrent convolutional neural network, December 2017
9. Ghosal, D., Kolekar, M.: Music genre recognition using deep neural networks and transfer learning, pp. 2087–2091, September 2018
10. Costa, Y., de Oliveira, L.S., Silla, C.: An evaluation of convolutional neural networks for music classification using spectrograms. Appl. Soft Comput. **52**, 28–38 (2017)
11. Guaus, E.: Audio content processing for automatic music genre classification: descriptors, databases, and classifiers. Ph.D. thesis, University Pompeu Fabra, Barcelona, Spain (2009)
12. Gatys, L.A., Ecker, A.S., Bethge, M.: A neural algorithm of artistic style. CoRR, abs/1508.06576 (2015)
13. Dai, S., Zhang, Z., Xia, G.G.: Music style transfer: a position paper. arXiv preprint arXiv:1803.06841 (2018)
14. Bitton, A., Esling, P., Chemla-Romeu-Santos, A.: Modulated variational autoencoders for many-to-many musical timbre transfer. CoRR, abs/1810.00222 (2018)
15. van den Oord, A., et al.: WaveNet: a generative model for raw audio. CoRR, abs/1609.03499 (2016)
16. Peeters, G., Giordano, B.L., Susini, P., Misdariis, N., McAdams, S.: The timbre toolbox: extracting audio descriptors from musical signals. J. Acoust. Soc. Am. **130**(5), 2902–2916 (2011)
17. Balazs, P., Dörfler, M., Jaillet, F., Holighaus, N., Velasco, G.: Theory, implementation and applications of nonstationary Gabor frames. J. Comput. Appl. Math. **236**(6), 1481–1496 (2011)
18. Mor, N., Wolf, L., Polyak, A., Taigman, Y.: A universal music translation network. CoRR, abs/1805.07848 (2018)
19. Engel, J., et al.: Neural audio synthesis of musical notes with WaveNet autoencoders. In: ICML (2017)
20. Jolicoeur-Martineau, A.: The relativistic discriminator: a key element missing from standard GAN (2018)
21. Lu, C.-Y., Xue, M.-X., Chang, C.-C., Lee, C.-R., Su, L.: Play as you like: timbre-enhanced multi-modal music style transfer (2018)
22. Brunner, G., Konrad, A., Wang, Y., Wattenhofer, R.: MIDI-VAE: modeling dynamics and instrumentation of music with applications to style transfer (2018)
23. Verma, P., Smith, J.O.: Neural style transfer for audio spectograms (2018)
24. Zwicker, E., Fastl, H.: Psychoacoustics: Facts and Models, vol. 22. Springer, Heidelberg (2013)
25. Parncutt, R.: Harmony: A Psychoacoustical Approach, vol. 19. Springer, Heidelberg (2012)
26. Ganchev, T., Fakotakis, N., Kokkinakis, G.: Comparative evaluation of various MFCC implementations on the speaker verification task. In: Proceedings of the SPECOM 2005, pp. 191–194 (2005)

27. Sturm, B.L.: The GTZAN dataset: its contents, its faults, their effects on evaluation, and its future use. arXiv preprint arXiv:1306.1461 (2013)
28. Sturm, B.L.: An analysis of the GTZAN music genre dataset. In: Proceedings of the Second International ACM Workshop on Music Information Retrieval with User-Centered and Multimodal Strategies, vol. 2012, pp. 7–12. Association for Computing Machinery. ACM Multimedia (2012)
29. Sturm, B.L.: A survey of evaluation in music genre recognition. In: Nürnberger, A., Stober, S., Larsen, B., Detyniecki, M. (eds.) AMR 2012. LNCS, vol. 8382, pp. 29–66. Springer, Cham (2014). https://doi.org/10.1007/978-3-319-12093-5_2
30. Ben-Tal, O., Harris, M.T., Sturm, B.L.: How music AI is useful: engagements with composers, performers, and audiences. In: Leonardo, pp. 1–13 (2020)
31. Sturm, B.L., et al.: Machine learning research that matters for music creation: a case study. J. New Music Res. **48**(1), 36–55 (2019)
32. Raffel, C., Ellis, D.P.W.: Large-scale content-based matching of midi and audio files. In: ISMIR, pp. 234–240 (2015)
33. Hawthorne, C., et al.: Enabling factorized piano music modeling and generation with the MAESTRO dataset. In: International Conference on Learning Representations (2019)
34. Raffel, C.: Learning-based methods for comparing sequences, with applications to audio-to-midi alignment and matching. Ph.D. thesis, Columbia University (2016)
35. Użytkownik: midi_man. The largest midi collection on the internet, collected and sorted diligently by yours truly (2019). https://www.reddit.com/r/WeAreTheMusicMakers/comments/3ajwe4/the_largest_midi_collection_on_the_internet/
36. Ezen-Can, A.: A comparison of LSTM and BERT for small corpus. arXiv preprint arXiv:2009.05451 (2020)
37. Ding, W., et al.: Audio and face video emotion recognition in the wild using deep neural networks and small datasets. In: Proceedings of the 18th ACM International Conference on Multimodal Interaction, pp. 506–513 (2016)
38. Wu, Y., et al.: Google's neural machine translation system: bridging the gap between human and machine translation. CoRR, abs/1609.08144 (2016)
39. Kingma, D.P., Ba, J.: Adam: a method for stochastic optimization. arXiv preprint arXiv:1412.6980 (2014)

Impact of ELM Parameters and Investment Horizon for Currency Exchange Prediction

Jakub Morkowski(✉)

Poznan University of Economics and Business, Poznan, Poland
jakub.morkowski@ue.poznan.pl

Abstract. The foreign exchange market is of the utmost importance for many sectors of the economy, therefore attempts to forecast changes in currency price levels are the research area of many practitioners and theorists. The article aims at examining the impact of settings of various neural network parameters on the results of currency forecasts. The three currency pairs the US dollar, British pound, and Swiss franc to EUR were selected for the analysis. The forecast results for different network settings are examined with three different indicators: forecast error, the ratio of correctly forecasted changes in the course direction and the potential profit generated. The neural network used for the study is Extreme Learning Machine and the forecast horizons taken into account are in the range of one to ten days. The better-quality forecasts based on price levels than on rates of return was shown and good quality forecasts for two out of three currency pairs was obtained in the study. The article also presents the relationship between the results generated by the neural network and the settings of these networks - in particular, the impact of the number of delays on forecast errors and the number of hidden nodes on all three assessment parameters.

Keywords: Neural networks · Currency · Forecasting

1 Introduction

In the era of open economy and highly developed globalization, one of the key factors influencing the economy of individual countries is the situation of the national unit of currency. The exchange rate is considered by many as a factor reflecting the current situation and condition of the economy of a given country [12]. Due to the specific nature of the currency exchange markets, the level of currency rates should be constantly monitored because of possible fluctuations in the market rates. In international trade, the exchange rate plays a key role and has a real impact on the profits and/or losses generated by companies. In the exchange between counterparties using different currencies, the size of cash flow is determined by the current exchange rate of currencies involved in the transaction. This directly translates into profit or loss for the enterprise

L. Rutkowski et al. (Eds.): ICAISC 2021, LNAI 12854, pp. 133–144, 2021.
https://doi.org/10.1007/978-3-030-87986-0_12

or other institution. Exchange rates can also be a direct profit (loss) factor by trading them [13]. By proper forecasting of the change in the currency market, the investor is able to obtain positive cash flows on their account.

Referring to the above characteristics of currency pairs, it should be noted that both practitioners and scientists are interested in creating the most effective model or method of forecasting future values of the time-series patterns of exchange rates. The main objective of forecasting should be to reduce the risk associated with making decisions which affect the work of enterprises, financial organizations, and private investors. The aim of the following study was to check the effectiveness of using neural networks in investments, to ensure positive cash flows using currency pairs as an underlying asset. For this purpose, a study was conducted to verify the ability of ELM (Extreme Learning Machine) to generate profit when investing in US$, British Pound, and Swiss Franc currency pairs in relation to Euro (USD/EUR, GBP/EUR, CHF/EUR). The study took into account different ELM neural network settings and different investment horizons from 1 to 10 days. Various settings of the neural network were introduced in order to search for dependencies between them and to analyze it for the optimal selection of settings depending on the expected investment horizon.

Forecasting methods used in financial markets fall into two trends or categories. One of them is the use of statistical measures such as the ARMA (Autoregressive moving average) or the ARIMA (Autoregressive integrated moving average) model and the later introduced ARCH (Autoregressive conditional heteroskedasticity) and GARCH (Generalized autoregressive conditional heteroskedasticity) models. The models listed above may be less effective due to the non-linear nature of the relationship between assets listed on stock exchanges.

In a recent literature review, among the methods belonging to the first stream, research on forecasting the risk of the US dollar exchange rate under volatile conditions can be distinguished [5] with the use of estimated value at risk (VaR), so that is possible to forecast significant declines in the US dollar (USD) price levels. The literature also includes studies on the volatility of the Euro and USD (EUR/USD) currency pair, where various GARCH and ARCH models are considered, estimated using the maximum likelihood method with different variants and using different tests (normality, Student's t, etc.) [9]. Therefore, research is often undertaken on the possibilities of modeling volatility using the ARCH and GARCH models recently, e.g., for the economy of Turkey [16] or Tanzania, where the impact of previous information about the exchange rate on the current course value. The above indicates that volatility in the previous day's rate may affect the current volatility of the exchange rate [10]. The above-mentioned models are also used in the course forecasting research. Some of the articles show better results in forecasting the British pound against the US dollar for the ARFIMA model than for the ARIMA model [17]. The reasons for such an advantage can be found in showing the property of long memory by series.

In the area of forecasting, much more dynamic development can be observed in the case of methods from the latter group, i.e., those based on soft computational methods. In the literature, authors [7] indicate that the largest share of

MLP (Multi-Layer Perception Network) and FLANN (Functional Link Artificial Neural Network) among artificial neural networks used for forecasting time series. Very often scientists use MLP (Multilayer Perceptron) neural networks for the prediction of time series [23] and [18]. Many scientists indicate that it is this type of neural network that works best for forecasting, among others, currency exchange rates, but there are indications to look for more effective ways than that of MLP. These premises result from overmatching, a large number of iterations and thus long computing time, as well as slow convergence, and local minimum of the MLP network [14]. In the case of ELM models, scientists point to the high speed of learning and good efficiency of generalization. However, they also point to the disadvantages of this type of network, which is often not resistant to badly conditioned data, which may cause learning errors such as, for instance, overfitting of the neural network [22]. In literature, there are many other examples of correct results of currency price prediction using models implementing the ELM neural network. [6] used ELM with the multi-population search scheme of Jaya optimization technique to forecast two Indian Rupee currency pairs in relation to the US$ USD/IND and the euro against US$ USD/EUR. ELM is also used together with other methods, such as empirical mode decomposition (EMD) and phase space reconstruction (PSR) responsible for the initial processing of data, which is then predicted by the ELM neural network [21].

In literature, there is a great interest in combining [4] both of these methods, creating hybrid models [15]. Hybrid methods are designed to reduce the limitations stemming from statistical methods [3]. The approach combining parametric methods, such as neural and nonparametric networks, shows promising results than using these methods separately [1,2,19].

The rest of the paper is organized as follows: the second part contains a description of the data sets, the third part describes the methodology, including the configuration of neural network parameters and methods of assessing the quality of the forecasts made. Part four is an experimental analysis while part five concludes the article presenting conclusions derived from the research.

2 Dataset Description

The exchange rate of two currencies characterizes the price at which one currency is convertible to another. All currencies have a buying and selling price. In this study, three currency pairs were taken into consideration - the US$, the British Pound and the Swiss Franc - for which the buying price of 1 Euro was analysed. The study was conducted for ten different time horizons ranging from one to ten days. The data has been divided into training data and test data. The data come from the period between 01.01.2015 and 31.12.2019. The learning set always contained the last hundred observations up to day $t-1$, where t is the day when the investment decision was made. Therefore, the size of the training set was always the same - one hundred observations, while the range changed depending on the date of making the investment decision. The study was conducted for price levels and for rates of return. In the first attempt, the empirical results will compare the prices of past currencies, that have actually been recorded on the stock exchange, to the prices of the currencies derived from the prediction. In the case of return rates, the study will compare the direction of change forecasted by neural networks with the direction of change recorded on the stock exchange, for a given period.

3 Methodology

Forecasts are made on the basis of historical data, on which the neural network learns. Due to the 1-day, 2-day, etc. investment horizons with data of daily frequency. The selected type of neural network to perform the test is ELM (i.e., feedforward neural networks). In this model, the biases in the hidden layer are selected randomly and the output weights are calculated based on them. This network is characterized by the lack of recursion [24]. Forecasting models from the nnfor package [25] were used in the R program for time series forecasting with Multilayer Perceptrons (MLP) and Extreme Learning Machines (ELM). What is more, the current version of this pack (version 0.9.6) was used.

At the preliminary stage of forecasting, both models provided by the abovementioned package were used. The first results indicated better results of forecasts for ELM currency pairs, therefore forecasting using this neural network was extended and is presented below.

In the following study, only one type of feedforward neural networks was taken into account for comparisons, however the parameters within the network were being changed and ten different forecast horizons were analysed. The study analyses three currency pairs: USD/EUR, CHF/EUR, and GBP/EUR. For each currency pair, forecasts were made on both rates of return and price levels, which gives two studies for each currency pair for forecasts from one to ten days.

For each length of the investment horizon, variances of two parameters were used: the number of hidden nodes (2, 5, 10, 15, 20, 25, 50) and the number of lags (1, 2, 3, 4, 5, 6, 7).

The number of hidden nodes was assumed not as consecutive natural numbers but keeping certain intervals in order to search for the most optimal orders of magnitude of nodes and to investigate the dependence of the forecast results on this parameter. Through such settings, the study answers the question whether increasing the number of hidden nodes has a positive effect on the results of neural networks, and if so, to what level it is profitable to increase them so that the benefits of improving the results are greater than the computational load resulting from the use of many hidden nodes. The use of the number of delays in the range from one to seven stems from the character of research on financial markets and the support for such an approach in the selection of this parameter can be found in many scientific studies [8]. Due to the research method used, it can be concluded that the size of the collected results is big. For all three currency pairs, forecasts were made at price levels and rates of return - for each of them, forecasts were made at ten different investment horizons (from one to ten days ahead). In addition, forty-nine different ELM network settings were used for forecasts resulting from a combination of two parameters (the number of hidden nodes and the number of delays). The study was carried out on such a large scale of various settings of the neural network, so to obtain an answer to the question on how individual parameters of the neural network affect the quality of prediction and in order to search for the relationship between the parameters of the neural network and the results of predictions in various forecast horizons.

The assessment of effectiveness of the forecasts for 1 to 10 days ahead, made with the use of neural networks on the rates of currency pairs, will be presented in the study using three categories. The first category includes the most common forecast errors in similar studies. Mean Absolute Error (MAE) will be used in this category. In order to determine the error for the entire method or model which was used, a measure called the mean error ME. In case of perfect fit of forecasts to the model, ME is 0. The smaller the value of ME, the better the model fit. Based on the default ME character, it can be concluded that when its sign is positive our forecasts are undervalued, and when its sign is negative, our forecasts are overvalued. It is believed that ME can be applied to evaluate the model used for forecasting and to guide researchers to the possibility of changing the model and/or removing some variables from the model to reduce forecast errors. However, this measure cannot be the only used for comparative evaluation between different models [11,20].

Additionally, the proposed categories to evaluate the forecast results are:

- the percentage of correct decisions made by the neural network in the total number of decisions made. A correct decision is understood as a correctly forecasted increase or decrease in the rate of a given currency pair, both for forecasting of returns and price levels. For return rates, the correct decision made by the network is confirmed by the same sign by the forecast and actual return rate for a given iteration. For the price level, the correct decision is when the difference between the rate on the investment start date and the rate on the investment completion date is characterized by the same sign as the difference between the rate on the day of starting the investment and the expected rate value on the day of its completion,
- profit - potential profit understood for each iteration as the absolute value of the differences between the exchange rate on the investment completion date and the exchange rate on the investment completion date. In this way, we get a potential profit with a well-forecasted direction of exchange rate change for each of the days covered by the study. The measure of profit is the sum of these values, where the values of the potential profit receive the "+" sign if the network has made a correct decision on the direction of the exchange rate change (the correct decision is understood here in the same way as in the category of the percentage of correct decisions made) and the sign "–" otherwise.

4 Experimental Analysis

Three currency pairs USD/EUR, GBP/EUR and CHF/EUR were used for forecasting, and the dataset spans over four years and includes over 1,290 sample market quotations. The model forecasts for 1, 2, 3, 4, 5, 6 and 7 days ahead using historical data price levels and rates of return. The analysis takes into account the influence of forecasting parameters on the methods of assessing the quality of these forecasts

The generalized results of three measures taken into account are presented in the Table 1 below. The first table shows the averaged results for all observations. These results were presented with the assumption of maximization of the profit and the number of correct direction predictions of change made by the network in the total number of attempts made. The table shows two measures of error - ME and MAE and how they look depending on the use of different types of data (price levels and rates of return) for the same currency pairs.

Table 1. Averaged results of profit, correct decisions and ME/MAE for three currency pairs when forecasting price levels (PR) and rates of return (RR)

	Profit PL	Cor. PL	ME/MAE PL	Profit RR	Cor. RR	ME/MAE RR
USD/EUR	0,9874	0,5464	0,0007/0,0094	0,1498	0,5067	−0,0000134/0,0036
CHF/EUR	0,3725	0,5353	−0,0004/0,0062	0,0573	0,5035	−0,00000631/0,0023
GBP/EUR	−0,0667	0,4802	−0,0024/0,0165	−0,5498	0,4843	−0,0000257/0,0004

As can be seen in the two tables above, forecasting on price levels is characterized by better results, which are understood as the possibility of generating a greater profit and a higher percentage of correct decisions in the total number of decisions than forecasting on rates of return. Later in the analysis of the results, the results for the price levels of the three currency pairs will be presented in order to support the reasons why forecasting on this data gives better results in the two parameters mentioned above. For the sake of clarity in the presentation of results, they will be presented in the form of graphs.

4.1 MAE

In further stages of the following article, for the sake of clarity, the graphs and results will be presented in narrower data ranges, considering the fact that conclusions drawn on these narrower ranges can be generalized for the entire study, taking into account all the applied settings of neural networks and three currency pairs. There are several conclusions that will be analysed in more detail later on in the discussion of the results. Those include:

- Noticeable arrangement of the MAE results for the shortest horizons closest to the centre and its systematic expansion along with the increase of the investment horizon (arrangement (in the order from the centre) in blue-yellow-red with various markers and ending in green symbolizing the longest ten-day prediction horizon).

- For the first four horizons, there is a very clear gradual proximity to the middle of the MAE results graph for the network settings where the investment horizon is equal to the set number of lags. This phenomenon can be observed especially well for forecasts for one day ahead, where for a setting with one delay, the markers are around the 0.004 line. Then they gradually increase for delays from two to seven, and again with a network setting with one delay, but with increased number of hidden nodes, get closer to the middle of the graph, to line 0.004.

  On the charts below, presenting the Mean Absolute Error values for currency pairs, it should be noticed that:
- For all three currency pairs, the shorter the investment horizon, the lower the MAE. The chart presented in Fig. 1 shows the MAE for USD/EUR for the number of hidden nodes one and two and for delays ranging from one to seven. The chart is only a representative one, because such a relationship can also be noticed for a higher number of hidden nodes and for all three currency pairs. (see Fig. 1) shows that the MAE levels range from 0.003 to 0.013, but each time for a given grid setting, the longer the investment horizon, the greater the MAE.

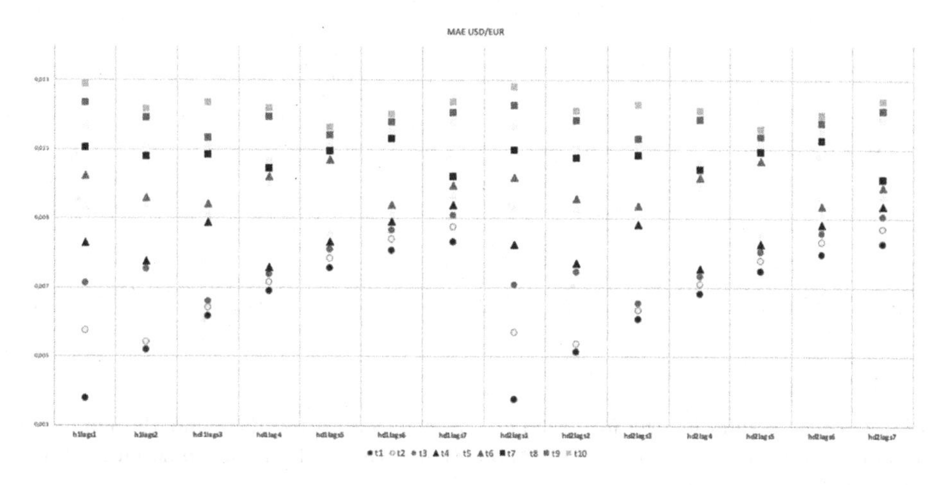

Fig. 1. Results of the MAE for the forecast horizon from one to ten days and the number of hidden nodes from one to two in combination with lags from one to seven for the CHF/EUR currency pair

Figure 1 shows the trend for the best results, i.e., the lowest MAE, for lags equal to the investment horizon with the same number of hidden nodes. In one-day ahead predictions for the first seven different network settings, the lowest MAE was obtained for the setting: one delay and one hidden node, for the next seven for two hidden nodes, and also one delay for the next three nodes and again one delay, etc. For the two days ahead prediction, the lowest MAE was obtained for the hdXlags2 settings (where X are consecutive natural numbers from 1 to 7) - i.e., the number of delays is equal to two and the number of hidden nodes changes. What is more, this relation applies to all currency pairs.

When deepening the analysis for the best results for the forecasts from a given forecast horizon from the previous point, it is visible that the value of MAE decreases together with the increase in the number of nodes. This means that the results of the neural network for the predictions for one, two and three days with the number of delays equal to the prediction horizon are characterized by the lower value of the forecast error, the greater the number of hidden nodes used. However, it should be also noted that the decrease in the value of the forecast errors together with the increase in the number of hidden nodes has an expiring nature.

The forecast error value increases along with the increase of the investment horizon. For horizons from one to three days, it ranges from 0.002 to 0.006. For investment horizons of more than seven days, it is higher than 0.006 for each combination of neural network settings.

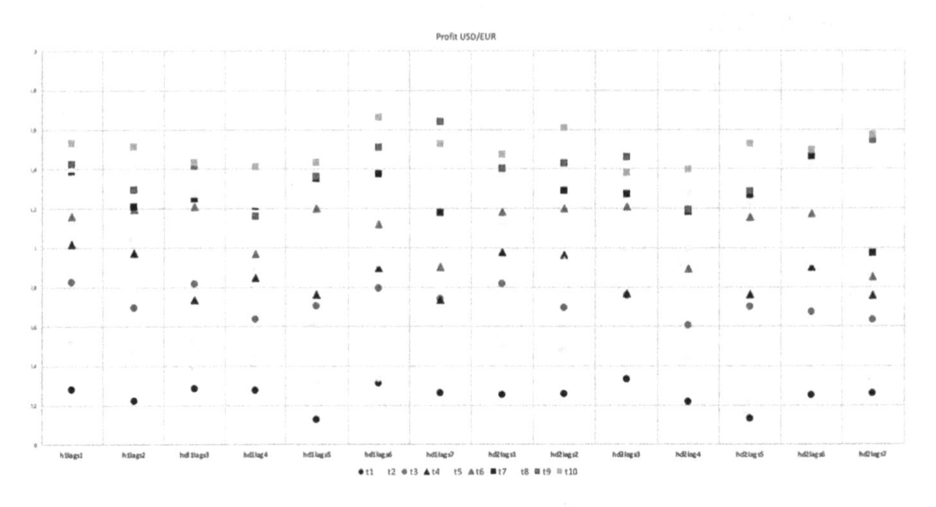

Fig. 2. Profit results for the forecast horizon from one to ten days and the number of hidden nodes from one to two in combination with lags from one to seven for the USD/EUR currency pair

4.2 Profit

The currency pair CHF/EUR and USD/EUR are two of the analysed currency pairs, which for all network settings and for all time horizons always show a positive profit, as for the assessment category described in the previous chapter. For the analysed currency pair, a similar trend can be noticed as in the case of errors in forecasts regarding the increase in profit together with the increase in the investment horizon - (see Fig. 2).

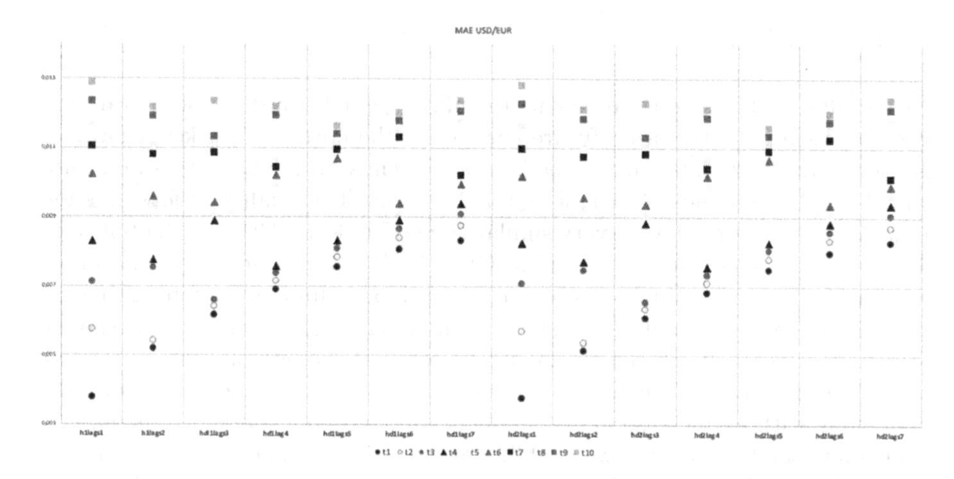

Fig. 3. Share of correct forecast decisions for the forecast horizon from one to ten days and for the number of hidden nodes from one to three in combination with lags from one to seven for the CHF/EUR currency pair

The forecast for more days ahead has a greater risk of making a mistake, but as can be seen in the Figure below, a greater risk is associated with the possibility of generating higher flows. In the case of profit, the relationships between the number of hidden layers and delays and the amount of profit are less visible. It seems that the forecasts for one day ahead for each of the 49 network settings have the lowest profit value compared to the forecasts for two to ten days. This is caused by smaller increases/decreases in the price of the currency for one day. The price fluctuations over a longer time horizon may be greater, and with an appropriate prediction, the profit ratio will be characterized by a higher level.

4.3 The Number of Correct Decisions

The percentage of correct decisions made by the neural network presents the number of correctly predicted directions of currency exchange rate changes to the total number of decisions made, which in this study were 1,184 cases. Results for all 49 network settings and for all 10 different prediction horizons for CHF/EUR and USD/EUR are higher than 50% of correct decisions. For forecasts for one to four days, one can observe an increasing number of correct decisions. For forecasts for more than four days, no clear trends or tendencies can be found. In Fig. 3 (see Fig. 3), one can see the best results for the ten-day prediction with the network setting – lags = 7 and regardless of the number of hidden nodes. For the remaining three currencies and with a greater number of hidden nodes, there is also no dependency. It is particularly well reflected in Fig. 3, which, in contrast to the MAE and profit charts, is characterized by a large variation in the order of the best results depending on the network settings.

5 Discussion

The results of the conducted research confirm the effectiveness of neural networks as used for forecasting future changes in the currency market. However, as shown by the results, their effectiveness depends on many factors and may vary depending on the settings of neural networks. The quality of forecasts for major currency pairs may be very similar (USD/EUR and CHF/EUR) but may also be characterized by worse results (GBP/EUR). Both researchers and practitioners find it very valuable to search for optimal solutions for forecasting on financial markets. Not only did the study show a few dependencies that should be further developed but also it confirmed some of the assumptions presented in the literature review, such as, for example, an increase in the quality of forecasts along with an increase in the number of hidden nodes, but this increase is smaller and smaller together with adding new nodes. In connection with this too many hidden nodes will significantly extend the time needed to conduct forecasts without significantly increasing the results generated by such a network.

Network setup results may play a valuable role in future research. They may minimize forecast errors (MAE) and maximize the number of correct decisions. The study presented in this article is a kind of introduction to further considerations on the possibility of using neural networks in forecasting the prices of currency pairs and the use of these forecasts for investments in derivatives. The desire of the networks to minimize forecast errors may be considered a flaw of neural networks used in this study. In further considerations on the possibilities of using this type of network for forecasting, which will then be used to invest in derivatives, the network's attempt to minimize forecast errors is not a fully desirable action. This is due to the fact that it is more important for investing in derivative instruments to predict the direction of changes, and not to minimize the forecast error. As also was mentioned in the preliminary literature review, the results presented in this study can be used as an introduction to the application of the hybrid method by applying the best settings of neural network parameters, e.g., fuzzy sets, in order to increase the effectiveness of their forecasts.

6 Conclusion

Extreme learning machines give satisfactory results in predicting future price levels of two key currency pairs - the US$ and the Swiss Franc in relation to the Euro, while the relation of the British Pound to Euro was slightly worse. Important for further research are answers to questions about the relationship between the settings of the neural networks and their results in various forecast horizons. The results that were noted, especially for short horizons, were confirmed for all three currencies. Regardless of the number of hidden nodes, the lowest forecast errors were found for the settings of the number of delays equal to the forecast horizon. It was also proven that there is a possibility of generating the highest profits for investments made with forecasts for longer periods (more than 5 days). However, it was also shown that for the same settings there is a higher risk than in the case of shorter forecasts (less than 5 days).

References

1. Abdelsalam, M., Hatem, A., Abdulwahed, W.F.: Evaluation of differential evolution and particle swarm optimization algorithms at training of neural network for prediction. IJCI Int. J. Comput. Inf. **3**(1), 2–14 (2014)
2. Abdual-Salam, M.E., Abdul-Kader, H.M., Abdel-Wahed, W.F.: Comparative study between differential evolution and particle swarm optimization algorithms in training of feed-forward neural network for stock price prediction. In: 2010 The 7th International Conference on Informatics and Systems (INFOS), pp. 1–8. IEEE (2010)
3. Agrawal, M., Khan, A.U., Shukla, P.K.: Stock price prediction using technical indicators: a predictive model using optimal deep learning. Learning **6**(2), 7 (2019)
4. Anastasakis, L., Mort, N.: Exchange rate forecasting using a combined parametric and nonparametric self-organising modelling approach. Expert Syst. Appl. **36**(10), 12001–12011 (2009)
5. Anjum, H., Malik, F.: Forecasting risk in the US$ exchange rate under volatility shifts. North Am. J. Econ. Financ. **54**, 101257 (2020)
6. Das, S.R., Mishra, D., Rout, M.: A hybridized ELM using self-adaptive multi-population-based Jaya algorithm for currency exchange prediction: an empirical assessment. Neural Comput. Appl. **31**(11), 7071–7094 (2018). https://doi.org/10.1007/s00521-018-3552-8
7. Dash, R., Dash, P.K., Bisoi, R.: A self adaptive differential harmony search based optimized extreme learning machine for financial time series prediction. Swarm Evol. Comput. **19**, 25–42 (2014)
8. De Myttenaere, A., Golden, B., Le Grand, B., Rossi, F.: Using the mean absolute percentage error for regression models. In: Proceedings, p. 113. Presses universitaires de Louvain (2015)
9. Dritsaki, C.: Modeling the volatility of exchange rate currency using GARCH model. Economia Internazionale/Int. Econ. **72**(2), 209–230 (2019)
10. Epaphra, M.: Modeling exchange rate volatility: application of the GARCH and EGARCH models. J. Math. Financ. **7**(1), 121–143 (2016)
11. Ferreira, T.A., Vasconcelos, G.C., Adeodato, P.J.: A new intelligent system methodology for time series forecasting with artificial neural networks. Neural Process. Lett. **28**(2), 113–129 (2008)
12. Kartono, A., Febriyanti, M., Wahyudi, S.T.: Predicting foreign currency exchange rates using the numerical solution of the incompressible Navier-Stokes equations. Physica A Stat. Mech. Appl. **560**, 125191 (2020)
13. Markova, M.: Foreign exchange rate forecasting by artificial neural networks. In: AIP Conference Proceedings, vol. 2164, no. 1, p. 060010. AIP Publishing LLC (2019)
14. Rout, A.K., Dash, P.K., Dash, R., Bisoi, R.: Forecasting financial time series using a low complexity recurrent neural network and evolutionary learning approach. J. King Saud Univ.-Comput. Inf. Sci. **29**(4), 536–552 (2017)
15. Rout, M., Majhi, B., Majhi, R., Panda, G.: Forecasting of currency exchange rates using an adaptive ARMA model with differential evolution based training. J. King Saud Univ.-Comput. Inf. Sci. **26**(1), 7–18 (2014)

16. Sekmen, F., Ravanoğlu, G.A.: The modelling of exchange rate volatility using Arch-Garch models: the case of Turkey. MANAS Sosyal Araştırmalar Dergisi **9**(2), 834–843 (2020)
17. Shittu, O.I., Yaya, O.S.: Measuring forecast performance of ARMA and ARFIMA models: an application to US$/UK pound foreign exchange rate. Eur. J. Sci. Res. **32**(2), 167–176 (2009)
18. Tahersima, H., Tahersima, M., Fesharaki, M., Hamedi, N.: Forecasting stock exchange movements using neural networks: a case study. In: 2011 International Conference on Future Computer Sciences and Application, pp. 123–126. IEEE (2011)
19. Wang, Y., Cai, Z., Zhang, Q.: Differential evolution with composite trial vector generation strategies and control parameters. IEEE Trans. Evol. Comput. **15**(1), 55–66 (2011)
20. Willmott, C.J., Matsuura, K.: Advantages of the mean absolute error (MAE) over the root mean square error (RMSE) in assessing average model performance. Climate Res. **30**(1), 79–82 (2005)
21. Yang, H.L., Lin, H.C.: Applying the hybrid model of EMD, PSR, and ELM to exchange rates forecasting. Comput. Econ. **49**(1), 99–116 (2017)
22. Yildirim, H., Özkale, M.R.: The performance of ELM based ridge regression via the regularization parameters. Expert Syst. Appl. **134**, 225–233 (2019)
23. Yu, L.Q., Rong, F.S.: Stock market forecasting research based on neural network and pattern matching. In: 2010 International Conference on E-Business and E-Government, pp. 1940–1943. IEEE (2010)
24. Zong, W., Huang, G.B., Chen, Y.: Weighted extreme learning machine for imbalance learning. Neurocomputing **101**, 229–242 (2013)
25. Kourentzes, N.: nnfor: Time Series Forecasting with Neural Networks. R package version 0.9.6 (2019). https://CRAN.R-project.org/package=nnfor

Spectroscopy-Based Prediction of In Vitro Dissolution Profile Using Artificial Neural Networks

Mohamed Azouz Mrad(✉), Kristóf Csorba, Dorián László Galata, Zsombor Kristóf Nagy, and Brigitta Nagy

Budapest University of Technology and Economics, Budapest műegyetem rkp. 3, Budapest 1111, Hungary
mmrad@edu.bme.hu

Abstract. In pharmaceutical industry, dissolution testing is part of the target product quality that are essentials in the approval of new products. The prediction of the dissolution profile based on spectroscopic data is an alternative to the current destructive and time-consuming method. Raman and near infrared (NIR) spectroscopies are two complementary methods, that provide information on the physical and chemical properties of the tablets and can help in predicting their dissolution profiles. This work aims to use the information collected by these methods by creating an artificial neural network model that can predict the dissolution profiles of the scanned tablets. The ANN models created used the spectroscopies data along with the measured compression curves as an input to predict the dissolution profiles. It was found that ANN models were able to predict the dissolution profile within the acceptance limit of the f_1 and f_2 factors.

Keywords: Artificial neural networks · Dissolution prediction · Raman spectroscopy · NIR spectroscopy

1 Introduction

In pharmaceutical industry, a target product quality profile is a term used for the quality characteristics that a drug product should process in order to satisfy the promised benefit from the usage and are essentials in the approval of new products or the post-approval changes. A target product quality profile would include different important characteristics, very often one of these is the in vitro (taking place outside of the body) dissolution profile [1]. A dissolution profile represents the concentration rate at which capsules, and tablets emit their drugs into bloodstream over the time. It is especially important in case of tablets that yield a controlled release into the bloodstream over several hours. That offers many advantages over immediate release drugs like reducing the side effects

L. Rutkowski et al. (Eds.): ICAISC 2021, LNAI 12854, pp. 145–155, 2021.
https://doi.org/10.1007/978-3-030-87986-0_13

due to the reduced peak dosage and better therapeutic results due to the balanced drug release [2]. In vitro dissolution testing has been a subject of scientific researches for several years and became a vital tool for accessing product quality performance [3]. However, this method is destructive since it requires immersing the tablets in a solution simulating the human body and time-consuming as the measurements usually take several hours. As a result, the tablets measured represent only a small amount of the tablets produced, also called batch. Therefore, there is a need to find different methods that do not have the limitations of the in vitro dissolution testing. The prediction of the dissolution profile based on spectroscopic data is an alternative on which many articles have been published and showed promising results. Raman and near infrared (NIR) spectroscopies are two complementary methods that are applied in the pharmaceutical industry. They offer the opportunity to obtain information on the physical and chemical properties of the tablets that can help predicting their dissolution profiles in few minutes without destroying them. Hence, Raman and NIR are recognized as a straight-forward, cost effective alternatives and non-destructive tools in the quality control process [4,5]. However, these spectroscopies produce a large amount of data as they consist of measurements of hundreds of wavelengths. This data can be filtered out or maintained depending on how much useful information can be extracted from it. This can be achieved using multivariate data analysis techniques such as Principal Component Analysis (PCA).

Several researchers have used the spectroscopies data along with the multivariate data analysis techniques in order to predict the dissolution profiles. Zannikos et al. worked on a model that permits hundreds of NIR wavelengths to be used in the determination of the dissolution rate [6]. Donoso et al. used the NIR reflectance spectroscopy to measure the percentage drug dissolution from a series of tablets compacted at different compressional forces using linear regression, non-linear regression and partial least square (PLS) models [7]. Freitas et al. created a PLS calibration model to predict drug dissolution profiles at different time intervals and for media with different pH using NIR reflectance spectra [8]. Hernandez et al. used PCA to study the sources of variation in NIR spectra and a PLS-2 model to predict the dissolution on tablets subjected to different levels of strain [9].

Artificial neural networks (ANNs) are very suitable for complex and highly nonlinear problems and have been used in pharmaceutical industry in many aspects, such as the prediction of chemical kinetics [10], monitoring a pharmaceutical freeze-drying process [11], solubility prediction of drugs [12]. ANN models have been also used for the prediction of the dissolution profile based on spectroscopic data. Ebube et al. trained an ANN model with the theoretical composition of the tablets to predict their dissolution profile [13]. Galata et al. developed a PLS model to predict the contained drotaverine (DR) and the hydroxypropyl methylcellulose (HPMC) content of the tablets which are respectively the drug itself and a jelling material that slows down the dissolution, based on both Raman and NIR Spectra, and used the predicted values along with the measured compression force as input to an ANN model in order to predict the dissolution profiles of the tablets defined in 53 time points [14]. Using NIR and

Raman spectra to predict the DR and HPMC content of the tablets then along with the concentration force predicting the dissolution profile is a fast method that require minimal amount of human labor and which makes it easier to evaluate a larger amount of the batch. However, using predicted values as input to the ANN model might limit the accuracy of the dissolution profile prediction. Furthermore, training a neural network model to map three inputs to the 53 time points of the dissolution profile, is still a heavy task that might also affect the accuracy of the prediction. Thus, our goal was to create ANN models with a different approach that do not use predicted values as input. Our aim was to extract the useful information directly from the NIR and RAMAN spectra using a multivariate data analysis technique. This information, along with the extracted information from the compression curve of the tablets were used as input of two ANN models created that predict the dissolution curve represented in the 53 points.

2 Data and Methods

In this section, the data used will be described and the methods used for the data pre-processing will be presented. The artificial neural networks created will be presented and finally the error measurement methods adopted to evaluate the results.

2.1 Data Description

We have been provided with the measurements of the NIR and RAMAN spectroscopy, along with the pressure curves extracted during the compression of the tablets. The data consists of the NIR reflection and transmission, Raman reflection and transmission spectra, the compression force - time curve and the dissolution profile of 148 tablets. The tablets were produced with a total of 37 different settings. Three parameters were varied: drotaverine content, HPMC content and the compression force. From each setting, four tablets were selected for analysis (37 * 4). The spectral range for NIR reflection spectra was 4000–10,000 cm^{-1},with a resolution of 8 cm^{-1}, which represents 1556 wavelength points. NIR transmission spectra were collected in the 4000–15,000 cm^{-1} wavenumber range with 32 cm^{-1} spectral resolution, which represents 714 wavelength points. Raman spectra were recorded in the range of 200–1890 cm^{-1} with 4 cm^{-1} spectral resolution for both transmission and reflection measurements which represents 1691 points. Two spectra were recorded for each tablet in both NIR and Raman. The pressure during the compression of the pill was recorded in 6037 time points. The dissolution profiles of the tablets were recorded using an in vitro dissolution tester. The length of the dissolution run was 24 h. During this period, samples were taken at 53 time points (at 2, 5, 10, 15, 30, 45 and 60 min, after that once in every 30 min until 1440 min).

2.2 Data Analysis

The collected data were visualized and analyzed using MATLAB and Excel in order to detect and fix missed and wrong values: Setting first point of the dissolution curves to zero, detecting missed values, and fixing negative values found due to error of calibration, etc. Specifically, the data is represented in matrices N_i^n for NIR transmission data and M_j^n for NIR Reflection data, where i = 1556, j = 714. R_k^n and Q_k^n respectively for Raman reflection and transmission data where k = 1691. C_l^n for the compression force data where l = 6037 and P_s^n for the dissolution profiles where s = 54. With n representing the number of samples which is equal to 296. All the different NIR, RAMAN and the compression force matrices have been standardized using scikit-learn preprocessing method: StandardScaler. StandardScaler fits the data by computing the mean and standard deviation and then centers the data following the equation $Stdr(NS) = (NS - u)/s$, where NS is the non-standardized data, u is the mean of the data to be standardized, and s is the standard deviation. All the different standardized NIR, RAMAN and the compression force matrices have been row-wise concatenated to form a new matrix D_m^n where n = 296 and m = i+j+2k+l=11686 as follow: $D_m^n = (N_i^n|M_j^n|R_k^n|Q_k^n|C_l^n)$. After standardization, PCA was applied to the different standardized matrices as well as the merged data D_m^n and in order to reduce the dimension of the data while extracting and maintaining the most useful variations. Basically, taking D_m^n as an example we construct a symmetric m*m dimensional covariance matrix Σ (where m = 11686) that stores the pairwise covariances between the different features calculated as follow:

$$\sigma_{j,k} = \frac{1}{n}\sum_{i=1}^{n}(x_j^{(i)} - \mu_j)(x_k^{(i)} - \mu_k) \tag{1}$$

With μ_j and μ_k are the sample means of features j and k. The eigenvectors of Σ represent the principal components, while the corresponding eigenvalues define their magnitude. The eigenvalues were sorted by decreasing magnitude in order to find the eigenpairs that contains most of the variances. Variance explained ratios represents the variances explained by every principal components (eigenvectors), it is the fraction of an eigenvalue λ_j and the sum of all the eigenvalues. The following plot (Fig. 1) shows the variance explained rations and the cumulative sum of explained variances.

It indicates that the first principal components alone accounts for 50% of the variance. The second component account for approximately 20% of the variance. The plot indicates that the seven first principal components combined explain almost 96% of the variance in D. These components are used to create a projection matrix W which we can use to map D to a lower dimensional PCA subspace D' consisting of less features:

$$D = [d_1, d_2, d_3, \dots d_m], d \in R^m \rightarrow D' = DW, W \in R^{m*v} \tag{2}$$

$$D' = [d_1, d_2, d_3, \dots d_m], d \in R^m \tag{3}$$

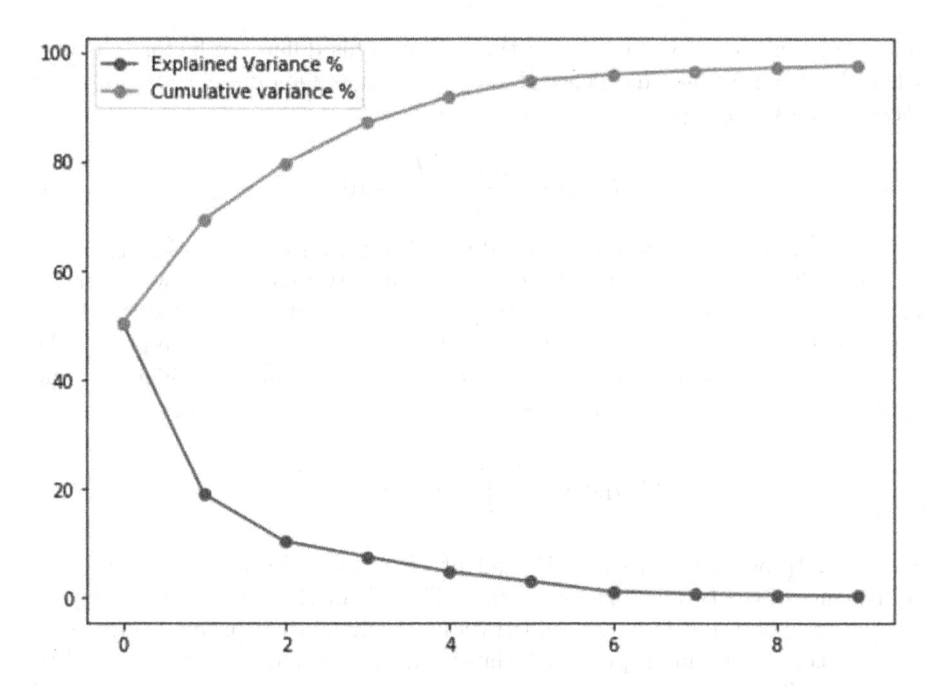

Fig. 1. PCA explained and cumulative variances.

2.3 Artificial Neural Networks

ANN models were used to predict the dissolution profiles of the tablets. The models were created using the Python library sklearn. Two ANN models were created, Model 1 and Model 2 respectively using $Input_1$ and $Input_2$ which are described later in this paper. The models used the rectified linear unit activation function referred to as ReLU on the hidden layers and the weights on the models were optimized using LBFGS optimizer. The mean-squared error (MSE), was the loss function used by the optimizer in both models. The training target for both models were the 53 dissolution curve points. The number of layers on the models and the number of neurons were optimized based on their performances. Regularization term has been varied in order to reduce overfitting. In each training, 16% of the training samples (49 samples) were selected randomly for testing. The accuracy of the models predictions was calculated by evaluating the similarity of the predicted and measured dissolution profiles using the f_2 and the f_1 values.

2.4 Error Measurement

Two mathematical methods are described in the literature to compare dissolution profiles [15]. A difference factor f_1 which is the sum of the absolute values of the vertical distances between the test and reference mean values at each dissolution time point, expressed as a percentage of the sum of the mean fractions released from the reference at each time point. This difference factor f_1 is zero when the mean profiles are identical and increases as the difference between the mean profiles increases.

$$f_1 = \frac{\sum_{t=1}^{n} |R_t - T_t|}{\sum_{t=1}^{n} |R_t|} * 100 \tag{4}$$

Where R_t and T_t are the reference and test dissolution values at time t.

The other mathematical method is the similarity function known as the f_2 measure, it performs a logarithmic transformation of the squared vertical distances between the measured and the predicted values at each time point. The value of f_2 is 100 when the test and reference mean profiles are identical and decreases as the similarity decreases.

$$f_2 = 50log_{10}[(1 + \frac{1}{n}\sum_{t=1}^{n}(R_t - T_t)^2]^{-0.5}) * 100 \tag{5}$$

Values of f_1 between zero and 15 and of f_2 between 50 and 100 ensure the equivalence of the two dissolution profiles. The two methods are accepted by the FDA (U.S. Food and Drug Administration) for dissolution profile comparison, however the f_2 equation is preferred, thus in this paper maximizing the f_2 will be prioritized. The average of f_1 and f_2 on the different samples was then calculated and returned.

3 Results and Discussions

In this section the results after the PCA dimensionality reduction will be discussed. The results and the performance of the Artificial Neural Network models created will be presented.

3.1 Dimensionality Reduction Using PCA

Principal component analysis transformation was applied in a first step to the standardized NIR and Raman spectra recorded in reflection and transmission mode (N_i^n, M_j^n, R_k^n, Q_k^n matrices) and the standardized compression force curve C_l^n, and in a second step on all the data merged in matrix D_m^n in order to investigate the effect of the transformation on the merged and the separate data (Fig. 2).

The resulting PCA decompositions, showed that in the case of NIR reflection, three principal components explaining 84.79%, 9.67% and 4.83% of the total variance in the data, respectively, leading to a cumulative explained variance of more than 99%. Four principal components explained more than 80% of the total variances of the NIR transmission data and 95% of the compression force data. However, for Raman transmission, the first principal component alone explains 99.69% of the variance in the data. The first two principal components explain 98.51% and 1.01% of the variance in the Raman Reflection data, respectively.

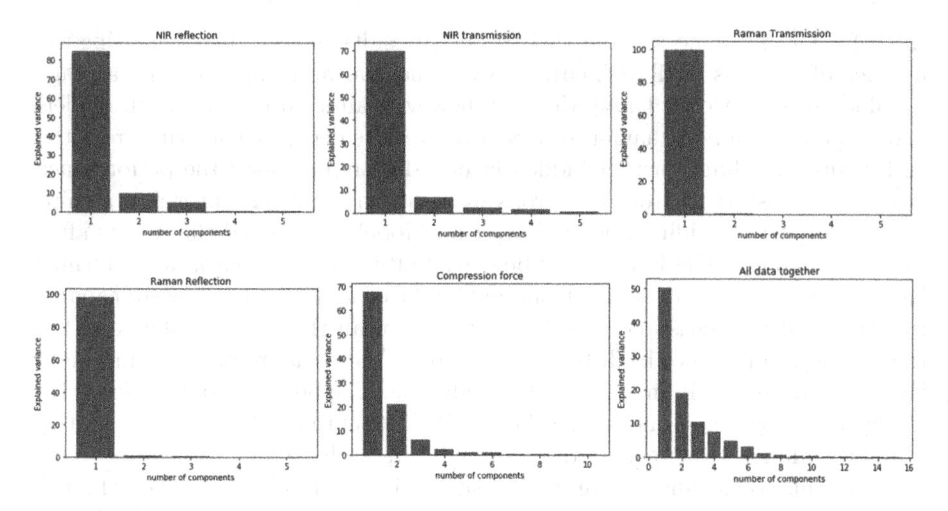

Fig. 2. Explained variance of spectral data, compression force, and all merged.

For matrix D_m^n, 7 principal components explain more than 95% of the variance and 33 explain more than 99% of the merged standardized data. These data resulting from the PCA decomposition were used as the inputs for the Artificial neural network models. Two different inputs were created for the ANN models. $Input_1$ was composed of the most important principal components of the different data. All components that explain less than 2% of the variance were eliminated, 12 principal components from the different data remained explaining a total average of 95% of the variance (Table 1). $Input_2$ maintain 99% of the variance in the merged data with 33 principal components.

Table 1. Inputs selected for the ANN models

Input name	Composition				
$Input_1$	NIR RE	NIR TR	Raman Tr	Raman Re	Compression
	3, 99.3%	3, 78.8%	1, 99.96%	1, 98.51%	4, 96.83%
$Input_2$	33 PCs, 99% of the variance of all data merged				

3.2 Predicting the Dissolution Profile by Artificial Neural Network

Two ANN models were created, Model 1 and Model 2 having $Input_1$ (12 components) and $Input_2$ (33 components) as inputs. For each setting, 100 trainings with randomly selected testing datasets were run. The average f_2 of the 100 trainings was recorded. The number of neurons was increased from one to 13 and results were recorded. A second layer and then a third layer having all the same number of neurons, were added in order to check the effect of deeper network on the behavior of the models and their results. Results showed that increasing the number of neurons until 10 neurons was beneficial and improved the average f_2 value on both Model 1 and Model 2. however, after 10 neurons both models maintained the same behavior and even the results decreased starting from the 12th neurons. Adding further hidden layers, slightly increased the performance of the models starting from 5 neurons when adding a second layer and a third layer on Model 1. Adding a second layer on Model 2 increased the results after the 5th neurons, while it was only beneficial after the 7th neuron for the third layer. Both models with at least one hidden layer and 7 neurons, were capable to predict all the measured dissolution profile of all the 49 test tablets within the acceptance range of both f_1 and f_2 factors. Adding more neurons and more layers was beneficial in improving the prediction. The best f_2 result achieved for Model 1 was by using one hidden layer with 10 neurons. However, the best f_2 result for Model 2 was achieved by three layers each having 10 neurons. This might be due to the higher number of features in the input used in Model 2 (33 features) compared to only 12 features in Model 1 inputs (Fig. 3 and Table 2).

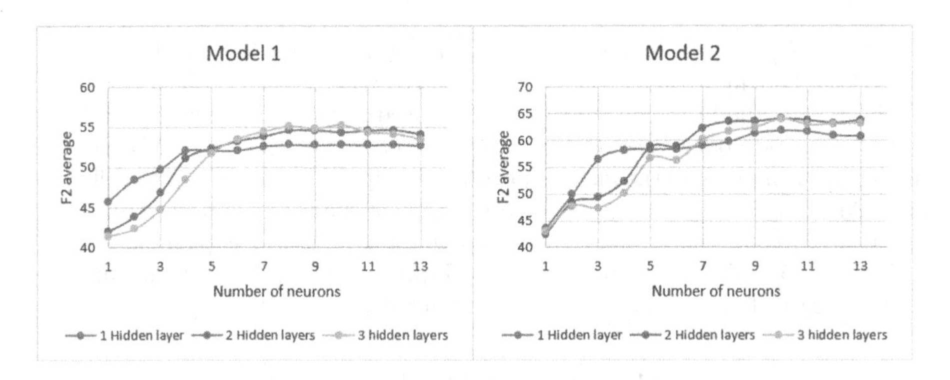

Fig. 3. Average f_2 values, using different layers and neuron numbers

Table 2. Results of best performing models

Model	f_2 value of the best ANN model	f_1 value of the best ANN model
Model 1	67.05	6.51
Model 2	70.35	5.29

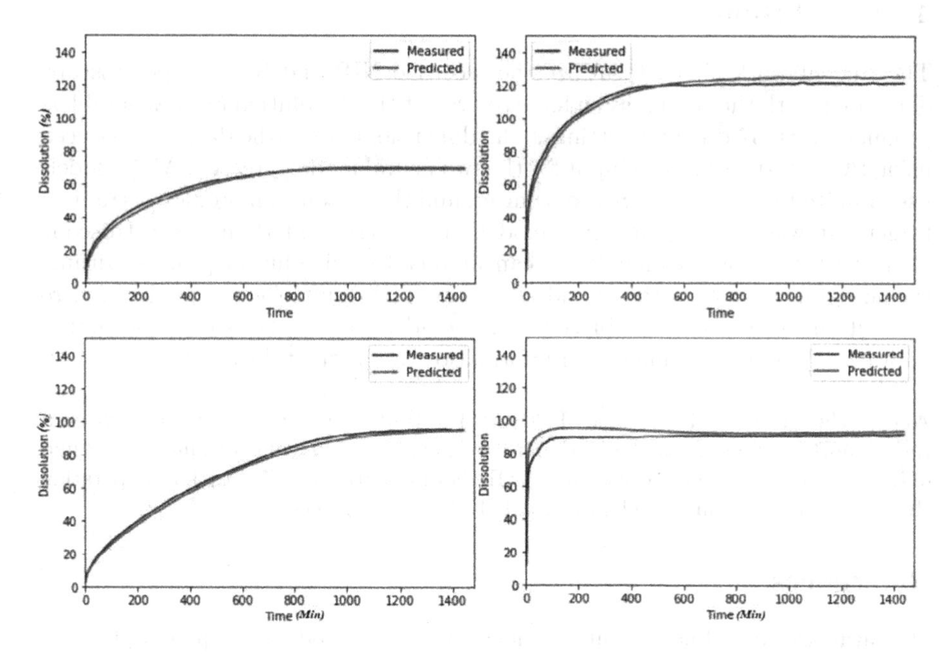

Fig. 4. Sample predicted dissolution profiles using Model 1

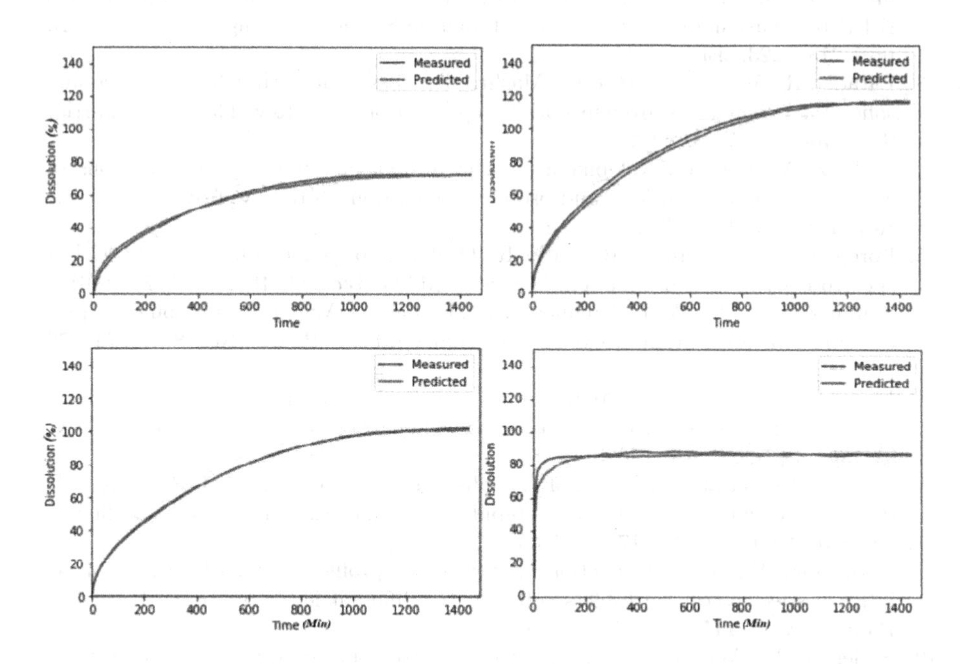

Fig. 5. Sample predicted dissolution profiles using Model 2

4 Conclusion

The current work aimed to utilize the recorded NIR and Raman spectroscopy data along with the compression force to predict the dissolution profiles of tablets produced with 37 different settings. The dimensionality of the data was reduced using PCA, then used as an input for the two neural models created. ANN models used Limited-memory BFGS as optimizer, and the dissolution profiles as training targets. It was found that ANN models using NIR and Raman spectroscopy along with the compression force, can predict the dissolution profiles withing the acceptance range of the f2 and f1 factors. The results show that the in vitro dissolution testing can be replaced by more advanced methods that use similar data providing a large amount of information about the tablets (Figs. 4 and 5).

Acknowledgments. Project no. FIEK_16-1-2016-0007 has been implemented with the support provided from the National Research, Development and Innovation Fund of Hungary, financed under the Centre for Higher Education and Industrial Cooperation Research infrastructure development (FIEK_16) funding scheme.

References

1. Lawrence, X.Y.: Pharmaceutical quality by design: product and process development, understanding, and control. Pharm. Res. **25**(4), 781–791 (2008)
2. Susto, G.A., McLoone, S.: Slow release drug dissolution profile prediction in pharmaceutical manufacturing: a multivariate and machine learning approach. In: 2015 IEEE International Conference on Automation Science and Engineering (CASE), pp. 1218–1223. IEEE (2015)
3. Patadia, R., Vora, C., Mittal, K., Mashru, R.: Dissolution criticality in developing solid oral formulations: from inception to perception. Crit. Rev. Ther. Drug Carrier Syst. **30**(6), 495–534 (2013)
4. Hédoux, A.: Recent developments in the Raman and infrared investigations of amorphous pharmaceuticals and protein formulations: a review. Adv. Drug Deliv. Rev. **100**, 133–146 (2016)
5. Porep, J.U., Kammerer, D.R., Carle, R.: On-line application of near infrared (NIR) spectroscopy in food production. Trends Food Sci. Technol. **46**(2), 211–230 (2015)
6. Zannikos, P.N., Li, W.-I., Drennen, J.K., Lodder, R.A.: Spectrophotometric prediction of the dissolution rate of carbamazepine tablets. Pharm. Res. **8**(8), 974–978 (1991)
7. Donoso, M., Ghaly, E.S.: Prediction of drug dissolution from tablets using near-infrared diffuse reflectance spectroscopy as a nondestructive method. Pharm. Dev. Technol. **9**(3), 247–263 (2005)
8. Freitas, M.P., et al.: Prediction of drug dissolution profiles from tablets using NIR diffuse reflectance spectroscopy: a rapid and nondestructive method. J. Pharm. Biomed. Anal. **39**(1–2), 17–21 (2005)
9. Hernandez, E., et al.: Prediction of dissolution profiles by non-destructive near infrared spectroscopy in tablets subjected to different levels of strain. J. Pharm. Biomed. Anal. **117**, 568–576 (2016)
10. Szaleniec, M., Witko, M., Tadeusiewicz, R., Goclon, J.: Application of artificial neural networks and DFT-based parameters for prediction of reaction Kinetics of ethylbenzene dehydrogenase. J. Comput. Aided Mol. Des. **20**(3), 145–157 (2006)

11. Drăgoi, E.N., Curteanu, S., Fissore, D.: On the use of artificial neural networks to monitor a pharmaceutical freeze-drying process. Drying Technol. **31**(1), 72–81 (2013)
12. Jouyban, A.G., Soltani, S., Asadpour, Z.K.: Solubility prediction of drugs in supercritical carbon dioxide using artificial neural network. Iranian J. Pharm. Res. (IJPR) **6**(4), 243 (2007)
13. Ebube, N.K., McCall, T., Chen, Y., Meyer, M.C.: Relating formulation variables to in vitro dissolution using an artificial neural network. Pharm. Dev. Technol. **2**(3), 225–232 (1997)
14. Galata, D.L., et al.: Fast, spectroscopy-based prediction of in vitro dissolution profile of extended release tablets using artificial neural networks. Pharmaceutics **11**(8), 400 (2019)
15. Moore, J., Flanner, H.: Mathematical comparison of dissolution profiles. Pharm. Technol. **20**(6), 64–74 (1996)

Possibilistic Classification Learning Based on Contrastive Loss in Learning Vector Quantizer Networks

Seyedfakhredin Musavishavazi, Marika Kaden, and Thomas Villmann(✉)

Saxon Institute for Computational Intelligence and Machine Learning (SICIM), University of Applied Sciences Mittweida, Mittweida, Germany
thomas.villmann@hs-mittweida.de

Abstract. Classification in a possibilistic scenario is a kind of multiple class assignments for data. One of the most prominent and interpretable classifier is the learning vector quantization (LVQ) realizing a nearest prototype classifier model. Figuring out the problem of classifying based on possibilistic or probabilistic class labels (assignments) leads to the use of likelihood ratio to organize a sustainable approach. To this end, we start with a special kind of probabilistic LVQ, known as Robust Soft LVQ, and propose a possibilistic extension to pave the way to our new method. Particularly, the proposed possibilistic variant takes positive and negative reasoning known from RSLVQ into account to secure a contrastive learning model in the end. In the paper we will explain the model and give the mathematical justification.

Keywords: Contrastive learning (CL) · Multiple classification · LVQ · RSLVQ · Interpretable models

1 Introduction

The challenge of secure classification is still subject of ongoing research, in particular if the classification learning has to deal with label noise, decision stability, etc. [2,4,5,15]. Recently, in [13] we proposed a mathematical framework which guarantees an optimum solution for trade-off between classification and rejection of a data point. In this method, we try to assign a crisp label given a probability class assignment, albeit if it is not rejected yet. But, in reality sometimes we face a situation that rejection of data as an option is absolutely off the table. In addition, instead of class probability assignment we have to deal with class possibility assignments. For example, in case of medical diagnosing patients can suffer from several diseases with high certainty. A probabilistic classification scheme would suggest respective diseases all with low probability which could lead to a classification reject because of ambiguous decision. Here, the importance of possibilistic classification emerges.

Hence, it is inevitable to deal with possibilities for classification and avoid the inherent uncertainty of possibilistic assignments. To clarify our concerns,

L. Rutkowski et al. (Eds.): ICAISC 2021, LNAI 12854, pp. 156–167, 2021.
https://doi.org/10.1007/978-3-030-87986-0_14

sometimes an event with high possibilities for several classes would lead to low probabilistic decisions, which may lead us to confusion. But, definitely the less the possibility of an event is the less is its probability [14,21,25].

Prototype based classifiers are known to be both interpretable and robust [2,17,24]. Among them, the family of learning vector quantizers (LVQ) plays a leading role [1,9,10,18]. A probabilistic LVQ based on cross-entropy loss was proposed in [22] whereas likelihood ratio loss was taken in robust soft LVQ (RSLVQ, [19]) to obtain a probabilistic class assignment.

In this paper we adopt the RSLVQ approach to obtain a possibilistic soft LVQ. For this purpose, the conventional log likelihood ratio in RSLVQ is replaced by a new ratio, defined based on possibilities. Finally, the resulting new ratio is able to investigate both positive and negative reasoning to establish a contrastive learning structure for a possibilistic labeling method which is not only interpretable but more comprehensive than some other common methods. Further, this positive and negative reasoning shows similarities to recently proposed probabilistic classifier learning scheme inspired by cognitive learning theory [16].

2 Problem and Model Description

The task is to classify a given data point $\boldsymbol{x} \in \mathbb{R}^n$ with a class assignment possibility $\boldsymbol{y}(\boldsymbol{x}) = (\boldsymbol{y}_1(\boldsymbol{x}), \ldots, \boldsymbol{y}_{N_c}(\boldsymbol{x}))$ where $\boldsymbol{y}_j(\boldsymbol{x}) \in [0,1]$ represents the possibility of assigning the class $j \in C = \{1, \ldots, N_c\}$ to the sample $\boldsymbol{x} \in \mathbb{R}^n$. For this purpose, we suppose a prototype-based scheme with a set of prototypes $\boldsymbol{W} = \{\boldsymbol{\omega}_i \in \mathbb{R}^n | i = 1, \ldots, N\}$ where each prototype $\boldsymbol{\omega}_i$ stands as a feature representative of a certain class such that at least one prototype is responsible for each class.

The starting point is to adopt the RSLVQ-model and modify it accordingly to handle multiple class assignments in a possibilistic setting adequately.

Remark 1. Our priority is to classify the sample data $\boldsymbol{x}$, regardless of its classification costs. Thus, $\boldsymbol{x}$ should be classified unless proven otherwise.

2.1 Likelihood Ratio and Predicting Labels – Preliminary Concepts and Notations

Let $\boldsymbol{P}(i|\boldsymbol{x})$ be the probability of choosing $i \in C$ as a label given the data point $\boldsymbol{x}$, which as a *posterior* to be predicted by the classifier. Due to the Bayes' theorem we have the relation

$$\boldsymbol{P}(i|\boldsymbol{x}) = \boldsymbol{P}(\boldsymbol{x}|i) \cdot \frac{p_i}{\boldsymbol{P}(\boldsymbol{x})} \tag{1}$$

to be valid with class priors p_i.

Additionally, we also consider the quantity

$$\boldsymbol{P}(\neg i|\boldsymbol{x}) = \sum_{j \neq i} \boldsymbol{P}(j|\boldsymbol{x})$$

as the probability of assigning any label to $\boldsymbol{x}$ except i. Both quantities are combined into the class dependent *likelihood ratio*:

$$\begin{aligned}\mathcal{L}_i(\boldsymbol{x}) &= \frac{\boldsymbol{P}(i|\boldsymbol{x})}{\boldsymbol{P}(\neg i|\boldsymbol{x})} \\ &= \frac{\boldsymbol{P}(i|\boldsymbol{x})}{\sum_{j\neq i} \boldsymbol{P}(j|\boldsymbol{x})} \\ &= \frac{\boldsymbol{P}(\boldsymbol{x}|i) \cdot \frac{p_i}{\boldsymbol{P}(\boldsymbol{x})}}{\sum_{j\neq i} \boldsymbol{P}(\boldsymbol{x}|j) \cdot \frac{p_j}{\boldsymbol{P}(\boldsymbol{x})}} \\ &= \frac{\boldsymbol{P}(\boldsymbol{x}|i) \cdot p_i}{\sum_{j\neq i} \boldsymbol{P}(\boldsymbol{x}|j) \cdot p_j},\end{aligned}$$

Next, we assume each class is represented by just one prototype, i.e. $N_c = N$ for simplicity. Further, if we suppose equal priors $p_i = \frac{1}{N}$ the likelihood ratio simplifies to

$$\mathcal{L}_i(\boldsymbol{x}) = \frac{\boldsymbol{P}(\boldsymbol{x}|i)}{\sum_{j\neq i} \boldsymbol{P}(\boldsymbol{x}|j)}. \tag{2}$$

which can be used as an alternative class predictor instead of the prediction function $\boldsymbol{P}(i|\boldsymbol{x})$ from (1). In case of, $p_i \neq \frac{1}{N}$ we get

$$\mathcal{L}_i(\boldsymbol{x}) = \frac{\boldsymbol{P}(\boldsymbol{x}, i)}{\sum_{j\neq i} \boldsymbol{P}(\boldsymbol{x}, j)} \tag{3}$$

for the likelihood ratio.

So far, we considered $\boldsymbol{P}(\boldsymbol{x}|i)$ as a probability density function. In other words, $0 \leq \boldsymbol{P}(\boldsymbol{x}|i) \leq 1$ and $\sum_i \boldsymbol{P}(\boldsymbol{x}|i) = 1$. Yet, we did not make use of the normalization condition.

Therefore, all considerations remain valid if we relax this normalization condition changing to possibility functions, i.e. allowing $\sum_i \boldsymbol{P}(\boldsymbol{x}|i) \neq 1$. In consequence, we can apply the same arguments for the *possibility* function and generalize the class dependent *likelihood ratio* $\mathcal{L}_i(\boldsymbol{x})$ to this scenario.

2.2 Likelihood Ratio Based Loss Function in Relation to RSLVQ – The Probabilistic Case

In this section we will introduce a local contrastive loss based on the class dependent likelihood ratios $\mathcal{L}_i(\boldsymbol{x})$ and relate this to the RSLVQ local loss

$$Loss_{RSLVQ}(\boldsymbol{x}) = \sum_i Loss_{RSLVQ}(i|\boldsymbol{x}) \tag{4}$$

with class dependent local losses

$$Loss_{RSLVQ}(i|\boldsymbol{x}) = \boldsymbol{y}_i(\boldsymbol{x}) \cdot \ln\left(\frac{\boldsymbol{P}(\boldsymbol{x}|i)}{\boldsymbol{P}(\boldsymbol{x}|i) + \sum_{j\neq i} \boldsymbol{P}(\boldsymbol{x}|j)}\right) \tag{5}$$

as defined in [19]. We will show that this local contrastive loss includes positive and negative reasoning, whereas the RSLVQ loss $Loss_{RSLVQ}(\boldsymbol{x})$ only includes positive reasoning.

Hence, in analogy to soft LVQ (SLVQ) as the initial setting for RSLVQ in [19], we introduce the class dependent loss

$$Loss(i|\boldsymbol{x}) = -\boldsymbol{y}_i(\boldsymbol{x}) \cdot \ln(\mathcal{L}_i(\boldsymbol{x})) \tag{6}$$

using the likelihood ratios $\mathcal{L}_i(\boldsymbol{x})$ and consider the local loss function $Loss(\boldsymbol{x}) = \sum_i Loss(i|\boldsymbol{x})$ as follows:

Without loss of generality, we assume both $\boldsymbol{y}_i(\boldsymbol{x})$ and $\boldsymbol{P}(\boldsymbol{x}|i)$, to be probability density functions $\forall i$ and prove that the $Loss(i|\boldsymbol{x})$ evaluates dissimilarities between them. For this purpose, we calculate

$$\begin{aligned}
Loss(i|\boldsymbol{x}) &= -\boldsymbol{y}_i(\boldsymbol{x}) \cdot \ln\left(\mathcal{L}_i(\boldsymbol{x})\right) \\
&= -\boldsymbol{y}_i(\boldsymbol{x}) \cdot \ln\left(\frac{\boldsymbol{P}(\boldsymbol{x}|i)}{\sum_{j\neq i} \boldsymbol{P}(\boldsymbol{x}|j)} \cdot \frac{\boldsymbol{y}_i(\boldsymbol{x})}{\boldsymbol{y}_i(\boldsymbol{x})}\right) \\
&= -\boldsymbol{y}_i(\boldsymbol{x}) \cdot \ln\left(\frac{\boldsymbol{P}(\boldsymbol{x}|i)}{\boldsymbol{y}_i(\boldsymbol{x})}\right) + \boldsymbol{y}_i(\boldsymbol{x}) \cdot \ln\left(\frac{\sum_{j\neq i} \boldsymbol{P}(\boldsymbol{x}|j)}{\boldsymbol{y}_i(\boldsymbol{x})}\right).
\end{aligned}$$

Summing up both sides of the equation $\forall i$ we obtain

$$\begin{aligned}
\sum_{i=1}^{N_c} Loss(i|\boldsymbol{x}) &= \sum_{i=1}^{N_c} -\boldsymbol{y}_i(\boldsymbol{x}) \cdot \ln\left(\frac{\boldsymbol{P}(\boldsymbol{x}|i)}{\boldsymbol{y}_i(\boldsymbol{x})}\right) + \boldsymbol{y}_i(\boldsymbol{x}) \cdot \ln\left(\frac{\sum_{j\neq i} \boldsymbol{P}(\boldsymbol{x}|j)}{\boldsymbol{y}_i(\boldsymbol{x})}\right) \\
&= \sum_{i=1}^{N_c} -\boldsymbol{y}_i(\boldsymbol{x}) \cdot \ln\left(\frac{\boldsymbol{P}(\boldsymbol{x}|i)}{\boldsymbol{y}_i(\boldsymbol{x})}\right) - \sum_{i=1}^{N_c} -\boldsymbol{y}_i(\boldsymbol{x}) \cdot \ln\left(\frac{\sum_{j\neq i} \boldsymbol{P}(\boldsymbol{x}|j)}{\boldsymbol{y}_i(\boldsymbol{x})}\right)
\end{aligned}$$

yielding

$$Loss(\boldsymbol{x}) = D_{KL}(\boldsymbol{y}(\boldsymbol{x})\|\boldsymbol{P}(\boldsymbol{x})) - D_{KL}(\boldsymbol{y}(\boldsymbol{x})\|\neg\boldsymbol{P}(\boldsymbol{x})) \tag{7}$$

as a local loss function with the Kullback-Leibler divergences $D_{KL}(\boldsymbol{y}(\boldsymbol{x})\|\boldsymbol{P}(\boldsymbol{x}))$ and $D_{KL}(\boldsymbol{y}(\boldsymbol{x})\|\neg\boldsymbol{P}(\boldsymbol{x}))$. In fact, both divergences can be taken as positive and negative reasoning functions, respectively. By means of positive and negative reasoning it is obvious that the local loss $Loss(\boldsymbol{x})$ is a *contrastive* learning function or contrastive loss.

In the next step we relate the local loss $Loss(\boldsymbol{x})$ to that of RSLVQ given in (4). For this purpose we consider the class dependent local loss

$$
\begin{aligned}
Loss(i|\boldsymbol{x}) &= -\boldsymbol{y}_i(\boldsymbol{x}) \cdot \ln\left(\mathcal{L}_i(\boldsymbol{x})\right) \\
&= -\boldsymbol{y}_i(\boldsymbol{x}) \cdot \ln\left(\frac{\boldsymbol{P}(\boldsymbol{x}|i)}{\sum_{j\neq i}\boldsymbol{P}(\boldsymbol{x}|j)} \cdot \frac{\boldsymbol{P}(\boldsymbol{x}|i)+\sum_{j\neq i}\boldsymbol{P}(\boldsymbol{x}|j)}{\boldsymbol{P}(\boldsymbol{x}|i)+\sum_{j\neq i}\boldsymbol{P}(\boldsymbol{x}|j)}\right) \\
&= -\boldsymbol{y}_i(\boldsymbol{x}) \cdot \ln\left(\frac{\boldsymbol{P}(\boldsymbol{x}|i)}{\boldsymbol{P}(\boldsymbol{x}|i)+\sum_{j\neq i}\boldsymbol{P}(\boldsymbol{x}|j)}\right) \\
&\quad + \boldsymbol{y}_i(\boldsymbol{x}) \cdot \ln\left(\frac{\sum_{j\neq i}\boldsymbol{P}(\boldsymbol{x}|j)}{\boldsymbol{P}(\boldsymbol{x}|i)+\sum_{j\neq i}\boldsymbol{P}(\boldsymbol{x}|j)}\right) \\
&= Loss_{RSLVQ}(i|\boldsymbol{x}) + \boldsymbol{y}_i(\boldsymbol{x}) \cdot \ln\left(\frac{\sum_{j\neq i}\boldsymbol{P}(\boldsymbol{x}|j)}{\boldsymbol{P}(\boldsymbol{x}|i)+\sum_{j\neq i}\boldsymbol{P}(\boldsymbol{x}|j)} \cdot \frac{\boldsymbol{y}_i(\boldsymbol{x})}{\boldsymbol{y}_i(\boldsymbol{x})}\right) \\
&= Loss_{RSLVQ}(i|\boldsymbol{x}) + \boldsymbol{y}_i(\boldsymbol{x}) \cdot \ln\left(\frac{\sum_{j\neq i}\boldsymbol{P}(\boldsymbol{x}|j)}{\boldsymbol{y}_i(\boldsymbol{x})}\right) \\
&\quad - \boldsymbol{y}_i(\boldsymbol{x}) \cdot \ln\left(\frac{\boldsymbol{P}(\boldsymbol{x}|i)+\sum_{j\neq i}\boldsymbol{P}(\boldsymbol{x}|j)}{\boldsymbol{y}_i(\boldsymbol{x})}\right) \\
&= Loss_{RSLVQ}(i|\boldsymbol{x}) + \boldsymbol{y}_i(\boldsymbol{x}) \cdot \ln\left(\frac{\sum_{j\neq i}\boldsymbol{P}(\boldsymbol{x}|j)}{\boldsymbol{y}_i(\boldsymbol{x})}\right) + \boldsymbol{y}_i(\boldsymbol{x}) \cdot \ln \boldsymbol{y}_i(\boldsymbol{x}) \\
&\quad - \boldsymbol{y}_i(\boldsymbol{x}) \cdot \ln\left(\boldsymbol{P}(\boldsymbol{x}|i) + \sum_{j\neq i}\boldsymbol{P}(\boldsymbol{x}|j)\right)
\end{aligned}
$$

whereby we used the equalities

- $\sum_i \boldsymbol{y}_i(\boldsymbol{x}) \cdot \ln \frac{\sum_{j\neq i}\boldsymbol{P}(\boldsymbol{x}|j)}{\boldsymbol{y}_i(\boldsymbol{x})} = -D_{KL}(\boldsymbol{y}(\boldsymbol{x})\|\neg\boldsymbol{P}(\boldsymbol{x}))$,
- $\sum_i \boldsymbol{y}_i(\boldsymbol{x}) \cdot \ln \boldsymbol{y}_i(\boldsymbol{x}) = H(\boldsymbol{y}(\boldsymbol{x}))$,
- $\ln\left(\boldsymbol{P}(\boldsymbol{x}|i) + \sum_{j\neq i}\boldsymbol{P}(\boldsymbol{x}|j)\right) = 0$,

as well as (4) to simplify. Hence, we achieve the local loss as

$$Loss(\boldsymbol{x}) = Loss_{RSLVQ}(\boldsymbol{x}) - D_{KL}(\boldsymbol{y}(\boldsymbol{x})\|\neg\boldsymbol{P}(\boldsymbol{x})) + H(\boldsymbol{y}(\boldsymbol{x})). \tag{8}$$

which relates to the RSLVQ loss according to (7)

$$Loss_{RSLVQ}(\boldsymbol{x}) = D_{KL}(\boldsymbol{y}(\boldsymbol{x})\|\boldsymbol{P}(\boldsymbol{x})) - H(\boldsymbol{y}(\boldsymbol{x})). \tag{9}$$

In other words, the RSLVQ loss $Loss_{RSLVQ}(\boldsymbol{x})$, compared to our $Loss(\boldsymbol{x})$, includes only positive reasoning.

2.3 Generalization to the Possibilistic Case

So far, we considered $\boldsymbol{y}_i(\boldsymbol{x})$ and $\boldsymbol{P}(\boldsymbol{x}|i)$, $\forall i$, as probability density functions. In following steps we show that the local loss $Loss(\boldsymbol{x})$ as a *contrastive* learning function, in general, holds its properties when the normalization condition for probability densities is dropped. Accordingly, now we suppose

- $\boldsymbol{y}_i(\boldsymbol{x}) \in [0,1]$ and $\sum_{i=1}^{N_c} \boldsymbol{y}_i(\boldsymbol{x}) = \mathcal{Y}$ for the class assignments
- $\boldsymbol{P}(\boldsymbol{x}|i) \in [0,1]$ and $\sum_{i=1}^{N_c} \boldsymbol{P}(\boldsymbol{x}|i) = \mathcal{P}$ for the conditional probabilities.

We emphasize that generally $\mathcal{Y}, \mathcal{P} \neq 1$ is valid. To emphasize this distinction we replace the notation $Loss(i|\boldsymbol{x})$ by $Loss^g(i|\boldsymbol{x})$ in the following. We consider and rewrite

$$
\begin{aligned}
Loss^g(i|\boldsymbol{x}) &= -\boldsymbol{y}_i(\boldsymbol{x}) \cdot \ln\big(\frac{\boldsymbol{P}(\boldsymbol{x}|i)}{\sum_{j\neq i} \boldsymbol{P}(\boldsymbol{x}|j)} \cdot \frac{\boldsymbol{y}_i(\boldsymbol{x})}{\boldsymbol{y}_i(\boldsymbol{x})}\big) \\
&= -\boldsymbol{y}_i(\boldsymbol{x}) \cdot \ln\big(\frac{\boldsymbol{P}(\boldsymbol{x}|i)}{\boldsymbol{y}_i(\boldsymbol{x})}\big) + \boldsymbol{y}_i(\boldsymbol{x}) \cdot \ln\big(\frac{\sum_{j\neq i} \boldsymbol{P}(\boldsymbol{x}|j)}{\boldsymbol{y}_i(\boldsymbol{x})}\big) \\
&= -\boldsymbol{y}_i(\boldsymbol{x}) \cdot \ln\big(\frac{\boldsymbol{P}(\boldsymbol{x}|i)}{\boldsymbol{y}_i(\boldsymbol{x})}\big) - (\boldsymbol{y}_i(\boldsymbol{x}) - \mathcal{P}) \\
&\quad - [-\boldsymbol{y}_i(\boldsymbol{x}) \cdot \ln\big(\frac{\sum_{j\neq i} \boldsymbol{P}(\boldsymbol{x}|j)}{\boldsymbol{y}_i(\boldsymbol{x})}\big) - (\boldsymbol{y}_i(\boldsymbol{x}) - \mathcal{P})] \\
&= -\boldsymbol{y}_i(\boldsymbol{x}) \cdot \ln\big(\frac{\boldsymbol{P}(\boldsymbol{x}|i)}{\boldsymbol{y}_i(\boldsymbol{x})}\big) - (\boldsymbol{y}_i(\boldsymbol{x}) - \boldsymbol{P}(\boldsymbol{x}|i)) \\
&\quad + \sum_{j\neq i} \boldsymbol{P}(\boldsymbol{x}|j) - [-\boldsymbol{y}_i(\boldsymbol{x}) \cdot \ln\big(\frac{\sum_{j\neq i} \boldsymbol{P}(\boldsymbol{x}|j)}{\boldsymbol{y}_i(\boldsymbol{x})}\big) \\
&\quad - (\boldsymbol{y}_i(\boldsymbol{x}) - \sum_{j\neq i} \boldsymbol{P}(\boldsymbol{x}|j))] - \boldsymbol{P}(\boldsymbol{x}|i) \\
&= [-\boldsymbol{y}_i(\boldsymbol{x}) \cdot \ln\big(\frac{\boldsymbol{P}(\boldsymbol{x}|i)}{\boldsymbol{y}_i(\boldsymbol{x})}\big) - (\boldsymbol{y}_i(\boldsymbol{x}) - \boldsymbol{P}(\boldsymbol{x}|i))] \\
&\quad + [-\boldsymbol{y}_i(\boldsymbol{x}) \cdot \ln\big(\frac{\sum_{j\neq i} \boldsymbol{P}(\boldsymbol{x}|j)}{\boldsymbol{y}_i(\boldsymbol{x})}\big) - (\boldsymbol{y}_i(\boldsymbol{x}) - \sum_{j\neq i} \boldsymbol{P}(\boldsymbol{x}|j))] \\
&\quad + [-\boldsymbol{P}(\boldsymbol{x}|i) + \sum_{j\neq i} \boldsymbol{P}(\boldsymbol{x}|j)]
\end{aligned}
$$

Using the relations

- $D_{GKL}(\boldsymbol{y}(\boldsymbol{x})\|\boldsymbol{P}(\boldsymbol{x})) = \sum_i -\boldsymbol{y}_i(\boldsymbol{x}) \cdot \ln\big(\frac{\boldsymbol{P}(\boldsymbol{x}|i)}{\boldsymbol{y}_i(\boldsymbol{x})}\big) - (\boldsymbol{y}_i(\boldsymbol{x}) - \boldsymbol{P}(\boldsymbol{x}|i))$ for the generalized Kullback-Leibler divergence for the possibilitic setting known from [3] and [23]
- $D_{GKL}(\boldsymbol{y}(\boldsymbol{x})\|\neg\boldsymbol{P}(\boldsymbol{x})) = \sum_i -\boldsymbol{y}_i(\boldsymbol{x}) \cdot \ln\big(\frac{\sum_{j\neq i} \boldsymbol{P}(\boldsymbol{x}|j)}{\boldsymbol{y}_i(\boldsymbol{x})}\big) - (\boldsymbol{y}_i(\boldsymbol{x}) - \sum_{j\neq i} \boldsymbol{P}(\boldsymbol{x}|j))$ accordingly and
- $-\boldsymbol{P}(\boldsymbol{x}|i) + \sum_{j\neq i} \boldsymbol{P}(\boldsymbol{x}|j) = -2 \cdot \boldsymbol{P}(\boldsymbol{x}|i) + \mathcal{P}$,

and summing up $Loss^g(i|\boldsymbol{x})$ over all possible values of $i = 1, \ldots, N$ we finally get

$$
Loss^g(\boldsymbol{x}) = D_{GKL}(\boldsymbol{y}(\boldsymbol{x})\|\boldsymbol{P}(\boldsymbol{x})) - D_{GKL}(\boldsymbol{y}(\boldsymbol{x})\|\neg\boldsymbol{P}(\boldsymbol{x})) + (N-2) \cdot \mathcal{P} \tag{10}
$$

as the contrastive loss in the possibilistic setting.

To investigate the relation between $Loss(\boldsymbol{x})$ and $Loss_{RSLVQ}(\boldsymbol{x})$, first we introduce the quantities $\mathcal{L}_i^{gRSLVQ}(\boldsymbol{x})$, which have to be deviating from the RSLVQ likelihood ratios

$$\mathcal{L}_i^{RSLVQ}(\boldsymbol{x}) = \frac{\boldsymbol{P}(\boldsymbol{x}|i)}{\boldsymbol{P}(\boldsymbol{x}|i) + \sum_{j \neq i} \boldsymbol{P}(\boldsymbol{x}|j)}$$

to reflect the possibilistic setting. Thus we define

$$\mathcal{L}_i^{gRSLVQ}(\boldsymbol{x}) = \frac{\boldsymbol{P}(\boldsymbol{x}|i)}{\mathcal{P}}, \tag{11}$$

where $\mathcal{P} = \sum_j \boldsymbol{P}(\boldsymbol{x}|j)$. Accordingly, we consider the class dependent local likelihood ratios

$$\mathcal{L}_i^{g}(\boldsymbol{x}) = \frac{\boldsymbol{P}(\boldsymbol{x}|i)}{\sum_{j \neq i} \boldsymbol{P}(\boldsymbol{x}|j)}$$

for the possibilistic case and relate them to the class dependent local losses $Loss^g(i|\boldsymbol{x})$. We calculate

$$\begin{aligned}
Loss^g(i|\boldsymbol{x}) &= -\boldsymbol{y}_i(\boldsymbol{x}) \cdot \ln \mathcal{L}_i^g(\boldsymbol{x}) \\
&= -\boldsymbol{y}_i(\boldsymbol{x}) \cdot \ln \left(\frac{\boldsymbol{P}(\boldsymbol{x}|i)}{\sum_{j \neq i} \boldsymbol{P}(\boldsymbol{x}|j)} \cdot \frac{\mathcal{P}}{\mathcal{P}}\right) \\
&= -\boldsymbol{y}_i(\boldsymbol{x}) \cdot \ln \left(\frac{\boldsymbol{P}(\boldsymbol{x}|i)}{\mathcal{P}}\right) + \boldsymbol{y}_i(\boldsymbol{x}) \cdot \ln \left(\frac{\sum_{j \neq i} \boldsymbol{P}(\boldsymbol{x}|j)}{\mathcal{P}}\right) \\
&= -\boldsymbol{y}_i(\boldsymbol{x}) \cdot \ln \mathcal{L}_i^{gRSLVQ}(\boldsymbol{x}) + \boldsymbol{y}_i(\boldsymbol{x}) \cdot \ln \left(\frac{\sum_{j \neq i} \boldsymbol{P}(\boldsymbol{x}|j)}{\mathcal{P}} \cdot \frac{\boldsymbol{y}_i(\boldsymbol{x})}{\boldsymbol{y}_i(\boldsymbol{x})}\right) \\
&= Loss^{gRSLVQ}(i|\boldsymbol{x}) + \boldsymbol{y}_i(\boldsymbol{x}) \cdot \ln \left(\frac{\sum_{j \neq i} \boldsymbol{P}(\boldsymbol{x}|j)}{\boldsymbol{y}_i(\boldsymbol{x})}\right) + \boldsymbol{y}_i(\boldsymbol{x}) \cdot \ln \left(\frac{\boldsymbol{y}_i(\boldsymbol{x})}{\mathcal{P}}\right) \\
&= Loss^{gRSLVQ}(i|\boldsymbol{x}) + \boldsymbol{y}_i(\boldsymbol{x}) \cdot \ln \left(\frac{\sum_{j \neq i} \boldsymbol{P}(\boldsymbol{x}|j)}{\boldsymbol{y}_i(\boldsymbol{x})}\right) + \Big(\boldsymbol{y}_i(\boldsymbol{x}) - \sum_{j \neq i} \boldsymbol{P}(\boldsymbol{x}|j)\Big) \\
&\quad - \Big(\boldsymbol{y}_i(\boldsymbol{x}) - \sum_{j \neq i} \boldsymbol{P}(\boldsymbol{x}|j)\Big) + \boldsymbol{y}_i(\boldsymbol{x}) \cdot \ln \left(\frac{\boldsymbol{y}_i(\boldsymbol{x})}{\mathcal{P}}\right),
\end{aligned}$$

and make use of the obvious relations

- $\sum_i Loss^{gRSLVQ}(i|\boldsymbol{x}) = Loss^{gRSLVQ}(\boldsymbol{x})$,
- $\sum_i \boldsymbol{y}_i(\boldsymbol{x}) \cdot \ln \left(\frac{\sum_{j \neq i} \boldsymbol{P}(\boldsymbol{x}|j)}{\boldsymbol{y}_i(\boldsymbol{x})}\right) + \left(\boldsymbol{y}_i(\boldsymbol{x}) - \sum_{j \neq i} \boldsymbol{P}(\boldsymbol{x}|j)\right) = -D_{KL}(\boldsymbol{y}(\boldsymbol{x}) \| \neg \boldsymbol{P}(\boldsymbol{x}))$,
- $\sum_i \boldsymbol{y}_i(\boldsymbol{x}) \cdot \ln \boldsymbol{y}_i(\boldsymbol{x}) = H^g(\boldsymbol{y}(\boldsymbol{x}))$,

to derive

$$Loss^g(\boldsymbol{x}) = Loss^{gRSLVQ}(\boldsymbol{x}) - D_{KL}(\boldsymbol{y}(\boldsymbol{x}) \| \neg \boldsymbol{P}(\boldsymbol{x})) + H^g(\boldsymbol{y}(\boldsymbol{x})) - \mathcal{Y} + (N-1) \cdot \mathcal{P} - \mathcal{Y} \cdot \ln \mathcal{P},$$

as the desired relation. Moreover, using the equality (10) we have

$$Loss^{gRSLVQ}(\boldsymbol{x}) = D_{GKL}(\boldsymbol{y}(\boldsymbol{x}) \| \boldsymbol{P}(\boldsymbol{x})) - H^g(\boldsymbol{y}(\boldsymbol{x})) - \mathcal{P} + \mathcal{Y} \cdot \ln \mathcal{P} \tag{12}$$

in terms of the generalized Kullback-Leibler divergence. As we can see, this modified local loss $Loss^{gRSLVQ}(\boldsymbol{x})$ for the possibilistic setting is like in the probabilistic case for RSLVQ, i.e. it only takes the positive reasoning into account.

2.4 Discussion on Class Possibility Assignment

In the beginning, we mentioned that taking $\boldsymbol{y}_i(\boldsymbol{x})$ as a class possibility assignment has some disadvantages. To tackle this problem, we introduce the entropic function

$$f(\boldsymbol{y}_i(\boldsymbol{x})) = -[\boldsymbol{y}_i(\boldsymbol{x}) \ln \boldsymbol{y}_i(\boldsymbol{x}) + \left(1 - \boldsymbol{y}_i(\boldsymbol{x})\right) \ln \left(1 - \boldsymbol{y}_i(\boldsymbol{x})\right)], \tag{13}$$

which is inspired by Shannon entropy [20]. This function holds the following properties:

- for $\boldsymbol{y}_i(\boldsymbol{x}) \in (0, \frac{1}{2})$: if $\boldsymbol{y}_i(\boldsymbol{x}) \to \frac{1}{2}$ then $f(\boldsymbol{y}_i(\boldsymbol{x})) \to 1$, i.e. the uncertainty *increases*,
- $f(\frac{1}{2}) = 1$, i.e. at the vertex we have the maximum uncertainty,
- for $\boldsymbol{y}_i(\boldsymbol{x}) \in (\frac{1}{2}, 1)$: if $\boldsymbol{y}_i(\boldsymbol{x}) \to 1$ then $f(\boldsymbol{y}_i(\boldsymbol{x})) \to 0$, i.e. the uncertainty *decreases*.

which can be easily verified.

Using this entropic function, the local losses $Loss^g(i|\boldsymbol{x})$ are obtained as

$$Loss^g(i|\boldsymbol{x}) = -f(\boldsymbol{y}_i(\boldsymbol{x})) \cdot \ln(\mathcal{L}_i(\boldsymbol{x})). \tag{14}$$

Comparing this equation with the SLVQ local losses in (6), we can interpret the entropic function $f(\boldsymbol{y}_i(\boldsymbol{x}))$ as a measure of *fuzziness*. Yet, compared to $\boldsymbol{y}_i(\boldsymbol{x})$, it is a better choice because it takes both factors $\boldsymbol{y}_i(\boldsymbol{x})$ and $1 - \boldsymbol{y}_i(\boldsymbol{x})$ into account and resembles the functionality of $\mathcal{L}_i(\boldsymbol{x})$.

Moreover, in practice it does not have any drastic changes in compare to the corresponding $\boldsymbol{y}_i(\boldsymbol{x})$ but it has the drawback not to sum up to one, i.e. $\sum_i f(\boldsymbol{y}_i(\boldsymbol{x})) \neq 1$. To overcome this difficulty we apply the softmax transformation known from multi-layer perceptrons to achieve a probabilistic output [6,7]. Thus, we replace the entropic function $f(\boldsymbol{y}_i(\boldsymbol{x}))$ by following *softmax* function:

$$S(f(\boldsymbol{y}_i(\boldsymbol{x}))) = \frac{\exp\left(f(\boldsymbol{y}_i(\boldsymbol{x}))\right)}{\sum_j \exp\left(f(\boldsymbol{y}_j(\boldsymbol{x}))\right)}, \tag{15}$$

such that

$$Loss^g(i|\boldsymbol{x}) = -S(f(\boldsymbol{y}_i(\boldsymbol{x}))) \cdot \ln(\mathcal{L}_i(\boldsymbol{x})) \tag{16}$$

is obtained.

2.5 Specifications for Practical Use

Up to this point, we have not specified the class conditional probabilities $\boldsymbol{P}(\boldsymbol{x}|i)$ for $\boldsymbol{x}$ to be needed for calculation of $\boldsymbol{P}(i|\boldsymbol{x})$ according to the Bayes' relation (1). Taking the idea of stochastic neighbor embedding [8] we determine the probabilities as $\boldsymbol{P}(\boldsymbol{x}|i)$, i.e. they are defined based on proximity between the point $\boldsymbol{x}$ and any prototype assigned to label i. Accordingly, we propose $\boldsymbol{P}(\boldsymbol{x}|i) = \exp\left(-\lambda \cdot d^2(\boldsymbol{x}, \boldsymbol{\omega}_i)\right)$. We emphasize that $\boldsymbol{P}(\boldsymbol{x}|i) \to 0$ is valid, if $\lambda \to \infty$ and

$\boldsymbol{P}(\boldsymbol{x}|i) \to 1$, for the values of λ close enough to zero, i.e. it can act as a possibility function.

Further, we should remember that $\boldsymbol{P}(\boldsymbol{x}|i)$ is not a *probability* function. But, in the special case $\lambda = \frac{1}{2\cdot\sigma^2}$ it acts as a standard normal distribution function, $\mathcal{N}(\boldsymbol{x}; 0, \lambda)$. To clarify our claim we consider

$$\begin{aligned}
\mathcal{L}_i^g(\boldsymbol{x}) &= \frac{\boldsymbol{P}(\boldsymbol{x}|i)}{\sum_{j\neq i} \boldsymbol{P}(\boldsymbol{x}|j)} \\
&= \frac{\exp\big(-\lambda \cdot d^2(\boldsymbol{x}, \boldsymbol{\omega}_i)\big)}{\sum_{j\neq i} \exp\big(-\lambda \cdot d^2(\boldsymbol{x}, \boldsymbol{\omega}_j)\big)} \\
&= \frac{\exp\big(-\frac{1}{2\cdot\sigma^2} \cdot d^2(\boldsymbol{x}, \boldsymbol{\omega}_i)\big)}{\sum_{j\neq i} \exp\big(-\frac{1}{2\cdot\sigma^2} \cdot d^2(\boldsymbol{x}, \boldsymbol{\omega}_j)\big)} \cdot \frac{\sqrt{2\cdot\pi}\cdot\sigma}{\sqrt{2\cdot\pi}\cdot\sigma} \\
&\equiv \frac{\mathcal{N}_i(\boldsymbol{x}; 0, \lambda)}{\sum_{j\neq i} \mathcal{N}_j(\boldsymbol{x}; 0, \lambda)} \\
&= \mathcal{L}_i(\boldsymbol{x}).
\end{aligned}$$

which demonstrates the appropriate choice.

3 Cost Function and Learning Step

For a machine learning model approach we consider the overall loss

$$L = \sum_{\boldsymbol{x}} Loss(\boldsymbol{x})$$

such that stochastic gradient descent learning can be applied for minimizing L using the gradients of the local losses $Loss(\boldsymbol{x})$. In our model, the parameters to be adjusted during learning are the prototype vectors $\boldsymbol{\omega}_i(t)$. Hence, we get

$$\boldsymbol{\omega}_i(t+1) = \boldsymbol{\omega}_i(t) + \alpha(t) \cdot \frac{\partial Loss(\boldsymbol{x})}{\partial \boldsymbol{\omega}_i} \tag{17}$$

as the learning update for the prototype $\boldsymbol{\omega}_i(t)$ given a randomly chosen training sample pair $(\boldsymbol{x}, \boldsymbol{y}(\boldsymbol{x}))$. The learning rate is $0 < \alpha(t) \ll 1$ typically slowly decreasing during time.

Since, the entropic function $f(\boldsymbol{y}_i(\boldsymbol{x}))$ and its *softmax* version $S(f(\boldsymbol{y}_i(\boldsymbol{x})))$ do not depend on $\boldsymbol{\omega}_i$ we can conclude that the proportionalities

$$\frac{\partial Loss(\boldsymbol{x})}{\partial \boldsymbol{\omega}_j} \propto \frac{\partial Loss(i|\boldsymbol{x})}{\partial \boldsymbol{\omega}_j} \propto \frac{\partial\big(\mathcal{L}_i(\boldsymbol{x})\big)}{\partial \boldsymbol{\omega}_j},$$

hold. Therefore, we can take an arbitrary choice and, hence, we continue with the exponential-based $\mathcal{L}_i(\boldsymbol{x})$

$$\ln \mathcal{L}_i(\boldsymbol{x}) = -\lambda \cdot d^2(\boldsymbol{x}, \boldsymbol{\omega}_i) - \ln\left(\sum_{j\neq i} \exp\left(-\lambda \cdot d^2(\boldsymbol{x}, \boldsymbol{\omega}_j)\right)\right). \tag{18}$$

for convenience. Since

$$\frac{\partial \ln \left(\sum_{j \neq i} \exp \left(-\lambda \cdot d^2(\boldsymbol{x}, \boldsymbol{\omega}_j) \right) \right)}{\partial \boldsymbol{\omega}_i} = 0$$

is valid, which holds due to its independence of $\boldsymbol{\omega}_i$, we get

$$\begin{aligned} \frac{\partial \ln \mathcal{L}_i(\boldsymbol{x})}{\partial \boldsymbol{\omega}_i} &= \frac{\partial \big(-\lambda \cdot d^2(\boldsymbol{x}, \boldsymbol{\omega}_i) \big)}{\partial \boldsymbol{\omega}_i} \\ &= -2 \cdot \lambda \cdot d(\boldsymbol{x}, \boldsymbol{\omega}_i) \cdot \frac{\partial \big(d(\boldsymbol{x}, \boldsymbol{\omega}_i) \big)}{\partial \boldsymbol{\omega}_i}, \end{aligned}$$

whereas for $\frac{\partial \ln \mathcal{L}_i(\boldsymbol{x})}{\partial \boldsymbol{\omega}_k}$ with $k \neq i$

$$\begin{aligned} \frac{\partial \ln \mathcal{L}_i(\boldsymbol{x})}{\partial \boldsymbol{\omega}_k} &= \frac{\partial \big(-\ln \sum_{j: j \neq i} \exp \left(-\lambda \cdot d^2(\boldsymbol{x}, \boldsymbol{\omega}_j) \right) \big)}{\partial \boldsymbol{\omega}_k} \\ &= -\frac{-2 \cdot \lambda \cdot d(\boldsymbol{x}, \boldsymbol{\omega}_k) \cdot \exp \big(-\lambda \cdot d^2(\boldsymbol{x}, \boldsymbol{\omega}_k) \big)}{\sum_j \exp \big(-\lambda \cdot d^2(\boldsymbol{x}, \boldsymbol{\omega}_j) \big)} \cdot \frac{\partial \big(d(\boldsymbol{x}, \boldsymbol{\omega}_k) \big)}{\partial \boldsymbol{\omega}_k} \\ &= 2 \cdot \lambda \cdot S \left(\exp \big(-\lambda \cdot d^2(\boldsymbol{x}, \boldsymbol{\omega}_k) \big) \right) \cdot d(\boldsymbol{x}, \boldsymbol{\omega}_k) \cdot \frac{\partial \big(d(\boldsymbol{x}, \boldsymbol{\omega}_k) \big)}{\partial \boldsymbol{\omega}_k} \end{aligned}$$

is obtained.

A careful initial choice of λ together with an appropriate decreasing scheme supports good convergence behavior of the learning process [11,12].

4 Conclusions

In this paper we propose a machine learning approach for possibilistic classification learning such that fuzzy class assignments are predicted, which do not sum up to one. Thus it is a kind of gradual multiple class learning. The approach is based on prototype learning paradigm known to be interpretable and robust.

The mathematical framework is motivated and explained in detail. Future work is to realize the model computationally and show successful numerical simulations for illustrative and real world problems.

Acknowledgement. M.K. was supported by a grant of the European Social Fund (ESF).

References

1. Biehl, M., Hammer, B., Villmann, T.: Prototype-based models in machine learning. Wiley Interdiscip. Rev. Cogn. Sci. **2**, 92–111 (2016)

2. Chen, C., Li, O., Tao, D., Barnett, A., Rudin, C., Su, J.: This looks like that: deep learning for interpretable image recognition. In: Wallach, H., Larochelle, H., Beygelzimer, A., d'Alché-Buc, F., Fox, E., Garnett, R. (eds.) Advances in Neural Information Processing Systems, vol. 32, pp. 8930–8941. Curran Associates, Inc. (2019). https://proceedings.neurips.cc/paper/2019/file/adf7ee2dcf142b0e11888e72b43fcb75-Paper.pdf
3. Cichocki, A., Zdunek, R., Phan, A., Amari, S.I.: Nonnegative Matrix and Tensor Factorizations. Wiley, Chichester (2009)
4. Devarakota, P., Mirbach, B., Ottersten, B.: Confidence estimation in classification decision: a method for detecting unseen patterns. In: International Conference on Advances Pattern Recognition (ICaPR), pp. 136–140 (2007)
5. Frénay, B., Verleysen, M.: Classification in the presence of label noise: a survey. IEEE Trans. Neural Netw. Learn. Syst. **25**(5), 845–869 (2014)
6. Goodfellow, I., Bengio, Y., Courville, A.: Deep Learning. MIT Press, Cambridge (2016)
7. Haykin, S.: Neural Networks - A Comprehensive Foundation. IEEE Press, New York (1994)
8. Hinton, G., Roweis, S.: Stochastic neighbor embedding. In: Advances in Neural Information Processing Systems, vol. 15, pp. 833–840. The MIT Press, Cambridge (2002)
9. Kohonen, T.: Learning vector quantization. Neural Netw. **1**(Supplement 1), 303 (1988)
10. Kohonen, T.: Self-Organizing Maps. SSINF, vol. 30. Springer, Heidelberg (1995). https://doi.org/10.1007/978-3-642-97610-0. (Second Extended Edition 1997)
11. Lange, M., Zühlke, D., Holz, O., Villmann, T.: Applications of l_p-norms and their smooth approximations for gradient based learning vector quantization. In: Verleysen, M. (ed.) Proceedings of European Symposium on Artificial Neural Networks, Computational Intelligence and Machine Learning (ESANN 2014), pp. 271–276. i6doc.com, Louvain-La-Neuve, Belgium (2014)
12. Montavon, G., Orr, G.B., Müller, K.-R. (eds.): Neural Networks: Tricks of the Trade. LNCS, vol. 7700. Springer, Heidelberg (2012). https://doi.org/10.1007/978-3-642-35289-8
13. Musavishavazi, S., Mohannazadeh Bakhtiari, M., Villmann, T.: A mathematical model for optimum error-reject trade-off for learning of secure classification models in the presence of label noise during training. In: Rutkowski, L., Scherer, R., Korytkowski, M., Pedrycz, W., Tadeusiewicz, R., Zurada, J.M. (eds.) ICAISC 2020. LNCS (LNAI), vol. 12415, pp. 547–554. Springer, Cham (2020). https://doi.org/10.1007/978-3-030-61401-0_51
14. Provost, F., Fawcett, T.: Robust classification for imprecise environments. Mach. Learn. **42**, 203–231 (2001)
15. Ravinchandran, J., Kaden, M., Saralajew, S., Villmann, T.: Variants of DropConnect in learning vector quantization networks for evaluation of classification stability. Neurocomputing **403**, 121–132 (2020). https://doi.org/10.1016/j.neucom.2019.12.131
16. Saralajew, S., Holdijk, L., Rees, M., Asan, E., Villmann, T.: Classification-by-components: probabilistic modeling of reasoning over a set of components. In: Proceedings of the 33rd Conference on Neural Information Processing Systems (NeurIPS 2019), pp. 2788–2799. MIT Press (2019)
17. Saralajew, S., Holdijk, L., Villmann, T.: Fast adversarial robustness certification of nearest prototype classifiers for arbitrary seminorms. In: Proceedings of the 34th

Conference on Neural Information Processing Systems (NeurIPS 2020). MIT Press (2020, in press)

18. Sato, A., Yamada, K.: Generalized learning vector quantization. In: Touretzky, D.S., Mozer, M.C., Hasselmo, M.E. (eds.) Advances in Neural Information Processing Systems 8. Proceedings of the 1995 Conference, pp. 423–9. MIT Press, Cambridge (1996)
19. Seo, S., Obermayer, K.: Soft learning vector quantization. Neural Comput. **15**(7), 1589–1604 (2003)
20. Shannon, C.: A mathematical theory of communication. Bell Syst. Tech. J. **27**, 379–432 (1948)
21. Villmann, A., Kaden, M., Saralajew, S., Hermann, W., Biehl, M., Villmann, T.: Reliable patient classification in case of uncertain class labels using a cross-entropy approach. In: Verleysen, M. (ed.) Proceedings of the 26th European Symposium on Artificial Neural Networks, Computational Intelligence and Machine Learning (ESANN 2018), Bruges (Belgium), pp. 153–158. i6doc.com, Louvain-La-Neuve, Belgium (2018)
22. Villmann, A., Kaden, M., Saralajew, S., Villmann, T.: Probabilistic learning vector quantization with cross-entropy for probabilistic class assignments in classification learning. In: Rutkowski, L., Scherer, R., Korytkowski, M., Pedrycz, W., Tadeusiewicz, R., Zurada, J.M. (eds.) ICAISC 2018. LNCS (LNAI), vol. 10841, pp. 724–735. Springer, Cham (2018). https://doi.org/10.1007/978-3-319-91253-0_67
23. Villmann, T., Haase, S.: Divergence based vector quantization. Neural Comput. **23**(5), 1343–1392 (2011)
24. Villmann, T., Saralajew, S., Villmann, A., Kaden, M.: Learning vector quantization methods for interpretable classification learning and multilayer networks. In: Sabourin, C., Merelo, J., Barranco, A., Madani, K., Warwick, K. (eds.) Proceedings of the 10th International Joint Conference on Computational Intelligence (IJCCI), Sevilla, pp. 15–21. SCITEPRESS - Science and Technology Publications, Lda., Lissabon (2018). ISBN: 978-989-758-327-8
25. Zadeh, L.: Fuzzy Sets as a Basis for a Theory of Possibility. North Holland Publishing Company, Amsterdam (1978)

Convolutional Autoencoder Based Textile Defect Detection Under Unconstrained Setting

Deepak Nagaraj(✉), Pramod Vadiraja, Oliver Nalbach, and Dirk Werth

AWS-Institute for Digitized Products and Processes, Saarbrücken, Germany
deepak.nagaraj@aws-institut.de

Abstract. Automated visual defect detection on textile products under unconstrained setting is a much sought-after, and at the same time a challenging problem. In general, textile products are structurally complex and highly varied in design, which makes the development of a generalized approach using conventional image processing methods impossible. Deep supervised machine learning models have been very successful on similar problems but cannot be applied in this use-case due to lack of annotated data. This paper demonstrates a novel automated approach which still leverages on the ability of deep learning models to capture complex features on the textured and colored fabric, but in an unsupervised manner. Specifically, deep autoencoders are applied to capture the complex features, which are further processed by image processing techniques like thresholding and blob detection, subsequently leading to detection of defects in the images.

Keywords: Fabric defect detection · Unsupervised learning · Dimensionality reduction · Autoencoder

1 Introduction

Defects in textile industry can occur during various phases of products' lifecycle. Such defects have been identified into more than 70 different kinds by the textile industry [1]. They can occur either because of faulty raw-material and/or because of malfunctioning of the machines involved in the textile production. Accordingly, various kinds of defects can be attributed to type of raw-material, product-type and the production facility. Quality control process in textile industry is very essential as the price of a defect ridden product can be reduced by 45% to 65% [2]. The process is repetitive and costs unreasonable duration per item if done manually. Certain improvements have been made in recent years by capturing the previous "lessons learnt" in terms of machine learning models and utilizing the learnt models to predict and correspondingly to avoid quality problems. One such technique is predictive quality assurance (PreQA) [3], where the idea is to learn the relationship between product features starting from design until manufacturing and the corresponding quality related problems. The information (machine learning model) thus learnt is later used in the product development phase to predict possible quality related problems, and if necessary, to suggest design alternatives to overcome the problems.

L. Rutkowski et al. (Eds.): ICAISC 2021, LNAI 12854, pp. 168–181, 2021.
https://doi.org/10.1007/978-3-030-87986-0_15

As explained in [3], a PreQA system gathers and integrates information about the product from both structured and unstructured data. The structured data can be for example from enterprise resource planning (ERP) software, bill of materials (BOM) and quality check logs. Unstructured data here is in the form of images of defective products, taken either by the manufacturer during quality check process, or directly by the customers during returning. Conventionally the visual defect detection has been carried out manually and offline with many inherent restrictions in efficiency and efficacy because of carelessness, optical illusion etc. Given that the textile products can be visually and structurally widely varying ranging from plain fabric to multi-colored and multi-textured textiles, it is challenging to automate visual defect detection in textile industry. In addition, as the sources of images can be varying, the required automated defect detection technique must be robust with respect to view of the product, lighting conditions and zoom level of the image. In this regard, an automated fabric defect detection system which functions in unconstrained environment is of high importance for the efficacy of the PreQA system. With respect to analysis on unstructured data, the article [4] demonstrates a novel image processing pipeline to obtain structured information from the images of defective product and accordingly to enrich the solution obtained using structured data. However, the pipeline lacked a method for automated detection of abnormalities in the image. This article demonstrates a novel deep learning-based framework for automated anomaly detection in images of textile products.

Given the infeasibility of composing a labeled dataset covering all possible types of defects, we opt for an unsupervised approach based on convolutional neural network (CNN)-based autoencoders (CAE) which is able to detect visual defects without imposing restrictions on how an image has been captured. The remainder of this article is organized as follows: Sect. 2 reviews previous studies on fabric defect detection, followed by a brief introduction to relevant machine learning algorithms (Sect. 3). Section 4 presents the proposed framework and explains its individual components. Section 5 evaluates the framework with respect to different criteria. Finally, Sect. 6 concludes the article with summary and future work.

2 Previous Work

In this section, we briefly review previous work applied in related problem scenarios. We focus on both conventional image processing and machine learning-based methods.

2.1 Image Processing Techniques

The field of automated defect detection has seen abundant research mainly focusing on machine vision based techniques. Core image processing techniques have been historically widely studied and applied for anomaly detection in fabrics. [5] proposed Scale-Invariant features (SIFT) between equivalent images, which are being continuously used in image processing techniques quite often. The studies [6] and [7] have demonstrated the application of SIFT features for defect detection in textiles and steel surfaces respectively. Image processing techniques that propose usage of filters are also widely studied for this application. One such commonly used filter for texture analysis is the Gabor filter used for analyzing the presence of various frequency contents in specific directions in the image. [8] proposed a technique to automatically choose an optimal Gabor filter for a given application from a bank of filters - the scales and orientations of which are preset. This optimal filter is then used for isolating the defect also in textured fabrics. However, the preset parameters in the bank of filters may not be well generalized for diversified and outlier input fabrics. A similar study [9] proposed a strategy that involves usage of Gabor Filter banks along with Kernelized-Principal Component Analysis (KPCA) [10] for feature reduction followed by Median Filtering and thresholding using OTSU's [11] algorithm. However, as the parameters for the Gabor filter banks are chosen by experience, they might not be representative of all possible unseen defects in the unconstrained environment. [12] proposed a pipeline that can be briefly compiled into three steps; Contrast enhancement, that enhances contrast between defective regions and textured backgrounds in the fabrics; Pattern extraction, that extracts periodic patterns in the fabric, texture and median filtering of the considered windowed regions; and finally cluster the filtered regions to isolate the defects. While this kind of framework expects the textile to be plain or has regular patterns, a fabric or an apparel might not always be plain or have periodic patterns in them (especially if the framework is supposed to function in an unconstrained environment). Hence, the suggested pipelined steps might not perform well for fabrics that have non-uniform regions in them.

2.2 Machine Learning Techniques

The studies discussed so far attempt to detect defects by employing or developing techniques from the conventional image processing field. Although sometimes they are effective, the techniques pose difficulties while considering fabrics that were obtained under a non-controlled environment. Along with this problem, it is also challenging to determine the optimal parameters for processing the images through experiences, as the fashion industry is dynamic and widely varying. Hence for techniques that involve presetting parameters (e.g.: for Filters, Thresholds) as the first step of processing, certain expertise is needed to determine optimal parameters in each case to obtain relatively better results.

In this respect, learning based methods which learn from real-world datasets and therefore generalize the solution, gain prominence. Because of non-availability of annotated data, supervised learning methods are ruled out for the task of defect detection in fabrics. Unsupervised ML methods which do not anticipate labelling of the data are relevant for this task. The Principal Component Analysis (PCA) [13] is one such technique that can be leveraged for defect detection. PCA, is a kind of dimensionality reduction method, which reduces data points from a higher space to a lower space while retaining the major portion of variance in the data. This is done by performing eigen decomposition of the covariance matrix of the multivariate dataspace, followed by using the top-k eigenvectors to transform the data points from the original space to the reduced space. [14] suggests a technique to leverage PCA to isolate defects. Initially PCA is used to model the training images that are non-defective, these are portions of the image that do not contain the defects. Post this, PCA is once again used to model the test image that contains the defect. The PCA model of the test image is projected onto the train images' model and a score matrix is constructed that acts as a mask for the test image to isolate the defects. However, this approach requires pre-definition of the defective and non-defective versions of the same image/fabric which might not always be possible.

Deep Learning Techniques. With recent advancements in the computer hardware and corresponding computation speed, deep learning is gaining more importance in computer vision. The deep learning approaches are relatively more promising as they capture complex non-linear relationships between the input and output variables of a system, that might not have been possible before. In this regard, autoencoders are deep neural network based dimensionality reduction method; they compress the given data and consequently try to reconstruct the original representation of data from the compressed representation. More detail about autoencoder can be found in Sect. 3. [15] utilized a Convolutional Autoencoder (CAE) with an inception-block like layer to identify images that deviates from the rest of the lot. They pretrain the CAE on a set of images, which makes the process of anomaly detection restricted to the dataset and fails to generalize to the real-world defective images. [16] proposed an approach for defect segmentation using autoencoders that uses the Structural Similarity [17] metric for computing the per-pixel reconstruction error. The approach was tested only on grayscale versions of non-woven fabric textures and a dataset of nanofibrous materials. Similar to their approach, our approach leverages on the CAE, and additionally on the image processing techniques like thresholding algorithms and blob detections, which make it relatively robust against uncertainties posed by unconstrained environment.

3 Background

Autoencoders are artificial neural networks that attempt to compress and reconstruct input data in an unsupervised manner. As such, the aim of an autoencoder is to learn a representation of the input dataset (from original space) in a reduced dimensional space. By doing so, it generally ignores the extremities and noise in the dataset. As shown in Fig. 1, it has two functional parts, namely the encoder and the decoder. The encoder encodes/projects the input data onto a lower-dimensional space called latent space. The decoder tries to generate a representation as close as possible to input data from the reduced encoding.

Given a dataset of images represented as $\mathbb{R}^{h \times w \times k}$, the encoder function maps from the original image (x) to the latent space, $\varphi: \mathbb{R}^{h \times w \times k} \to \mathbb{R}^{j}$ and the decoder function tries to generate a representation ($\hat{x}$) as close as possible to original image form the latent space vector, $\psi: \mathbb{R}^{j} \to \mathbb{R}^{h \times w \times k}$; where h, w, k are height, width and color specification (channels) respectively and j is the dimensionality of the latent space. The number of channels k can be different depending on the color space, we have chosen RGB scheme, accordingly an image has three ($k = 3$) channels: red, green, and blue. Generally, $j \ll h \times w \times k$, this facilitates the encoder to extract most essential and meaningful features from the input images followed by accurate reconstruction by the decoder. The non-essential part, which the autoencoder neglects (fails to capture) is expected to be an anomaly in the given image. The complete process of compression and consequent reconstruction can be given as: $\hat{x} = \psi(\varphi(x)) = \psi(z)$; where z is the latent space vector.

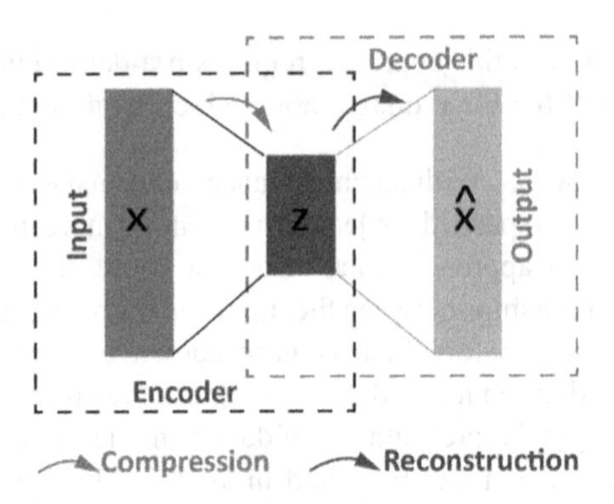

Fig. 1. Autoencoder with basic components

Both encoder and decoder functions can be built using different kinds of neural network architectures depending on the application. In general, they can be either multi-layer perceptron (MLP) [18], or convolutional neural network (CNN) [19] or recurrent neural network (RNN) [20]. This paper discusses the application of MLP and CNN as encoder/decoder components for anomaly detection. An MLP is a classical feed-forward neural network which consists of at least three layers of nodes: an input layer, a hidden layer and an output layer. Except for the nodes in input layer, each node is a neuron that gets inputs from all the neurons of the previous layer, processes the weighted sum of these inputs using nonlinear activation function, and sends this function value to all the neurons of the next layer. This fully-connectedness of MLP make them prone to overfitting to the data. Also, the number of trainable parameters increases drastically with an increase in the size of the image.

CNNs are a class of neural networks, typically applied to analyzing visual data. A CNN typically consists of three types of layers: Convolutional layers, pooling layers and fully connected layers. Convolutional layers compile the presence of structural features of a given image. They systematically apply different kinds of filters on to the input image to extract feature maps corresponding to the filter. These layers stacked in a deep model capture features hierarchically, starting from simpler features like lines in the layers close to the input to complex features like objects and shapes in the deeper layers. The feature maps thus obtained would be very sensitive to the position of the features in the input image. Pooling layers, which effectively down-sample these feature maps, are responsible for making the CNNs more robust to any change of position (translation or rotation) of any feature in the input images. Generally, one pooling layer will be placed after each convolutional layer, which effectively reduces the feature map generated by convolutional layer by a factor of 'p', i.e. a feature map of size $n * n$ (n^2 pixels) will be reduced to $n/p * n/p$ (n^2/p^2 pixels). It preserves the spatial structure of the images using small squares of input data called kernels. Thus, it can identify the object, its location, as well as its relationship with other objects in an image, accurately. Because of these qualities CNNs are proven to be very useful in computer vision applications.

4 Automated Textile Defect Detection

The input to our approach is an image of a textile product taken in an uncontrolled setting. We do not make any assumptions on the particular view of the product (e.g., full product, part of the product or close-up view), the color and texture (light, dark, strips, checkered, complex texture, etc.), the type of product (shirt, jeans, jacket, etc.) or the specific lighting conditions. The output of our approach consists of one or multiple regions in the image, described by bounding boxes, in which an irregularity which usually corresponds to a defect is suspected. To keep the method simple, we assume that the object (i.e. piece of clothing) of interest has already been to be additionally given a foreground mask for the input image, i.e., that it is known which pixels correspond to the object of interest and which are background pixels. Unsupervised foreground-background segmentation for images of textile products has been demonstrated in the literature [4]. More specifically, for all images presented in this paper, the publicly available software Remove.bg [21] was used to perform the background removal without any manual input. Furthermore, we resize all input images to be at most 512 pixels in either dimension, while keeping the original aspect ratio.

The computational flow of our method is depicted in Fig. 2. The method consists of two major steps: training of an autoencoder network to represent (and reconstruct) patches in the original image (Fig. 2, top row) and a detection of connected regions with high per-patch reconstruction errors (Fig. 2, bottom). In the following, we will describe both computational steps in more detail.

More specifically, we compare two neural network architectures for the autoencoder: one based on basic hidden layers, such that the encoder and decoder part of the autoencoder correspond to multilayer perceptron (MLP), and a second one based on convolutional and deconvolutional layers in the encoder and decoder parts, respectively. Both architectures are depicted briefly in Fig. 3. As it can be seen, the first hidden layer of encoder part of MLP-based autoencoder (MAE) has 128 activations units (functions) and decreasing thereafter in steps of half of the previous, the decoder part is symmetric

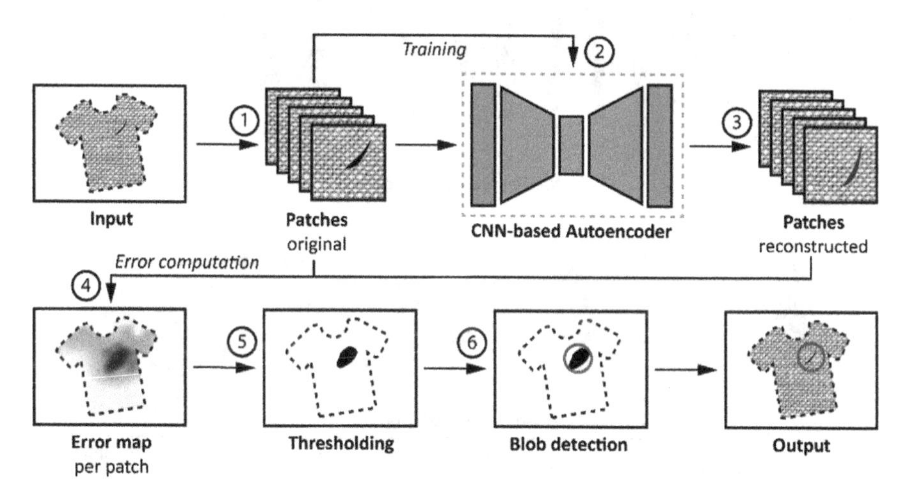

Fig. 2. Computational flow of the proposed defect detection method on textile products

to the encoder part. In CAE, doubling the number of kernels 'k' in successive layers leads to down-sampling (pooling) of the image in the encoder, followed by up-sampling of the image in the decoder part. Most notably, the MAE treats the input patches as $3 \cdot n \cdot n$ - dimensional vectors (3 being number of color channels and 'n' is the patch size), and the spatial ordering of pixels and presence of different color channels has no effect on neural links in the network. Whereas the CAE treats inputs as tensors of size 3 by n by n and is explicitly sensitive to the spatial order by only considering spatially nearby pixels or features, respectively, in its computations.

The optimal model architectures found empirically are detailed in Fig. 3, with the following specifics pertaining to CAE:

- Encoder contains blocks of convolution, activation (PReLU [22]) and max-pooling.
- Convolutions of 3×3 are applied at all the convolution layers and max-pooling is applied after convolution layer.
- Number of convolution kernels/output features is set to f for the first block and is doubled with each block. We use $f = 2$.
- Decoder part is symmetric with respect to Encoder in terms of output features, only difference is that max-pool is replaced by up-sampling.

Dimensionality reduction using either autoencoder network corresponds to training the autoencoder on the set of extracted input patches. As our method is intended to be fully automated, the training process must be automated, too. To this end, we train both network architectures using the Adam optimizer [23] which autonomously adjusts the parameter learning rates in the training process to achieve stable learning and fast convergence. The training process is automatically terminated using an early stopping criterion which checks if the relative improvement of the reconstruction error of the autoencoder on the validation part of the dataset, measured using the mean squared error

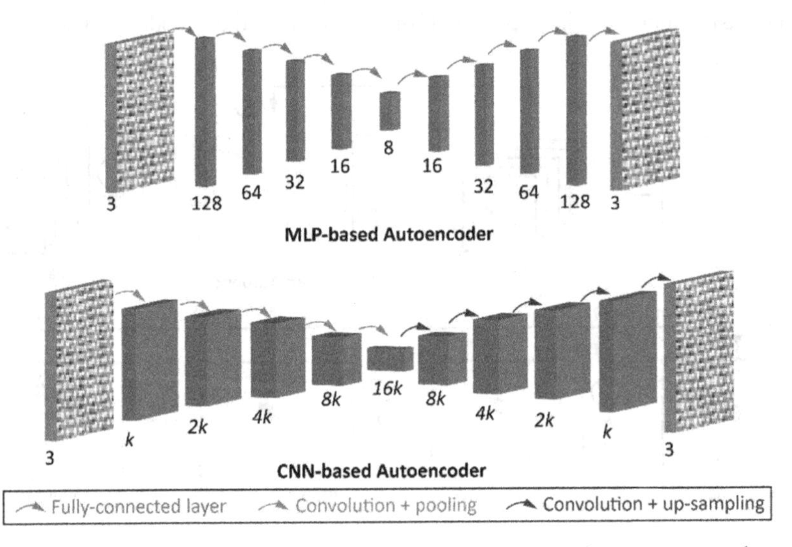

Fig. 3. The two autoencoder architectures used in our experiments. compared

(MSE), was larger than 2% in the last epoch of training. While the required number of epochs to reach this point can vary between input images in theory, in practice it was experienced that training was automatically terminated after around three epochs for most input images. After training the autoencoder, we process all input patches by the network to obtain the reconstructions corresponding to their dimensionally reduced representations (Fig. 2, step 3).

4.1 Dimensionality Reduction Using Deep Learning

The actual anomaly detection step is applied to a grayscale image in which the value of each pixel is equal to the mean squared error between the original patch centered around it and that patch's reconstruction (Fig. 2, step 4). Figure 4 shows several examples in the second column. For easier interpretation, a color map has been applied to the error maps. Intuitively, regions of higher reconstruction error, which consequently appear brighter in Fig. 4, are not captured as well in the autoencoder's learnt representation which is usually due to the fact that they correspond to outliers compared to other image patches.

To automatically detect and localize these regions in the error map, we apply a thresholding (Fig. 2, step 5; Fig. 4, third col.) to binarize the image and differentiate between low and high error pixels. For the thresholding, we use Yen's method [24] which autonomously selects a suitable threshold based on the histogram of the error map. We apply a slight Gaussian blur to the error map before applying the thresholding to reduce the effects of noise. In the final step, we apply a simple blob detection method which identifies connected regions in the thresholded error map (Fig. 2, step 6). We use the implementation provided by the image processing library OpenCV [25]. The method filters out regions/blobs which are either very small or large (area less than 300 or more than 5000 pixels) or very elongate. The output of the blob detection is a list of key-points defined by a center and radius (cf. Fig. 4) indicating the location and estimated scale of the anomalies in our input images.

Extensions of the approach could perform an additional classification step in which detected anomalies, which can be localized using the key-points, are classified as expected anomalies (e.g. singular features of a piece of clothing like a single button visible on an image of a jeans) and actual defects, respectively. This kind of classification is already being considered in the past work [4].

4.2 Comparison of MLP and CNN-Based Autoencoders

This subsection compares the efficacy of MAE and CAE architectures (Fig. 3) applied to 3 different images. We employed generic architectures for both the types. The encoder and decoder parts are symmetric. Figure 4 compares the output of the individual computational steps for the two types of neural network architectures we experimented with. More specifically, except of the step of error map production (Fig. 2, step 2; Fig. 4, first column), rest of the pipeline stays same for the two cases. Clearly CAE performs much better than MAE. As explained in Subsect. 4.1, convolutions pertaining to CAE focus computations on local neighborhood, leading to structures and relationship between objects being captured better. As seen, in exemplar 1 of Fig. 4, unlike CAE, MAE fails to associate the region near the collar to its surroundings. The error map in exemplar

2 from CAE shows clear distinction of defect region from the normal region. Despite better performance, due to the facts outlined in Sect. 3, CAE has much lesser trainable parameters compared to MAE; CAE has 380,707 parameters, while MAE has 812,320 parameters.

5 Evaluation

As expected, and as proved by the previous section, CAE performed much better than MAE. Therefore, we take CAE as our neural network and evaluate the complete automated pipeline with respect to diverse set of input images depicting defective or flawed textile products. The dataset includes images showing different kinds of flaws (e.g., holes, stains, seam defects), different types of products (shirts, jeans, and other fabrics), of different color and texture as well as seen from varying views and in varying lighting conditions. Most of the images were taken from user reviews on shopping websites. To increase the variety, we however also manually added defects to few images of non-defective pieces of clothing by means of photo manipulation.

5.1 Qualitative Analysis

Figure 5 shows the final detection results for different common types of product flaws. For each type of flaw, three examples for drastically different situations (according to the properties specified in the preceding paragraph) are given. The approach has shown satisfactory performance for a variety of input image types. The model successfully differentiates between the local textures on the fabric and the actual defect, which is particularly evident from the defect type "Stain" and "Far" view.

For the example corresponding to "Stain" and "Close" view although multiple abnormalities like pen, brand-tag and wrinkles are present, the defective region is clearly highlighted, this can be attributed to the highly contrasting defect region. The approach has worked decently also for defects which are needle-shape. This can be seen in the "Seam/stitching defect" row (especially in the "Far" view). The approach also proved to be effective in capturing and isolating defects for detailed views of inputs (even with the presence of textures in the input), which can be seen in the "Detail" column of our findings.

Along with evaluating the error-map resulted from the proposed CAE, the usage of image processing techniques like the thresholding algorithms and the blob detection proved to be effective in isolating multiple defect locations from the same piece of fabric. This can be seen in the "Miscellaneous" row's "Far" column where multiple defects were isolated which otherwise might have been difficult to be captured. With the application of Blob detection technique, we were able to pin-point the exact defect location better or provide us with a potential defective region. The defect is thus parameterized by the circle's (shown red in the figure) center and its radius. This kind of automated parameterization further improves the efficacy of the PreQA approach.

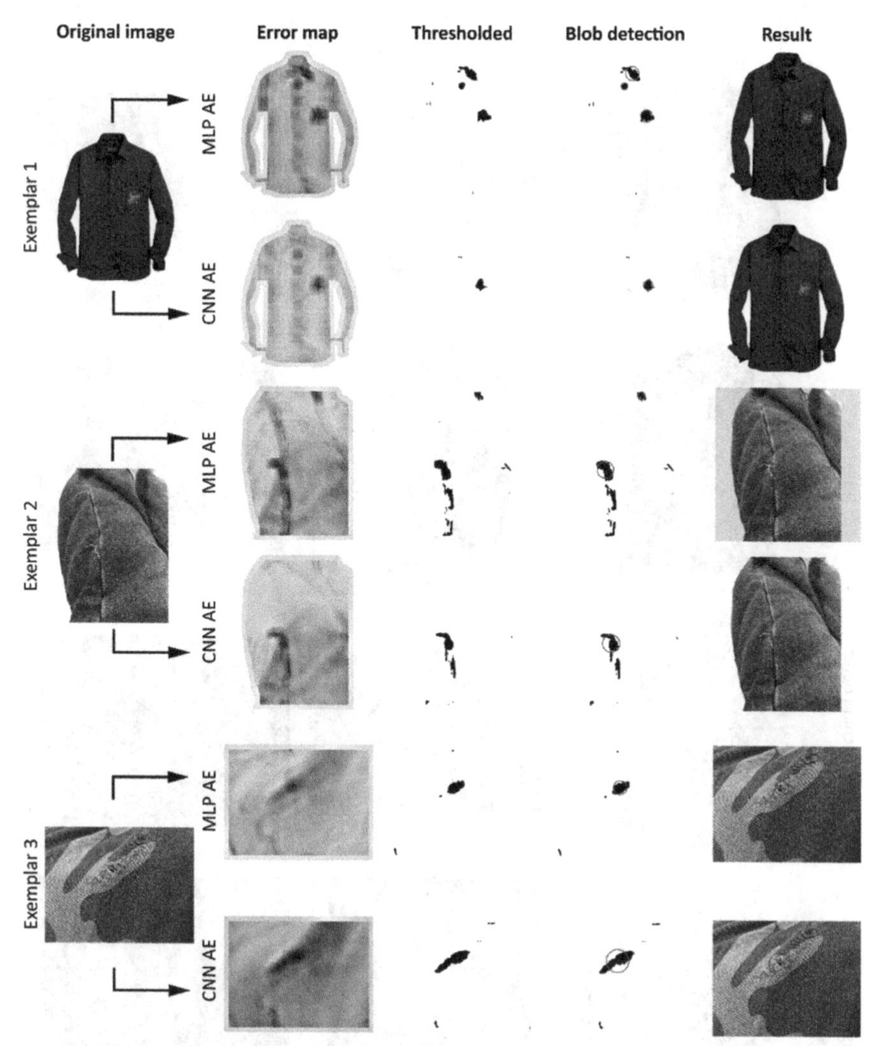

Fig. 4. Comparison of MAE and CAE for anomaly detection on images of defective textile products. The MAE based approach resulting in similarly looking heatmaps at first glance, however the CAE can capture the distribution of patches better, which can be seen from the error maps in which patterns like that of exemplar 1 or the camouflage design of exemplar 3 are hardly visible anymore. For MLPs, more structures of the original image remain visible in the heatmap, leading to less precise results. The gray borders in the error maps highlight regions in which it was not possible to form patches without background pixels. These regions do not contribute to the training set of the autoencoder.

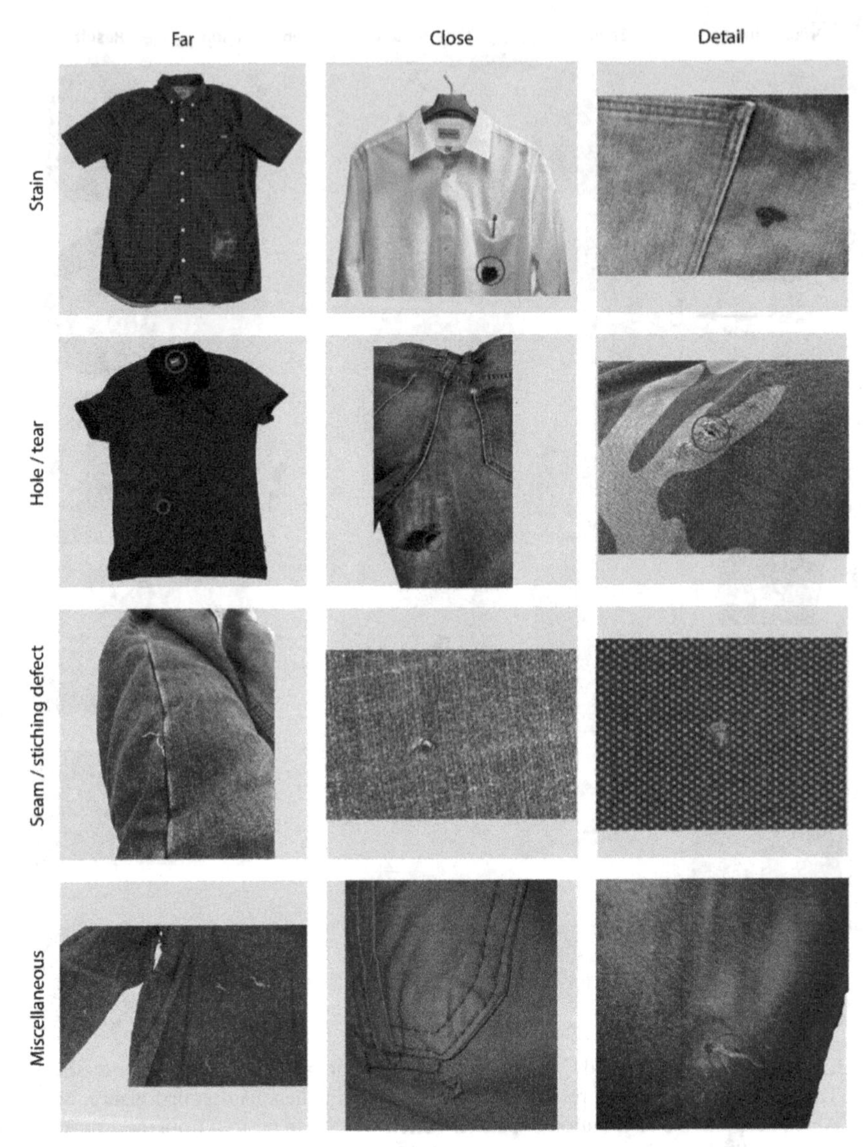

Fig. 5. Final detection results using the CAE for various types of flaws and products. The red circles visualize the center and radius of the identified blobs. (Color figure online)

5.2 Limitations

The approach demonstrated in this paper relies on anomaly detection to identify likely defective areas in images of diverse textile products taken in uncontrolled settings. However, on its own merits, it cannot judge whether an identified anomaly corresponds to a defect or not. For example, for the image in the second row and first column of Fig. 5, which shows a polo shirt in full, not only the actual defect but also the brand-tag is identified by the method because it, too, is an irregularity with respect to the rest of the image and therefore has a high reconstruction error by the autoencoder.

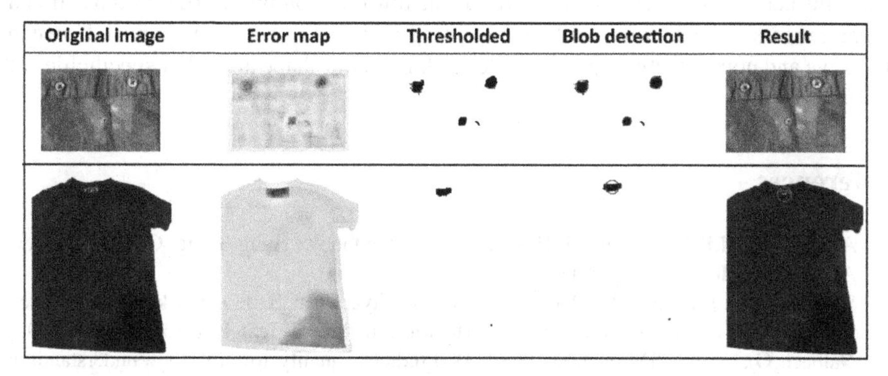

Fig. 6. Not every anomaly also corresponds to a defect.

Two more such examples are depicted in Fig. 6. As depicted in the first row, the actual defect, a small hole has been identified by the method. However, as the image only shows part of the object and only two buttons are visible in the images, these, too, have been considered irregular compared to the rest of the image. On the other hand, as shown in the second row, the brand-tag is identified as an abnormality, but the actual defect (here, color-fade) is ignored. One can argue that it might be difficult for even naked eye to detect this defect. If we take closer look into the corresponding "Error map", although the defective region has been captured, but the CAE reconstruction error is highest near brand-tag, the second is the wet region and finally the defective region. While, in general, defects like color variations or small tears are often also used as design elements which makes the problem of defect detection impossible to solve with full certainty without additional contextual information, combining our anomaly detection approach with a classification method [4] could enable a distinction between true defects and false positives in most cases like the one from Fig. 6.

6 Conclusion and Future Work

This paper demonstrates a general method that can spot anomalies in images of textile products. A deep CAE is leveraged in an unsupervised manner, together with image processing techniques, for the detection of anomalies under mostly unconstrained setting. Because of this fact, the method can be directly applied to any image of textile products without any preprocessing. The CAE performed much better compared to MAE.

Evaluation of CAE based pipeline yielded in decent results for various kinds of defects, products and views. While the approach is relatively robust against unconstrained setting, it has certain inherent drawbacks as well. Mainly, the distinction between defective and non-defective anomalies is not well considered, accordingly the method described in this paper is not sufficient by itself.

However, when combined with the complete approach as described in [4], this distinction can be feasible. It would also result in localization as well as classification of anomalies on any image of a textile products. Common to any other machine learning technique, the performance of our CAE model could be further optimized empirically. Also, further research need to be carried out towards combining other conventional image processing techniques and deep learning methods to better distinguish between defective and non-defective anomalies depending on the shape and other specificities of the anomalies.

References

1. Association H.K.C.S, Council H.K.P: Textile handbook. Hong Kong Cotton Spinners Association, Hong Kong (2001)
2. Srinivasan, K., Dastoor, P.H., Radhakrishnaiah, P., Jayaraman, S.: FDAS: a knowledge-based framework for analysis of defects in woven textile structures. J. Text. Inst. **83**, 431–448 (1992)
3. Nalbach, O., Linn, C., Derouet, M., Werth, D.: Predictive quality: towards a new understanding of quality assurance using machine learning tools. In: Abramowicz, W., Paschke, A. (eds.) Business Information Systems. BIS 2018. Lecture Notes in Business Information Processing, vol. 320. Springer, Cham (2018). https://doi.org/10.1007/978-3-319-93931-5_3
4. Nalbach, O., Derouet, M., Swamy, A.K.E., Werth, D.: Leveraging Unstructured Image Data for Product Quality Improvement. Wirtschaftsinformatik 2019 (2019)
5. Low, D.G.: Distinctive image features from scale-invariant keypoints. Int. J. Comput. Vis. **60**, 91–110 (2004)
6. Zhao, X., Zhu, X., Xue, T., LIU, K., Deng, Y.: SIFT-based textile defects detection by using adaptive neural-fuzzy inference system (2018)
7. Suvdaa, B., Ahn, J.-H., Ko, J.-G.: Steel Surface Defects Detection and Classification Using SIFT and Voting Strategy (2012)
8. Kumar, A., Pang, G.K.H.: Defect detection in textured materials using Gabor filters. IEEE Trans. Ind. Appl. **38**, 425–440 (2002)
9. Jing, J., Fan, X., Li, P.: Automated fabric defect detection based on multiple Gabor filters and KPCA. Int. J. Multimed. Ubiquitous Eng. **11**, 93–106 (2016)
10. Mika, S., Schölkopf, B., Smola, A., Müller, K.R., Scholz, M., Rätsch, G.: Kernel PCA and de-noising in feature spaces. Adv. Neural Inf. Process. Syst. **11**, 536–542 (1999)
11. Smith, P., Reid, D.B., Environment, C., Palo, L., Alto, P., Smith, P.L.: Otsu_1979_otsu_method. IEEE Trans. Syst. Man. Cybern. C **100**, 62–66 (1979)
12. Hamdi, A.A., Sayed, M.S., Fouad, M.M., Hadhoud, M.M.: Unsupervised patterned fabric defect detection using texture filtering and K-means clustering. In: Proceedings of the 2018 International Conference on Innovative Trends in Computer Engineering ITCE 2018. 2018-March, pp. 130–144 (2018)
13. Jolliffe, I.T.: Principal Component Analysis and Factor Analysis. In: Principal Component Analysis, pp. 115–128. Springer New York, New York, NY (1986). https://doi.org/10.1007/978-1-4757-1904-8_8

14. Shanbhag, P.M.: Fabric defect detection using principal component analysis. **2**, 2863–2867 (2013)
15. Sarafijanovic-Djukic, N., Davis, J.: Fast distance-based anomaly detection in images using an inception-like autoencoder. In: Novak, P.K., Šmuc, T., Džeroski, S. (eds.) Discovery Science. DS 2019. Lecture Notes in Computer Science, vol. 11828. Springer, Cham (2019). https://doi.org/10.1007/978-3-030-33778-0_37
16. Bergmann, P., Löwe, S., Fauser, M., Sattlegger, D., Steger, C.: Improving unsupervised defect segmentation by applying structural similarity to autoencoders. In: VISIGRAPP 2019 - Proceedings of the14th International Joint Conference on Computer Vision, Imaging and Computer Graphics Theory and Applications, vol. 5, pp. 372–380 (2019)
17. Wang, Z., Bovik, A.C., Sheikh, H.R., Simoncelli, E.P.: Image quality assessment: from error visibility to structural similarity. IEEE Trans. Image Process. **13**, 600–612 (2004)
18. Hu, G., Huang, J., Wang, Q., Li, J., Xu, Z., Huang, X.: Unsupervised fabric defect detection based on a deep convolutional generative adversarial network. Text. Res. J. **90**, 247–270 (2020)
19. Mahdizadehaghdam, S., Panahi, A., Krim, H.: Sparse generative adversarial network. In: Proceedings - 2019 International Conference on Computer Vision Workshop ICCVW 2019, pp. 3063–3071 (2019)
20. Werbos, P.: Backpropagation through time: what it does and how to do it. Proc. IEEE. **78**, 1550–1560 (1990)
21. Remove.bg: Background removal tool. https://www.remove.bg/
22. He, K., Zhang, X., Ren, S., Sun, J.: Delving deep into rectifiers: Surpassing human-level performance on ImageNet classification. In: Proceedings of the IEEE International Conference on Computer Vision 2015 Inter, pp. 1026–1034 (2015)
23. Kingma, D.P., Ba, J.L.: Adam: a method for stochastic optimization. In: 3rd International Conference on Learning Representations ICLR 2015 - Conference Track Proceedings, pp. 1–15 (2015)
24. Yen, J.-C., Chang, F.-J., Chang, S.: A new criterion for automatic multilevel thresholding. IEEE Trans. Image Process. **4**, 370–378 (1995)
25. Bradski, G.: The OpenCV Library. Dr Dobb's J. Softw. Tools Prof. Program. **25**(11), 120–123 (2000)

Clustering-Based Adaptive Self-Organizing Map

Dominik Olszewski(✉)

Faculty of Electrical Engineering, Warsaw University of Technology, Warsaw, Poland
dominik.olszewski@ee.pw.edu.pl

Abstract. We propose an improvement of the Self-Organizing Map (SOM). In our version of SOM, the neighborhood widths of the Best Matching Units (BMUs) are computed on the basis of the data density and scattering in the input data space. The density and scattering are expressed by the values of the inner-cluster variances, which are obtained after the preliminary input data clustering. The experiments conducted on the two real datasets evaluated the proposed approach on the basis of a comparison with the three reference data visualization methods. By reporting the superiority of our technique over the other tested algorithms, we confirmed the effectiveness and accuracy of the introduced solution.

Keywords: Self-Organizing Map · Adaptive self-organizing map · clustering · Neighborhood width · Gaussian kernel · Visualization

1 Introduction

The Self-Organizing Map (SOM) [3] is a type of an artificial neural network architecture, however, at the same time, it may be recognized as a data visualization technique. The term data visualization refers in our research to a linear or non-linear projection from an original input high-dimensional space onto a resulting output 2- or 3-dimensional data space. Consequently, any data visualization formulated in the following way can be treated as a particular case of a dimensionality reduction problem, where the output number of dimensions is 2 or 3, typically 2. The SOM technique generates a 2-dimensional map structure. The location of points in 2-dimensional grid aims to reflect the similarities between the corresponding samples in an input multidimensional space. Therefore, the SOM algorithm allows for visualization of relationships between samples in multidimensional space.

1.1 Our Proposal

We propose an adaptive rule for the determination of the neighborhood widths of the SOM's BMUs during the SOM's training. The rule is formulated on the basis

L. Rutkowski et al. (Eds.): ICAISC 2021, LNAI 12854, pp. 182–192, 2021.
https://doi.org/10.1007/978-3-030-87986-0_16

of the preliminary data clustering in the input space. After forming of the input data clusters, the inner-cluster variances for each of the generated clusters are calculated and utilized afterward as a basis for the SOM's BMUs' neighborhood widths computation.

In our method, the neighborhood widths are determined independently for each BMU neuron in the SOM grid. The neighborhood of BMU is mathematically described using the Gaussian kernel function, where the radius of the Gaussian kernel is calculated as a result of initial data clustering in the input space. This is achieved in this way that the inner-cluster variances for each of the input data clusters are utilized as the basis for the radius of the Gaussian kernel.

2 Related Work

The SOM visualization technique has been extensively studied, and numerous improvements and extensions have been developed, including the Growing Hierarchical SOM (GHSOM) [12], the asymmetric SOM [4,7,10], and the adaptive SOM [1,5,9,11,13], to name a few. Naturally, the adaptive SOM versions are of particular interest for the purposes of our research.

An approach allowing to gain a control over the neurons' neighborhood widths in SOM delivered in the paper [13] is the magnification control approach. The issue is thoroughly studied by the authors of [13], where the three learning rule modifications for SOM are considered, namely, the localized learning, the winner-relaxing learning, and the concave-convex learning. The one closest to our research is the localized learning modification leading to inserting the local learning step size in the SOM weights update formula, in this way, affecting the SOM's BMUs' neighborhood widths. The local learning step size depends on the stimulus density of the weight vectors (prototypes) of SOM. As it is noticed in [13], a major drawback of the approach is that one has to estimate the generally unknown data distribution corresponding to the mentioned stimulus density, which may lead to numerical instabilities of the control mechanism [13]. Such a drawback does not concern the proposal of the present paper, because in our method, there is no necessity of any data distribution estimation. The second important difference between our technique and the localized learning is that in our method, preliminary data clustering and the resulting inner-cluster variances refer to the input samples in the SOM input space, whereas in case of the localized learning, the local learning step size is determined on the basis of the stimulus density of the weight vectors of SOM, i.e., on the basis of the intrinsic SOM structural information.

In the articles [9,11], an adaptive version of the SOM technique is introduced, in which, the frequency information about the input dataset is employed in the adaptive training process of SOM, i.e., in the adaptive form of the SOM's update formula. Strictly speaking, the frequency of occurrences of data samples in the input data space is utilized as the basis for the SOM's BMUs' neighborhood widths determination. The distinction between the approach from [9,11] and our research is that in the present paper, the information on the density and scattering of the input data samples in the input data space is included in the adaptive training process of SOM, whereas in the study from [9,11], the input data samples' frequencies of occurrences are taken into account, when SOM is being trained, and its lattice is being constructed.

Finally, the paper [1] proposes a Local Adaptive Receptive Fields Self-Organizing Map (LARFSOM). Local models are built by calculating between the output associated with the winning node and the difference vector between the input vector and the weight vector. These models are combined by using a weighted sum to yield the final approximate value. The topology is adapted in a self-organizing way, and the weight vectors are adjusted in a modified unsupervised learning algorithm for supervised problems.

3 Traditional SOM Method

The SOM algorithm provides a non-linear mapping from an original input high-dimensional data space onto a resulting output 2-dimensional map of neurons.

Besides the classical algorithmic description of the SOM method, which is well-known in the existing literature (see, e.g., [3]), an additional mathematical scaffolding has been presented in [4].

According to [4], the results obtained by the SOM method are equivalent to the results delivered by minimizing the following error function with respect to the prototypes w_r and w_s:

$$\begin{aligned} e &= \sum_r \sum_{x_i \in V_r} \sum_s h_{rs} d^2_{\text{Euc}}\left(x_i,\ w_s\right) \\ &\approx \sum_r \sum_{x_i \in V_r} d^2_{\text{Euc}}\left(x_i,\ w_r\right) \\ &+ K \sum_r \sum_{s \neq r} h_{rs} d^2_{\text{Euc}}\left(w_r,\ w_s\right) , \end{aligned} \tag{1}$$

where x_i, $i = 1, \ldots, N$ is the ith input sample in high-dimensional input space, N is the total number of input samples; w_r, $r = 1, \ldots, M$ and w_s, $s = 1, \ldots, M$ are the prototypes of input samples in the grid (the different indeces r and s are used in order to compute the sum of distances between neurons within the SOM grid, including the values of the function h_{rs}); M is the total number of prototypes/neurons in the grid; h_{rs} is a neighborhood function (e.g., the Gaussian kernel) that transforms non-linearly the neuron distances (see [3] for other choices of neighborhood functions); $d_{\text{Euc}}\left(\cdot,\ \cdot\right)$ is the Euclidean distance; and V_r is the Voronoi region corresponding to prototype w_r.

The width of the kernel h_{rs} is adapted in each iteration of the algorithm using the rule proposed by [5], i.e.:

$$\sigma\left(t\right) = \sigma_m \left(\frac{\sigma_f}{\sigma_m}\right)^{\frac{t}{N_{\text{iter}}}} , \tag{2}$$

where $\sigma_m \approx \frac{M}{2}$ is typically assumed in the literature (e.g., in [3]), σ_f is the parameter that determines the smoothing degree of the principal curve generated by the SOM algorithm [5], and N_{iter} is the total number of iterations during the SOM training process.

4 A Novel Clustering-Based Adaptive SOM Method

The main proposal of this paper is a method for adaptive SOM training, which is based on the preliminary data analysis, i.e., clustering of the input data samples in the input data space. After the clustering process, one calculates the inner-cluster variances for each of the generated clusters. These values represent the density and scattering of the input data samples, which is, as we claim in our research, the crucial information for the proper adaptation and adjustment of the SOM's lattice to the properties and characteristics of the input dataset. Therefore, the inner-cluster variances are subsequently included in the SOM's exponential update formula (2) from the work [5], and consequently, these values are employed in the SOM's unsupervised training process. In our research, we formulate an assertion that the introduced modification to (2) results in a higher SOM's performance and accuracy, and therefore, it can be recognized as the traditional SOM's enhancement.

The main idea behind the proposed method is the utilization of the information about the data density and scattering in the input data space for improving the training of SOM by making it adaptive and intelligent. This information is included during the setting of the neighborhood widths of SOM's BMUs', in this way, constituting a novel adaptive rule for the training of SOM, which is the main contribution of our paper. The information about the data density and scattering in the input data space is obtained from the measurements of the inner-cluster variance possible to determine after a preliminary data clustering in the input data space.

The entire proposal of the extension to the traditional NeRV method is presented completely and formally in Procedure 1.

Procedure 1. *The clustering-based and density-preserving adaptive SOM method proposed in current paper proceeds as follows:*

Step 1. Perform a clustering of the analyzed dataset in the input high-dimensional space.

Step 2. Compute the inner-cluster variance for each of the clusters according to the following formula:

$$\nu_k = \frac{1}{N_k} \sum_{i=1}^{N_k} d\left(c_k,\, x_i\right), \tag{3}$$

where ν_k is the inner-cluster variance of the kth cluster, $k = 1, \ldots, K$; K is the number of clusters; N_k is the number of data samples in the kth cluster; $d\left(\cdot,\, \cdot\right)$ is a given suitable dissimilarity measure in the input *high-dimensional space;* c_k *is the centroid of the kth cluster; and* x_i *are the data samples in the input high-dimensional space.*

Step 3. *Assign the inner-cluster variances* ν_k *to each data sample in the input space:*

$$\nu_i = \nu_k \text{ , such that } x_i \in C_k \text{ ,} \tag{4}$$

where ν_i *is the inner-cluster variance of the ith data sample in the input space,* $i = 1, \ldots, N$*;* N *is the total number of data samples in the input space;* C_k *is the kth cluster in the input space,* $k = 1, \ldots, K$*; and the rest of the notation has been explained previously in this paper.*

Step 4. *Include the inner-cluster variances* ν_i *in the exponential update formula (2):*

$$\sigma_i (\nu_i, t) = (1 + \nu_i) \sigma_m \left(\frac{\sigma_f}{\sigma_m} \right)^{\frac{t}{N_{\text{iter}}}} , \tag{5}$$

where σ_i *is the width of the Gaussian kernel used during the training of SOM for the ith data sample in the input space and the BMU in the SOM grid corresponding to that ith data sample.*

Step 4. *Minimize the error function (1) utilizing the novel form of the neighborhood function* h_{rs} *including the novel adaptive exponential update formula (5) for* $\sigma_i (\nu_i, t)$*.*

5 Experiments

In our experimental research, we aimed to verify the effectiveness of the approach introduced in the current paper. All of the experiments have been carried out in two phases, i.e., the input data visualization itself and the *a posteriori* output data clustering, i.e., clustering of the visualized data, or in other words, clustering of the data projected on the SOM grid. The data clustering within the visualization space has been conducted using the weight vectors (prototypes) attached to the neurons in the SOM grid. The results of the *a posteriori* SOM projection clustering have served us as the basis of the comparisons between the introduced technique and the three selected reference data visualization methods.

The experiments have been conducted on real data in the two different research fields: in the field of words visualization and clustering and in the field of speakers visualization and clustering. The first part of the experimental study has been carried out on the large dataset of high-dimensionality (Subsect. 5.3), whereas the second part has been conducted on smaller dataset, but also of high-dimensionality (Subsect. 5.4).

As a result, the scalability of our approach is presented, i.e., the ability to effectively operate on datasets of significantly different size. Note that our method does not increase essentially the computational complexity of the traditional SOM, and it adds only the computational demand of the initial clustering, which is run only once, and it is not repeated during the training of SOM.

Hence, in case of the DBSCAN clustering algorithm, the complexity of the initial clustering is $\mathcal{O}(N \log N)$, does not constraint the scalability property of our technique.

As the reference methods in our empirical study, we have chosen the standard SOM algorithm and the two modified versions of the conventional SOM, i.e., the Time Adaptive SOM (TASOM) and the data visualization approach proposed in [9,11], which will be called throughout this paper as the Frequency-Based SOM (FBSOM).

In case of the speakers' dataset, a graphical illustration of the generated SOMs is provided, whereas in case of the "Bag of Words" dataset, no such illustration is given, because of the high number of data samples in this dataset, which would make such images unclear and unreadable.

5.1 Evaluation Criteria

As the basis of the comparisons between the investigated methods, i.e., as the clustering evaluation criteria, we have used the accuracy rate [6,7] and the uncertainty degree [4,7].

Hence, the following two evaluation criteria have been used:

1. **Accuracy rate.** This evaluation criterion determines the number of correctly assigned samples divided by the total number of samples. Hence, for the entire dataset, the accuracy rate is determined as follows:

$$q = \frac{N_c}{N}, \tag{6}$$

   where N_c is the number of correctly assigned samples, and N is the total number of samples in the entire dataset.
   The accuracy rate q assumes values in the interval $\langle 0, 1 \rangle$, and naturally, greater values are preferred.
2. **Uncertainty degree.** This evaluation criterion determines the number of overlapping samples divided by the total number of samples in a dataset. The samples belonging to the overlapping area are determined on the basis of the ratio of dissimilarities between them and the two nearest clusters centroids. If this ratio is in the interval $\langle 0.9, 1.1 \rangle$, then the corresponding sample is said to be in the overlapping area.
   The uncertainty degree is determined as follows:

$$U_d = \frac{N_o}{N}, \tag{7}$$

   where N_o is the number of overlapping samples in the dataset, and N is the total number of samples in the dataset.
   The uncertainty degree assumes values in the interval $\langle 0, 1 \rangle$, and, smaller values are desired.

5.2 Experimental Setup

In our experimental research, we have utilized the DBSCAN clustering algorithm, because of the important and significant advantages of the algorithm from the point of view of our data analysis, i.e., the automatic clusters' number determination and the capability to handle the non-linearly separable data.

The output data *a posteriori* clustering in the SOM visualization space has been conducted using the standard k-means clustering algorithm.

Each of the investigated methods has been run 50 times, because all of the methods are non-deterministic, and by repeating their executions, we obtain results, which may be recognized as more reliable. The randomness of the methods exists in both phases of our data analysis and processing, i.e., in the data visualization and in the following data clustering.

The values of the accuracy rates and uncertainty degrees in Tables 1 and 2 are computed as the arithmetic averages over all the executed runs of the evaluated methods.

Feature extraction of the textual data investigated in the part of our empirical study demonstrated in Subsect. 5.3 was carried out using the term frequency – inverse document frequency (*tf-idf*) approach.

Features of the speakers' sound signals considered in Subsect. 5.4 have been extracted using a method based on the Discrete Fourier Transform (DFT), which is described in details in [8].

5.3 Words Visualization and Clustering

In the first part of our experimental research, we have utilized excerpts from the "Bag of Words" dataset from the UCI Machine Learning Repository [2].

Our dataset consists of five text collections: Enron E-mail Collection, Neural Information Processing Systems (NIPS) full papers, Daily KOS Blog Entries, New York Times News Articles, PubMed Abstracts. The total number of analyzed words was approximately 10,868,000. On the visualizations generated by the investigated methods, five clusters representing those five text collections in the "Bag of Words" dataset were formed.

Experimental Results. The results of this part of our experiments are reported in Table 1, where the accuracy rates and uncertainty degrees corresponding to each of the evaluated methods are given.

Table 1. Accuracy rates and uncertainty degrees of the words visualization and clustering.

	q	U_d
SOM & k-means	8,175,273/10,868,000 = 0.7522	2,459,195/10,868,000 = 0.2263
TASOM & k-means	8,389,009/10,868,000 = 0.7719	2,304,016/10,868,000 = 0.2120
FBSOM & k-means	9,183,460/10,868,000 = 0.8450	1,523,471/10,868,000 = 0.1402
Proposed SOM & k-means	**9,262,276/10,868,000 = 0.8523**	**1,248,863/10,868,000 = 0.1149**

The results obtained in this part of our experiments have shown a superiority of our approach over the other three data visualization algorithms considered as the benchmark methods. The solution introduced in this paper produced the highest value of the accuracy rate and the lowest value of the uncertainty degree among all the investigated methods.

5.4 Speakers Visualization and Clustering

The speakers visualization and clustering experiment has been conducted on the dataset of sound signals gathered independently by the author of this work.

In this part of our experiments, we considered four clusters representing four speakers. Each speaker was represented by 40 speeches, i.e., sound signals (time series). This kind of clustering can be regarded as the speaker recognition based on the sound signals.

Four clusters representing four different speakers have been formed. Each speaker has been represented by 40 10-s sound signals sampled with the 44.1 kHz frequency. Our dataset is composed of 160 sound signals. Feature extraction was carried out according to the DFT-based algorithm, as it was written in Subsect. 5.2. The dataset for the speakers visualization and clustering has been collected autonomously be the author of this research.

Experimental Results. The results of this part of our experiments are reported in Figs. 1a, 1b, 2a, 2b, and in Table 2, which has the same form as Table 1 in Subsect. 5.3. Figures 1a, 1b, 2a, and 2b show the map structures of four considered SOM versions. The points in the 2-dimensional space of these SOM variants' visualizations are the projections of the input data samples from the input data space. In each of the figures, the clusters, generated in the output data *a posteriori* clustering, are indicated and marked with different colors. The colors are assigned to the clusters randomly, therefore, a given cluster may have different colors assigned in different runs of the same algorithm. Hence, the clusters themselves are important, and not their particular colors. Each of Figs. 1a, 1b, 2a, and 2b presents SOM graphics for a single execution of a given SOM version.

Table 2. Accuracy rates and uncertainty degrees of the speakers visualization and clustering.

	q	U_d
SOM & k-means	$134/160 = 0.8375$	$15/160 = 0.0938$
TASOM & k-means	$137/160 = 0.8563$	$16/160 = 0.1000$
FBSOM & k-means	$145/160 = 0.9063$	$10/160 = 0.0625$
Proposed SOM & k-means	$\mathbf{154/160 = 0.9625}$	$\mathbf{5/160 = 0.0313}$

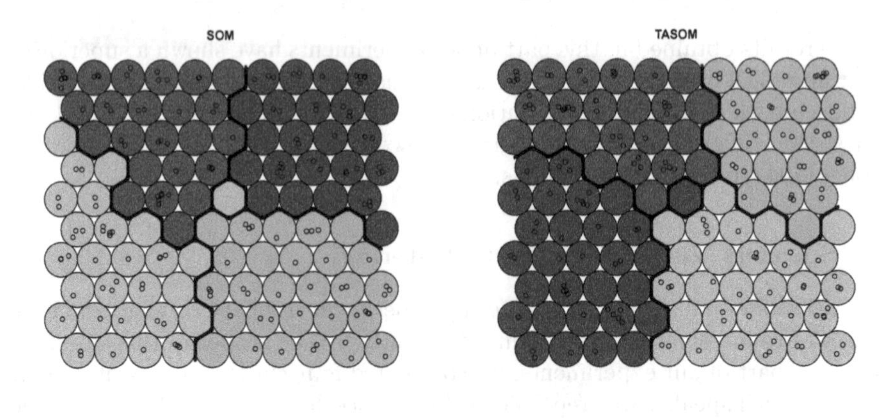

(a) Conventional SOM method. (b) TASOM method.

Fig. 1. Results of speakers visualization and clustering using the conventional SOM method and the TASOM method.

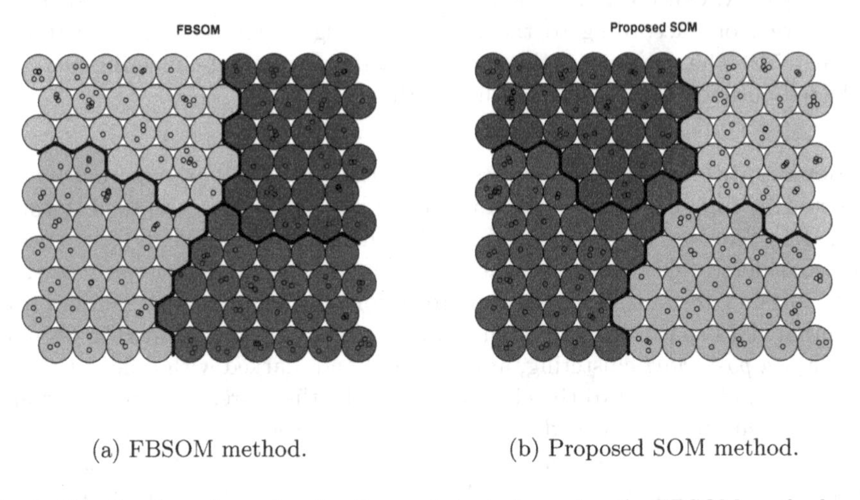

(a) FBSOM method. (b) Proposed SOM method.

Fig. 2. Results of speakers visualization and clustering using the FBSOM method and the proposed SOM method.

The outcome of the second part of our experiments verified and confirmed the effectiveness and usefulness of our proposed method by indicating that it outperforms all the other tested variants and improvements of SOM. The approach developed in our work returned the higher accuracy rate than all the other algorithms being a subject of evaluation, and furthermore, it allowed for obtaining the lowest value of uncertainty degree, when compared to the reference data visualization methods.

6 Summary

In this paper, a novel version of SOM has been proposed. The novelty in our extension to the traditional SOM was a concept of establishing a relationship between the SOM's BMUs' neighborhood widths and the density and scattering of the data in the input data space. In other words, the SOM's BMUs' neighborhood widths have been determined on the basis of the information about the data density and scattering in the input data space. Precisely speaking, the density and scattering properties of the input data have been numerically represented and conveyed by the quantities of inner-cluster variances obtained as a result of a preliminary input data clustering.

References

1. Ferreira, P.H.M., Araújo, A.F.R.: Growing self-organizing maps for nonlinear time-varying function approximation. Neural Process. Lett. **51**(2), 1689–1714 (2020). https://doi.org/10.1007/s11063-019-10168-9
2. Frank, A., Asuncion, A.: UCI machine learning repository (2010). http://archive.ics.uci.edu/ml
3. Kohonen, T.: Self-Organizing Maps, 3rd edn. Springer, Heidelberg (2001). https://doi.org/10.1007/978-3-642-56927-2
4. Martín-Merino, M., Muñoz, A.: Visualizing asymmetric proximities with SOM and MDS models. Neurocomputing **63**, 171–192 (2005)
5. Mulier, F., Cherkassky, V.: Self-organization as an iterative kernel smoothing process. Neural Comput. **7**(6), 1165–1177 (1995)
6. Olszewski, D.: Asymmetric *k*-means algorithm. In: Dobnikar, A., Lotrič, U., Šter, B. (eds.) ICANNGA 2011. LNCS, vol. 6594, pp. 1–10. Springer, Heidelberg (2011). https://doi.org/10.1007/978-3-642-20267-4_1
7. Olszewski, D.: An experimental study on asymmetric self-organizing Map. In: Yin, H., Wang, W., Rayward-Smith, V. (eds.) IDEAL 2011. LNCS, vol. 6936, pp. 42–49. Springer, Heidelberg (2011). https://doi.org/10.1007/978-3-642-23878-9_6
8. Olszewski, D.: *k*-means clustering of asymmetric data. In: Corchado, E., Snášel, V., Abraham, A., Woźniak, M., Graña, M., Cho, S.-B. (eds.) HAIS 2012. LNCS (LNAI), vol. 7208, pp. 243–254. Springer, Heidelberg (2012). https://doi.org/10.1007/978-3-642-28942-2_22
9. Olszewski, D.: An improved adaptive self-organizing map. In: Rutkowski, L., Korytkowski, M., Scherer, R., Tadeusiewicz, R., Zadeh, L.A., Zurada, J.M. (eds.) ICAISC 2014. LNCS (LNAI), vol. 8467, pp. 109–120. Springer, Cham (2014). https://doi.org/10.1007/978-3-319-07173-2_11
10. Olszewski, D., Kacprzyk, J., Zadrożny, S.: Time series visualization using asymmetric self-organizing map. In: Tomassini, M., Antonioni, A., Daolio, F., Buesser, P. (eds.) ICANNGA 2013. LNCS, vol. 7824, pp. 40–49. Springer, Heidelberg (2013). https://doi.org/10.1007/978-3-642-37213-1_5
11. Olszewski, D., Kacprzyk, J., Zadrożny, S.: An improved adaptive self-organizing map. In: De Tré, G., Grzegorzewski, P., Kacprzyk, J., Owsiński, J.W., Penczek, W., Zadrożny, S. (eds.) Challenging Problems and Solutions in Intelligent Systems. SCI, vol. 634, pp. 75–102. Springer, Cham (2016). https://doi.org/10.1007/978-3-319-30165-5_5

12. Rauber, A., Merkl, D., Dittenbach, M.: The growing hierarchical self-organizing map: exploratory analysis of high-dimensional data. IEEE Trans. Neural Netw. **13**(6), 1331–1341 (2002)
13. Villmann, T., Claussen, J.C.: Magnification control in self-organizing maps and neural gas. Neural Comput. **18**(2), 446–469 (2006)

Applying Machine Learning Techniques to Identify Damaged Potatoes

Aleksey Osipov[1(✉)], Andrey Filimonov[2], and Stanislav Suvorov[2]

[1] Financial University under the Government of the Russian Federation, Shcherbakovskaya. 38, 105187 Moscow, Russian Federation

[2] Moscow Polytechnic University, Bolshaya Semyonovskaya 38, 107023 Moscow, Russian Federation

Abstract. This paper examines the problem of detecting potatoes with mechanical damage using machine learning techniques.

In this article, the authors proposed an algorithm for detecting damaged potato tubers on a conveyor belt that is characterized by speed and accuracy of recognition.

The distinctive features of the algorithm are combining the methods of Viola-Jones and the convolutional networks, the application of two complementary classifiers, working in the usual gray color and inverted color. Also, the distinguishing feature is that the identified tubers are processed by the classifiers only once, regardless of the time in front of the video camera.

The Viola-Jones method was used to identify individual tubers on the conveyor belt, and the convolutional networks were only used to recognize damaged tubers. Moreover, two complementary networks were used for classification, one of which worked in gray gradation and the other in inverted color.

The algorithm was implemented using the OpenCV library in Python. Testing was carried out in conditions close to the conditions of potato storage at vegetable bases.

The percentage of properly-recognized damaged tubers was 92,1%.

Keywords: Neural networks · Identify defects · Potato classification · Fast detection

1 Introduction

Potatoes are one of the most common crops in the world. According to the Food and Agriculture Organization of the United Nations (FAO), 368 million tons of potatoes from a total area of 17.6 million hectares were harvested worldwide for 2018 [1].

At such volumes of harvest, the problem of preserving the harvest in winter becomes especially urgent. Crop losses caused by diseases in potato production account for 22% annually. With purchase prices for potatoes at the beginning of 2020, according to the resource east-fruit.com $0.19–0.31/kg damage can be more than $25 billion per year!

There are mechanical damages [2] and damage caused by diseases and pests [3–5].

L. Rutkowski et al. (Eds.): ICAISC 2021, LNAI 12854, pp. 193–201, 2021.
https://doi.org/10.1007/978-3-030-87986-0_17

In this article, the authors looked at mechanical damage, because of all the variety of damage, the most important is mechanical damage to tubers. Organisms such as fungi and bacteria cannot penetrate the intact peel and gain access to tuber tissues only in mechanical damage. Therefore, infection depends on the presence of mechanical damage, and resistance to the latter protects tubers from disease.

It should be pointed out mechanical damage leads to additional losses in the form of increased waste when using potatoes for canteens. Therefore, potatoes with mechanical damage very quickly lose weight due to increased evaporation from the damaged surface and rapid breathing. According to some data, weight loss after 4 months of storage is: in cut potatoes - up to 15%, with pulp damage - up to 12%, and in undamaged potatoes - up to 7–10%. Mechanical damage can be divided into two groups: external (superficial) and internal. External damage includes damage that can be determined by external inspection. These are rippings, scratches, cracks, dents, cuts, crushed tubers, etc. Internal damage: darkening of pulp, internal cracks, damage to vascular beams [6].

Incisions and cuts of tubers during cleaning are caused mainly by incorrect adjustment of the colter depth.

Peeling of the peel occurs when tubers slide over the surface of the working organs in the presence of a relatively high coefficient of friction.

Damages such as dents, crushes, cracks, breaks arise from pressure on tubers at static loads.

But the greatest amount of damage such as serious as cracks, dents, damage to vascular beams, darkening of the pulp is caused exclusively by dynamic loads - impacting tubers with working organs.

Besides, damage to tubers may also occur as a result of improper storage, for example, due to the storage under direct sunlight, potatoes are green, or, as another example, the freezing of tubers. In this case, the flesh of the tubers gets a pinkish tint (Fig. 1).

Regular monitoring of the appearance of potatoes will allow to quickly apply measures that exclude further damage to tubers, and, as a result, reduce the damage.

In this paper, the authors propose a fast algorithm of visual control of the appearance of potatoes, allowing to assess the quality of the appearance of large volumes presented for potato analysis.

Over the past decade, the successful use of convolutional neural networks (CNN) in the field of imaging has significantly improved the performance of computer vision tasks [7–11].

The authors of the article [7] suggested using this method to determine the market price of potatoes. The whole process is divided into four phases. In the first stage, the convolutional neural network divided all tubers into 6 different classes of healthy, damaged, greening, black dot, scab, and black plaque) in the work was offered a solution to the problem when classes intersect. In the second and third phases, defects were localized with the help of defect activation cards (DAM). The severity of the damage was assessed in the fourth phase.

The authors of the article [8] showed that the CNN method, and in particular ResNet18, could work as a detector to detect potato diseases at the stage of growth, which would allow for several pre-emptive measures to reduce crop damage. However,

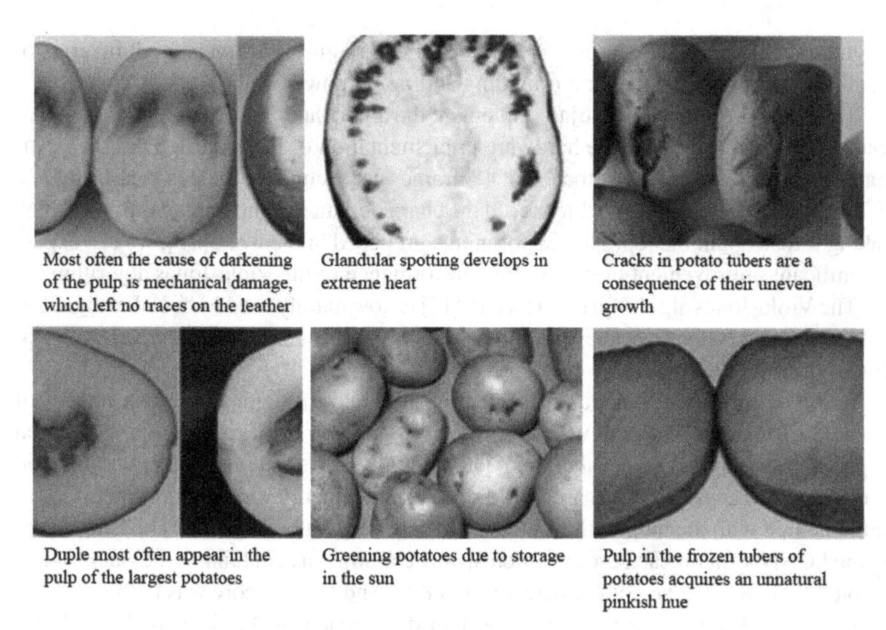

Fig. 1. Examples of various potato damage due to improper storage.

the complexity of the video-fixing system, special requirements for the quality of the image does not allow using it in masse.

The authors of the article [9] used reverse laser scattering, which was processed by the neural network (CNN) to detect defects in apples. The filter system, segmentation, and selection of an area of 150 to 150 pixels were used to localize the defect. The stated computing power of the computer and the large image at the entrance of the neural network did not allow to make enough for high accuracy training sample (accuracy was 90%).

The authors of the article [10] for the detection of apples during the harvest proposed to use the fastest modification of the convolutional neural network YOLOv3. Experiments are underway to install this detection device on an apple-picker robot.

However, as the authors of the work [12] neural networks, designed to solve the problems of understanding high-level images are not suitable for implementation onboard intelligent portable devices. Large cores (e.g. 7×7 or 9×9) or a large number of layers [13] are required to gain acceptable susceptibility to the field. Both of these schemes lead to a very significant slowdown of the system. To achieve acceptable image processing time on low-productivity portable devices, most existing systems are limited to smaller images of 41×41 (pixel). And the processing time of each such frame can be up to several seconds. In a continuously moving pipeline, this is unacceptable.

In these circumstances, algorithms with a favorable ratio of speed/resources and power [14] deserve special interest. For this reason, several hardware optimization methods [15, 16] have recently been proposed.

The authors of the work [15] proposed to implement a deep neural network (DNN) to use GPUs. This increases the computing capabilities of the system by an order of magnitude but will make it more expensive and will require additional, both in terms of installing additional devices and providing additional power.

The authors of the work [16] to implement the algorithm of facial recognition Viola-Jones [17] proposed to use the hardware implementation of FPGA. The DE2–115 estimated board provided a performance of 4.4 frames per second for images measuring 320 × 240 pixels. These indicators far exceed the characteristics of input images of the CNN. Taking into account the reprocessing of inputs proposed in the work [18], we can expect a significant improvement in the existing performance of the Viola-Jones algorithm.

The Viola-Jones algorithm described in [17] is now mainly used for facial recognition tasks. However, several articles have appeared in the press indicating the versatility of this algorithm [19, 20].

In the work [19] automatic assessment of the yield of melons based on images of the field of melons. The task was to solve three stages: melon recognition, extraction of geometric features, and evaluation of individual weight. The first stage was divided into two sub-stages. First, an area on the field, presumably containing a melon (was implemented with the help of the Viola-Jones detector) was selected. Then, using a pre-defined CNN system, each area of interest was classified as containing melon or not. It is noted that the accuracy of the detector was 82% and the F1 score was 0.85.

In the work [20] with the help of a hybrid approach of the detector Viola-Jones and multi-pattern comparison successfully identified varieties of flower anthurium. The authors noted the high speed of the algorithm and assumed that the methodology offered by them may be useful for other purposes of identification.

2 Description of the Method

As has been shown above, the use of convolutional neural networks to recognize sick and damaged potatoes is not justified for rapid diagnosis in large volumes, such as when the potato is moving on the conveyor belt.

This is due to the large number of computing operations required for the operation of the network.

The authors propose to divide the diagnostic procedure into separate phases, allowing to speed up the process of identifying individual tubers in a video stream and direct analysis. Below, each of the phases will be discussed in more detail.

The first phase: identifying individual tubers in the image. Given the fact that the potato is moving on a conveyor belt and that a normal image capture by a video camera requires a frequency of at least 16 fps, and better than 22 fps and above, the most appropriate method of identifying tubers is the Viola-Jones method.

This method uses the so-called Haar primitives to identify the characteristics inherent in the objects identified.

Examples of Haar's primitives are presented in Fig. 2.

To properly operate the algorithm and its acceleration, it was suggested to abandon the color representation of potatoes in favor of representation in gray gradations with subsequent binarization. Also, we looked at the image of the potato both in normal form and inverted (Fig. 4).

Fig. 2. Haar's primitives

Fig. 3. Original image

Fig. 4. Image translated into gray gradations

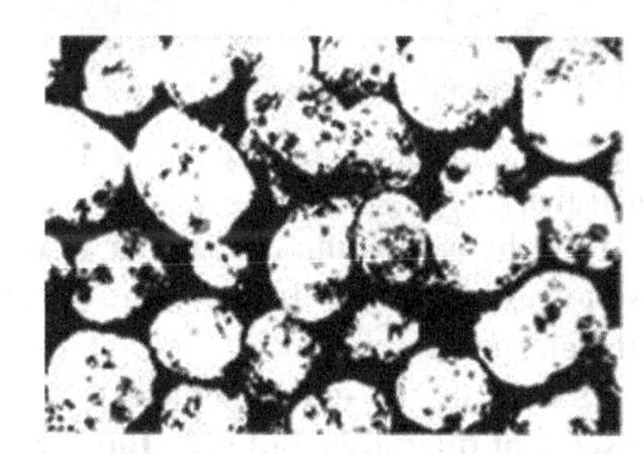

Fig. 5. Threshold binarization (Otsu method) has been applied

Figure 5 shows that the tubers in this view are difficult to distinguish as separate objects, but if we knew in advance the location of each tuber in the image, the damage to the tubers themselves is very noticeable. Moreover, it is easy to calculate the area of damage.

On this basis, the authors proposed to determine the location of tubers in the image, not in the usual color but inverted with subsequent binarization (Figs. 7, 8).

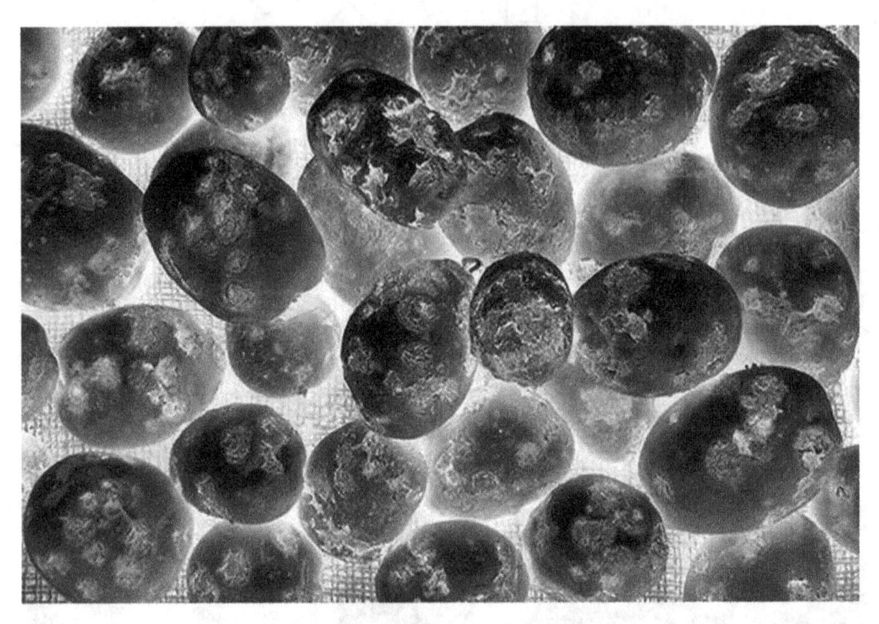

Fig. 6. Inverted image

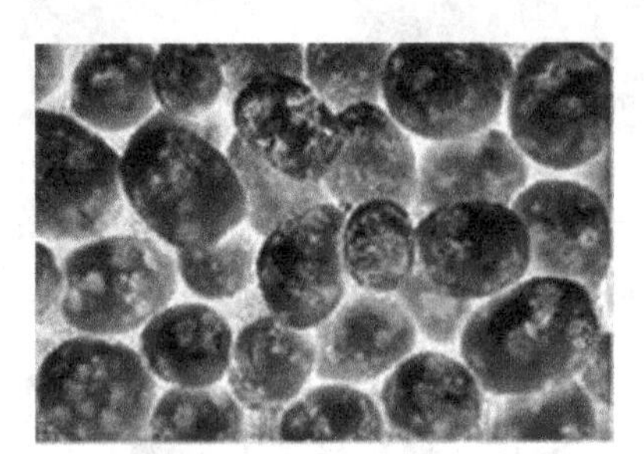

Fig. 7. Inverted image translated into gray gradation

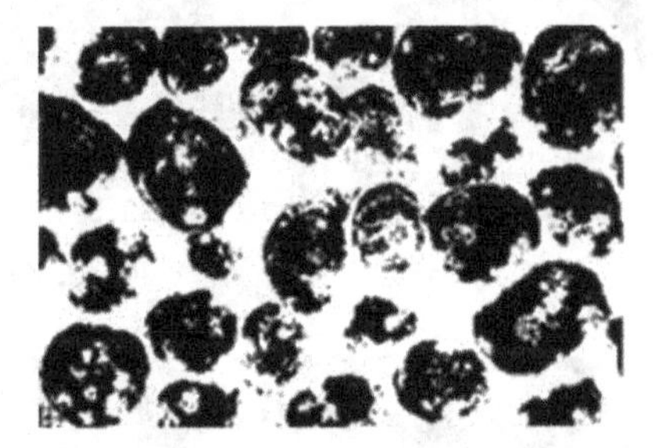

Fig. 8. Threshold binarization (Otsu method) has been applied

Thus, by comparing the two views of tubers in Figs. 3 and 6, it is possible in one case to easily identify the locations of tubers, and in the other case to identify tubers with damage.

That is, at the end of Phase 1, not all images of tubers are received on the subsequent diagnosis, but only those where there is damage.

Also, knowing the speed of the conveyor and the frequency of the video camera, you can discard the need to re-analyze the same tuber while it is in the frame. Roughly speaking, the algorithm singled out the tuber in the frame once it was processed and marked as processed. In this case, the tuber will not be called in the re-analysis frame when the tuber is shifted. There will only be an escort. **This is also an optimization element that allows you to speed up the algorithm by several orders of magnitude.**

Phase two: classification of identified tubers. In this article, the authors used a binary classification: damaged tuber or undamaged.

Given that the light in the potato warehouse is not good, that part of the tubers can be partially covered by other tubers, it is proposed to use not one classifier, but two.

The first classifier uses the location information of the tubers obtained during the first phase but works with the original image presented in the gray gradations.

The second classifier uses an inverted image.

The results of both classifiers complement each other.

CNN is used as classifiers (Fig. 9).

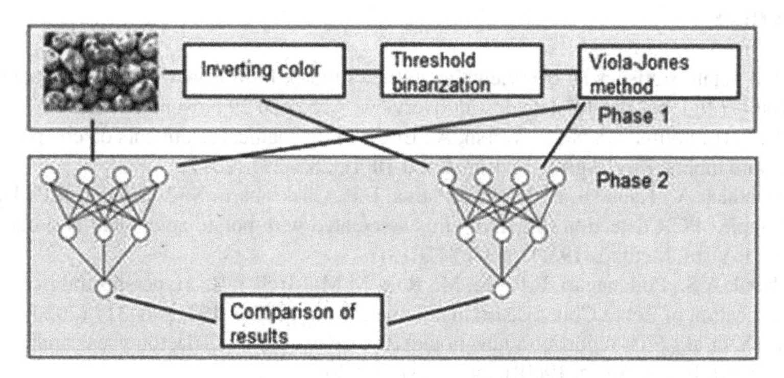

Fig. 9. A generalized diagram of the algorithm

The algorithm was implemented using the OpenCV library in Python. Testing was carried out in conditions close to the conditions of potato storage at vegetable bases.

The percentage of properly-recognized damaged tubers was 87%. At the same time, the accuracy of recognition was greatly influenced by lighting and overlapping tubers to each other.

If the lighting lamps are located directly above the conveyor belt near the recording video camera, the accuracy of recognition increased to 91.3%.

The overlap of tubers can be partially eliminated by increasing the speed of the conveyor. In this case, the accuracy of the recognition increased to 92.1%.

However, given the recommended speed of conveyors for the transport of root vegetables, given in Table 1, it is clear that a significant increase in the accuracy of recognition due to the cleaning of the tuber overlap, there is no need to expect.

3 Conclusion

In this article, the authors proposed an algorithm for detecting damaged potato tubers on a conveyor belt that is characterized by speed and accuracy of recognition.

Table 1. Recommended conveyor speeds for root vegetables.

Tape speed v, m/s, at tape width B, mm							
300–500	650	800	1000	1200	1400	1600	2000
0,8	0,8	1	1	1	1	1	1

The distinctive features of the algorithm are combining the methods of Viola-Jones and the convolutional nets, the application of two complementary classifiers, working in the usual gray color and inverted color. Also, the distinguishing feature is that the identified tubers are processed by the classifiers only once, regardless of the time in front of the video camera.

References

1. Global potato statistics an overview. Potato news today. https://www.potatonewstoday.com/2020/08/12/global-potato-statistics-an-overview. Accessed 29 November 2020
2. Rady, A.H., Soliman, S.N., El-Wersh, A.: Effect of mechanical treatments on creep behavior of potato tubers. Eng. Agric. Environ. Food **10**(4), 282–291 (2017)
3. Jeevalathaa, A., Kaundal, P., Venkatasalama, E.P., Chakrabarti, S.K., Singha, B.P.: Uniplex and duplex PCR detection of geminivirus associated with potato apical leaf curl disease in India. J. Virol. Methods **193**(1), 62–67 (2013)
4. Garhwal, A.S., Pullanagari, R.R., Li, M., Reis, M.M., Archer, R.: Hyperspectral imaging for identification of Zebra Chip disease in potatoes. Biosyst. Eng. **197**, 306–317 (2020)
5. Song, S, et al.: First report of a new potato disease caused by Galactomyces candidum F12 in China. J. Integr. Agric. **19**(10), 2470–2476 (2020)
6. Wrigley, C.W.: Definitions of terms for the food- and grain-processing industries. Ref. Module Food Sci. (2016)
7. Marino. S., Beauseroy. P., Smolarz, A.: Weakly-supervised learning approach for potato defects segmentation. Eng. Appl. Artif. Intell. **85**, 337-346 (2019)
8. Afonso, M., Blok, P.M., Polder, G., van der Wolf, J.M., Kamp, J.: Blackleg detection in potato plants using convolutional neural networks. IFACOnLine **52**(30), 6–11 (2019)
9. Wu, A., Zhu, J., Ren, T.: Detection of apple defect using laser-induced light backscattering imaging and convolutional neural network. Comput. Electr. Eng. **81**, 106454 (2020)
10. Kuznetsova, A., Maleva, T., Soloviev, V.: Using YOLOv3 algorithm with pre- And post-processing for apple detection in fruit-harvesting robot (Article). Agronomy **10**(7), 1016 (2020)
11. Kuznetsova, A., Maleva, T., Soloviev, V.: Detecting apples in Orchards using YOLOv3. In: Gervasi, O., et al. (eds.) ICCSA 2020. LNCS, vol. 12249, pp. 923–934. Springer, Cham (2020). https://doi.org/10.1007/978-3-030-58799-4_66
12. Yin, H., Gong, Y., Qiu, G.: Fast and efficient implementation of image filtering using a side window convolutional neural network, Signal Process. **176**, 107717 (2020)
13. Shen, X., Chen, Y.-C., Tao, X., Jia, J.: Convolutional neural pyramid for image processing. arXiv:1704.02071v1 [cs.CV] (2017)
14. Spagnoloa, F., Perrib, S., Corsonello, P.: Design of a real-time face detection architecture for heterogeneous systems-on-chips. Integration **74**, 1–10 (2020)
15. Feng, X., Jiang, Y., Yang, X., Du, M., Li, X.: Computer vision algorithms and hardware implementations: a survey. Integration **69**, 309-320 (2019)

16. Irgens, P., Bader, C., Lé, T., Saxena, D., Ababei, C.: An efficient and cost effective FPGA based implementation of the Viola-Jones face detection algorithm. HardwareX **1**, 68–75 (2017)
17. Viola, P., Jones, M.: Rapid object detection using a boosted cascade of simple features. December 2001. In: IEEE Conference on Computer Vision and Pattern Recognition (CVPR), pp. 511–518 (2001)
18. Shia, W., Alawieha, M.B., Lib, X., Yud, H.: Algorithm and hardware implementation for visual perception system in autonomous vehicle: a survey. Integration **59**, 148–156 (2017)
19. Kalantar, A., Edan, Y., Gur, A., Klapp, I.: A deep learning system for single and overall weight estimation of melons using unmanned aerial vehicle images. Comput. Electr. Agric. **178**, 105748 (2020)
20. Soleimanipour, A., Chegini, G.R.: A vision-based hybrid approach for identification of Anthurium flower cultivars of Anthurium flower cultivars. Comput. Electr. Agric. **174**, 105460 (2020)

Quantifying the Severity of Common Rust in Maize Using Mask R-CNN

Nelishia Pillay[1(✉)], Mia Gerber[1], Katerina Holan[2], Steven A. Whitham[2], and Dave K. Berger[3]

[1] Department of Computer Science, University of Pretoria, Pretoria, Gauteng, South Africa
nelishia.pillay@up.ac.za
[2] Plant Pathology and Microbiology, Iowa State University, Ames, Iowa, USA
[3] Department of Plant and Soil Sciences, Forestry and Agricultural Biotechnology Institute (FABI), University of Pretoria, Pretoria, Gauteng, South Africa
https://www.cs.up.ac.za
https://www.plantpath.iastate.edu/
https://www.up.ac.za/faculty-of-natural-agricultural-sciences

Abstract. The second sustainable development goal defined by the United Nations focuses on achieving food security and supporting sustainable agriculture. This paper focuses on one such initiative contributing to attaining this goal, namely, the identification or prediction of disease in crops. More specifically the paper examines the automated quantification of the severity of common rust in maize. Previous work has focused on using standard image processing algorithms for this problem. This is the first study, to the knowledge of the authors, employing machine learning techniques to determine the severity of common rust disease in maize. Quantifying the severity of common rust is achieved by counting the number of pustules on maize leaves and determining the surface area of the leaf covered by pustules. In this study a Mask R-CNN is used to determine this. Both the standard image processing algorithms and the Mask R-CNN were evaluated on a real-world dataset created from images of maize leaves grown in a greenhouse. The Mask R-CNN was found to outperform the standard image processing algorithms in terms of counting the number of pustules, calculation of the pustule surface area and the average pustule size. These results were found to be statistically significant at a 5% level of significance. One of the challenges with Mask R-CNN is finding suitable parameter values, which is time consuming. Future work will examine automating parameter tuning for the Mask R-CNN.

Keywords: Plant disease severity quantification · Image processing · Mask R-CNN

1 Introduction

One of the sustainable development goals defined by the United Nations focuses on food security and sustainable agriculture [6]. Some of the initiatives towards

L. Rutkowski et al. (Eds.): ICAISC 2021, LNAI 12854, pp. 202–213, 2021.
https://doi.org/10.1007/978-3-030-87986-0_18

attaining this goal include automated irrigation, precision agriculture, and automated detection of pests and plant disease. Global crop production is constantly under threat by pests and pathogens, and a recent study quantified average global losses of maize, wheat and rice from these biotic threats to be 22.5%, 21.5% and 30%, respectively [17]. In addition to crop disease diagnostics there is a need to develop high-throughput methods for quantifying the severity of crop disease symptoms, for example the area of lesions on a maize leaf caused by a pathogen [9].

Common rust caused by the fungus *Puccinia sorghi Schwein* is a widespread disease of maize in North and South America, Africa and Asia, and it can account for significant yield losses [18]. Quantification of the severity of common rust disease is done by counting rust (red-brown) lesions, referred to as *pustules*, and measuring the lesion surface area over a time course after artificial inoculation [2]. An example of a greenhouse maize leaf with pustules is illustrated in Fig. 1.

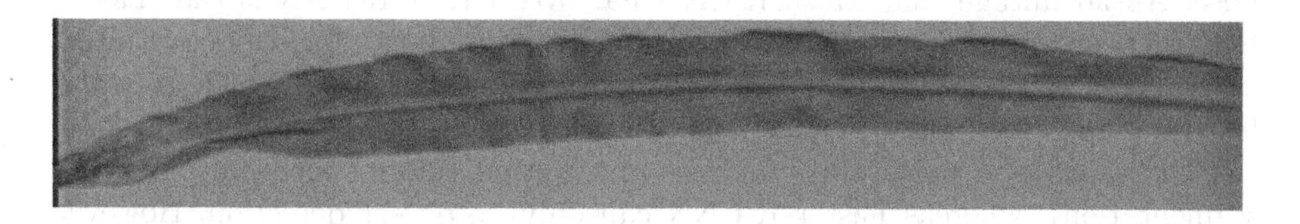

Fig. 1. Pustules on a maize leaf caused by common rust disease

These rust disease metrics are measured at different stages of the infection by plant pathologists or plant breeders in maize improvement projects either: (i) to test rust control measures such as fungicides or biological control agents;, or (ii) to screen different varieties of maize for genetic resistance to rust [2]. The study presented in this paper examines automating the process of quantifying the severity of common rust in maize. In initial work a standard image processing algorithm was used to determine the pustule count and the area covered by pustules. This study examines the use of neural networks, namely, a Mask R-CNN [7], for this purpose. The performance of the neural network is compared to that of the standard image processing algorithm that was previously applied. The Mask R-CNN was found to outperform the image processing algorithm.

The following section provides an overview of neural networks used for plant disease diagnosis and quantification. Section 3 presents the standard image processing algorithms used for determining the pustule count and surface area covered and Sect. 4 describes the Mask R-CNN employed. Section 5 presents the experimental setup used to assess the performance of the approaches. A comparison of the performance of the standard image processing algorithms and the Mask R-CNN is discussed in Sect. 6. Section 7 provides a summary of the findings of the study and proposes future extensions of the work.

2 Related Work

Deep neural networks have proven to be effective for predicting plant disease and determining severity from images [11,23]. The research into the use of deep neural networks can be categorised into two areas. The first involves classifying images as having the disease or not. This is essentially a binary classification problem with the classes being infected and not infected or a multiclass classification problem with the classes being the different levels of severity of the disease [11]. Convolutional neural networks that have proven to be effective for this include VGG16, VGG19, Inception-v3 and ResNet50.

The second area is image segmentation for quantifying disease severity. This essentially involves segmenting the image into areas so as to count the spots on a leaf. This area has not been as well researched as plant disease diagnosis. Deep neural networks have also been effective for this area [15,22]. More recently Faster R-CNN and Mask R-CNN [3,16] have been used for this purpose. An advantage that Mask R-CNN has over Faster R-CNN is that Faster R-CNN is only able to generate rectangular bounding boxes for regions of interest whereas Mask R-CNN can generate binary polygon masks for regions in the feature map, which align perfectly on a pixel level. Stated differently, Mask R-CNN is capable of performing image segmentation (more specifically instance segmentation), whereas Faster R-CNN can only do object detection. However Mask R-CNN is not the only neural network that has been applied to the task of image segmentation, other neural networks such as YOLO and U-Net are also popular choices. Mask R-CNN has however been shown in previous work to outperform both YOLO [4,5] as well as U-Net [1,14,25]. YOLO specifically would not be suited to the task of pustule identification as it is known to struggle with identification of small objects as well as objects that overlap [20]. Although Mask R-CNN has not before been applied to the task of common rust pustule identification on Puccinia sorghi, it has performed well in similar applications to that investigated in this paper [3], specifically it has been used for problems such as leaf counting [24] and leaf blight phenotyping [21]. Mask R-CNN appears to be the most suitable for the application at hand, namely, counting pustules and the surface area covered by pustules from the images of greenhouse leaves. The neural network produces a mask allowing for the number of pustules to be counted as well as the surface area covered to be determined from the mask.

In this study we compare the performance of Mask R-CNN to that of the standard image processing algorithms previously used to count the number of pustules and determine the surface area covered by pustules on a maize leaf. The following section presents the standard image processing algorithms used.

3 Image Processing Algorithms

The image processing pipeline, employing standard image processing algorithms, employed to determine the pustule count and surface area covered by pustules, is specified in Algorithm 1.

Algorithm 1. Image Processing Pipeline

1: Divide the image into three RGB components using Algorithm 2 below.
2: Find the edges of the pustules using the Sobel edge detection algorithm [8].
3: Apply the Watershed transformation [19] to the edges of pustules and red channel markers (identified in Step 1).
4: Fill holes with a square connectivity equal to one.
5: Discard regions of the image that are greater than 10000 pixels.
6: Label the remaining regions as pustules.

Algorithm 2 describes the process of dividing the image into the RGB components which is the first step of the image processing pipeline. The first two steps of Algorithm 2 prevents leaf areas that are simply chlorotic or browning from being highlighted as pustules.

Algorithm 2. Algorithm for Dividing the Image into RGB Components

1: For the blue component, if a pixel has a value greater than 140, that pixel has its value set to 0.
2: For the green component, if a pixel has a value greater than 130, that pixel has its value set to 255.
3: The maximum pixel intensity value for each X,Y coordinate is calculated.
4: All pixel values less than the maximum pixel intensity value are set to zero.
5: The three channels are merged back together for subsequent processing steps.

The next section describes the Mask R-CNN employed to determine the pustule count and surface area covered by pustules.

4 Mask R-CNN

The Matterport [10] version of Mask R-CNN is employed in this study. The ResNet 101 convolutional neural network forms the backbone of the Mask R-CNN. ResNet 101 is pretrained on ImageNet. The parameter values for Mask R-CNN was determined empirically. These parameter values are listed in Table 1.

The Mask R-CNN essentially performs supervised learning. The label for each image in the training set is comprised of:

- The coordinates for each pustule.
- The surface area of the leaf covered by the pustules.
- The size of each pustule. This is used to determine the average pustule size.

The Mask R-CNN produces a mask from which the number of pustules, surface area covered by pustules and the average pustule size is determined. The performance of the Mask R-CNN is assessed using the following measures:

Table 1. Parameter Values for Mask R-CNN

Parameter	Value
Number of epochs	150
Learning rate	0.00001
Ground truth instances	400
Detection instances	400
Region of interest instances	400
Non-maximal suppression threshold	0.75

- PCAcc - This measures the pustule count accuracy. This measure is calculated by firstly taking the difference between the number of pustules specified in the label for the training instance and the number of pustules identified by the Mask R-CNN at the same coordinates. This value is expressed as a percentage of the total number of pustules specified for the training instance and averaged over all the training instances in the training set.
- SAAcc -This is a measure of the accuracy of the surface area covered by pustules determined by the Mask R-CNN. The surface area is measured in terms of the number of pixels. For each training instance the difference in the number of pixels is calculated and expressed as a percentage of the surface area specified in the label of the training instance. This percentage is then averaged over all the training instances.
- APSAcc - This measure assesses the average pustule size accuracy. The difference in the average pustule size specified in the training instance and that produced by the Mask R-CNN is calculated and expressed as a percentage. This percentage is averaged over training instances.

The following section describes the experimental setup used to assess the performance of the image processing algorithms and the Mask R-CNN.

5 Experimental Setup

This section describes the experimental setup used to evaluate the performance of the standard image processing algorithms and the Mask R-CNN. The main aim of the research presented is to compare the performance of the standard image processing algorithms and Mask R-CNN. The Mann Whitney-U test has been used to ascertain the statistical significance of the results obtained. Section 5.1 describes the dataset used to evaluate the approaches and Sect. 5.2 presents the technical specifications of the machines used to run experiments.

5.1 Dataset of Common Rust Images

The dataset was created from maize plants grown in a greenhouse. The maize plants were inoculated by spraying the leaves with common rust urediniospores.

Disease symptoms were monitored over a two week period. At different stages of infection, maize leaves were removed and scanned on a flatbed scanner at 1200 DPI. Common rust exhibits different stages of symptom development, and it was hypothesised that it may be more difficult to quantify the severity of common rust in one stage than another. To test this hypothesis the images were divided into two datasets:

- Dataset 1 - This dataset contains leaves from the early stage of common rust.
- Dataset 2 - This dataset contains leaves from the later stage of common rust.

A total of 1040 images were created, with 400 images in Dataset 1 and 640 images in Dataset 2.

One of the characteristics of maize leaves is that they are longer than they are wider. Large images can slow down the training of neural networks. To work around this problem the individual images were sliced into 8 smaller, equal sided images. This had the added benefit of increasing the size of the dataset without requiring that additional scans be taken. The slices were individually inspected and if a slice was found to not contain any pustules it was removed from the dataset.

Manual annotation using Fiji was performed to specify "ground truth" labels for each of the images (pustule number, pustule area). Rules for annotation were established with the help of a domain expert. The details of the labels are presented in Sect. 4.

For Dataset 1 320 of the 400 images were used for training and 80 for testing. For Dataset 2 512 of the 640 images were used for training and 128 for testing.

5.2 Technical Specifications

All the algorithms were coded in Python. A multicore cluster was used to run simulations. Approximately 144 cores were used for training and 24 cores for testing.

6 Results and Discussion

This section firstly presents the results obtained by applying the Mask R-CNN to Dataset 1 and Dataset 2. A comparison between the standard image processing algorithms and the Mask R-CNN is then presented.

6.1 Mask R-CNN Performance

This section discusses the performance of the Mask R-CNN in predicting the pustule count, surface area covered by pustules and the average pustule size. Table 2 presents the training results and Table 3 the test results.

The neural network appears to have performed better for Dataset 1, i.e. images from the early stage of common rust, than for Dataset 2 which contains

Table 2. Mask R-CNN training performance

Dataset	PAAcc	SAAcc	APSAcc
Dataset 1	74.02%	69.89%	79.68%
Dataset 2	58.54%	58.97%	69.76%

Table 3. Mask R-CNN test performance

Dataset	PAAcc	SAAcc	APSAcc
Dataset 1	73.69%	69.12%	78.88%
Dataset 2	58.04%	58.24%	69.16%

the late stage common rust images. A potential explanation for this discrepancy is illustrated by looking at the difference between the two datasets. Figure 2 shows a typical image taken from the Dataset 1 and Fig. 3 shows a typical image taken from Dataset 2.

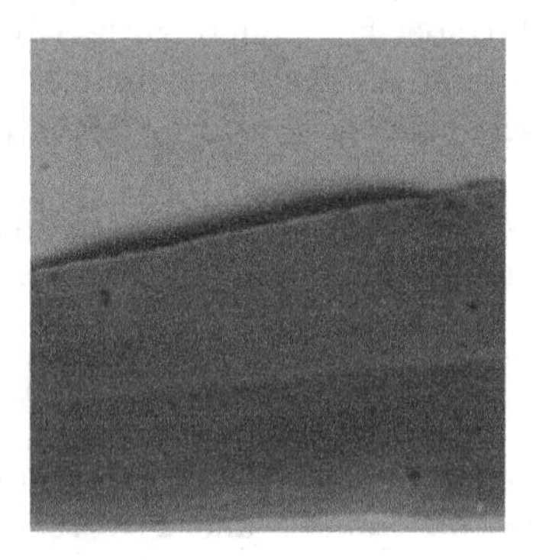

Fig. 2. Dataset 1 example

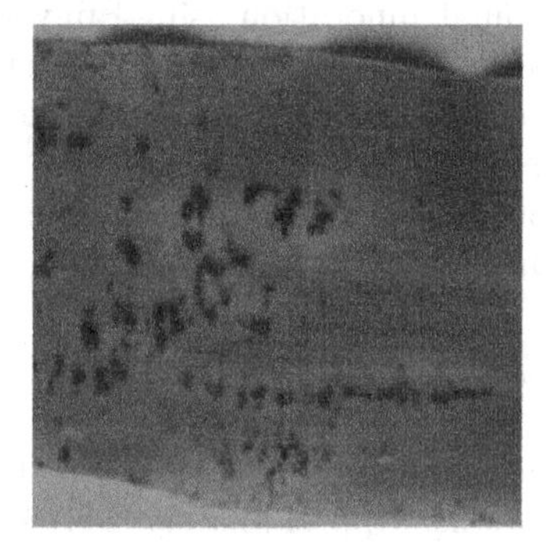

Fig. 3. Dataset 2 example

The leaf in Fig. 2 not only has far less pustules, but they are also more clearly defined and well separated. The leaf in Fig. 3 has far more pustules, in addition to the quantity of pustules, certain pustules are starting to coalesce with other pustules in close proximity. The coalescing of pustules can make it difficult for even a human domain expert to correctly count the number of pustules. The other potential issue with Fig. 3 is the presence of spores on the leaf surface from pustules that have burst open. Even though these spores have not been annotated as pustules during dataset creation, they add additional noise that could affect accuracy. Future work will investigate techniques for addressing coalescing pustules and noise.

6.2 Qualitative Results

This subsection presents some qualitative results in the form of images with their masks overlayed for both Dataset 1 and Dataset 2. Three example leaves from Dataset 1 are given in Fig. 4, the ground truth annotations for these leaves are shown in Fig. 5 and the masks produced for the leaves by the neural network are shown in Fig. 6. Three example leaves from Dataset 2 are given in Fig. 7, the ground truth annotations for these leaves are shown in Fig. 8 and the masks produced for the leaves by the neural network are shown in Fig. 9.

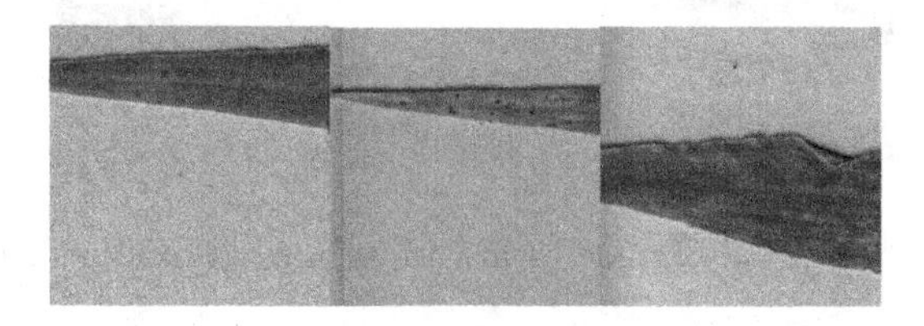

Fig. 4. Dataset 1 example - leaf without ground truth annotations

Fig. 5. Dataset 1 example - leaf with ground truth annotations

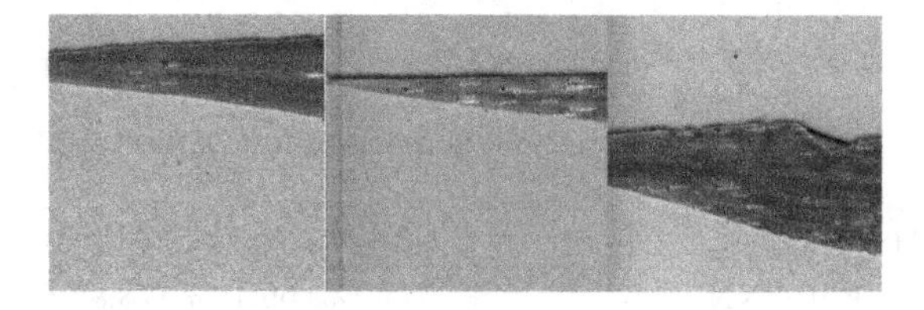

Fig. 6. Dataset 1 example - leaf with neural network predictions

Fig. 7. Dataset 2 example - leaf without ground truth annotations

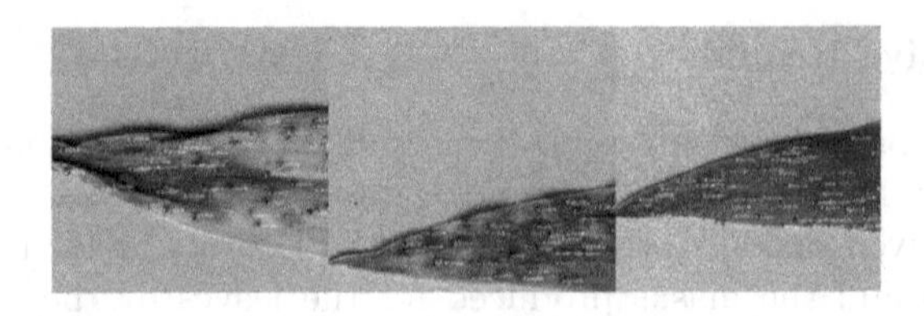

Fig. 8. Dataset 2 example - leaf with ground truth annotations

Fig. 9. Dataset 2 example - leaf with neural network predictions

6.3 Performance Comparison

This section compares the performance of the standard image processing algorithms to that of the Mask R-CNN. The performance comparison for Dataset 1 is presented in Table 4 and for Dataset 2 in Table 5.

From Table 4 and Table 5 it is evident that the Mask R-CNN has outperformed the standard image processing algorithms for both Dataset 1 and Dataset 2. These results were found to statistically significant a 5% level of significance.

As can be anticipated the standard image processing algorithms have lower runtimes than the mask R-CNN. Table 6 lists the runtimes for all the experiments.

Table 4. Performance comparison for Dataset 1

Approach	PAAcc	SAAcc	APSAcc
Mask R-CNN	73.69%	69.12%	78.88%
Image processing algorithms	37.24%	30.71%	36.07%

Table 5. Performance comparison for Dataset 2

Approach	PAAcc	SAAcc	APSAcc
Mask R-CNN	58.04%	58.24%	69.16%
Image processing algorithms	31.22%	16.97%	31.26%

Table 6. Runtime comparison

Dataset	Mask R-CNN	Mask R-CNN	Algorithms
	Training	Testing	Algorithms
Dataset 1	4 days 10 h 21 min	12 min 8 s	1 min 34 s
Dataset 2	4 days 18 h 5 min	14 min 34 s	1 min 35 s

The training of the Mask R-CNN can be conducted offline and the testing runtimes are reasonable to use the model in realtime.

7 Conclusion and Future Work

The research presented in this paper compares the performance of standard image processing algorithms and the Mask R-CNN in quantifying the severity of common rust on maize leaves. This is the first study applying machine learning techniques to the quantification of the severity of common rust in maize. The study revealed that the Mask R-CNN is more effective than the standard image processing algorithms in quantifying the severity of common rust in maize. This result was found to be statistically significant at a 5% level of significance. As expected the runtimes for Mask R-CNN are higher than that for the standard image processing algorithms. However, the testing times are reasonable for real-time use, especially given the improvement in accuracy. The study also revealed that quantifying the severity of common rust in later stages of the disease proved to be more challenging than quantifying the severity in early stages. It is hypothesised that the possible reason for this is noise and coalescing pustules in the late stage common rust. Future work will investigate techniques for addressing this.

One of the challenges with Mask R-CNN is finding effective parameter values. Future work will investigate the automating the process of parameter tuning. Previous work has shown the effectiveness of selection perturbative hyper-heuristics [13] and evolutionary algorithms for parameter tuning [12], hence future research will investigate this for Mask R-CNN.

Acknowledgements. This work is based on the research supported wholly/in part by the National Research Foundation of South Africa (Grant Numbers 46712). Opinions expressed and conclusions arrived at, are those of the author and are not necessarily to be attributed to the NRF.

References

1. Alfaro, E., Fonseca, X.B., Albornoz, E.M., Martínez, C.E., Ramrez, S.C.: A brief analysis of U-Net and mask R-CNN for skin lesion segmentation. In: 2019 IEEE International Work Conference on Bioinspired Intelligence (IWOBI), pp. 000123–000126. IEEE (2019)

2. Bade, C., Carmona, M.: Comparison methods to assess severity of common rust caused by Puccinia Sorghi in maize. Trop. Plant Pathol. **36**, 264–266 (2011)
3. Bharati, P., Pramanik, A.: Deep learning techniques—R-CNN to mask R-CNN: a survey. In: Das, A.K., Nayak, J., Naik, B., Pati, S.K., Pelusi, D. (eds.) Computational Intelligence in Pattern Recognition. AISC, vol. 999, pp. 657–668. Springer, Singapore (2020). https://doi.org/10.1007/978-981-13-9042-5_56
4. Burić, M., Pobar, M., Ivašić-Kos, M.: Object detection in sports videos. In: 2018 41st International Convention on Information and Communication Technology, Electronics and Microelectronics (MIPRO), pp. 1034–1039. IEEE (2018)
5. Dorrer, M., Tolmacheva, A.: Comparison of the YOLOv3 and mask R-CNN architectures' efficiency in the smart refrigerator's computer vision. J. Phys. Conf. Ser. **1679**, 042022 (2020)
6. Guterres, A.: The sustainable development goals report. Technical report, United Nations (2018)
7. He, K., Gkioxari, G., Dollar, P., Girshick, R.: Mask R-CNN. In: Proceedings of the 2017 IEEE International Conference on Computer Vision (ICCV) (2017). https://doi.org/10.1109/ICCV.2017.322
8. Kanopoulos, N., Vasanthavada, N., Baker, R.: Design of an image edge detection filter using the Sobel operator. IEEE J. Solid-State Circuits **23**(2), 358–367 (1988)
9. Korsman, J., Meisel, B., Kloppers, F.J., Crampton, B., Berger, D.: Quantitative phenotyping of grey leaf spot disease in maize using real-time PCR. Eur. J. Plant Pathol. **133**(2), 461–471 (2012). https://doi.org/10.1007/s10658-011-9920-1
10. Abudulla, W.: Mask R-CNN for object detection and instance segmentation on keras and tensorflow (2017). https://github.com/matterport/Mask_RCNN
11. Mohanty, S., Hughes, D., Salathe, M.: Using deep learning for image-based plan disease detection. Front. Plant Sci. **7**(1419), 1–10 (2016)
12. Nyathi, T., Pillay, N.: Comparison of a genetic algorithm to grammatical evolution for automated design of genetic programming classification algorithms. Expert Syst. Appl. **104**, 213–234 (2018)
13. Pillay, N., Qu, R.: Conclusions and future research directions. In: Hyper-Heuristics: Theory and Applications. NCS, pp. 99–101. Springer, Cham (2018). https://doi.org/10.1007/978-3-319-96514-7_13
14. Quoc, T.T.P., Linh, T.T., Minh, T.N.T.: Comparing U-Net convolutional network with mask R-CNN in agricultural area segmentation on satellite images. In: 2020 7th NAFOSTED Conference on Information and Computer Science (NICS), pp. 124–129. IEEE (2020)
15. Rahnemoonfar, M., Sheppard, C.: Deep count: fruit counting based on deep simulated learning. Sensors **17**(4), 905 (2017). https://doi.org/10.3390/s17040905
16. Ren, S., He, K., Girshick, R., Sun, J.: Faster R-CNN: towards real-time object detection with region proposal networks. IEEE Trans. Pattern Anal. Mach. Intell. **39**(6), 1137–1149 (2017). https://doi.org/10.1109/TPAMI.2016.2577031
17. Savary, S., Williocquet, L., Pethybridge, S., Esker, P., McRoberts, N., Nelson, A.: The global burden of pathogens and pests on major food crops. Nat. Ecol. Evol. **3**(3), 430–439 (2019)
18. Shah, D., Dillard, H.: Yield loss in sweet corn caused by Puccinia Sorghi: a meta-analysis. Plant Dis. **90**(11), 1413–1418 (2006)
19. Soille, P., Ansoult, M.: Automated basin delineation from digital elevation models using mathematical morphology. Signal Process. **20**(2), 171–182 (1990)

20. Sommer, L., Schumann, A., Schuchert, T., Beyerer, J.: Multi feature deconvolutional faster R-CNN for precise vehicle detection in aerial imagery. In: 2018 IEEE Winter Conference on Applications of Computer Vision (WACV), pp. 635–642. IEEE (2018)
21. Stewart, E.L., et al.: Quantitative phenotyping of northern leaf blight in UAV images using deep learning. Remote Sens. **11**(19) (2019). https://doi.org/10.3390/rs11192209. https://www.mdpi.com/2072-4292/11/19/2209
22. Ubbens, J., Cieslak, M., Prusinkiewicz, P., Stavness, I.: The use of plant models in deep learning: an application of leaf counting in rosette plants. Plant Methods **14**(6) (2018). https://doi.org/10.1186/s13007-018-0273-z
23. Wang, G., Sun, Y., Wang, J.: Automatic image-based plant disease severity estimation using deep learning. Comput. Intell. Neurosci. (2017). https://doi.org/10.1155/2017/2917536
24. Xu, L., Li, Y., Sun, Y., Song, L., Jin, S.: Leaf instance segmentation and counting based on deep object detection and segmentation networks. In: 2018 Joint 10th International Conference on Soft Computing and Intelligent Systems (SCIS) and 19th International Symposium on Advanced Intelligent Systems (ISIS), pp. 180–185. IEEE (2018)
25. Zhao, T., Yang, Y., Niu, H., Wang, D., Chen, Y.: Comparing u-net convolutional network with mask R-CNN in the performances of pomegranate tree canopy segmentation. In: Multispectral, Hyperspectral, and Ultraspectral Remote Sensing Technology, Techniques and Applications VII, vol. 10780, p. 107801J. International Society for Optics and Photonics (2018)

Federated Learning Model with Augmentation and Samples Exchange Mechanism

Dawid Połap[1(✉)], Gautam Srivastava[2], Jerry Chun-Wei Lin[3], and Marcin Woźniak[1]

[1] Faculty of Applied Mathematics, Silesian University of Technology, Kaszubska 23, 44-101 Gliwice, Poland
{Dawid.Polap,Marcin.Wozniak}@polsl.pl
[2] Department of Mathematics and Computer Science, Brandon University, Brandon, MB R7A 6A9, Canada
SrivastavaG@brandonu.ca
[3] Western Norway University of Applied Sciences, Bergen, Norway
JerryLin@ieee.org

Abstract. The use of intelligent solutions often comes down to the use of already trained classifiers, which is caused by one of their biggest drawbacks. It is the accuracy or effectiveness of artificial intelligence methods, which are algorithms called data-hungry. It means that it depends on the number of samples in the database, and the quality of the classifier could be better if their number is high and the samples are different. In this paper, we propose a solution based on the idea of federated learning in an application for intelligent systems. The proposed solution consists not only in the division of the database among workers but also in the quality of the samples and their possible exchange. Exchanging samples for a particular worker means labeling difficult to classify samples. These samples are used to expand the sets using the generative adversarial network. The mathematical model of a proposal is described, then the experimental results are shown and discussed with the comparison to the classic approach.

Keywords: Internet of Things · Artificial intelligence · Federated learning · Generative adversarial network · Convolutional neural network

1 Introduction

The last few years brought 5G networks [9,12], which is still implemented around the world today. 5G allows using solutions in the field of the Internet of Things in everyday life. These solutions bring benefits such as the improvement and monitoring of activities or operations, and even the availability of analysis results. The idea is that certain things can pick up signals, process, and share them with

L. Rutkowski et al. (Eds.): ICAISC 2021, LNAI 12854, pp. 214–223, 2021.
https://doi.org/10.1007/978-3-030-87986-0_19

others. The acquisition of signals is done with the help of sensors that allow performing certain information-producing operations. With the help of technological solutions, especially machine learning said information is processed or sent to others. Communication is a key element due to additional data from other areas that may influence the present ones.

Information processing most often requires artificial intelligence methods that classify information, or even make predictions. Unfortunately, the use of artificial intelligence methods involves taking over all the problems or disadvantages associated with them. One such drawback is the training process to a certain level of classifier effectiveness. Using methods such as artificial neural networks, obtaining a high level of effectiveness is associated with the need for a large amount of data in the training database [3]. In addition to requiring a large number of samples, their diversity is important for a classifier to be trained on different variants in terms of given classification classes. This problem is particularly acute in the case of using a specific database under test conditions and then using the classifier in a new environment [7,14,15]. The classifier may not be able to correctly process the new data.

One of the latest solutions which modified the approach to training artificial intelligence methods is a collaborative way called federated learning [23]. The idea is to divide the existing database into many workers, where each worker trains the classifier on its part of the database. Then, the trained configuration is sent to the server, whose purpose is to create one, general configuration based on all received ones. This solution allows implementing Internet of Things solutions using artificial neural networks, where objects collect information from the environment and train their classifier. After some time, they send the classifier information to the central server which generates a new configuration. The main advantage of this solution is that each object can process its data and train classifiers with private data, and after some time it gets a more generalized version, which is again adapted to private data.

Federated learning idea was analyzed in [10], where the main challenges and future directions were mainly described and discussed. The authors focused among other things on expensive communication between workers and servers, privacy, and security. These two aspects are important from the point of practical applications. As the future directions, there were indicated some ideas like Pareto frontier and novel asynchronous models. In terms of security, the most important solutions are blockchain [13]. Again in [22], the researchers described a secure and verifiable federated learning model. The aspect of security was also described in [2], where different attacks were shown and examined. The presented research shows that each worker has a big impact on the aggregated model. Again, communications are mainly analyzed in wireless networks [20]. The authors defined this idea as a FEDL optimization problem, which takes into account communication delays at the communication level through the accuracy of classifiers and computing power. In [18], the sparse ternary compression method was proposed which was developed for federated learning requirements during transmission. A similar optimization task was considered in [17], where the idea of federated learning involves edge computing. The authors proposed a solution that brings the heavy computing operation to edge nodes.

The topic of communication and learning techniques can be also defined as one optimization problem which involves joint learning methods, wireless resource allocation and user selection [4]. Nowadays, collaborative learning has many applications like mobile keyboard prediction [6], where recurrent neural networks were used to predict the next-word in smartphones. These solutions show the advantages of the main assumptions of the idea, i.e. keeping private data, but using them to create a more general model (using many different users' smartphones). The industrial application needs efficient and safe solutions, and an example of such solutions with orthographic security was proposed in [5]. Practical application may also be subjected to various cyber-attacks not only on the model of operation but also on devices [16]. The authors of the analysis proposed a scheme for detecting intrusion into devices in the Internet of Things based on the collaborative learning model. It is also an important aspect in medical purpose, where the federated learning idea could be spread to many hospitals, and in each of them, some devices will train a classifier with patients' data. It is crucial to secure this data [19]. To create effective tools and secure all the data, the researchers try to develop a new mechanism and improve existing ones. Another aspect that modifies federated learning is to introduce adaptivity like control algorithms which determine some aspects [21].

In this paper, we propose an alternative construction of a federated learning mechanism, where each worker has their private database, but the samples with the best classification results can be sent to the server for changing on other samples if certain conditions are met. The server collects the samples, that were sent back as having the best classifier result, and in return receive a new augmented sample via the generative adversarial network (GAN). This solution guarantees continuous improvement in the effectiveness of individual workers.

2 Proposed Architecture

Federated learning has to train one, aggregated learning model θ which is holded on server. A training database D is split into N (where $N > 0$) subsets D_i which satisfy the condition $D = \bigcup_{i=1}^{N} D_i$. Mark workers as $\{\xi_1, \xi_2, \ldots, \xi_N\}$. In original idea, each set $D_i$is assigned to worker ξ_i, in our proposition, a database is split into $N+1$ subsets, so $D = \bigcup_{i=1}^{N+1} D_i$. The subset marked as D_{N+1} is assigned to server.

2.1 Worker

The size of a private dataset of i-th worker can be marked as $|\xi_i| \stackrel{\circ}{=} |x_i|$. In the beginning, the worker gets its database and waits to receive the initial model θ^0 from the server. Then, the worker split his datasets into two subsets – training and validating. The first subset is used to retrain the classifier with model θ^0 by T_w iterations. After retraining, all samples from the validating subset are

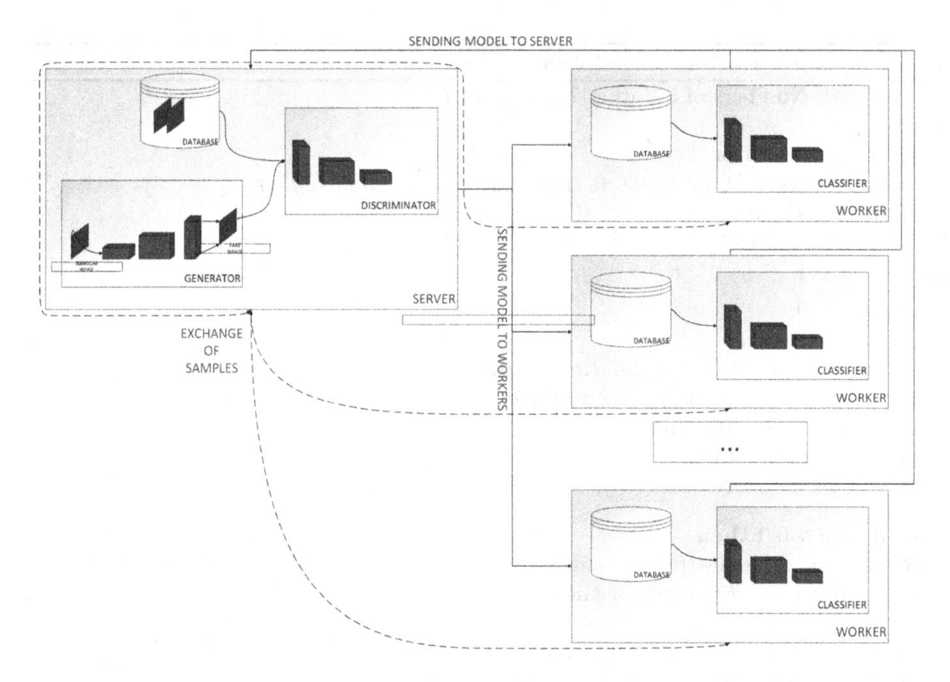

Fig. 1. Visualization of the operation of the proposed modification of the federated learning concept.

evaluated by this model. The samples with the best class prediction can be sent to the server for exchange with another sample. It is done if the accuracy of prediction of that sample is higher than 90%, which can be defined as

$$l_j(\theta) > 0.9, \tag{1}$$

where $l_j(\cdot)$ is loss function for j-th function using θ model.

After this process, i-th worker calculate loss value of model using D_i dataset according to

$$L_i(\theta) = \frac{1}{|D_i|} \sum_{j \in \xi_i} l_j(\theta), \tag{2}$$

where $l_j(\cdot)$ is the value of function on given data sample. Then send the model with the value of $L_i(\theta)$ to the server. And then the operations like receiving a new model, retraining, and sending it are repeated by K rounds of federated learning. A pseudocode of the worker's operation is presented in Algorithm 1.

2.2 Server

In the beginning, the server creates an initial model θ^0 which is sent to all workers. This model is created by training a discriminator in GAN using private

Algorithm 1: Worker operation in federated learning.

```
  Input: Numbers of iteration T_w, dataset D_i
1 t := 0;
2 Get model θ from server;
3 Split dataset D_i into subsets trainingSet and validatingSet;
4 for t < T_w do
5 |   Train model;
6 |_  t++;
7 tmp := 0;
8 sampleNumber := 0;
9 for each sample j in validatingSet do
10 |   Evaluate sample j using retrained model θ;
11 |   if tmp < l_i(θ) then
12 |   |   tmp = l_i(θ);
13 |_  |_  sampleNumber = j;
14 if tmp > 0.9 then
15 |   Send a request for exchange sample j;
16 |   if a request is accepted then
17 |   |   Send sample j;
18 |   |   Delete sample j from dataset D_i;
19 |_  |_  Get a new sample and add it to dataset D_i;
20 Send trained model to the server;
```

dataset D_{N+1} by T_s iterations. After training this model, a server sent generated θ^0. During waiting for all models from workers, a server trains GAN, especially generators for new samples. The idea of that is the purpose for extending subset D_{N+1} with new samples that are passed by discriminator and used for possible exchange with workers if there is such a need. The exchange is made only when in dataset D_{N+1}, there are samples which has accuracy lower than 0.5 on the discriminator with the current model

$$l_j(\theta) < 0.5. \tag{3}$$

If there are no such samples, then the exchange does not take place.

After getting all models from workers in j round, the server creates a new model θ^j, that is used in discriminator and resent it to all workers for the next round of federated learning. The federated learning process can be defined as minimizing task

$$\min_{\theta} \hat{L}(\theta) = \frac{1}{N}\sum_{j=1}^{N} L_j(\theta), \tag{4}$$

whose goal is to find the optimal model by minimizing the average value of the loss functions of all workers. The server operation is presented in Algorithm 2, and the whole idea is visualized in Fig. 1.

Algorithm 2: Server operation in federated learning.

```
   Input: Numbers of federated learning round T_s, dataset D_{N+1}
 1 t := 0;
 2 for t < T_s do
 3     Send model θ^t to all workers;
 4     while all models were not received from workers do
 5         Train GAN;
 6         Generate new samples k with GAN;
 7         if l_k(θ) > 0.9 then
 8             Add sample k to dataset D_{N+1};
 9         if a request for a exchange has been received then
10             tmp := 1;
11             sampleNumber := 0;
12             for each sample j in D_{N+1} do
13                 if tmp > l_i(θ) then
14                     tmp = l_i(θ);
15                     sampleNumber = j;
16             if tmp < 0.5 then
17                 Get a new sample and add it to D_{N+1};
18                 Send j-th sample;
19                 Delete sample j from D_{N+1};
20     Calculate loss value using Eq. (4);
21     Update the existing model of discriminator by averaging all on server;
22     t++;
```

3 Experiments

In conducted experiments, we used a MNIST dataset [8], which contains 60000 : 10000 training to validating images for 10 different classes. All data were splitted equal for all workers and server. We examinated proposal for 2, 3, 4 workers, $T_w \in \{10, 15, 20\}$ training iteration, and $T_s \in \{10, 15, 20\}$ rounds of federated learning. The architecture of convolutional neural network (and discriminator) was as follows – convolutional 5×5 with $\tanh(\cdot)$ as activation function, max pooling 2×2, convolutional 5×5 with $\tanh(\cdot)$ as activation function, max pooling 2×2, fully-connected with 1024 neurons and $\tanh(\cdot)$ as activation function, and output layer with one neuron with $(1 + \exp(-x))^{-1}$ as activation function. Generator in GAN has reverse structure (max pooling was replaced with up sampling layer). As training algorithm, stochastic gradient descent [1,11] was chosen.

We examined third parameters – the average accuracy of the model after T_s rounds of federated learning, the sum of exchanges from all workers, and the number of generated samples by GAN which was added to the database. All obtained results are presented in Table 1, 2 and 3. In the case of accuracy measurements, the lowest value was achieved using only 2 workers (0.78 and

0.77), and the highest by the use of 3 workers ($T_s = 20$) 4 workers ($T_s = 10$, $T_s = 20$). The accuracy of the aggregated model increased with a higher number of workers. Of course, there are cases when it was also lower ($T_s = 20$ for 3 workers reached only 0, 81) which is due to the random distribution of samples to individual databases.

Table 1. Obtained results for $T_s = 10$ round of federated learning.

Number of workers	2			3			4		
T_w	10	15	20	10	15	20	10	15	20
Average accuracy	0,83	0,78	0,81	0,83	0,84	0,85	0,83	0,87	0,88
Number of exchanges	9	9	5	6	11	17	26	31	17
Number of generated samples	20	20	20	30	30	30	40	40	40

Table 2. Obtained results for $T_s = 15$ round of federated learning.

Number of workers	2			3			4		
T_w	10	15	20	10	15	20	10	15	20
Average accuracy	0,82	0,84	0,84	0,86	0,85	0,85	0,82	0,83	0,86
Number of exchanges	18	13	10	39	30	20	14	26	19
Number of generated samples	30	30	30	45	45	45	60	60	60

Table 3. Obtained results for $T_s = 20$ round of federated learning.

Number of workers	2			3			4		
T_w	10	15	20	10	15	20	10	15	20
Average accuracy	0,77	0,84	0,83	0,81	0,88	0,89	0,86	0,86	0,88
Number of exchanges	12	38	23	33	31	50	50	31	44
Number of generated samples	40	40	40	60	60	60	80	80	80

The number of exchanges was very chaotic, because of the distribution of samples in databases. In the case, when the number of federated learning rounds T_s was higher, the exchanges were much more frequent compared to a smaller number of T_s. In general, the exchange mechanism was in many cases very often used. This indicates that such a mechanism is important because a large number of exchanges also reached higher results at the accuracy level. Replacing a very well-classified sample with a new one makes training on a modified basis more diverse, and even increases the generalization of the model itself. The third parameter that was analyzed is the number of created samples by GAN on a server. This number was limited to $T_s \cdot workers$ due to the small number of exchanges. It is worth noting that the exchange took place after the full training process of the worker, so the exchange could be as many as federated learning round. We noticed that, in each case, GAN generated a maximum number of samples. Some example of created ones are presented in Fig. 2.

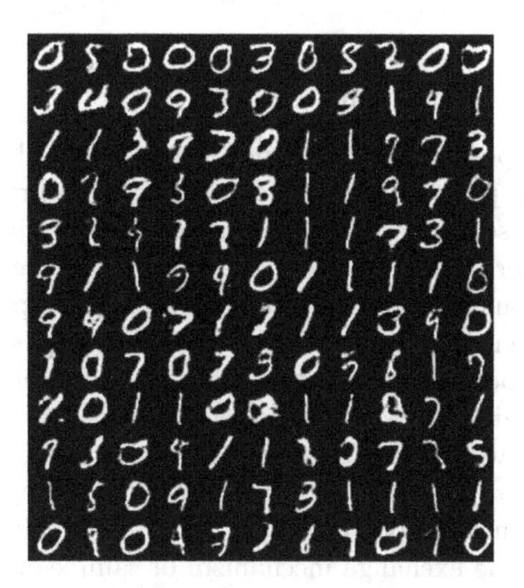

Fig. 2. Generated samples using GAN and added to the server's database.

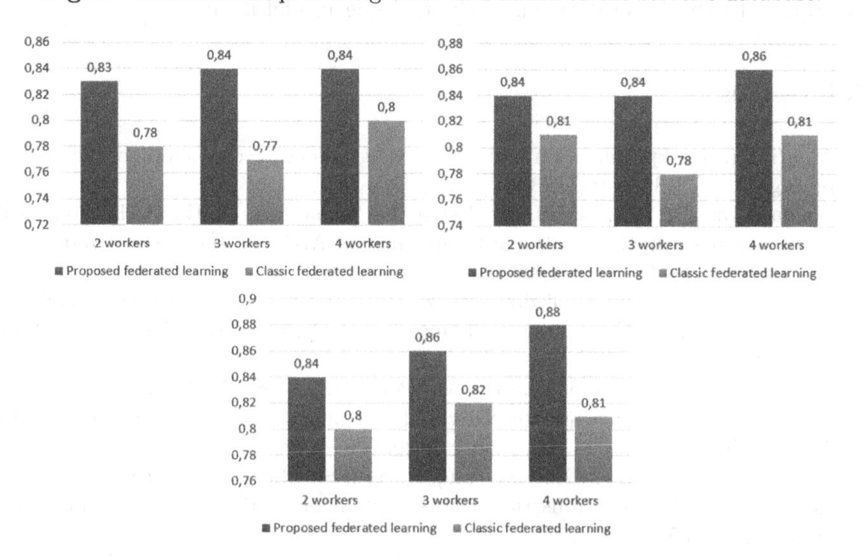

Fig. 3. Comparison classic concept to proposed one. In the first row for $T_s = 10$, $T_s = 15$, and in the second one $T_s = 20$.

In Fig. 3, the comparison of the proposed method with classic federated learning is presented. Described solutions have better results on average by over 5%. The main differences between both solutions are provided modification as augmentation by GAN and exchanging mechanism. Exchanging mechanism is not too burdensome for the algorithm itself, but only for data transfer. Generating new samples is a more computationally complex mechanism, especially when training GAN, and therefore it is located on the server. It is worth to notice, that GAN is trained until some maximum number of samples would be generated. If this number is small, then this process may be short.

4 Conclusion

In this paper, we propose a modified concept of federated learning for the image recognition task. We proposed two modifications. The first one was extending the operation of the server which in the original version just aggregate models and send it to workers. Here, the server has GAN, where the discriminator is trained by workers, and the server train generator. The purpose of GAN is to create a new, augmented sample. These samples can be sent to workers. This idea is the second modification which is an exchange mechanism. If the worker has very well-classified samples, it can be sent to the server, which will send back another sample with worse classification results.

This proposal was examined on the classic database, where the achieved results show the potential of practical application. Our solution achieved results higher by over 5%. The exchange mechanism of samples modifies the database by removing very well-classified samples, replacing them with worse ones, which contributes to the generalization of the training model itself.

References

1. Amari, S.i.: Backpropagation and stochastic gradient descent method. Neurocomputing **5**(4–5), 185–196 (1993)
2. Bagdasaryan, E., Veit, A., Hua, Y., Estrin, D., Shmatikov, V.: How to backdoor federated learning. In: International Conference on Artificial Intelligence and Statistics, pp. 2938–2948. PMLR (2020)
3. Barbedo, J.G.A.: Impact of dataset size and variety on the effectiveness of deep learning and transfer learning for plant disease classification. Comput. Electron. Agric. **153**, 46–53 (2018)
4. Chen, M., Yang, Z., Saad, W., Yin, C., Poor, H.V., Cui, S.: A joint learning and communications framework for federated learning over wireless networks. IEEE Trans. Wireless Commun. **20**, 269–283 (2020)
5. Hao, M., Li, H., Luo, X., Xu, G., Yang, H., Liu, S.: Efficient and privacy-enhanced federated learning for industrial artificial intelligence. IEEE Trans. Industr. Inf. **16**(10), 6532–6542 (2019)
6. Hard, A., et al.: Federated learning for mobile keyboard prediction. arXiv preprint arXiv:1811.03604 (2018)

7. Hyla, T., Wawrzyniak, N.: Identification of vessels on inland waters using low-quality video streams. In: Proceedings of the 54th Hawaii International Conference on System Sciences, p. 7269 (2021)
8. LeCun, Y.: The MNIST database of handwritten digits (1998). http://yann.lecun.com/exdb/mnist/
9. Li, S., Da Xu, L., Zhao, S.: 5G Internet of Things: a survey. J. Ind. Inf. Integr. **10**, 1–9 (2018)
10. Li, T., Sahu, A.K., Talwalkar, A., Smith, V.: Federated learning: challenges, methods, and future directions. IEEE Signal Process. Mag. **37**(3), 50–60 (2020)
11. Li, Y., Liang, Y.: Learning overparameterized neural networks via stochastic gradient descent on structured data. In: Advances in Neural Information Processing Systems, pp. 8157–8166 (2018)
12. Liu, Y., Peng, M., Shou, G., Chen, Y., Chen, S.: Toward edge intelligence: multiaccess edge computing for 5G and internet of things. IEEE Internet Things J. **7**(8), 6722–6747 (2020)
13. Lu, Y., Huang, X., Dai, Y., Maharjan, S., Zhang, Y.: Blockchain and federated learning for privacy-preserved data sharing in industrial IoT. IEEE Trans. Industr. Inf. **16**(6), 4177–4186 (2019)
14. Nourani, V., Foroumandi, E., Sharghi, E., Dabrowska, D.: Ecological-environmental quality estimation using remote sensing and combined artificial intelligence techniques. J. Hydroinf. **23**(1), 47–65 (2021)
15. Polap, D., Wlodarczyk-Sielicka, M.: Classification of non-conventional ships using a neural bag-of-words mechanism. Sensors **20**(6), 1608 (2020)
16. Rahman, S.A., Tout, H., Talhi, C., Mourad, A.: Internet of things intrusion detection: centralized, on-device, or federated learning? IEEE Network **34**, 310–317 (2020)
17. Ren, J., Wang, H., Hou, T., Zheng, S., Tang, C.: Federated learning-based computation offloading optimization in edge computing-supported Internet of Things. IEEE Access **7**, 69194–69201 (2019)
18. Sattler, F., Wiedemann, S., Müller, K.R., Samek, W.: Robust and communication-efficient federated learning from non-iid data. IEEE Trans. Neural Netw. Learn. Syst. **31**, 3400–3413 (2019)
19. Sheller, M.J., et al.: Federated learning in medicine: facilitating multi-institutional collaborations without sharing patient data. Sci. Rep. **10**(1), 1–12 (2020)
20. Tran, N.H., Bao, W., Zomaya, A., Nguyen, M.N.H., Hong, C.S.: Federated learning over wireless networks: optimization model design and analysis. In: IEEE INFOCOM 2019-IEEE Conference on Computer Communications, pp. 1387–1395. IEEE (2019)
21. Wang, S.S., et al.: Adaptive federated learning in resource constrained edge computing systems. IEEE J. Sel. Areas Commun. **37**(6), 1205–1221 (2019)
22. Xu, G., Li, H., Liu, S., Yang, K., Lin, X.: VerifyNet: secure and verifiable federated learning. IEEE Trans. Inf. Forensics Secur. **15**, 911–926 (2019)
23. Yang, Q., Liu, Y., Chen, T., Tong, Y.: Federated machine learning: concept and applications. ACM Trans. Intell. Syst. Technol. (TIST) **10**(2), 1–19 (2019)

Flexible Data Augmentation in Off-Policy Reinforcement Learning

Alexandra Rak[1(✉)], Alexey Skrynnik[1,2], and Aleksandr I. Panov[1,2]

[1] Moscow Institute of Physics and Technology, Moscow, Russia
rakalexandra@mail.ru

[2] Federal Research Center Computer Science and Control of the Russian Academy of Sciences, Moscow, Russia

Abstract. This paper explores an application of image augmentation in reinforcement learning tasks - a popular regularization technique in the computer vision area. The analysis is based on the model-free off-policy algorithms. As a regularization, we consider the augmentation of the frames that are sampled from the replay buffer of the model. Evaluated augmentation techniques are random changes in image contrast, random shifting, random cutting, and others. Research is done using the environments of the Atari games: Breakout, Space Invaders, Berzerk, Wizard of Wor, Demon Attack. Using augmentations allowed us to obtain results confirming the significant acceleration of the model's algorithm convergence. We also proposed an adaptive mechanism for selecting the type of augmentation depending on the type of task being performed by the agent.

Keywords: Reinforcement learning · Image augmentation · Rainbow · Regularization

1 Introduction

One of the best techniques that can significantly increase the generalizing property of a machine learning model is increasing the amount of data for the training. In practice, this approach often cannot be applied due to the limited amount of available data. One of the ways to solve this problem is to form an extension of the source data based on some knowledge about the problem being solved and some known requirements for model invariance. For example in the image classification problem, such requirement is model invariance to a wide variety of transformations, such as reflection, resizing, adding noise, or changing color. In speech recognition problems such properties are model invariance to adding noise, changing the volume, and changing the speed of the audio track. This approach is called data augmentation and it allows us to generate new data-target pairs using the described transformations and thus obtain more data for learning [15]. However, we should be careful about using different types of augmentations, since such data transformations may affect the label of the true

L. Rutkowski et al. (Eds.): ICAISC 2021, LNAI 12854, pp. 224–235, 2021.
https://doi.org/10.1007/978-3-030-87986-0_20

class of the object being classified. For example, vertical reflection can change the label of a true class for images that represent numbers or letters [5].

The described augmentation methods have proven their effectiveness in the fields of Computer Vision and speech recognition, where knowledge about model invariance can be easily used for data augmentation. However, such techniques weren't studied a lot in the field of reinforcement learning. At the same time image provides a representation of the agent's environment state, it is often one of the main information components received from the environment by the agent as an observation. This work is devoted to exploring the image augmentation influence on the quality of the off-policy model-free model in reinforcement learning. In this work, we investigate the effect of applying a simple idea of image augmentation representing the current and subsequent state of the environment, which are extracted from the replay buffer along with the rest of the data during the operation of the model algorithm. It is assumed that due to the limited size of the buffer this technique will work as regularization and will help to increase the generalization property of the model's algorithm.

Thus, the contribution of the work is as follows:

- studying the influence of popular Computer Vision augmentation techniques in model-free off-policy reinforcement learning algorithms,
- conducting experiments with several Atari environments and identifying augmentation techniques that affect the final quality and speed up the model's algorithm convergence,
- results generalization and interpretation of the augmentation effect for the off-policy algorithms.

2 Related Works

Image augmentation is actively researched in the field of Computer Vision. Successful attempts to solve image classification problem using image augmentation are made in [11]. The efficiency of training data extension using simple augmentation methods, such as cropping, rotating, and flipping input images is demonstrated widely. In this work, the authors restricted data access to a small subset of the ImageNet dataset and evaluated the results for each conversion method. One of the most successful strategies to increase the size of training data in the work is the traditional transformations mentioned above. This work also experiments with generative models for creating images of various styles and suggests a method called Neural Augmentation, which allows the neural network to study transformations that are most suitable for classification problems. In the article [2] a new algorithm was used to automatically search for the best image augmentation technique for each individual dataset. The proposed automatic augmentation selection mechanism showed high validation accuracy on the target dataset and achieved state-of-art accuracy on CIFAR-10, CIFAR-100, Svhn, and ImageNet (without using additional data). The augmentation algorithms obtained on the ImageNet dataset can be transferred to another algorithm and can be used for other datasets such as Oxford Flowers, Caltech-101, Oxford-IIT Pets, FGVC Aircraft, and Stanford Cars.

In Reinforcement Learning (RL) the topic of augmentation has not been studied a lot, but more interesting papers on this topic have recently appeared. In [9] authors presented a plug-and-play method that can improve any RL algorithm. Authors demonstrated how random cutout, color jitter, patch cutout, and random convolution can allow simple algorithms to perform the same way or even outperform complex modern methods by common criteria in terms of efficiency and communication capability. It is argued that only a variety of data can cause agents to focus on meaningful information from multidimensional observations without any changes in the training method. The results are demonstrated in DeepMind environments, where state-of-art quality is shown for 15 environments.

The article [8] offers a simple method that can be applied to standard model-free reinforcement learning algorithms, allowing to get more stable learning of the model directly from images without introducing auxiliary loss functions or doing pre-training. The approach uses augmentations commonly used in Computer Vision tasks to transform input data and as a result dramatically improves the Soft Actor-Critic algorithm's performance, enabling it to reach state-of-the-art performance on the DeepMind control suite. The proposed algorithm, which they dub DrQ: Data-regularized Q, can be combined with any model-free reinforcement learning algorithm. It was also demonstrated by applying it to DQN algorithm and significantly improve its data-efficiency on the Atari 100k benchmark.

In other work [12] the method UCB-DrAC was proposed. This new method is used for automatically finding effective data augmentation for RL tasks. It enables the principled use of data augmentation with actor-critic algorithms by regularizing the policy and value functions with respect to state transformations. It was shown that UCB-DrAC avoids the theoretical and empirical pitfalls typical in naive applications of data augmentation in RL. The method improves training performance by 16% and test performance by 40% on the Procgen benchmark, relative to standard RL methods such as Proximal Policy Optimization [14].

We can also mention works in which task-oriented augmentation was carried out [16,17]. Episodes of solving one of the tasks available to the agent were replenished by episodes in which another goal was achieved. In our work, we focus on studying the impact of data augmentation specifically for off-policy algorithms, for which such research has not been conducted before.

3 Model Description

Q-Learning. Unlike classical algorithms and machine learning methods, Reinforcement Learning is a class of models that do not receive direct information about object-response pairs. Instead of this the agent learns to act in some environment in a way that maximizes some scalar value of the reward. At each discrete step $t = 0, 1, 2, ...$, the environment presents an observation S_t to the agent, the agent reacts by selecting an action A_t, after which it receives a new reward value from the environment R_{t+1} and the next state S_{t+1}. This interaction is formalized by the concept of MDP or Markov decision - making process represented by a tuple $< S, A, T, r, \gamma >$, where S is a finite state space, A is a finite set of actions, $T(s, a, s') = P[S_{t+1} = s'|S_t = s, A_t = a]$ is a stochastic transition function between states, $r(s, a) = E[R_{t+1}|s_t = s, a_t = a]$ - reward function, and $\gamma \in [0, 1]$ - discount coefficient.

Reinforcement Learning uses the assumption that future rewards are discounted with a coefficient of γ for each step. Then the total discounted reward at time t is defined as

$$R_t = \sum_{t'=t}^{T} \gamma^{t'-t} r_{t'}$$

where T is understood as the moment when the game ends. Here it is important to define the so-called action-value function $Q^*(s, a)$ as the maximum expectation of the reward received for any policy π after the agent visits the state s and performs the action a:

$$Q^*(s, a) = \max_{\pi} E[R_t|S_t = s, A_t = a, \pi]$$

The π policy determines the probability distribution of actions for each of the states. The optimal Q-function, in this case, obeys the *equation of Bellman.*

$$Q^*(s, a) = E_{s' \sim \epsilon} \left[r + \gamma \max_{a'} Q^*(s', a')|s, a \right]$$

The basic idea of many reinforcement learning algorithms is to evaluate the action-value function using the Bellman equation in an iterative algorithm:

$$Q_{i+1}(s, a) = E \left[r + \gamma \max_{a'} Q_i(s', a')|s, a \right]$$

Such iterative algorithms converge to the optimal action-value function $Q_i \to Q^*$ for $t \to \infty$ In practice, this basic approach is not very applicable, since the Q-function is evaluated separately for each sequence of steps, without any generalization. Instead, an approximation is used to estimate the Q-function:

$$Q(s, a; \theta) \approx Q^*(s, a)$$

This can be a linear approximator function, or a non-linear approximator can be used instead, including a neural network with θ-weights. Such a Q-neural network can be trained by minimizing the sequence of loss functions $L_i(\theta_i)$:

$$L_i(\theta_i) = E_{s,a \sim p(.)} \left[(y_i - Q(s, a; \theta_i))^2 \right]$$

where y_i is the value of the desired function on the $i$iteration:

$$y_i = E_{s' \sim \epsilon} \left[r + \gamma \max_{a'} Q^*(s', a'; \theta_{i-1})|s, a \right]$$

The gradient of such a function will be equal to:

$$\nabla_{\theta_i} L_i(\theta_i) = E_{s,a\sim p(.);s'\sim\epsilon}\left[\left(r + \gamma \max_{a'} Q(s',a';\theta_{i-1}) - Q(s,a;\theta_i)\right)\nabla_{\theta_i} Q(s,a;\theta_i)\right]$$

Instead of calculating the total expectation in the gradient above, it is often computationally appropriate to optimize the loss function using stochastic gradient descent. If the model weights update each step and replace the expectation with a sample from the distributions p and the emulator ϵ, respectively, then we come to the familiar Q-learning algorithm [20]. This algorithm is a model-free algorithm, as well as an off-policy - it learns using an ϵ-greedy strategy, choosing an action with the maximum value of the Q-function with probability ϵ and with probability $1 - \epsilon$ taking a random action. This strategy allows the model to adjust its own estimates if necessary [10].

Deep Q-Network. Deep Q-Network is a successful generalization of combining convolutional neural networks and reinforcement learning to approximate the Q-function for s_t in the t-step. In this case, the state is fed as input to the neural network in the form of a sequence of pre-processed pixel frames. At each step of the algorithm, depending on the current state, the agent selects the next action using the ϵ-greedy strategy described above and also adds a tuple $(s_t, a_t, r_{t+1}, s_{t+1})$ to a special playback buffer called the replay buffer. The parameters of the neural network are optimized using stochastic gradient descent to minimize the loss of the function:

$$L = \left(r + \gamma \max_{a'} Q_{\hat{\theta}}(s',a') - Q_{\theta}(s,a)\right)^2$$

In this case, the gradient of the loss function is considered only for *online* – a neural network that is also used to select the optimal action. The parameters $\hat{\theta}$ represent *target*, a neural network that is a periodic copy of the *online* network. *Target* – the network is not directly optimized. Using a replay buffer and a target network allows for greater stability of model training and leads to good results for many reinforcement learning tasks.

Rainbow. This algorithm is an extension of the DQN algorithm described above and uses the following 6 improvements: Double Q-Learning [18], Prioritized replay [13], Dueling networks [19], Multi-step learning [3], Distributed RL [1], Noisy Nets [4]. This modification of the DQN was used for research on the use of augmentation in reinforcement learning. The described algorithm was chosen because it exceeds the standard DQN in terms of convergence rate, and also preserves the model-free and off-policy properties.

4 Augmentation in Reinforcement Learning

Data augmentation in Computer Vision problems is considered as a good method not only for increasing the amount of source data but also for increasing the diversity of these data [15]. The most popular image augmentation techniques are horizontal reflection, multi-pixel shifts, rotations, and other transformations. However, not all of these types of augmentation can be applied in reinforcement learning tasks. For example, rotation and horizontal reflection can significantly affect the optimal action that will maximize the agent's reward at a particular step of the game. Therefore, for the model under study, we have chosen augmentations with respect to which the model is invariant. This property means that after the data augmentation steps the optimal action at a particular step does not change.

4.1 Augmentation Types

Random Erase. To augment an arbitrary frame of the game a random point is selected within the borders of the image and a rectangle is constructed from it with side lengths distributed randomly from 0 to 20. The built shape is colored gray.

Random Crop. The original image is expanded by 4 pixels on each side and filled in with black. Then the resulting frame is randomly cut from the resulting image. The size of the final frame corresponds to the size of the initial image without expansion.

Random Contrast. Before augmentation, the regularization coefficient for contrast is randomly selected as a sample from the normal distribution $N(1, 0.5)$. Then the image contrast is changed with this coefficient. In such transformation, the value $k = 1$ corresponds to the absence of changes, and $k = 2$ corresponds to a double increasing of the contrast.

Random Augmentation. One of the three augmentations described above is applied to the image with equal probability.

4.2 General Framework for Action-Value Function Regularization

In DQN-algorithm deep networks and reinforcement learning were successfully combined by using a convolutional neural net to approximate the action-value function Q for a given state S_t, which is fed as input to the network in the form of a stack of raw pixel frames. Applying any augmentation to such frames can be considered as some regularization of the value function. Here we can define the general framework for such regularization by adding augmentation function f, which will apply some exact type of state augmentation. Generalizing, we can say, that augmentation function f can apply to the exact state any transformation which preserves invariance property of the model. Invariant state transformation function can be defined:

$$f : S \times T \to S$$

as a mapping that preserves the Q-values

$$Q(s,a) = Q(f(s,\nu),a)$$

for all $s \in S, a \in A$ and $\nu \in T$. where ν are the parameters of f, drawn from the set of all possible parameters T.

In DQN-algorithm a non-linear approximator can be used for Q-value function approximation, including a neural network with θ-weights. Applying a new value-function regularization framework such Q-neural network can be trained by minimizing the sequence of loss functions $L_i(\theta_i)$:

$$L_i(\theta_i) = E_{s,a \in p(.)}\left[(y_i - Q(f_i(s,\nu),a;\theta_i))^2\right]$$

where y_i is the value of the desired function on the i-iteration.

The choice of a particular regularization function strongly depends on the current problem being solved. The choice of function can be influenced by the specifics of the game, the agent's strategy for getting the maximum reward, and other features. Thus, the task of selecting the f function is non-trivial and requires additional experiments. We need to find such a function, and to choose its parameters to minimize the loss function described above:

$$\hat{f} = argmin_{f \in F, \nu \in T} L(\hat{\theta})$$

where F is a family of functions with mapping $f : S \times T \rightarrow S$ that preserves the property of model invariance.

Due to the complexity of the optimization problem for such a family of functions, this problem can be simplified and reduced to the problem of finding some approximation of a function $\hat{f}$. An important property of such an approximation function is adaptability, the ability of the function to take into account the features of the problem being solved.

In further experiments, we consider f as one of the image augmentation types described above. Set of all possible parameters T can be understood as all possible values of such augmentation parameters. For example among T will be all possible probability distributions of different augmentations for random augmentation.

Each state of the environment can be associated with a certain frame from the game. However, in this case, we can say little about the direction in which, for example, the ball is moving in a Breakout game and at what speed it does so. This raises the question of fulfilling the Markov property of the model, that is, if only one frame is used to represent a certain state of the environment, this property is violated. To solve this problem instead of a single frame of the game to characterize the state, several consecutive frames are used (in our case, 4). This allows you to get more information about the environment, and we can draw conclusions about the direction of the ball and its speed from two frames of the game, and about its acceleration – by three.

In all evaluated games, the source frames are reduced to the size of 84×84 and converted to the black and white format.

An augmentation procedure is added to the Rainbow algorithm, which is performed every time an image is sampled from the replay buffer. In this case, the randomness property should be performed only for different pairs of (s_t, s_{t+1}). Within each such pair, the augmentation is exactly the same for s_t and s_{t+1}. This is necessary in order to preserve the integrity of each observation of the environment, that is, the tuple $(s_t, a_t, r_{t+1}, s_{t+1})$ is an integral, indivisible unit for a reinforcement learning algorithm.

5 Experiments

5.1 Data Augmentation for Off-Policy RL

Currently, there is a lot of environments for solving RL tasks. We decided to explore the impact of augmentation regularization using the OpenAI Gym library, which provides APIs for simulating a large number of virtual environments, including Atari games. In this section, we compare four types of augmentation for the following environments: Berzerk, Breakout, Demon Attack, Space Invaders, Wizard of Wor. As a baseline model, we used the original Rainbow algorithm with the parameters specified in the original paper [6]. Table 1 reports summary results for whole 5 games. Learning curves for Berzerk and Space Invaders environments presented in Fig. 1.

Breakout. For Breakout, several augmentations showed a result that exceeds the result of the original algorithm. Augmentations not only increased the convergence rate of the algorithm but also resulted in a higher reward after convergence. The best result was shown by Random Augmentation, which increased the model's reward by 17%. Random Crap and Random Erase augmentations also accelerated convergence and were able to increase the reward by 10% and 8%, respectively. Random Contrast augmentation did not show results significantly different from baseline.

Space Invaders. The game requires more training time compared to other tested games, so we got significant results only after a large number of training steps. We obtained higher reward results for all types of augmentations in this game. The best type of augmentation is Random Erase, which increases the model's reward by rather 400%. In our experiment, Random Augmentation and Random Contrast are also successful with the result of more than 250% reward increasing. The model with Random Crop augmentation got 47% higher score than a Baseline model.

Wizard of Wor. For Wizard of Wor the best types of augmentation were Random Augmentation and Random Erase with 45% and 20% higher reward result. These two augmentations also increased the convergence rate of the algorithm. For this environment, we got worse results for Random Crop and Random Contrast augmentation with reward decreasing by 17% and 32%.

Berzerk. For this game, we obtained the highest reward results for Random Crop augmentation, which increases the model's reward after 30M training steps by 202%. In our experiment, Random Augmentation is also successful with a result of more than 173% reward increasing. The model with Random Erase on average shows the same score as a baseline model. The worse results were provided by the model with Random Contrast augmentation with 20% reward decreasing.

Demon Attack. For Demon Attack all augmentation types provided good results. But here we got more acceleration of the convergence rate and less increase of models reward in the end. The best augmentations for this game are Random Crop and Random Augmentation with over 15% reward increasing after 30M training steps. For these two augmentation types, we got over 160% increasing of convergence from 10M to 20M training steps. Random Erase provided 90% increasing of convergence from 10M to 20M training steps and 6% reward increasing in the end. For Random Contrast we got 15% increasing in convergence and nearly the same results as a baseline model in the end.

Table 1. Final results for each type of the augmentation for five Atari games. The column for each augmentation shows the final cumulative reward and its percentage relative to the Rainbow without augmentations (Original).

	Steps	Original		Random		Erase		Crop		Contrast	
Environment	$\times 10^6$	Score	%	Score	%	Score	%	Score	%	Score	%
Berzerk	40	2527	0	7301	189	2821	12	**11841**	**369**	2298	−9
Breakout	20	353	0	**389**	**10**	370	5	368	4	338	−4
Demon Attack	40	107648	0	**124237**	**15**	115004	7	124057	**15**	106936	−1
Space Invaders	40	2663	0	9406	253	**13253**	**398**	3926	47	10021	276
Wizard of Wor	30	8261	0	**12746**	**54**	10196	23	6175	−25	5027	−39

5.2 Behavioral Cloning

Augmentations work fine for such tasks as classification. In the following series of experiments, we decided to study the effects of different augmentation types on the behavioral cloning model. The main idea of this section is to explore how different proportions of image transformations can influence the final quality of a model trained to copy the behavior of an expert. We took the model with the best results for the Wizard of Wor game from the previous section. The selected model was trained during 50 m steps using the Rainbow algorithm with Random Augmentation regularization. We launched inference for this model and saved 10,000 state-action transitions to use it as input data for the behavioral cloning model. The input data was augmented using the Random Augmentation method with different proportions of Random Crop, Random Erase, Random Contrast, and Original Data augmentations (data without changes). As a model, we took a simple classifier with Cross-Entropy Loss. To iterate through various augmentation ratios for each of the augmentations in Random Augmentation we used discretization of [0, 33, 66, 100] followed by normalization to obtain the final discrete

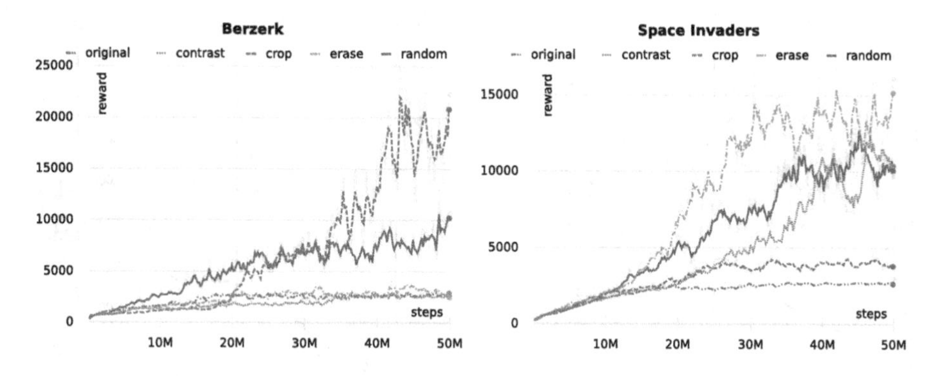

Fig. 1. Learning curves for Berzerk **(left)** and Space Invaders **(right)** environments. The correct choice of augmentation leads to a significant improvement in the convergence of the algorithm. The Crop augmentation shows superior results for Berzerk environment, but for the Space Invaders environment, the result is almost the same as the original Rainbow.

distribution $\{RandomCrop, RandomErase, RandomContrast, OriginalData\}$. This distribution is used to sample an augmentation for each item of a batch during the training process.

We tested model results with augmentation discretization above. Figure 2 shows the results for the top 30 and worst 30 sets from 256 experiments. The trained model for each ratio set tested over 100 evaluation episodes in the Wizard of Wor environment. The best cumulative reward 8650 showed the following augmentation ratio: contrast 33%, crop 100%, erase 100%, original 66% (normalized: 0.11, 0.33, 0.33, 0.22). From these results, one can conclude that a properly augmented model can achieve a better reward score than the original model while testing in the environment. Also, different proportions of augmentation types lead to different model quality. For example, for the Wizard of Wor environment, we got that the higher rate of Random Contrast and low rates of other augmentations leads to lower model quality. We also inspect the correlation between the obtained reward score and model accuracy. For all 256 experiments model accuracy varies not much (in the interval: 0.83 .. 0.86). Despite this, the model reward varies a lot, so the small perturbation in model accuracy leads to a significant difference in the reward score. Thus in our experiment, the mixture of all discussed augmentations types with a higher proportion of Erase and Crop Augmentations and a lower proportion of Contrast showed the best result.

Behavioral cloning experiments can be considered as a part of the search for an approximation of an adaptable regularization function described in Sect. 4.2. The experiment can be continued with subsequent model training using Reinforcement Learning algorithms and applying a new adaptable regularization function. For example, the best augmentations distribution found in this section can be applied to DQN from Demonstrations (DQfD) algorithm [7], where some expert's demonstrations produced by human or another well-trained agent also can be used in DQN algorithm and could significantly speed up the training

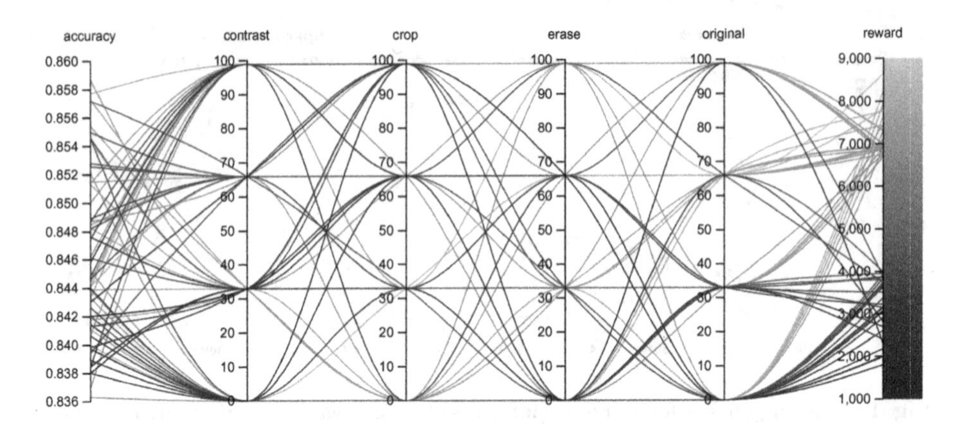

Fig. 2. Hyperparameter search for behavioral cloning on Wizard of Wor environment. We grid search over the ratio of frames for different augmentation type in training batch, considering the following percentages: 0%, 33%, 66%, 100%. For visibility purposes, this chart shows the results only for the top 30 and worst 30 sets from 256 experiments. The trained model for each ratio set tested over 100 evaluation episodes in the environment. The best cumulative reward 8650 showed the following augmentation ratio: contrast 33%, crop 100%, erase 100%, original 66% (normalized: 0.11, 0.33, 0.33, 0.22). The leftmost column reports validation accuracy on test data consisting of 20,000 frames.

process. Such experiments can include applying adaptable regularization function during pre-train or fine-tuning stages and will be the next step of our research.

6 Conclusion

Data augmentation methods have proven to be effective in image analysis. In this paper, we have applied a number of well-known augmentation techniques to the problem of Reinforcement Learning with image-based observations. We have developed an adaptive version of data augmentation for off-policy algorithms that use replay buffer as temporary memory. We have shown that augmentation improves the quality of one of the well-known state-of-the-art Rainbow algorithm. We conducted an experimental study on the selection of hyperparameters for our method of augmented data mixing. In future work, we plan to develop this method and conduct experiments with algorithms that use demonstrations.

Acknowledgements. The reported study was supported by the Ministry of Education and Science of the Russian Federation, project No. 075-15-2020-799.

References

1. Bellemare, M.G., Dabney, W., Munos, R.: A distributional perspective on reinforcement learning. arXiv preprint arXiv:1707.06887 (2017)

2. Cubuk, E.D., Zoph, B., Mane, D., Vasudevan, V., Le, Q.V.: Autoaugment: Learning augmentation policies from data. arXiv preprint arXiv:1805.09501 (2018)
3. De Asis, K., Hernandez-Garcia, J.F., Holland, G.Z., Sutton, R.S.: Multi-step reinforcement learning: A unifying algorithm. arXiv preprint arXiv:1703.01327 (2017)
4. Fortunato, M., et al.: Noisy networks for exploration. arXiv preprint arXiv:1706.10295 (2017)
5. Goodfellow, I., Bengio, Y., Courville, A.: Deep Learning. MIT Press (2016). http://www.deeplearningbook.org
6. Hessel, M., et al.: Rainbow: Combining improvements in deep reinforcement learning. arXiv preprint arXiv:1710.02298 (2017)
7. Hester, T., et al.: Deep q-learning from demonstrations. arXiv preprint arXiv:1704.03732 (2017)
8. Kostrikov, I., Yarats, D., Fergus, R.: Image augmentation is all you need: Regularizing deep reinforcement learning from pixels. arXiv preprint arXiv:2004.13649 (2020)
9. Laskin, M., Lee, K., Stooke, A., Pinto, L., Abbeel, P., Srinivas, A.: Reinforcement learning with augmented data. arXiv preprint arXiv:2004.14990 (2020)
10. Mnih, V., Kavukcuoglu, K., Silver, D., Graves, A., Antonoglou, I., Wierstra, D., Riedmiller, M.: Playing atari with deep reinforcement learning. arXiv preprint arXiv:1312.5602 (2013)
11. Perez, L., Wang, J.: The effectiveness of data augmentation in image classification using deep learning. arXiv preprint arXiv:1712.04621 (2017)
12. Raileanu, R., Goldstein, M., Yarats, D., Kostrikov, I., Fergus, R.: Automatic data augmentation for generalization in deep reinforcement learning. arXiv preprint arXiv:2006.12862 (2020)
13. Schaul, T., Quan, J., Antonoglou, I., Silver, D.: Prioritized experience replay. arXiv preprint arXiv:1511.05952 (2015)
14. Schulman, J., Wolski, F., Dhariwal, P., Radford, A., Klimov, O.: Proximal policy optimization algorithms. arXiv preprint arXiv:1707.06347 (2017)
15. Shorten, C., Khoshgoftaar, T.M.: A survey on image data augmentation for deep learning. J. Big Data **6**(1), 60 (2019)
16. Skrynnik, A., Staroverov, A., Aitygulov, E., Aksenov, K., Davydov, V., Panov, A.I.: Forgetful experience replay in hierarchical reinforcement learning from expert demonstrations. Knowledge-Based Systems 218, 106844 (2021), https://linkinghub.elsevier.com/retrieve/pii/S0950705121001076 https://arxiv.org/abs/2006.09939
17. Skrynnik, A., Staroverov, A., Aitygulov, E., Aksenov, K., Davydov, V., Panov, A.I.: Hierarchical Deep Q-Network from imperfect demonstrations in Minecraft. Cognitive Syst. Res. **65**, 74–78 (2021). https://arxiv.org/pdf/1912.08664.pdfwww.sciencedirect.com/science/article/pii/S1389041720300723?via%3Dihubwww.scopus.com/record/display.uri?eid=2-s2.0-85094320898&origin=resultslistlinkinghub.elsevier.com/retrieve/pii/S138904172
18. Van Hasselt, H., Guez, A., Silver, D.: Deep reinforcement learning with double q-learning. arXiv preprint arXiv:1509.06461 (2015)
19. Wang, Z., Schaul, T., Hessel, M., Hasselt, H., Lanctot, M., Freitas, N.: Dueling network architectures for deep reinforcement learning. In: International Conference on Machine Learning, pp. 1995–2003 (2016)
20. Watkins, C.J., Dayan, P.: Q-learning. Mach. Learn. **8**(3–4), 279–292 (1992)

Polynomial Neural Forms Using Feedforward Neural Networks for Solving Differential Equations

Toni Schneidereit(✉) and Michael Breuß

Brandenburg University of Technology Cottbus-Senftenberg,
Platz der Deutschen Einheit 1, 03046 Cottbus, Germany
{Toni.Schneidereit,breuss}@b-tu.de

Abstract. Several neural network approaches for solving differential equations employ trial solutions with a feedforward neural network. There are different means to incorporate the trial solution in the construction, for instance one may include them directly in the cost function. Used within the corresponding neural network, the trial solutions define the so-called neural form. Such neural forms represent general, flexible tools by which one may solve various differential equations. In this article we consider time-dependent initial value problems, which requires to set up the trial solution framework adequately.

The neural forms presented up to now in the literature for such a setting can be considered as first order polynomials. In this work we propose to extend the polynomial order of the neural forms. The novel construction includes several feedforward neural networks, one for each order. The feedforward neural networks are optimised using a stochastic gradient descent method (ADAM). As a baseline model problem we consider a simple yet stiff ordinary differential equation. In experiments we illuminate some interesting properties of the proposed approach.

Keywords: Feedforward neural networks · Initial value problem · Trial solution · Differential equations

1 Introduction

Over the last decades several neural network approaches for solving differential equations have been developed [1–3]. The application and extension of these approaches is a topic of recent research, including work on different network architectures like Legendre [4] and polynomial neural networks [5] as well as computational studies [6,7].

One of the early proposed methods [8] introduced a trial solution (TS) in order to define a cost function using one feedforward neural network. The TS is supposed to satisfy given initial or boundary values by construction. It is also referred to as neural form in this context [8,9]. Let us note that such neural forms

L. Rutkowski et al. (Eds.): ICAISC 2021, LNAI 12854, pp. 236–245, 2021.
https://doi.org/10.1007/978-3-030-87986-0_21

represent a general tool that enable to solve ordinary ordinary differential equations (ODEs), partial differential equations (PDEs) and systems of ODEs/PDEs alike. We will refer here to this approach as the trial solution method (TSM).

Later, the initial method from [8] has been extended by a TS with two feedforward neural networks, which allows to deal with boundary value problems for irregular boundaries [10] and yields broader possibilities for constructing the TS [9]. In the latter context, let us also mention [11] where an algorithm is proposed in order to create a TS based on grammatical evolution.

A technique related to TSM that avoids the explicit construction of trial solutions has been proposed in [12]. The given initial or boundary values from the underlying differential equation are included in the cost function as additional terms, so that the neural form can be set to equal the neural network output. We will refer to this approach as modified trial solution method (mTSM).

The fact that the neural network output computation resembles a linear combination of basis functions leads to a network architecture (for PDEs) presented in [13]. In this work one hidden layer incorporates two sets of activation functions, one of which is supposed to satisfy the PDE and the second dealing with boundary conditions. The basis function coefficients are set to be the connecting weights from the hidden layer to the output, and the sum over all basis functions and coefficients makes up the TS.

Motivated by the construction principle of collocation methods in numerical analysis, we propose in this paper a novel extension of the TS approach. Our extension is based on the observation, that the neural form using one feedforward neural network as employed in [8] may be interpreted as a first order collocation polynomial. We propose to extend the corresponding polynomial order of the neural form. The novel construction includes several feedforward neural networks, one for each order. Compared to a collocation method from standard numerics, the networks take on the role of coefficients in the collocation polynomial expansion. In computational experiments we illuminate some interesting properties of our extension.

2 Setting up the Neural Form

In this section, we first recall the TSM and its modified version mTSM, respectively, compare [8,12]. Then we proceed with details on the feedforward neural networks we employ here, followed by a description of the novel set-up.

2.1 Construction of the Neural Form

Consider an initial value problem written in a general form as

$$G\left(x, u(x), \frac{d}{dx}u(x)\right) = 0, \quad u(x_0) = u_0, \quad x \in D \subset \mathbb{R} \tag{1}$$

with given initial value $u(x_0) = u_0$. In order to connect G with a neural network, several approaches introduce a TS as a differentiable function $u_t(x, \vec{p})$, where $\vec{p}$

contains the network weights. With the collocation method we discretise the domain D by a uniform grid with n grid points x_i, so that G is transformed into

$$G\left(x_i, u_t(x_i, \vec{p}), \frac{\partial}{\partial x} u_t(x_i, \vec{p})\right) = \vec{0} \tag{2}$$

Let us note that, in a slight abuse of notation, we identify Eq. (2) with the vector of corresponding entries, since this enables to give many formula a more elegant, compact notation.

In order to satisfy the given initial value, TSM [8] employs the TS as a sum of two terms

$$u_t(x, \vec{p}) = A(x) + F(x, N(x, \vec{p})) \tag{3}$$

where $A(x)$ is supposed to match the initial condition (with the simplest choice to be $A(x) = u_0$), while $F(x, N(x, \vec{p}))$ is constructed to eliminate the impact of $N(x, \vec{p})$ at x_0. The choice of $F(x, N(x, \vec{p}))$ determines the influence of $N(x, \vec{p})$ over the domain.

Since the TS as used in this work satisfies given initial values by construction, we define the corresponding cost function incorporating Eq. (3) as

$$E^{TSM}[\vec{p}] = \frac{1}{2}\left\|G\left(x_i, u_t(x_i, \vec{p}), \frac{\partial}{\partial x} u_t(x_i, \vec{p})\right)\right\|_2^2 \tag{4}$$

Let us now turn to the mTSM approach after [12]. The mTSM approach chooses the TS to be equivalent to the neural network output directly

$$u_t(x, \vec{p}) = N(x, \vec{p}) \tag{5}$$

Since no condition is imposed by the initial value on the TS in this way, the conditions are added to the cost function for Eq. (5):

$$E^{mTSM}[\vec{p}] = \frac{1}{2}\left\|G\left(x_i, u_t(x_i, \vec{p}), \frac{\partial}{\partial x} u_t(x_i, \vec{p})\right)\right\|_2^2 + \frac{1}{2}\left\|N(x_0, \vec{p}) - u_0\right\|_2^2 \tag{6}$$

2.2 Neural Network Architecture

Our neural network architecture is depicted in Fig. 1. With one hidden layer, five hidden layer neurons, one bias neuron in the input layer and a linear output layer neuron, the neural network output reads

$$N(x, \vec{p}) = \sum_{j=1}^{5} v_j \sigma(z_j) \tag{7}$$

Thereby $\sigma(z_j) = 1/(1 + e^{-z_j})$ represents the sigmoid activation function with the weighted sum $z_j = w_j x + u_j$. Here, w_j, u_j and v_j, $j = 1, \ldots, 5$, denote the weights which are stored in the weight vector $\vec{p}$. The input layer passes the domain data x, weighted by w_j and u_j, to the hidden layer for processing. The neural network output $N(x, \vec{p})$ is again a weighted sum of the values $v_j \sigma(z_j)$. With $N(x, \vec{p})$ given, the trial solutions and cost functions in Eqs. (4), (6), are obtained.

The cost function gradient is used to update $\vec{p}$ in order to find a (local) minimum in the weight space. One training cycle is called an epoch and consists of a full iteration over all training data points. If an update is performed after every single training data computation, we call this method single batch training (SBtraining) here. An alternative proceeding, performing the weight update after a complete iteration with all training data points, averaging the cost function gradient, is denoted here as full batch training (FBtraining).

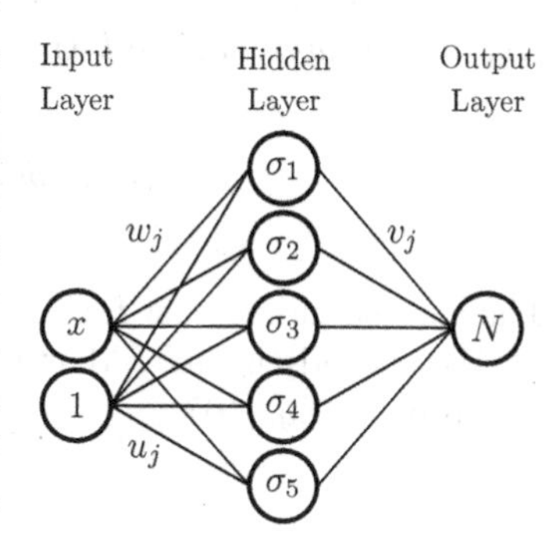

Fig. 1. Architecture for $N(x, \vec{p})$

For optimising the cost function we consider here ADAM (adaptive moment estimation) [14] which is a stochastic gradient descent method, using adaptive learning for every weight.

2.3 The Novel Neural Form

Setting for simplicity $x_0 = 0$ and $u(0) = 1$ (otherwise one may modify corresponding terms in the TS), a suitable choice for the TS is

$$u_t(x, \vec{p}) = 1 + N(x, \vec{p}) \cdot x \tag{8}$$

as mentioned in [8]. Compared to a first order polynomial $q_1(x) = a_0 + a_1 x$ with $a_0 = 1$ one may find similarities in the structure. Motivated by the expansion of an m-th order collocation function polynomial [15]

$$q_m(x) = a_0 + \sum_{k=1}^{m} a_k x^k \tag{9}$$

we are lead to set up our collocation-based TS approach for TSM:

$$u_t^C(x, \mathbf{P}_m) = 1 + \sum_{k=1}^{m} N_k(x, \vec{p}_k) x^k \tag{10}$$

The weight vector is denoted by $\vec{p}_k$ and we define the matrix $\mathbf{P}_m$ of m weight vectors $\mathbf{P}_m = (\vec{p}_1, \ldots, \vec{p}_m)$.

Let us observe, that the neural nets take on the roles of coefficient functions for the monomials x^k. We conjecture at this point that this construction makes sense since in this way several possible multipliers (not only x as in (8)) are included for neural form construction, which may contribute to total accuracy. Let us remark that the new trial solution construction (10) fulfills the initial condition.

Let us stress that the proposed ansatz (10) includes m neural networks, where $N_k(x, \vec{p}_k)$ represents the k-th neural network

$$N_k(x, \vec{p}_k) = \sum_{j=1}^{5} v_{j,k}\sigma(w_{j,k}x + u_{j,k}) \tag{11}$$

The corresponding cost function is then given as in Eq. (4).

We extend the mTSM method in a similar way as we obtained the TSM extension in Eq. (10):

$$u_t^C(x, \mathbf{P}_m) = N_1(x, \vec{p}_1) + \sum_{k=2}^{m} N_k(x, \vec{p}_k)x^{k-1} \tag{12}$$

The first neural network $N_1(x, \vec{p}_1)$ is set to learn the initial condition in the same way as stated in Eq. (6).

From now on we will refer to the number of neural networks in the neural form as the trial solution order (TSo).

3 Experiments and Results

We will perform experiments on the well known initial value problem

$$u'(x) - \lambda u(x) = 0, \quad u(0) = 1 \tag{13}$$

with $\lambda \in \mathbb{R}$, $\lambda < 0$, which has the exact solution $u(x) = e^{\lambda x}$. The Eq. (13) involves a damping mechanism, making this a simple model for stiff phenomena [16].

The numerical error Δu shown in subsequent diagrams is defined as the l_1-norm of the difference between the exact solution and the corresponding trial solution

$$\Delta u = \left\| u(x_i) - u_t^C(x_i, \mathbf{P}_m) \right\|_1$$

If we do not say otherwise, the fixed computational parameters in the subsequent experiments are: 1 input layer bias, 1 hidden layer with 5 sigmoid neurons, 1e5 training epochs, 10 training data, the ODE parameter $\lambda = -5$, $x \in [0,2]$ and the weight initialisation values which are $\vec{p}_{const}^{\,init} = -10$ and $\vec{p}_{rnd}^{\,init} = [-10.5, -9.5]$.

Weight initialisation with $\vec{p}_{const}^{\,init}$ applies to all corresponding neural networks so that they use the same initial values. In contrast, increasing the TSo for the initialisation with $\vec{p}_{rnd}^{\,init}$ works systematically. For TSo = 1, a set of random weights for the neural network is generated. For TSo = 2 (now with two neural networks), the first neural network is again initialised with the generated weights from TSo = 1, while for neural network number two, a new set of weights is generated. This holds for all TSo for higher orders, subsequently, in all experiments. The diagrams only display every hundredth data point. ADAM parameters are fixed as well with, as employed in [14], $\alpha = 1\text{e}{-3}$, $\beta_1 = 9\text{e}{-1}$, $\beta_2 = 9.99\text{e}{-1}$ and $\epsilon = 1\text{e}{-8}$.

Weight Initialisation. Let us comment in some more detail on weight initialisation. The weight initialisation plays an important role and determines the starting point for gradient descent. Poorly chosen, the optimisation method may fail to find a suitable local minimum.

The initial neural network weights are commonly chosen as small random values [17]. Let us note that this is sometimes considered as a computational characteristic of the stochastic gradient descent optimisation. Another option is to choose the initialisation to be constant. This method is not commonly used for the optimisation of neural networks since random weight initialisation may lead to better results. However, constant initialisation returns reliably results of reasonable quality if the computational parameters in the network remain unchanged. As previous experiments have documented [7,8,12], both TSM and mTSM are able to solve differential equations up to a certain degree of accuracy.

However, an example illustrating the accuracy of five computations with random weights $\vec{p}^{\,init}_{rnd}$ respectively constant weights $\vec{p}^{\,init}_{const}$ shows that the quality of approximations may vary considerably, see Table 1. As observed in many experiments, even a small discrepancy in the initialisation with several sets of random weights in the same range, may lead to a significant difference in accuracy. On the other hand, the network initialisation with constant values very often gives reliable results by the proposed novel approach. This motivates us to study in detail the effects of constant network initialisations.

Table 1. Results for five different realisations during optimisation (mTSM, TSo = 2)

No	$\Delta u(\vec{p}^{\,init}_{rnd})$	$\Delta u(\vec{p}^{\,init}_{const})$
1	5.7148e−6	2.6653e−6
2	7.5397e−6	2.6653e−6
3	3.7249e−5	2.6653e−6
4	1.1894e−5	2.6653e−6
5	7.7956e−6	2.6653e−6

3.1 Experiment 1: Number of Training Epochs

The first experiment shows for different TSo how the numerical error Δu behaves depending on the number of training epochs.

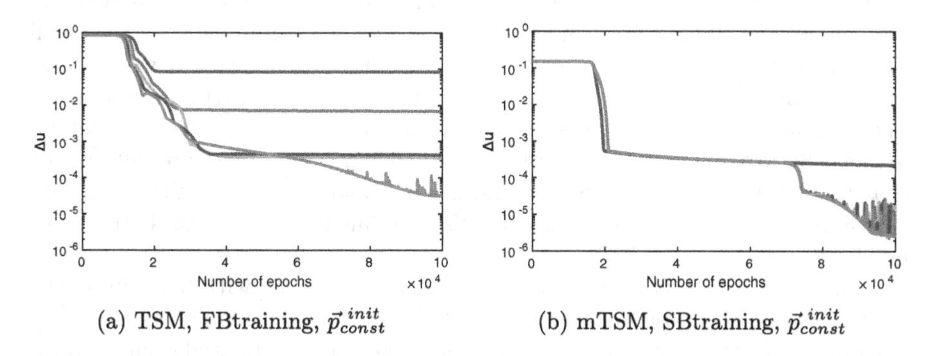

(a) TSM, FBtraining, $\vec{p}^{\,init}_{const}$ (b) mTSM, SBtraining, $\vec{p}^{\,init}_{const}$

Fig. 2. Experiment 1 Number of Training Epochs, (blue) TSo = 1, (orange) TSo = 2, (yellow) TSo = 3, (purple) TSo = 4, (green) TSo = 5 (Color figure online)

In Fig. 2a with TSM and $\vec{p}_{const}^{\,init}$ results for TSo = 1 (blue) do not provide any approximation, independent of the batch training method selected. With a second neural network for TSo = 2 (orange) in the trial solution, Δu approximately lowers by one order of magnitude so that we now obtain a solution which can be considered to rank at the lower end of reliability. However, the most interesting result in Fig. 2a is TSo = 5 (green) with the best accuracy at the end of the optimisation process but with the drawback of occurring oscillations. These may arise by the chosen optimisation method.

For mTSM with SBtraining and $\vec{p}_{const}^{\,init}$, already TSo = 1 converges to a solution accuracy that can be considered reliable. However, we observe within Fig. 2b that only the transition from TSo = 1 (blue) to TSO = 2 (orange) affects Δu with increasing accuracy, while heavy oscillations start to occur.

In the not documented results with $\vec{p}_{rnd}^{\,init}$, TSo has only minor influence on the accuracy. Especially FBtraining for mTSM shows the same trend for both initialisation methods with only minor differences in the last epochs.

Let us note that the displayed results show the best approximations using constant or random initialisation. This means, we obtain the best results for TSM with FBtraining, TSo = 5 (green) and for mTSM with SBtraining, TSo$\geq$2, respectively.

Concluding this experiment, we were able to get better results with $\vec{p}_{const}^{\,init}$ over $\vec{p}_{rnd}^{\,init}$. Increasing the TSo to at least order five seems to be a good option for TSM and FBtraining, whereas further TSo may provide even better approximations. For mTSM we can not observe benefits for TSo above order 2.

Moreover, we see especially that the increase in the order of the trial solution in (10) appears to have a similar impact on solution accuracy as the discretisation order in classical numerical analysis.

3.2 Experiment 2: Domain Size Variation

Investigating the methods concerning different domain sizes provides information on the reliability of computations on larger domains. The domains in this experiment read as $D = [0, x_{end}]$ and we directly compare in this experiment $\vec{p}_{const}^{\,init}$ with $\vec{p}_{rnd}^{\,init}$.

In Figs. 3a, 3b we observe TSM from around $x_{end} = 3.5$ to incrementally plateau to unreliable approximations. Increasing TSo improves Δu on small domains and shifts the observable step-like accuracy degeneration towards larger domains. However, even with TSo = 5 (green) the results starting from domain size $x_{end} = 3.5$ towards larger sizes are unreliable. Previous to the first plateau higher TSo provide significant better Δu for $\vec{p}_{const}^{\,init}$, while there are only minor changes for $\vec{p}_{rnd}^{\,init}$ for the TSM method. This holds for both SBtraining and FBtraining, and one can say that in this experiment TSM works better with $\vec{p}_{rnd}^{\,init}$, even without increasing TSo.

Turning to the mTSM extension, we observe in Fig. 3c with SBtraining the existence of a certain point from where different TSo return equal values, where FBtraining returns (close to) equal results for all the investigated domain sizes. However, we see some evidence for the use of TSo = 2 (orange) over TSo = 1

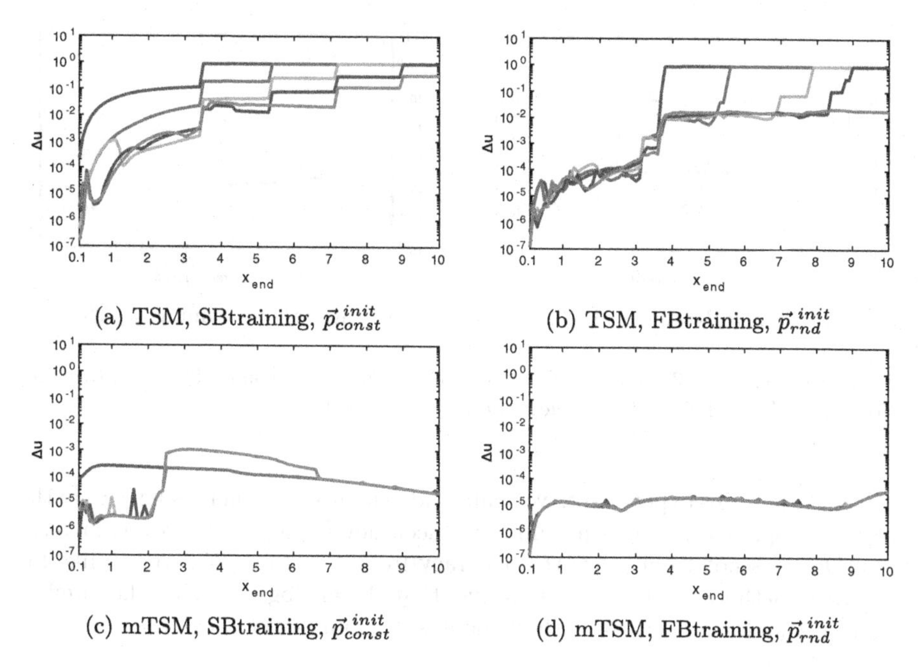

(a) TSM, SBtraining, $\vec{p}^{\,init}_{const}$ (b) TSM, FBtraining, $\vec{p}^{\,init}_{rnd}$

(c) mTSM, SBtraining, $\vec{p}^{\,init}_{const}$ (d) mTSM, FBtraining, $\vec{p}^{\,init}_{rnd}$

Fig. 3. Experiment 2 Domain Size Variation, (blue) TSo = 1, (orange) TSo = 2, (yellow) TSo = 3, (purple) TSo = 4, (green) TSo = 5

(blue) to show an overall good performance. A further increase of TSo is not necessary with this approach, confirming results from Experiment 3.1.

Let us also note that, with mTSM we find that a small domain seems to favour $\vec{p}^{\,init}_{const}$ which then provides better results than $\vec{p}^{\,init}_{rnd}$.

3.3 Experiment 3: Number of Training Data Variation

The behaviour of numerical methods highly depend on the chosen amount of grid points, so that in this experiment we analogously investigate the influence of the amount of training data. In every computation, the domain D is discretised by equidistant grid points.

As in the previous experiments, the TSo shows a major influence on the results with TSM, and the best approximations are provided by $\vec{p}^{\,init}_{const}$ with TSo = 5 (green) as seen in Fig. 4a. An interesting behaviour (observed also in a different context in Fig. 2a) is the equivalence between TSo = 3 (yellow) and TSo = 4 (purple). Both converge to almost exactly the same Δu, where one may assume a saturation for the TSo. However, another increase in the order decreases the numerical error again by one order of accuracy.

Turning to mTSM in Fig. 4b we again find a major increase in accuracy after a transition from TSo = 1 (blue) to TSo = 2. For ntD = 50, values for TSo $\geq$ 2 converge to the same results as provided by TSM with TSo = 5.

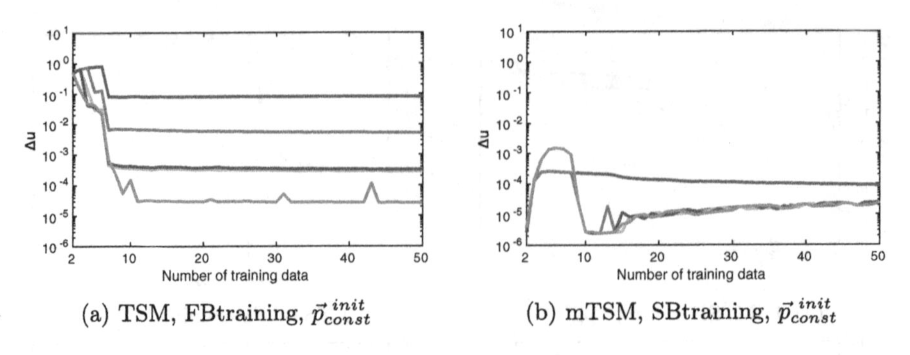

(a) TSM, FBtraining, $\vec{p}_{const}^{\,init}$ (b) mTSM, SBtraining, $\vec{p}_{const}^{\,init}$

Fig. 4. Experiment 3 Number of Training Data Variation, (blue) TSo = 1, (orange) TSo = 2, (yellow) TSo = 3, (purple) TSo = 4, (green) TSo = 5

Concluding this experiment, we again find evidence that increasing TSo in the proposed approach provides an improved accuracy for $\vec{p}_{const}^{\,init}$. However, increasing ntD seems not to improve the accuracy from a certain point on, unlike for numerical methods. But one could argue, that the analogy between the number of grid points for numerical methods here is the number of epochs.

4 Conclusion and Future Work

The proposed trial solution approach merging collocation polynomial basis functions with neural networks shows benefits over the previous trial solution constructions.

We have studied in detail the constant weight initialisation which seems to have some benefits for both the proposed TSM and mTSM extensions, depending on the batch learning methods. For the TSM collocation method this effect is more significant than observed for the mTSM extension.

Focusing on the mTSM collocation approach, using two neural networks, one for learning the initial value and one multiplied by x, seems to have some advantages. Considering approximation quality as most imperative, we find mTSM with TSo = 2 to provide the overall best results for the investigated initial value problem.

Future research may include work on other collocation functions and computational parameters. Especially with mTSM we still see space for improvement. Let us also note that the proposed method uses more than one neural network, so that there is the possibility to use very different computational parameter setups for each neural network.

Acknowledgement. This publication was funded by the Graduate Research School (GRS) of the Brandenburg University of Technology Cottbus-Senftenberg. This work is part of the Research Cluster Cognitive Dependable.

References

1. Yadav, N., Yadav, A., Kumar, M.: An Introduction to Neural Network Methods for Differential Equations. SpringerBriefs in Applied Sciences and Technology, Netherlands (2015). https://doi.org/10.1007/978-94-017-9816-7
2. Maede, A.J., Jr., Fernandez, A.A.: The numerical solution of linear ordinary differential equations by feedforward neural networks. Math. Comput. Model. **19**(12), 1–25 (1994). https://doi.org/10.1016/0895-7177(94)90095-7
3. Dissanayake, M.W.M.G., Phan-Thien, N.: Neural-network-based approximations for solving partial differential equations. Commun. Numer. Methods Eng. **10**(3), 195–201 (1994). https://doi.org/10.1002/cnm.1640100303
4. Mall, S., Chakraverty, S.: Application of Legendre Neural Network for solving ordinary differential equations. Appl. Soft Comput. **43**, 347–356 (2016). https://doi.org/10.1016/j.asoc.2015.10.069
5. Zjavka, L., Pedrycz, W.: Constructing general partial differential equations using polynomial and neural networks. Neural Netw. **73**, 58–69 (2016). https://doi.org/10.1016/j.neunet.2015.10.001
6. Famelis, I.T., Kaloutsa, V.: Parameterized neural network training for the solution of a class of stiff initial value systems. Neural Comput. Appl. **33**(8), 3363–3370 (2020). https://doi.org/10.1007/s00521-020-05201-1
7. Schneidereit, T., Breuß, M.: Solving ordinary differential equations using artificial neural networks - a study on the solution variance. In: Proceedings of the Conference Algoritmy, pp. 21–30 (2020)
8. Lagaris, I.E., Likas, A., Fotiadis, D.I.: Artificial neural networks for solving ordinary and partial differential equations. IEEE Trans. Neural Networks **9**(5), 987–1000 (1998). https://doi.org/10.1109/72.712178
9. Lagari, P.L., Tsoukalas, L.H., Safarkhani, S., Lagaris, I.E.: Systematic construction of neural forms for solving partial differential equations inside rectangular domains, subject to initial, boundary and interface conditions. Int. J. Artif. Intell. Tools **29**(5), 2050009 (2020). https://doi.org/10.1142/S0218213020500098
10. Lagaris, I.E., Likas, A., Papageorgiou, D.G.: Neural-network methods for boundary value problems with irregular boundaries. IEEE Trans. Neural Networks **11**(5), 1041–1049 (2000). https://doi.org/10.1109/72.870037
11. Tsoulos, I.G., Gavrilis, D., Glavas, E.: Solving differential equations with constructed neural networks. Neurocomputing **72**(10), 2385–2391 (2009). https://doi.org/10.1016/j.neucom.2008.12.004
12. Piscopo, M.L., Spannowsky, M., Waite, P.: Solving differential equations with neural networks: applications to the calculation of cosmological phase transitions. Phys. Rev. D **100**(1), 016002 (2019). https://doi.org/10.1103/PhysRevD.100.016002
13. Rudd, K., Ferrari, S.: A constrained integration (CINT) approach to solving partial differential equations using artificial neural networks. Neurocomputing **155**, 277–285 (2015). https://doi.org/10.1016/j.neucom.2014.11.058
14. Kingma, D.P., Ba, J.: ADAM: A Method for Stochastic Optimization. arXiv preprint:1412.6980 (2017)
15. Antia, H.M.: Numerical Methods for Scientists and Engineers, 1st edn. Hindustan Book Agency, New Delhi (2012)
16. Dahlquist, G.G.: G-stability is equivalent to A-stability. BIT Numer. Math. **18**(4), 384–401 (1978). https://doi.org/10.1007/BF01932018
17. Fernández-Redondo, M., Hernández-Espinosa, C.: Weight initialization methods for multilayer feedforward. ESANN, pp. 119–124 (2001)

Quantum-Hybrid Neural Vector Quantization – A Mathematical Approach

Thomas Villmann(✉) and Alexander Engelsberger

Saxon Institute for Computational Intelligence and Machine Learning (SICIM) at the University of Applied Sciences, Mittweida, Germany
thomas.villmann@hs.mittweida.de

Abstract. The paper demonstrates how to realize neural vector quantizers by means of quantum computing approaches. Particularly, we consider self-organizing maps and the neural gas vector quantizer for unsupervised learning as well as generalized learning vector quantization for classification learning. We show how quantum computing concepts can be adopted for these algorithms. The respective mathematical framework is explained in detail.

1 Introduction

Machine learning usually is a complex and time consuming task for many applications, e.g. in molecular genetics, speech processing, or geo- and astrophysics. Thus, the development of specific hardware as well as of computational alternatives are still challenging. One area contributing to both aspects is quantum computing based on the fundamental theory and realization of quantum phenomena [8]. Although the mathematical concepts of quantum computing are well-known [17,22], realizations of machine learning algorithms by quantum routines frequently is non-trivial [18]. Several approaches for supervised and unsupervised machine learning models and neural networks are known including quantum variants of k-means [12,32], quantum support vector machines (qSVM) [20], Hopfield and recurrent neural networks [3,19] and other. For an overview regarding supervised methods we refer to [26]. One disadvantage of quantum computing for machine learning are the limited hardware resources, which are currently available [18]. Thus hybrid approaches gain attraction integrating quantum routines into machine learning algorithms (some times denoted as quantum-enhanced machine learning [6]) or providing machine learning based parameter adaptation for the quantum approach [23].

An important subset of quantum approaches deals with supervised and unsupervised learning methods based on the nearest neighbor or nearest prototype principle closely related to classical vector quantization approaches based on the winner-takes-all rule (WTA). To this group of algorithms belong the a-fore mentioned q-means as quantum counterpart to classical k-means, vector quantization by means of the Schrödinger equation [9,11] as well as other distance

A. E. is supported by an ESF PhD grant.

L. Rutkowski et al. (Eds.): ICAISC 2021, LNAI 12854, pp. 246–257, 2021.
https://doi.org/10.1007/978-3-030-87986-0_22

based methods [1,2]. Supervised approaches are the already referenced qSVM as well as the quantum-inspired learning vector quantizer (QuI-LVQ) [30]. Both approaches make use of the underlying Hilbert-space theory in quantum computing, which relates kernels to quantum states from a mathematical perspective [25,29]. However, in QuI-LVQ the data vectors are mapped in to the Hilbert space of quantum bits (qubits) without entanglement and, hence, quantum parallelism due to qubit entanglement is not realized.

In the present contribution we consider quantum-hybrid neural vector quantizers including self-organizing maps (SOM), neural gas for unsupervised learning and learning vector quantization for supervised (classification) learning [13,14,16] and the respective possibilities for quantum computing realization For this purpose, data vectors as well as prototypes have to be transformed such that they represent quantum states and can be stored in a quantum register of quantum bits (qubits). We show that different mathematical description schemes are possible to establish respective quantum-hybrid computing solutions. Exemplary numerical simulations for classification learning are given.

2 Basic Notation and Concepts in Quantum Computing

A qubit is the counterpart to a classical bit defined by its state $|\psi\rangle = \alpha_1 |0\rangle + \alpha_2 |1\rangle$ with amplitudes $\alpha_k \in \mathbb{C}$ such that $|\alpha_1|^2 + |\alpha_2|^2 = 1$ is valid and the basis is $B_1 = \{|0\rangle, |1\rangle\}$. The basis states are frequently identified with the vectors $|0\rangle = (1,0)^T$ and $|1\rangle = (0,1)^T$ such that the qubit state space is isomorphic to $\mathbb{C}^2 \cong \mathbb{H}$, i.e. being a Hilbert space $\mathbb{H}$. The notation $|\psi\rangle$ is called Dirac's bra-vector. Accordingly, $|\psi\rangle = (\alpha_1, \alpha_2)^T$ is valid and the corresponding ket-vector is $\langle\psi| = (\alpha_1, \alpha_2)$. The inner product in $\mathbb{H}$ for states $|\psi\rangle$ and $|\varphi\rangle = \beta_1 |0\rangle + \beta_2 |1\rangle$ is given by the combination of the bra- and ket-vectors according to $\langle\psi|\varphi\rangle = \alpha_1^* \beta_1 + \alpha_2^* \beta_2$.

The tensor product state

$$|\psi\rangle = |\psi_1\rangle \otimes \ldots \otimes |\psi_N\rangle \in \mathbb{H}^{\otimes N} \tag{1}$$

$$= |\psi_1 \psi_2 \ldots \psi_{N-1} \psi_N\rangle$$

represents N unentangeled qubits making use of the common shorthand notation of $|a\rangle \otimes |b\rangle = |ab\rangle$. Note, that the tensor product does not commute but is bilinear (fulfils associativity and the distributive law). Note that $\mathbb{H}^{\otimes N}$ also is an Hilbert space.

Generally, a state $|\psi\rangle \in \mathbb{H}^{\otimes N}$ can be written as

$$|\psi\rangle = \alpha_1 |000\ldots00\rangle + \alpha_2 |000\ldots01\rangle + \cdots + \alpha_{2^n} |111\ldots11\rangle \tag{2}$$

$$= \sum_{k=1}^{n=2^N} \alpha_k |k\rangle$$

where $\sum_{k=1}^{n=2^N} |\alpha_k|^2 = 1$ holds and $|k\rangle$ refers to the kth element of the so-called computational basis

$$B_n = \{|000\ldots00\rangle, |000\ldots01\rangle, \ldots, |111\ldots11\rangle\}$$

and $k-1$ is the value taking the sequence of zeros and ones in the kth basis element in binary representation. The vector $\boldsymbol{\alpha} = (\alpha_1, \ldots, \alpha_n)^T$ is denoted as amplitude vector of $|\psi\rangle$. The respective density matrix $\boldsymbol{\rho}(|\psi\rangle) = |\psi\rangle \otimes |\psi\rangle = |\psi\rangle\langle\psi|$ is explicitly given as the Hermitean rank-one-matrix

$$\boldsymbol{\rho}(|\psi\rangle) = \sum_{k,j=1}^{n} \alpha_k^* \alpha_j |k\rangle\langle j|$$

based on the computational basis B_n. For such a pure state, the density matrix is idempotent, $\boldsymbol{\rho}(|\psi\rangle) \cdot \boldsymbol{\rho}(|\psi\rangle) = \boldsymbol{\rho}(|\psi\rangle)$ and its von-Neumann-entropy $S(\boldsymbol{\rho}(|\psi\rangle))$ satisfies

$$S(\boldsymbol{\rho}(|\psi\rangle)) = -\sum_j \lambda_j \cdot \log(\lambda_j) = 0$$

with λ_i are the eigen values of the density matrix.

If we consider two pure states $|\psi\rangle$ and $|\varphi\rangle$ with occurrence probabilities p_ψ and p_φ, respectively, such that $p_\psi + p_\varphi = 1$ is satisfied, then the density matrix of the respective mixed state

$$\boldsymbol{\rho}_{\text{mixed}}(|\psi\rangle, |\varphi\rangle) = p_\psi \cdot \boldsymbol{\rho}(|\psi\rangle) + p_\varphi \cdot \boldsymbol{\rho}(|\varphi\rangle) \tag{3}$$

has a von-Neumann-entropy $S(\boldsymbol{\rho}_{\text{mixed}}) > 0$. Thus the von-Neumann-entropy $S(\boldsymbol{\rho})$ quantifies the deviation of the respective state from a pure state [22]. Generally, the decomposition $\boldsymbol{\rho}_{\text{mixed}} = |\psi\rangle\langle\psi|$ of a density matrix regarding a mixed state $|\zeta\rangle$ by means of a state amplitude vector $|\psi\rangle$ is not available. In this case, the qubits of the system $|\zeta\rangle$ are said to be entangled.

The space $\mathbb{M}_{\mathbb{H}^{\otimes N}}$ of density matrices can be equipped with the Frobenius-inner-product (FIP)

$$\langle \boldsymbol{\rho}_1, \boldsymbol{\rho}_2 \rangle_{\mathbb{F}} = \operatorname{tr}(\boldsymbol{\rho}_1^* \cdot \boldsymbol{\rho}_2) \tag{4}$$

implying the Frobenius norm

$$\|\boldsymbol{\rho}\|_{\mathbb{F}} = \sqrt{\operatorname{tr}(\boldsymbol{\rho}^* \cdot \boldsymbol{\rho})} \tag{5}$$

such that $\mathbb{M}_{\mathbb{H}^{\otimes N}}$ also is a Hilbert space. Immediately, it follows that $\operatorname{tr}(\boldsymbol{\rho}^* \cdot \boldsymbol{\rho}) = 1$ iff $\boldsymbol{\rho}$ is a pure state whereas $\operatorname{tr}(\boldsymbol{\rho}^* \cdot \boldsymbol{\rho}) < 1$ iff $\boldsymbol{\rho}$ is a mixed state.

3 Protype-Based Machine Learning Models – Neural Vector Quantization

Neural vector quantization models require data $\mathbf{x} \in \mathcal{X} \subseteq \mathbb{R}^n$ and a set $W = \{\mathbf{w}_1, \ldots, \mathbf{w}_M\} \subset \mathbb{R}^n$ of prototype vectors, which should represent the data [4]. Further, a (differentiable) dissimilarity measure $d(\mathbf{x}, \mathbf{w}_k)$ is supposed, frequently chosen to be the squared Euclidean metric.

3.1 Unsupervised Neural Vector Quantization

Unsupervised neural vector quantization is mainly influenced by the pioneering self-organizing map (SOM) model introduced by T. KOHONEN [14]. The SOM assumes an external neural grid $A \subset \mathbb{R}^{n_A}$ with nodes identified by their coordinates $\mathbf{r} \in \mathbb{R}^{n_A}$. More formally, A is assumed to be a discrete topological structure which can be embedded into $\mathbb{R}^{n_A}$. The prototypes are formally assigned to these nodes by a mapping $\mathbf{w}_k \mapsto \mathbf{w}_\mathbf{r}$ assuming the $|A| = M$ for the cardinality of A. Further, we assume a dissimilarity measure $d_A(\mathbf{r}, \mathbf{r}')$ for A. Frequently, the grid A is taken as a two-dimensional rectangular lattice and the grid dissimilarity $d_A(\mathbf{r}, \mathbf{r}') = (\mathbf{r} - \mathbf{r}')^2$ is the squared Euclidean distance.

SOM training takes place as stochastic learning: For a given data vector $\mathbf{x} \in X$ the best matching prototype $\mathbf{w}_\mathbf{s}$ is determined by the winner-take-all (WTA) rule

$$\mathbf{s} = \mathrm{argmin}_{\mathbf{r} \in A} d(\mathbf{x}, \mathbf{w}_\mathbf{r}) \tag{6}$$

and the prototype update is given by

$$\Delta \mathbf{w}_\mathbf{r} = -\varepsilon \cdot h_A(\mathbf{r}, \mathbf{r}', \sigma) \cdot \frac{\partial d(\mathbf{x}, \mathbf{w}_\mathbf{r})}{\partial \mathbf{w}_\mathbf{r}} \tag{7}$$

with learning rate $0 < \varepsilon \ll 1$. The so-called neighborhood function

$$h_A(\mathbf{r}, \mathbf{r}', \sigma) = \exp\left(-\frac{d_A(\mathbf{r}, \mathbf{r}')}{2\sigma^2}\right)$$

realizes in (7) a neighborhood-cooperative learning. In case of the squared Euclidean distance the update (7) reads as

$$\mathbf{w}_\mathbf{r} \longleftarrow (1 - \varepsilon \cdot h_A(\mathbf{r}, \mathbf{r}', \sigma)) \mathbf{w}_\mathbf{r} + \varepsilon \cdot h_A(\mathbf{r}, \mathbf{r}', \sigma) \cdot \mathbf{x} \tag{8}$$

constituting a convex sum for $\mathbf{w}_\mathbf{r}$ and $\mathbf{x}$ with the shift parameter $\varsigma_{SOM} = \varepsilon \cdot h_A(\mathbf{r}, \mathbf{r}', \sigma) \in (0, 1)$. Unfortunately, the SOM-dynamic does not follows a gradient descent scheme of any cost function [7]. Therefore, HESKES suggested the alternative winner determination rule

$$\mathbf{s} = \mathrm{argmin}_{\mathbf{r} \in A} \sum_{\mathbf{r}'} h_A(\mathbf{r}, \mathbf{r}', \sigma) \cdot d(\mathbf{x}, \mathbf{w}_{\mathbf{r}'})$$

as the local sensoric response [10]. Another alternative is to throw away the neural grid assumption but keeping a neighborhood cooperativeness. Let

$$\mathrm{rk}_k(\mathbf{x}, W) = \sum_{\mathbf{w}_l \in W} H(d(\mathbf{x}, \mathbf{w}_k) - d(\mathbf{x}, \mathbf{w}_l))$$

be the winning rank of the prototype where $H(z)$ is the Heaviside function yielding the value one for $z > 0$ and being zero elsewhere. Then the prototype dynamic is given as

$$\Delta \mathbf{w}_k = -\varepsilon \cdot h_W(\mathrm{rk}_k(\mathbf{x}, W)) \cdot \frac{\partial d(\mathbf{x}, \mathbf{w}_k)}{\partial \mathbf{w}_k} \tag{9}$$

with the redefined neighborhood function

$$h_W\left(\mathrm{rk}_k\left(\mathbf{x},W\right),\sigma\right)=\exp\left(-\frac{\mathrm{rk}_k\left(\mathbf{x},W\right)}{\sigma}\right)$$

based on the winning ranks. It realizes a prototype dynamic of a diffusing gas if the prototypes are interpreted as gas particles [16]. As for the SOM, the Euclidean update is obtained as

$$\mathbf{w}_k \longleftarrow \left(1-\varepsilon\cdot h_W\left(\mathrm{rk}_k\left(\mathbf{x},W\right),\sigma\right)\right)\mathbf{w}_k+\varepsilon\cdot h_W\left(\mathrm{rk}_k\left(\mathbf{x},W\right),\sigma\right)\cdot\mathbf{x} \tag{10}$$

realizing also a convex sum for $\mathbf{w}_k$ and $\mathbf{x}$ with the shift parameter $\varsigma_{NG}=\varepsilon\cdot h_W\left(\mathrm{rk}_k\left(\mathbf{x},W\right),\sigma\right)\in(0,1)$.

Generally, NG performance much better than standard c-means and is insensitive with respect to initialization. Further, it should be emphasized that both, the prototype adaptation for SOM and NG, can be interpreted as an prototype attraction scheme in case of the squared Euclidean distance for $d\left(\mathbf{x},\mathbf{w}_k\right)$.

3.2 Supervised Classification Learning by Neural Vector Quantizers

Based on the SOM, T. Kohonen suggested a classifier model approximating a Bayes classification system known a learning vector quantization [13], if the data have to be assigned to classes $C=\{1,\ldots,N_C\}$. For this purpose, training data $\mathbf{x}\in X$ are equipped with class labels $c\left(\mathbf{x}\right)\in C$. Analogously, the prototypes are assigned to the classes $c\left(\mathbf{w}_k\right)\in C$ such that at least one prototype is responsible for each class. We partition the set W into subsets $W_c=\{\mathbf{w}_k\in W|c\left(\mathbf{w}_k\right)=c\}$ of class-responsible prototypes. The approximated classification accuracy can be optimized by a stochastic gradient descent learning dynamic according to

$$\Delta\mathbf{w}^{\pm}=-\varepsilon\cdot\Psi_f\left(\mathbf{w}^{\pm}\right)\cdot\frac{\partial d\left(\mathbf{x},\mathbf{w}^{\pm}\right)}{\partial\mathbf{w}_k} \tag{11}$$

with

$$\Psi_f\left(\mathbf{w}^{\pm}\right)=\frac{\partial f\left(\mu\left(\mathbf{x}\right)\right)}{\partial\mu\left(\mathbf{x}\right)}\cdot\frac{\partial\mu\left(\mathbf{x}\right)}{\partial d^{\pm}\left(\mathbf{x}\right)}$$

where f is a non-negative monotonically increasing sigmoid function, frequently chosen as $\mathrm{sgd}_\theta\left(z\right)=1/\left(1+\exp\left(\theta z\right)\right)$. The function

$$\mu\left(\mathbf{x}\right)=\frac{d^{+}\left(\mathbf{x}\right)-d^{-}\left(\mathbf{x}\right)}{d^{-}\left(\mathbf{x}\right)+d^{+}\left(\mathbf{x}\right)} \tag{12}$$

is the classifier function yielding negative values for correct classification. Here, $d^{+}\left(\mathbf{x}\right)$ is the distance of the best matching correct prototype $\mathbf{w}_k\in W_c$ for the data sample $\mathbf{x}$ with given class label $c\left(\mathbf{x}\right)$, i.e. $\mathrm{rk}_k\left(\mathbf{x},W_c\right)=0$, denoted as $\mathbf{w}^{+}$ whereas $d^{-}\left(\mathbf{x}\right)$ is the distance of matching prototype $\mathbf{w}^{-}$ among all other prototypes $\mathbf{w}_k\in W\setminus W_c$, i.e. $\mathrm{rk}_l\left(\mathbf{x},W\setminus W_c\right)=0$. Thus $\Psi_f\left(\mathbf{w}^{+}\right)>0$

whereas $\Psi_f(\mathbf{w}^-) < 0$ is valid.This adaptation scheme is known as generalized LVQ (GLVQ, [21]). In case of the squared Euclidean distance, the updates (11) read as

$$\mathbf{w}^+ \longleftarrow \left(1 - \varepsilon \cdot \Psi_f\left(\mathbf{w}^+\right)\right) \cdot \mathbf{w}^+ + \varepsilon \cdot \Psi_f\left(\mathbf{w}^+\right) \cdot \mathbf{x} \tag{13}$$

and

$$\mathbf{w}^- \longleftarrow \left(1 + \varepsilon \cdot \left|\Psi_f\left(\mathbf{w}^-\right)\right|\right) \cdot \mathbf{w}^- - \varepsilon \cdot \left|\Psi_f\left(\mathbf{w}^-\right)\right| \cdot \mathbf{x} \tag{14}$$

realizing an attraction-repulsing-scheme (ARS) with attraction for $\mathbf{w}^+$ and repulsion for $\mathbf{w}^-$.

The better f approximates the Heaviside function the better the classification accuracy is approximated by the underlying cost function [21]. Further, we remark that only the attraction scheme for $\mathbf{w}^+$ realizes a convex sum for $\mathbf{w}^+$ and $\mathbf{x}$ with the shift parameter $\varsigma^+_{GLVQ} = \varepsilon \cdot \Psi_f(\mathbf{w}^+) \in (0,1)$. whereas for the repulsion $1 + \varsigma^+_{GLVQ} > 1$ is valid for the shift parameter $\varsigma^-_{GLVQ} = \varepsilon \cdot |\Psi_f(\mathbf{w}^-)| \in (0,1)$.

A probabilistic variant of LVQ was proposed in [28] based on cross-entropy learning, which is also able to handle multiple class assignments.

4 Quantum-Hybrid Vector Quantization

Quantum-hybrid vector quantization (QhVQ) is proposed to be a (neural) vector quantization model which is at least partially motivated or realized by quantum computing algorithms. For this purpose we have to distinguish four main steps, which should be considered for quantum algorithms

1. data preprocessing such that data as well as prototypes represent quantum states
2. dissimilarity calculations
3. winner determination
4. prototype updates

We will explain them in the next subsections.

4.1 Data Preprocessing for Quantum State Generation

If we suppose data $\mathbf{x} \in \mathcal{X} \subseteq \mathbb{R}^n$ with $n = 2^N$ and a random set $W = \{\mathbf{w}_1, \ldots, \mathbf{w}_M\} \subset \mathbb{R}^n$ as usual, the simple normalization $\mathbf{x} \mapsto \hat{\mathbf{x}}$ and $\mathbf{w}_k \mapsto \hat{\mathbf{w}}_k$ with $\|\hat{\mathbf{x}}\|^2 = \|\hat{\mathbf{w}}_k\|^2 = 1$ can be taken where the coefficients α_j of $\hat{\mathbf{x}}$ and β_j of $\hat{\mathbf{w}}_k$ fulfill the requirements $\sum_j |\alpha_j|^2 = \sum_j |\beta_j|^2 = 1$. We identify these (squared) coefficients as amplitudes of the quantum states $|\boldsymbol{\xi}\rangle$ and $|\boldsymbol{\omega}_k\rangle$ with N qubits and computational basis B_n, respectively, representing pure states. This procedure is known as *amplitude encoding*. We denote the corresponding density matrices by $\boldsymbol{\rho}_{\mathbf{x}}$ and $\boldsymbol{\rho}_k = \boldsymbol{\rho}_{\mathbf{w}_k}$ both belonging to the matrix space $\mathbb{M}_{\mathbb{H}^{\otimes N}}$. The mapping $\Phi : \mathcal{X} \longrightarrow \mathbb{M}_{\mathbb{H}^{\otimes N}}$ is denoted as the *data-encoding feature map*. As mentioned in [24], it can be realized by a quantum circuit $U(|\boldsymbol{\xi}\rangle)$ (unitary transformation).

Another possibility is to take $\mathcal{X} \subseteq \mathbb{R}^N$ with vector entries x_j of $\mathbf{x}$ as scalars and apply the Pauli-X-rotation gate $R_X(x_j) = \exp\left(-i \cdot \frac{x_j}{2} \cdot \boldsymbol{\sigma}_X\right)$ to each, where

$\sigma_X = \begin{pmatrix} 0 & 1 \\ 1 & 0 \end{pmatrix}$ is the Pauli-X-matrix [22]. The resulting N qubits $|\xi_j\rangle$ form a state vector using the tensor product according to $|\boldsymbol{\xi}\rangle = \xi_1 \otimes \ldots \otimes \xi_N \in \mathbb{H}^{\otimes N}$. This procedure is known as *angle encoding*. Of course, $|\boldsymbol{\xi}\rangle$ again can be identified with a density matrix $\boldsymbol{\rho}_{\mathbf{x}} \in \mathbb{M}_{\mathbb{H}^{\otimes N}}$. In the same manner, the density matrices $\boldsymbol{\rho}_k$ can be generated for the original prototypes $\mathbf{w}_k \in \mathbb{R}^N$.

Regardless from the particular realization, we can interpret the mappings $\Phi : \mathcal{X} \longrightarrow \mathbb{M}_{\mathbb{H}^{\otimes N}}$ and $\Phi : \mathcal{X} \longrightarrow \mathbb{H}^{\otimes N}$ as feature codings into the Hilbert spaces $\mathbb{H}^{\otimes N}$ and $\mathbb{M}_{\mathbb{H}^{\otimes N}}$, respectively [25,30]. As pointed out in [24], $\mathbb{M}_{\mathbb{H}^{\otimes N}}$ and $\mathbb{H}^{\otimes N}$ are isomorphic and we can see these mapping as kernel mappings into a reproducing kernel Hilbert space (RKHS, [27]). According to [24], the respective kernel is given as $\kappa_\Phi(\mathbf{x}, \mathbf{w}) = |\langle \boldsymbol{\rho}_{\mathbf{x}}, \boldsymbol{\rho}_{\mathbf{w}} \rangle_{\mathbb{F}}|^2$ using the FIP (4), which is also denoted as the Hilbert-Schmidt-inner-product in this context.

Generally speaking, data encoding is a crucial aspect for quantum-hybrid machine learning classifier systems and, therefore has to be handled carefully keeping in mind the particular context [15].

4.2 Dissimilarity Calculations

Assuming quantum states given by the amplitude vectors $|\boldsymbol{\xi}\rangle, |\boldsymbol{\omega}_k\rangle \in \mathbb{H}^{\otimes N}$, i.e. supposing pure states, their distance is given as $d_{\mathbb{H}^{\otimes N}}(|\boldsymbol{\xi}\rangle, |\boldsymbol{\omega}_k\rangle) = \sqrt{2 - 2 \cdot \langle \boldsymbol{\xi} | \boldsymbol{\omega}_k \rangle}$. Hence,the distance $d_{\mathbb{H}^{\otimes N}}(|\boldsymbol{\xi}\rangle, |\boldsymbol{\omega}_k\rangle)$ it is completely determined by the inner product $\langle \boldsymbol{\xi} | \boldsymbol{\omega}_k \rangle$. Yet, the calculation of the inner product $\langle \boldsymbol{\xi} | \boldsymbol{\omega}_k \rangle$ can be realized by quantum circuit, which is the controlled SWAP-gate (cSWAP) schematically depicted in Fig. 1.

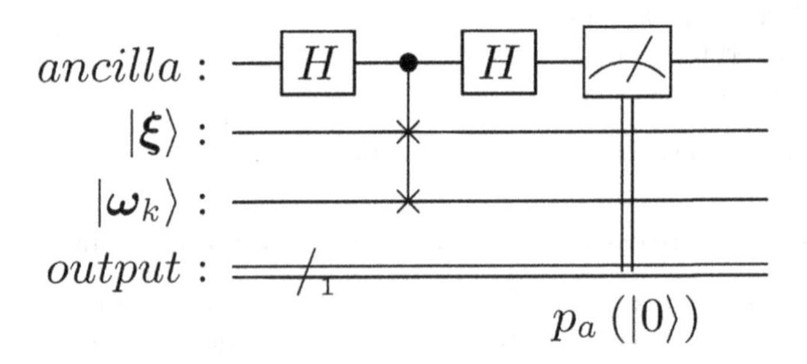

Fig. 1. Quantum circuit diagram of a cSWAP-gate.

In fact, the probability of the ancilla qubit to be in state $|0\rangle$ is $p_a(|0\rangle) = \frac{1}{2}\left(1 - |\langle \boldsymbol{\xi} | \boldsymbol{\omega}_k \rangle|^2\right)$ [17]. Thus we get the distance

$$d_{\mathbb{H}^{\otimes N}}(|\boldsymbol{\xi}\rangle, |\boldsymbol{\omega}_k\rangle) = \sqrt{2 - 2 \cdot \sqrt{2 \cdot p_a(|0\rangle) - 1}} \tag{15}$$

for this setting.

If we choose the density matrix representation, the distance between states is obtained as $d_{\mathbb{M}_{\mathbb{H}^{\otimes N}}}(\boldsymbol{\rho}_{\mathbf{x}}, \boldsymbol{\rho}_{\mathbf{w}}) = \sqrt{2 - 2 \cdot \langle \boldsymbol{\rho}_{\mathbf{x}}, \boldsymbol{\rho}_{\mathbf{w}} \rangle_{\mathbb{F}}}$. This procedure also works for density matrices $\boldsymbol{\rho}_{\mathbf{w}}$ representing mixed states.

4.3 Winner Determination

The winner determination requires first to calculate all distances $d_{\mathbb{M}_{\mathbb{H}^{\otimes N}}}(\boldsymbol{\rho}_{\mathbf{x}}, \boldsymbol{\rho}_k)$ or $d_{\mathbb{H}^{\otimes N}}(|\boldsymbol{\xi}\rangle, |\boldsymbol{\omega}_k\rangle)$. We assume that we collect them in a database D. Then the quantum Grover-algorithm allows to realize the minimum search as a quantum circuit [9]. Yet, an appropriate database coding is demanded, which, in fact, might be non-trivial to realize.

4.4 Prototype Update

For prototype update we focus on the density matrix representation. In this view, the convex updates (8), (10), and (14) for SOM, NG and GLVQ can be realized by means of the mixed state density calculation (3). Particularly, we get

$$\boldsymbol{\rho}_{\mathbf{w}} \longleftarrow (1 + \varsigma) \cdot \boldsymbol{\rho}_{\mathbf{w}} + \varsigma \cdot \boldsymbol{\rho}_{\mathbf{x}} \tag{16}$$

as the new density matrix representing in general a mixed state. Depending on the considered algorithm the choices ς_{SOM}, ς_{NG}, or ς^{+}_{GLVQ} have to be made for ς.

Unfortunately and as already mentioned, the repulsion update for GLVQ (14) is not a convex scheme and, hence, is not consistent with the quantum approach in this version. A heuristic could be to apply an attraction scheme for $\boldsymbol{\rho}_{\mathbf{w}^-}$ according to $\boldsymbol{\rho}_{\mathbf{w}^-} \longleftarrow (1 + \varsigma) \cdot \boldsymbol{\rho}_{\mathbf{w}^-} + \varsigma \cdot \boldsymbol{\rho}_{-\mathbf{x}}$. It is a heuristic, because this update would not longer be the gradient of the cost function.

5 Experiments

In this section we give preliminary experimental results. In particular, We considered the application of a quantum-hybrid GLVQ (Qh-GGLVQ) for the well-known IRIS data set. Compared to the usual GLVQ, the distance calculation were realized using an (ideal) cSWAP-test simulating a real quantum computer. Thus the distance is estimated via (15). The original data as well as the initial prototype settings were transformed into a quantum state representation by amplitude encoding. The quanWe show numerical experiments for quantum-hybrid GLVQ (QhGLVQ).

The prototype adaptation in QhGLVQ was realized as vector shifts of the amplitude prototype vectors. As pointed out in Sect. 3.2, vector shift operations are crucial because, in general, those shifts do not preserve the normalization. Hence, a renormalization after update is required to ensure the updated prototype to represent an quantum state. However, because of the small learning rate in stochastic gradient descent learning, the norm deviations after the update are small, as depicted in Fig. 2.

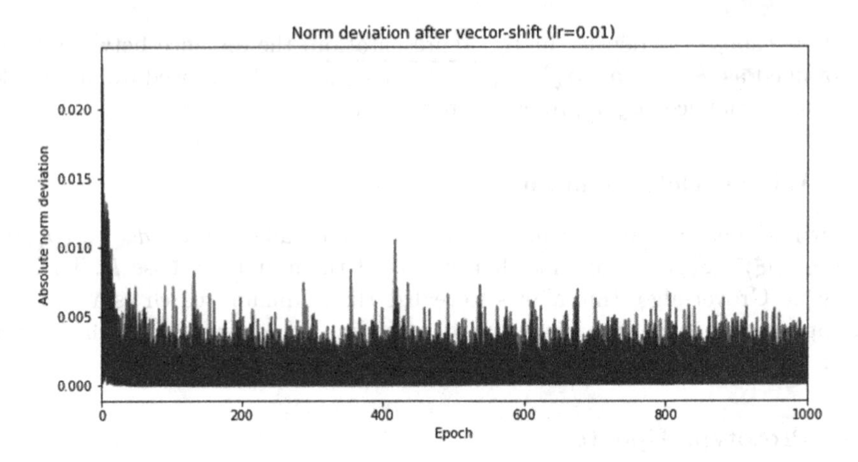

Fig. 2. Norm deviation of the prototypes after update.

As we can observe, the norm deiviations are roughly constant showing that the process is stable.

The distance calculation by means of the cSWAP is a probabilistic procedure, i.e. a repeated cSWAP-test and evaluating the respective statistics for the probabilistic outcome to estimate the inner product precisely. Mathematically, it can be seen as a Bernoulli-process such that we can give apriori information about the expected deviation of the estimated distance in dependence on the number of trials for the statistics. The confidence interval I_α can be estimated by

$$I_\alpha = \left[Q_\beta \left(1 - \frac{\alpha}{2}, p_0 N, 1 - p_0 + \frac{1}{N} \right), Q_\beta \left(1 - \frac{\alpha}{2}, N p_0 + \frac{1}{N}, (1 - p_0) N \right) \right]$$

according to [5] where Q_β is the quantile of the β-distribution. An approximation $\hat{I}_\alpha = [p_l, p_u]$ is given by

$$p_{l,u} = \frac{1}{1 + \frac{c_\alpha^2}{N}} \cdot \left(p_0 + \frac{c_\alpha^2}{N} \pm C \cdot \sqrt{\frac{p_0 \cdot (1 - p_0)}{N} + \frac{c_\alpha^2}{4 \cdot N^2}} \right)$$

where $c_\alpha = \Phi^{-1}\left(1 - \frac{\alpha}{2}\right)$ is the α-quantile of the standard normal distribution [31].

The resulting learning curves for 100 runs for the IRIS data, the accuracy curves, are depicted in Fig. 3.

As expected, the achieved accurracies are comparable to standard classifiers. After the initialization phase, the adaptation seems to be not changing before it starts to improving significantly after 30 epochs. This behavior is also known for standard GLVQ.

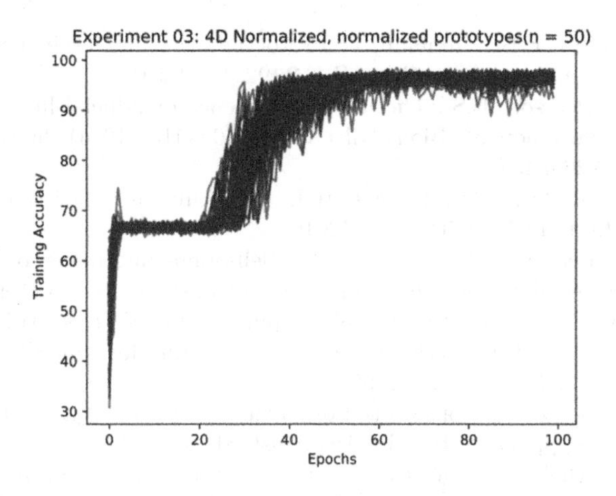

Fig. 3. Accuracy learning curves for IRIS-training.

6 Conclusions

In this contribution we considered the mathematical framework to describe neural vector quantizers by quantum computing paradigms. We have shown that dissimilarity calculations as well as winner determination can be realized by quantum algorithms which are the cSWAP-evaluation and the Grover algorithm, respectively. Yet, prototype adaptation using quantum computing concepts in neural vector quantization is more crucial. It turns out that prototype repulsion as used in GLVQ is incompatible with quantum computing. The attraction update, however, is consistent with quantum computing if the density matrix representation of quantum states is used, i.e. density representation of data. Thus, the successful neural vector quantizers SOM and NG can be adapted for quantum computing whereas GLVQ is not that easy to realize.

Future work will include full quantum computing simulations. A remaining difficult task, however, is to develop respective quantum circuites as well as realization of these concept on quantum computing hardware.

Acknowledgement. A.E. was supported by a grant of the European Social Fund (ESF).

References

1. Aïmeur, E., Brassard, Gr., Gambs, S.: Quantum clustering algorithms. In: Proceedings of the 24th International Conference on Machine Learning, vol. 518, pp. 1–8 (2007)
2. Aïmeur, E., Gr. Brassard, and S. Gambs. Quantum speed-up for unsupervised learning. Mach. Learn. 90(2), 261–287 (2013)
3. Bausch, J.: Recurrent quantum neural networks. In: Advances in Neural Information Processing Systems 33 (NIPS 2020), pp. 1–11. Curran Associates Inc. (2020)

4. Biehl, M., Hammer, B., Villmann, T.: Prototype-based models in machine learning. Wiley Interdisciplinary Rev. Cogn. Sci. **2**, 92–111 (2016)
5. Clopper, C., Pearson, E.S.: The use of confidence or fiducial limits illustrated in the case of the binomial. Biometrika **26**, S. 404–413 (1934). https://doi.org/10.1093/biomet/26.4.404
6. Dunjko, V., Taylor, J.M., Briegel, H.J.: Quantum-enhanced machine learning. Phys. Rev. Lett. **117**(130501), 1–6 (2016)
7. Erwin, E., Obermayer, K., Schulten, K.: Self-organizing maps: ordering, convergence properties and energy functions. Biol. Cyb. **67**(1), 47–55 (1992)
8. Feynman, R.P.: Quantum mechanical computers. Optics News **11**(2), 11–20 (1985)
9. Grover, L.K.: Quantum mechanics helps in searching for a needle in a haystack. Phys. Rev. Lett. **79**, 325–328 (1997)
10. Heskes, T.: Energy functions for self-organizing maps. In: Oja, E., Kaski, S. (eds.) Kohonen Maps, pp. 303–316. Elsevier, Amsterdam (1999)
11. Horn, D., Gottlieb, A.: Algorithm for data clustering in pattern recognition problems based on quantum mechanics. Phys. Rev. Lett. **88**(1), 1–4 (2002)
12. Kerenidis, I., Landman, J., Luongo, A., Prakash, A.: q-means: a quantum algorithm for unsupervised machine learning. In: Wallach, H., Larochelle, H., Beygelzimer, A., dAlché Buc, F., Fox, E., Garnett, R. (eds.) Advances in Neural Information Processing Systems 32 (NIPS 2019), pp. 4134–4144. Curran Associates Inc. (2019)
13. Kohonen, T.: Learning Vector Quantization. Neural Networks **1**(Supplement 1), 303 (1988)
14. Kohonen, T.: Self-Organizing Maps. Springer Series in Information Sciences, vol. 30. Springer, Berlin, Heidelberg (1995)
15. LaRose, R., Coyle, B.: Robust data encodings for quantum classifiers. Physical Review A (2020)
16. Martinetz, T.M., Berkovich, S.G., Schulten, K.J.: "Neural-gas" network for vector quantization and its application to time-series prediction. IEEE Trans. Neural Networks **4**(4), 558–569 (1993)
17. Nielsen, M.A., Chuang, I.L.: Quantum Computation and Quantum Information, 10th edn. Cambridge University Press, Cambridge (2016)
18. Preskill, J.: Quantum computing in the NISQ era and beyond. Quantum **2**(78), 1–20 (2018)
19. Rebentrost, P., Bromley, T.R., Weedbrook, C., Lloyd, S.: Quantum Hopfield neural network. Phys. Rev. A **98**(042308), 1–11 (2018)
20. Rebentrost, P., Mohseni, M., Lloyd, S.: Quantum support vector machines for big data classification. Physical Review Letters **113**(13050), 1–5 (2014)
21. Sato, A., Yamada, K.: Generalized learning vector quantization. In: Touretzky, D.S., Mozer, M.C., Hasselmo, M.E. (eds.) Advances in Neural Information Processing Systems 8. In: Proceedings of the 1995 Conference, pp. 423–9. MIT Press, Cambridge (1996)
22. Scherer, W.: Mathematics of Quantum Computing. Springer, Cham (2019). https://doi.org/10.1007/978-3-030-12358-1_9
23. Schuld, M.: Machine learning in quantum spaces. Nature **567**, 179–181 (2019)
24. Schuld, M.: Quantum machine learning models are kernel methods. arXiv, 2101.11020v1:1–26 (2021)
25. Schuld, M., Killoran, N.: Quantum machine learning in feature Hilbert spaces. Phys. Rev. Lett. **122**(040504), 1–6 (2019)
26. Schuld, M., Petruccione, F.: Supervised Learning with Quantum Computers. QST, Springer, Cham (2018). https://doi.org/10.1007/978-3-319-96424-9_9

27. Steinwart, I., Christmann, A.: Support Vector Machines. Information Science and Statistics. Springer, Heidelberg (2008)
28. Villmann, A., Kaden, M., Saralajew, S., Villmann, T.: probabilistic learning vector quantization with cross-entropy for probabilistic class assignments in classification learning. In: Rutkowski, L., Scherer, R., Korytkowski, M., Pedrycz, W., Tadeusiewicz, R., Zurada, J.M. (eds.) ICAISC 2018. LNCS (LNAI), vol. 10841, pp. 724–735. Springer, Cham (2018). https://doi.org/10.1007/978-3-319-91253-0_67
29. Villmann, T.: Quantum-inspired learning vector quantization - basic concepts and beyond. Machine Learning Reports, 14(MLR-02-2020):29–32 (2020). ISSN:1865-3960. www.techfak.uni-bielefeld.de/~fschleif/mlr/mlr_02_2020.pdf
30. Villmann, T., Engelsberger, A., Ravichandran, J., Villmann, A., Kaden, M.: Quantum-inspired learning vector quantizers for prototype-based classification. Neural Comput. Appl., 1–10 (2020). https://doi.org/10.1007/s00521-020-05517-y
31. Wilson, E.B.: Probable inference, the law of succession, and statistical inference. J. Am. Stat. Assoc. **22**, 209–212 (1927)
32. Xiao-Yan, Z., Xing-Xing, A., Wen-Jie, L., Fu-Gao, J.: Quantum k-means algorithm based on the minimum distance. J. Chinese Comput. Sci. **38**(5), 1059–1062 (2017)

A Graphic CNN-LSTM Model for Stock Price Predication

Jimmy Ming-Tai Wu[1], Zhongcui Li[1], Youcef Djenouri[2], Dawid Polap[3], Gautam Srivastava[4], and Jerry Chun-Wei Lin[5](✉)

[1] Shandong University of Science and Technology, Qingdao, China
wmt@wmt35.idv.tw
[2] SINTEF Digital, Department of Mathematics and Cybernetics, Oslo, Norway
youcef.djenouri@sintef.no
[3] Faculty of Applied Mathematics, Silesian University of Technology, Kaszubska 23, 44-100 Gliwice, Poland
Dawid.Polap@polsl.pl
[4] Department of Math and Computer Science, Brandon University, Brandon, Canada
srivastavag@brandonu.ca
[5] Department of Computer Science, Electrical Engineering and Mathematical Sciences, Western Norway University of Applied Sciences, Bergen, Norway
jerrylin@ieee.org

Abstract. In this paper, we presented a novel model that combines Convolution Neural Network (CNN) and Long Short-term Memory Neural Network (LSTM) for better and accurate stock price prediction. We then developed a model called stock sequence array convolutional LSTM (SACLSTM) that builds both a sequence array of the historical data and leading indicators (i.e., futures and options). This built array is then considered as the input data of the CNN model, thus specific feature vectors via convolutional and pooling layers are then extracted for being the input vector of the LSTM model. Based on this flowchart, the stock price can be better predicted, that can be seen from the conducted experiments in 10 stocks data from USA and Taiwan stock markets. Results also indicated that the designed model is better than the existing models.

Keywords: Convolution neural network · Long short-term memory neural network · Stock price prediction · Leading indicators

1 Introduction

Financing is a term that refers to the economic operation process, which includes several financial products, i.e., stock, saving, and bonds, among others. To obtain good benefits from those items, many prediction models have been investigated [14], particular in the stock markets. In stock market, several factors should be considered to influence the stock prices, i.e., economic structure, political issue, economic tread, industry development, and economic cycles, among others.

L. Rutkowski et al. (Eds.): ICAISC 2021, LNAI 12854, pp. 258–268, 2021.
https://doi.org/10.1007/978-3-030-87986-0_23

Several generic models i.e., statistical analysis, logistic regression, optimization model, and artificial intelligence. As the rapid growth of computer techniques, artificial neural network (ANN) has been employed into many domains and applications [9,20]. ANN is an simulation model that is to mimic the behaviors of human brain for decision making. ANN is a non-liner system which is to process the distributed information parallelly based on the concept of weighted neurons [5]. After several iterations, the weight of each neuron can be adjusted as an optimized value, thus achieving good performance for further prediction. In general, ANN can be considered as the characteristics with distributed processing, self-learning, distributed memory, and even self organization, which is able to process the large-scale and parallel procedures. In general, if several abilities used in ANN can be adapted in the progress for decision making, thus a better predictive model can thus be obtained and maintained. For instance, it is very useful to build the ANN model used in the financial market or time-series applications particular in predicting the short-term closing price of the stock. Besides of the generic ANN model, the deep neural network (DNN) [7] has also become a very hot issue in recent decades since it is used to make a better prediction results regarding the depth of the neural networks (i.e., the number of layers in the DNN model). Many ANN-extended models such as convolutional neural network (CNN), lstm long short term memory (LSTM), and residual neural network (ResNet) are respectively presented and discussed in recent years since they have achieved good performance compared to the generic ANN model. In CNN, it uses convolutional units and unit of pooling to process the neural network models [13]. For the LSTM model [4], it uses both the recurrent unit [16] and long-short term memory unit [11] to process the neural network models. Several well-known models such as deep multilayer perceptron (MLP), [18], Restricted Boltzmann Machine (RBM) [3], and autoencoder (AE) [2] are respectively presented to enhance the capability of the ANN model to process more complex and complicated data.

Multilayer neural network considers to map the features into the values, which is normally based on selection operation manually. First, the signal is considered as the input of the deep learning model. The features are then extracted and the expected value is then considered as the output afterward. The well-known CNN and LSTM models adapt this procedure in the network development. CNN is widely used in many domains and applications since it can solve the limitations of generic deep network (e.g., the number of parameters is huge for the training step). CNN considers the local receptive fields and shared weights to reduce the number of parameters used in the network. The local receptive field is the input data of the network model that is represented as a multi-dimensional vector. Neurons of the next layer is connected to the input neurons. In addition, weight sharing shows $N \times N$ hidden layer neurons are then connected to the input layer, and the parameters of neurons in the hidden layers are not different, which indicates that the corresponding hidden layer neurons of different windows share the same settings of the parameters. Based on the capability of CNN model, many studies used CNN to make the prediction of stock task

[10]. Recurrent neural network (RNN) is mainly used to process the sequence data particular in solving the problem in natural language process (NLP) since the meaning of a specific word in a sentence could be different based on the words in prefix or postfix positions. LSTM is a varied extension of RNN, which applies to solve the issues regarding gradient disappearance and explosion while the training step of long sequence. LSTM has better performance compared to the ordinary RNN model, and LSTM has achieved good results regarding the financial-market applications [8]. Siripurapu [15] considered stock candlestick chart as an input image used in the input layer in LSTM. Hoseinzade and Haratizadeh [12] mapped the historical data of the market to its future volatility. To avoid the overfitting issue in neural network model, Di Persio and Honchar presented a modified CNN by using one-dimensional input to make a better prediction [6]. This model uses the history of closing price without considering the other variable (e.g., technical indicators) for the prediction task. Gunduz *et al.* [10] presented another CNN-based model by adapting the technical indicators of each sample to solve the above limitation. Di Persio and Honchar used the historical data of the closing price of the S&P 500 index as the input for CNN, MLP and LSTM, respectively and the results showed that the LSTM and CNN have achieved good performance than that of the MLP. Zhang and Tan [19] developed a model that can predict the future returns of stock rankings. Azzouni and Pujolle [1] applied the LSTM for several tasks including classification, prediction of time series, processing, and learning from experience. Ghosh *et al.* [8] developed a model that can forecast and analyze the future growth of the company by LSTM. In this paper, we then consider the CNN and LSTM to make the stock price prediction. The major contributions are stated below.

1. As long as the various factors considering in the stock market, it is important to collect the sufficient indicators as the reference in the developed model. In the developed model, the CNN and LSTM is considered together, and the leading indicators of the stock and the historical data are the considered together as the input in the developed model to predict the stock price.
2. Two-dimensional vector is then established to simulate the image data as the input in the designed prediction model, thus the higher prediction results can thus be achieved by the developed model.
3. Experimental results showed that the developed model obtains good performance compared to the existing models.

The reminding of this manuscript is stated below. Second section shows the data preprocessing and environment setting regarding the designed model. Section 3 shows the experimental results of the developed model compared to the existing models. In Sect. 4, the conclusion of the developed model is then provided and the further works are discussed and studied.

2 Data Preprocessing and Environment Setting

Several settings include data sets, network parameters, evaluation models, and baseline approaches are then discussed and studied in this section. The optimized model designed in this paper is also studied and developed in the following subsections.

2.1 Dataset

The datasets used in the designed model is 10 stocks from 2 markets, which are MSFT, AMZN, AAPL, FB, and IBM from American market, and DJO, IJO, CDA, DVO, and CFO are from Taiwan market. Each sample includes some variables (e.g., futures, options, and historical data) used in the developed model.

2.2 Optimization Framework

In the designed model, the information of stock index vector in 30 days is then collected as the input image. An illustrated example is then indicated in Fig. 1. In Fig. 1, the x-axis is the data the of the continuous cycle of the input image, and y-axis shows the index of the historical data of stocks along with the dates in the input layer of the image data.

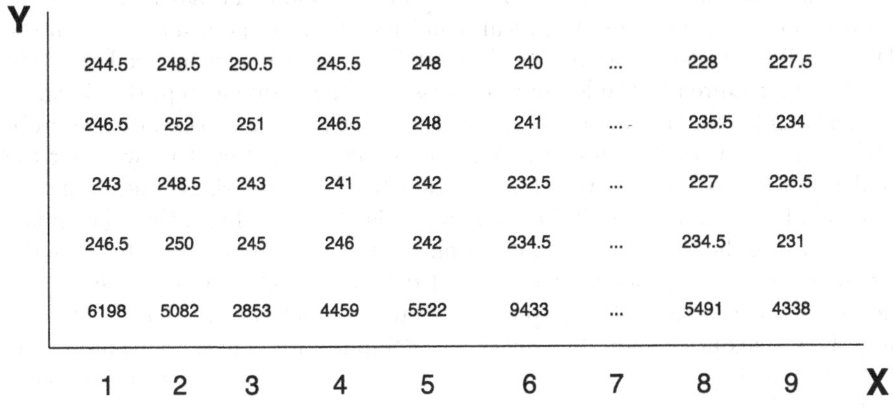

Y									
	244.5	248.5	250.5	245.5	248	240	...	228	227.5
	246.5	252	251	246.5	248	241	...	235.5	234
	243	248.5	243	241	242	232.5	...	227	226.5
	246.5	250	245	246	242	234.5	...	234.5	231
	6198	5082	2853	4459	5522	9433	...	5491	4338
	1	2	3	4	5	6	7	8	9	X

Fig. 1. An used example of the input layer in the designed model.

The width sequence of a sliding window of pre-determined herein stock index is set as 39 days in the evaluation process. Each window produces an input image. To get the next image of the current window, the sliding window can be moved to the next date. Thus, the designed model will obtain a series of images as the input data. Two adjacent images showed that their sliding window is not the same way to place the day.

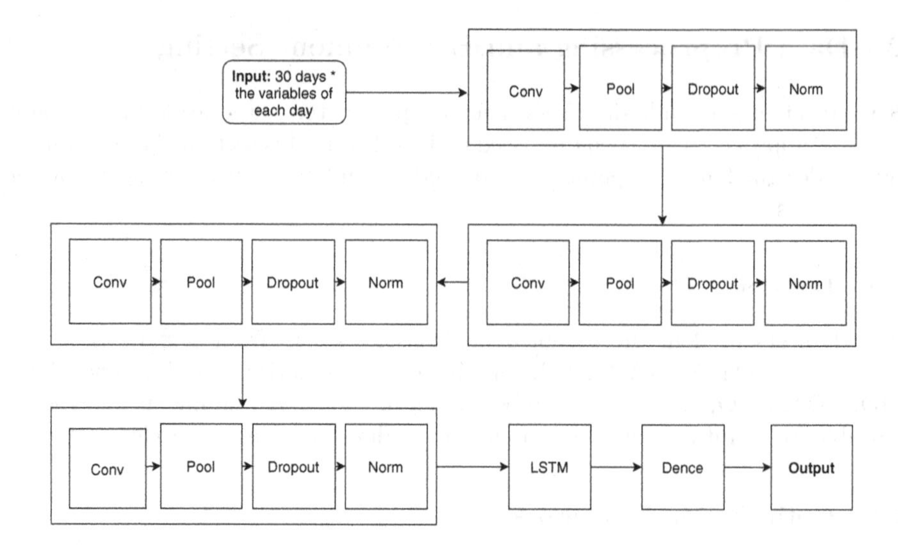

Fig. 2. A framework to improve the forecasting accuracy of stock prices.

As mentioned before, the designed model combines both CNN and LSTM together to improve the performance. The features od CNN model [10,15] are the extracted and converted into the images of the input data in the network. Different to the operations used in LSTM and the pooling mechanism, the developed model considers more technical operations (i.e., dropout and norm in the DNNs). The reason to adapt the dropout is because it can be used to avoid much data required in the learning process. For the training step, the designed model samples the parameters of the weight layer randomly based on the probability model. Also, the sub-network is considered as the target network for the updating progress. If a network has n parameters, thus the size of the subnets is calculated as 2^n. Moreover, if the number of n is extremely huge, thus the updating process of the subnet for each iteration would no update repeatly because the overfitting issue is considered to be avoided in the training step by the current network. The designed also convert the vector of stock index value for 30 days into the image data, which is then used in the input layer. Thus, the result for the stock forecasting is then produced, and the main framework is then illustrated in Fig. 2.

Figure 3 shows the procedure of the designed model in this paper, in which we can see that the stock data is basically divided into two datasets for the further processing, e.g., training and testing datasets. The designed model then produces a trading strategy, and the detailed algorithm in then described in Algorithm 1.

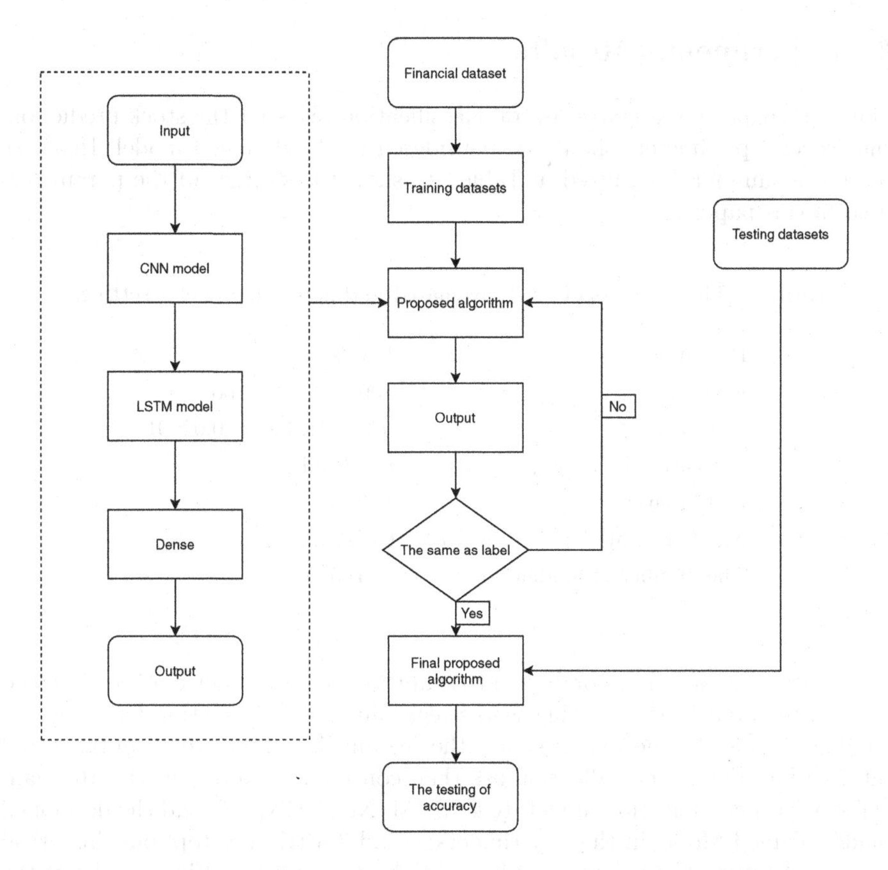

Fig. 3. The flowchart of the developed model in this paper.

Algorithm 1. The developed model

Require: b is the data of training; c is the data of testing; I is the number of iteration; B is batch size; Algorithm SGD is named *Adam*.

Ensure: the train model n; evaluation result *accuracy*

1: Initialize algorithm
2: $b \leftarrow$ *Initialize algorithm*
3: $P \leftarrow$ (split b in equal parts of B)
4: **for** each round $t = 1, 2, ..., z$ **do**
5: $\{verify, train\} \leftarrow \{P_t, P - P_t\}$
6: $(tf, vf) \leftarrow$ (generate feature of $train$ and $verify$)
7: $n_t \leftarrow$ modelFit$(Adam, tf)$
8: $r_t \leftarrow$ modelEvaluate(n_t, vf)
9: **end for**
10: $n \leftarrow$ bestModel
11: $c \leftarrow n$
12: $accuracy \leftarrow$ modelEvaluate$(n, test)$

---

3 Experimental Results

Since this paper investigates several classification tasks for the stock prediction, thus several parameters should be considered in the designed model. Here, we present a summary organized in Table 1 to show the settings of the parameters used in this paper.

Table 1. The numbers of levels evaluated in different parameter settings

Parameters	Levels
Epochs	200, 300, ..., 700
Learning rate	0.1, 0.01, 0.001, 0.00001
Activation functions	relu/tanh
LSTM layers	1, 2, 3
Number of hidden layer neurons	64, 128, ..., 512
The number of hidden layer	3, 4, 5

Experiments are then conducted to evaluate the performance of the designed model compared to the existing works regarding stock prediction based on the trading signals. In the experiments, the leading indicators (e.g., options and futures) and market classification are then considered. Moreover, the 10 financial stocks, 4 classification models (e.g., SVM, NN, CNNpred, and the developed model (named Model in the experiments)), and 3 attributes (options, historical data, and futures) are then considered in the experiments. Figure 4 shows the results for prediction, and from the results, it can be seen that the algorithms produce good performance by considering all the indices, and the accuracy is acceptable to be applied into the real-world stock markets. In general, the designed model has achieved the best results compared to the other approaches in the experiments.

To show that this combined model has achieved good performance, it then compares with the existing two models by CNN [17] and LSTM. 3 different time windows (e.g., 1, 3 and 7 days) are then used in the experiments, and we can observe the prediction performance regarding varied time slots. The results are then stated in Table 2. From the results, we can see that the developed model has achieved the best performance among 3 time slits, and the developed model combing CNN and LSTM outperforms the existing two models.

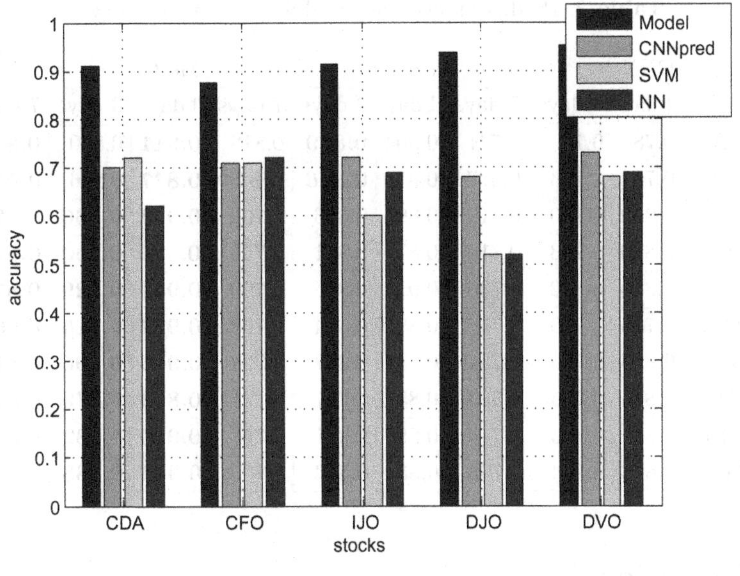

(a) Taiwanese stocks

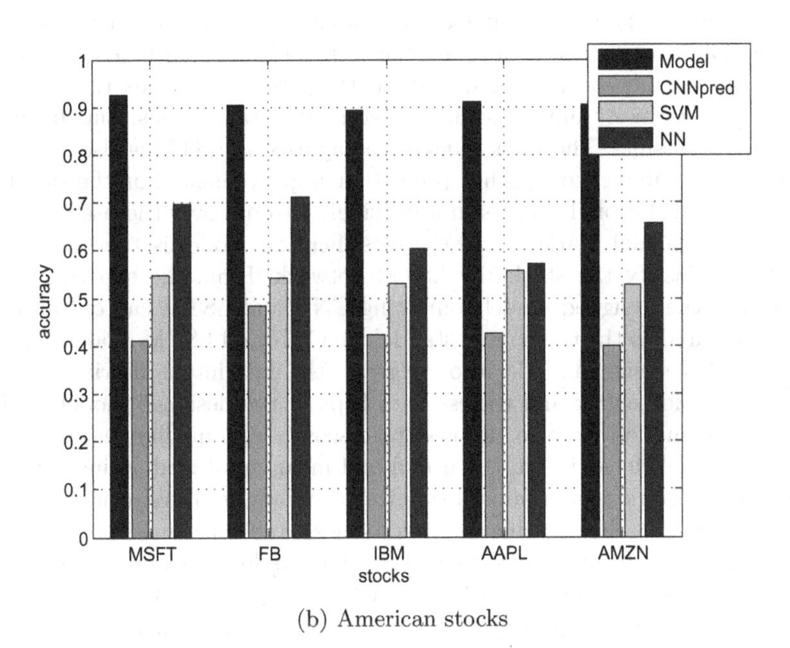

(b) American stocks

Fig. 4. The results of prediction accuracy for 4 models by using three indices.

Table 2. Prediction accuracy in different time windows

Stocks	CNN			LSTM			Model		
	1 day	3 days	7 days	1 day	3 days	7 days	1 day	3 days	7 days
CDA	0.78	0.734	0.731	0.904	**0.889**	0.818	**0.911**	0.870	**0.859**
CFO	0.791	0.773	0.677	0.874	**0.846**	**0.814**	**0.877**	0.826	0.795
IJO	0.840	0.740	0.719	0.894	0.826	0.762	**0.913**	**0.866**	**0.836**
DJO	0.800	0.763	0.719	0.874	0.821	0.755	**0.936**	**0.900**	**0.828**
DVO	0.880	0.842	0.736	0.910	0.875	0.790	**0.951**	**0.929**	**0.890**
MSFT	0.785	0.770	0.768	0.855	0.824	0.798	**0.927**	**0.915**	**0.904**
FB	0.886	0.746	0.734	0.886	0.862	0.794	**0.905**	**0.896**	**0.879**
IBM	0.815	0.725	0.583	0.857	0.857	0.786	**0.893**	**0.879**	**0.877**
AAPL	0.861	0.733	0.710	0.860	0.821	0.778	**0.910**	**0.882**	**0.892**
AMZN	0.818	0.764	0.795	0.852	0.833	0.805	**0.905**	**0.888**	**0.883**

4 Conclusion

Since the characteristics of nonlinear and noise of price in stock markets showed that it is not an easy task to forecast the trends of financial market, and more variables should be considered to make better prediction, thus the developed model considers many indices such as historical data, options, and futures to achieve better results. The stock sequence array used in LSTM is also considered in the designed model for further prediction improvement. For the developed model, the convolutional layer is first considered to discover the useful features used in the financial market, and the classification model is then adapted to predict and classify the stocks by LSTM network. From the experiments, we can see that the designed model combining CNN and LSTM can obtain better results compared to the traditional standalone CNN and LSTM models for stock prediction. The designed model also integrates the data into a matrix, which can be used to avoid scatter and useless data kept in the designed model, and the convolutional model can also be used to discover high-quality features in the developed model. In addition, the useful and meaningful leading indicators are considered in the designed model for better accuracy improvement. In general, the developed model has achieved effectiveness and efficiency for stock price predication. In the future works, we will consider to develop a model by utilizing the designed model to figure out the rise or fall point in the stock market, which is possible to make an expert system for the valuable investment.

Acknowledgment. This research is supported by Shandong Provincial Natural Science Foundation (ZR201911150391).

References

1. Azzouni, A., Pujolle, G.: A long short-term memory recurrent neural network framework for network traffic matrix prediction. arXiv preprint arXiv:1705.05690 (2017)
2. Bao, W., Yue, J., Rao, Y.: A deep learning framework for financial time series using stacked autoencoders and long-short term memory. PloS one **12**(7), e0180944 (2017)
3. Cai, X., Hu, S., Lin, X.: Feature extraction using restricted Boltzmann machine for stock price prediction. In: IEEE International Conference on Computer Science and Automation Engineering, vol. 3, pp. 80–83 (2012)
4. Chen, K., Zhou, Y., Dai, F.: A lstm-based method for stock returns prediction: a case study of china stock market. In: IEEE International Conference on Big Data, pp. 2823–2824 (2015)
5. Chen, Q.a., Li, C.D.: Comparison of forecasting performance of ar, star and ann models on the Chinese stock market index. In: International Symposium on Neural Networks, pp. 464–470 (2006)
6. Di Persio, L., Honchar, O.: Artificial neural networks architectures for stock price prediction: comparisons and applications. Int. J. Circuits Syst. Signal Process. **10**(2016), 403–413 (2016)
7. Ding, X., Zhang, Y., Liu, T., Duan, J.: Deep learning for event-driven stock prediction. In: Twenty-Fourth International Joint Conference on Artificial Intelligence (2015)
8. Ghosh, A., Bose, S., Maji, G., Debnath, N., Sen, S.: Stock price prediction using lstm on indian share market. In: The International Conference on Computer Applications in Industry and Engineering, vol. 63, pp. 101–110 (2019)
9. Graves, A., Mohamed, A.R., Hinton, G.: Speech recognition with deep recurrent neural networks. In: IEEE International Conference on Acoustics, Speech and Signal Processing, pp. 6645–6649 (2013)
10. Gunduz, H., Yaslan, Y., Cataltepe, Z.: Intraday prediction of borsa istanbul using convolutional neural networks and feature correlations. Knowl.-Based Syst. **137**, 138–148 (2017)
11. Hochreiter, S., Schmidhuber, J.: Long short-term memory. Neural Comput. **9**(8), 1735–1780 (1997)
12. Hoseinzade, E., Haratizadeh, S.: Cnnpred: Cnn-based stock market prediction using a diverse set of variables. Expert Syst. Appl. **129**, 273–285 (2019)
13. LeCun, Y., Bottou, L., Bengio, Y., Haffner, P.: Gradient-based learning applied to document recognition. Proc. IEEE **86**(11), 2278–2324 (1998)
14. Rani, S., Sikka, G.: Recent techniques of clustering of time series data: a survey. Int. J. Comput. Appl. **52**(15) (2012)
15. Siripurapu, A.: Convolutional networks for stock trading. Stanford Univ. Dep. Comput. Sci., 1–6 (2014)
16. Williams, R.J., Zipser, D.: A learning algorithm for continually running fully recurrent neural networks. Neural Comput. **1**(2), 270–280 (1989)
17. Wu, J.M.T., Li, Z., Srivastava, G., Tasi, M.H., Lin, J.C.W.: A graph-based convolutional neural network stock price prediction with leading indicators. Practice and Experience, Software (2020)
18. Yong, B.X., Rahim, M.R.A., Abdullah, A.S.: A stock market trading system using deep neural network. In: Asian Simulation Conference, pp. 356–364. Springer (2017)

19. Zhang, X., Tan, Y.: Deep stock ranker: a lstm neural network model for stock selection. In: International Conference on Data Mining and Big Data, pp. 614–623 (2018)
20. Zhao, Z., Zhang, X., Zhou, H., Li, C., Gong, M., Wang, Y.: Hetnerec: heterogeneous network embedding based recommendation. Knowl.-Based Syst. **204**, 106218 (2020)

Applying Convolutional Neural Networks for Stock Market Trends Identification

Ekaterina Zolotareva(✉)

Data Analysis and Machine Learning Department, Financial University under the Government of the Russian Federation, 38 Shcherbakovskaya St., Moscow 105187, Russia
elzolotareva@fa.ru

Abstract. In this paper we apply a specific type ANNs - convolutional neural networks (CNNs) - to the problem of finding start and endpoints of trends, which are the optimal points for entering and leaving the market. We aim to explore long-term trends, which last several months, not days. The key distinction of our model is that its labels are fully based on expert opinion data. Despite the various models based solely on stock price data, some market experts still argue that traders are able to see hidden opportunities. The labelling was done via the GUI interface, which means that the experts worked directly with images, not numerical data. This fact makes CNN the natural choice of algorithm. The proposed framework requires the sequential interaction of three CNN submodels, which identify the presence of a changepoint in a window, locate it and finally recognize the type of new tendency - upward, downward or flat. These submodels have certain pitfalls, therefore the calibration of their hyperparameters is the main direction of further research. The research addresses such issues as imbalanced datasets and contradicting labels, as well as the need for specific quality metrics to keep up with practical applicability. This is the reduced version of the research, full text will be submitted to arxiv.org.

Keywords: CNN · Stock market trends · Expert opinion · Image recognition

1 Introduction

An ability to identify stock market trends has obvious advantages for investors. Buying stock on an upward trend (as well as selling it in case of downward movement) results in profit, which makes predicting stock markets a highly attractive topic both for investors and researchers. Despite the long history, this field, as stated in [1] is still a promising area of research mostly because of the arising opportunities of artificial intelligence.

Modern machine learning technologies are presented by a number of various algorithms, but in general, there are three main classes of models used for the prediction of stock markets - artificial neural networks (ANNs), support vector machines (SVM/SVRs) and various decision tree ensembles (e.g., Random forests). As of 2017, the hegemony of ANNs and SVM/SVRs has been observed - the articles based on these models accounted for 86% of articles researched in [1]. Among twenty scientific papers on stock market forecasting published in the period between 2018 and early 2021, at least eight studies

L. Rutkowski et al. (Eds.): ICAISC 2021, LNAI 12854, pp. 269–282, 2021.
https://doi.org/10.1007/978-3-030-87986-0_24

[2–9] exploited ANNs as the only main algorithm, another two used ANN in the ensemble (or stack) with other algorithms of equal importance [10, 11]. In four studies ANNs were used to compare performance with other algorithms, chosen as main [12–15].

It should be noted, that despite the common main method, different researches have sufficient variations in model structure, problem formalization, and features and labels accordingly. Models also may be applied to different markets, assets and prediction horizons. A certain variation also exists in the selection of performance measures which should ensure the comparability of models.

The absolute majority of researches, which apply ANNs as the main algorithm (or part of an ensemble or stack) to stock market forecasting [2, 3, 6–10] are concentrated on predicting the direction of the stock market, thereby solving a classification problem ("direction type" studies). Fewer predict prices [4, 11, 15] by solving a regression problem ("price type" studies). In all cases the ground truth variable is extracted from the historical price series.

All the "price type" studies and most of the "direction type" studies focus on daily basis predictions and only in three papers the time horizon varies from one week [2, 7] to one month [10].

Another important difference between the suggested models is the choice of features. Though all of the researches used market data (e.g., prices, volumes and the values derived from them - technical analysis indicators, correlations, volatilities and returns) as input variables, in some studies they were supplemented by text features [7, 11] or fundamental macroeconomic variables [6, 10].

In this paper we apply a specific type of ANNs - convolutional neural networks (CNNs) - to the problem of finding start and endpoints of trends. CNNs have appeared and developed largely due to the increased need in solving computer vision problems. In 2011 the AlexNet convolutional network [16] led to a breakthrough in the field of image classification. Subsequently, CNNs were used not only in the problem of classification, but also in image detection (for example, the YOLO methodology [17], as well as the RCNN, Fast-RCNN, and Faster-RCNN algorithms [18]), therefore their application to stock market forecasting is less common compared to the traditional fully connected networks. Here we consider a mathematical model based on a convolutional neural network, which is used not for the classical problem of image classification and detection, but for recognizing the state of the financial market (upward trend, downward trend, or flat) and predicting future moments of a trend reversal.

Among the reviewed recent papers on stock market forecasting, only three have applied CNNs. [2] suggests an integrated framework, which fuses market and trading information for price movement prediction. CNN is used to extract trading features from the transaction number matrix, the buying volume matrix, and the selling volume matrix of investors, clustered by their trading behaviour profile. The output of CNN is concatenated with the market information weighted by stock correlation and then is fed into another deep neural network algorithm to obtain predictions for the several following trading days.

[6] apply CNN to capture correlations among different variables for extracting combined features from a diverse set of input data from five major U.S. stock market indices,

as well as fundamental macroeconomic variables (e.g., currency rates, commodity prices, etc.) to predict next-day price movements.

[13] address the problem of stock preselection for portfolio optimization. The authors consider the performance of five machine learning algorithms - random forest, support vector regression (SVR) and three neural networks (LSTM neural network, deep multilayer perceptron and CNN) - inputting past 60 days' daily returns to predict the next day's return.

Unlike the majority of other researchers, we aim to explore long-term trends, which last several months, not days. The start and endpoints of such trends accordingly are the optimal points for entering and leaving the market. Despite the various technical analysis algorithms and econometrics studies based solely on stock data, some market experts still argue that traders are able to see opportunities of making money (i.e., detecting trends or turning points) that cannot be formally expressed. Thus, using computer science algorithms to learn from successful traders' decisions (and not only stock data) is likely to improve financial market models.

The key distinction of our model is that its ground truth vector is fully based on expert opinion data, provided by one major investment company. The basis for building a mathematical model is the historical data of the financial market, divided by experts into markup windows, each of which corresponds to some unchanged market state. Unlike other researchers, we do not use a mathematical formula to define a trend, instead, it is defined by an expect as a potentially profitable (or unprofitable) pattern in price dynamics. The major difficulty of this approach is that it is exposed to subjective judgements of the experts. On the other hand, if the experts are successful traders in a certain investment company, it gives the employer the chance to 'digitalize' their exceptional skills and obtain a machine learning algorithm no one else on the market can employ.

The labelling of market states was performed in a specially designed graphical interface, where the experts marked certain consequent periods as "Trend", "Flat" or N/A. They worked directly with images, not numerical data. This fact makes CNN the natural choice for the case, as we need to extract implicit patterns from images which is exactly the original mission of CNN.

The model requires only raw historical price data as input. From one point of view, it can be considered a limitation, since we ignore fundamental factors and news feed. On the other hand, it makes the model unpretentious in production, since stock data is easily obtainable and can be downloaded into the company's informational systems or directly fed into the model via API. Besides, the concept of using prices as only input features corresponds with the hypothesis of market efficiency (EMH) in the sense that prices reflect all the available market information [19].

Technically we are solving a classification problem (it is a "direction type" research), but the standard quality metrics (accuracy, AUC and F1-Score) turn out to be inapplicable because of imbalanced datasets and contradicting labels. For these reasons, the returns of simulated trading are used as the main performance indicator. This study is related to [20], that shares the same dataset but introduces a different prediction algorithm.

2 The Proposed Framework

The initial markup contained three types of windows - trend, flat or unknown state. Approximately 90% of identified trends and flats last from 40 to 600 business days, which is in line with medium- and long-term trends. Initially, the task was set as recognizing the point of a change in the market state (the transition from trend to flat and vice versa, regardless of the trend direction) with a minimum time lag - for example, identifying the beginning of a 200-day trend 20 days after its start. However, as the study progressed, it also became necessary to distinguish between the direction of trend (upward and downward) to be able to assess the returns and compare various modifications of the model.

The practical purposes of the study impose several restrictions. First, the constructed model must be universal, that is, it must not be limited to a specific asset, market or time period. Another significant issue is the exclusion of future data in calculations. Violation of this condition, as well as a significant increase in time lag in changepoint identification, makes the simulation results inapplicable in practice.

2.1 Model Structure

The Submodels. The proposed framework for predicting future changepoints in the market state includes three components.

- ChangePoints_classifier (abbreviated to ChP-c), the classification model that determines the presence or absence of the changepoint at a certain time interval. For the output the model returns the value 1 ("The trend changed at the specified interval") and 0 ("The trend did not change at the specified interval").
- ChangePoints_regression (abbreviated to ChP-r), the regression model that determines the position of the last changepoint at a certain time interval, provided that at least one change of state has occurred. The position is determined by a real number in the range from 0 to 1 (proportion of the width of the time interval), where 0 means that the last changepoint was recorded at the beginning of the period, and 1 - at the end. Knowing the start date and the width of the time interval, we can convert the value obtained from the ChP-r model to a specific date.
- Trend_or_Flat (abbreviated as TF), the classification model that determines the type of trend observed in a given markup window. For the output the model returns the value 1 ("Upward trend"), -1 ("Downward trend") or 0 ("No trend, flat").

It was mentioned above that one of the practical requirements for the model is to ensure its universality (that is, the independence of a specific financial instrument, and, in particular, of the range of changes in its prices or trading volumes in any time interval). It is also unacceptable to "look into the future", which makes it impossible to normalize quantitative indicators with, for example, annual or monthly highs and lows - they are unknown at the time of calculation. One of the possible options for constructing a mathematical model under such conditions is to interpret the quotes of a financial instrument not as a set of quantitative indicators, but as an image, for example, a chart with Japanese candlesticks and corresponding labels. Image detection assumes recognition of both the type of object (classification problem) and determination of the boundaries of its location on a snapshot containing several types of objects of different sizes (regression problem). In our case, the problem of detection can be interpreted as determining the type of trend and its boundaries (beginning and end) in the time window which plays the role of a snapshot. Thus, the input data in all three submodels (ChP-c, ChP-r, TF) are matrices of digitalized images corresponding to the quotes charts on certain time intervals - data slices.

The Interaction of Submodels. The interaction of submodels in real-time simulation (that is when new information about quotes arrives every day) is shown in Fig. 1. The ChP-c model, which is responsible for determining the presence of changepoints in a slice with a width of n working days (for example, n_days = 25 corresponds to approximately 5 weeks, and n_days = 75 to 15 weeks), is fed a digital image - a quote chart for the selected data slice. To reduce the computation time, data slices can be taken not every day, but with a skip step (for example, skip = 5 days), which gives a non-critical error. If the ChP-c model detects the presence of a changepoint in the window (value 1), then the ChP-r model is activated, which determines exactly where in this data slice the last trend started (win_srt). If the ChP-c model returns the value 0 (there was no trend change), then the trend start point (win_srt) is set to the value determined from the previous slice (if it is absent, then win_srt is equal to the beginning of the quotes history). By default, the end of the trend (win_end) is the current date on which the ChP-c model is launched. A digital chart of stock quotes taken for the dates between win_srt and win_end is transmitted to the TF model, which determines the type of trend (uptrend/downtrend or flat) and the corresponding recommendation - to open and hold a long/short position or close the position.

2.2 Data Preparation

The dataset to explore consists of 1389 files labelled by 2 experts. The data contains quotes of 700 stocks included in the S&P index, covering the period from 2005–01-28 to 2017–09-13. Each file contains on average around 2600 daily quotes (Date, Open, High, Low, Close) for a certain period and stockname, labelled by a certain expert.

Image Processing. As noted above, the input data in all three submodels (ChP-c, ChP-r, TF) are matrices of digital images corresponding to the quotes charts on certain time intervals - data slices. One of the most common chart types that accounts for all price data (open price, close price, high and low) is the Japanese candlestick chart.

Logarithms. During the markup, the experts used the logarithmic scale, which gives a slightly different visual effect, therefore, in this study, all the quotes were replaced by their natural logarithms.

Duplicates, Contradictions and Imbalanced Datasets. Different experts could have marked the same “data points”, but their markup does not always coincide. In our case, the number of duplicates with the same “Type” field value is 1,246,726 (37.8%) “data

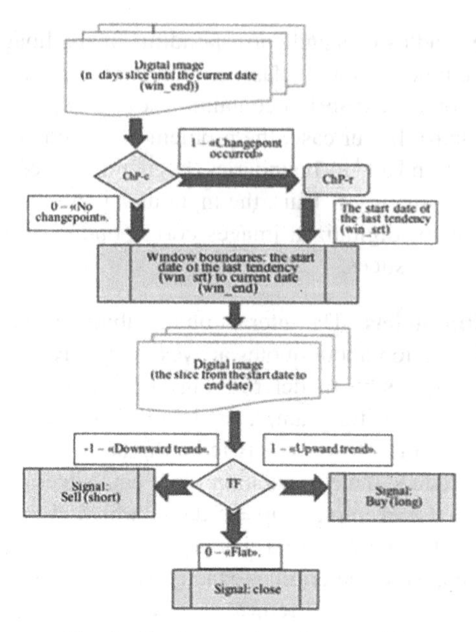

Fig. 1. The interaction of submodels.

points". The opposite situation (expert opinions for the same "data point" does not coincide) leads to contradictions. The number of such records is 397,321, which is approximately 24% of 1,660,441 - the number of unique "data points". Reasons, possible options for dealing with contradictions, their advantages and disadvantages are discussed [20] which uses the same dataset.

It should be noted that the problem of duplicates and contradictions is more significant for the models that determine the changepoints (ChP-c and ChP-r) than for the TF model: almost for every "data point" labelled as changepoint, there exists a "no changepoint" label obtained from another expert. Within the framework of this study, we corrected only technical blots, which eliminated 8% of contradictions.

Another issue is that the dataset is imbalanced. In the ChP-c model, most observations do not contain changepoints and the imbalance increases from 3:1 to 10:1 as n_days, the width of the data slice, decreases from 75 to 25. As for the TF model, more than half of the windows in both train and test set are labelled as flat, and there are about 2 times more upward trends than downward.

Ideally, the proportion of classes in the binary classification model should be close to 1:1, since as the imbalance increases, observations of the majority class begin to "suppress" observations of the minority class. In this study, we modified the standard loss function of CNN to deal with the imbalanced classes.

Train and Test Samples. To assess the generalizing ability of the model, the data array is traditionally divided into training and test samples. We considered three possible options: standard random split (inapplicable), split by source and split by date. Our final choice is to use 70% of older data as a train set and the remaining 30% as a test set. The threshold date separating the test and training samples is set to October 17, 2014.

3 The General Scheme for CNN

The general scheme for building a CNN includes several stages. At each stage, it is necessary to fix some parameters (more precisely, hyperparameters), on which the result of training will largely depend. The actual machine learning algorithm is implemented in Python 3.5/3.6 using the CNTK library from Microsoft (v.2.5.1) [21], Pandas 0.22.0, Numpy 1.14.2 and Matplotlib 2.2.2.

3.1 Extracting Features and Labels

At this stage, it is necessary to fix the following data preprocessing settings:

- image characteristics: resolution, number of channels. In the present study, colour images with a resolution of dpi = 10 were used to train the ChP-c and ChP-r models. For the TF model, which had a smaller sample size, the 20 dpi and 60 dpi options were also tested. It did not reveal any significant quality benefits but resulted in increased data volume and processing time.
- the procedure for dealing with duplicates and contradictions.
- the width of the data slice (n_days) and the step (skip) with which they are taken - for the models that determine the changepoints (ChP-c and ChP-r). In the designed models, the options n_days = 25 and n_days = 75 were used, which is slightly longer than monthly and quarterly intervals. The skip step is a technical hyperparameter that limits the size of the resulting sample (reaches several gigabytes) and, accordingly, the time of its generation (can take tens of hours).

The generated sets of features and labels are saved in a special text format CTF, compatible with the CNTK library.

3.2 CNN Structure

The convolutional neural network model can have three types of layers (convolutional, pooling and fully connected), characterized by different sets of hyperparameters. A multidimensional array containing input variables (features) is fed as the input of the neural network. For the output the network returns labels.

Figure 2 presents the final structure of the ChP-c model. The total number of parameters (weights) of the model in eight blocks is 32 117. The parameters f (filter size), n (number of filters), s (stride) depend on the original resolution of images. The ChP-r structure is similar except for the output layer – it uses linear activation function instead of sigmoid. TF models have two first convolutional layers instead of three and, of course, a different fully-connected layer with a vector output which is fed into softmax operator.

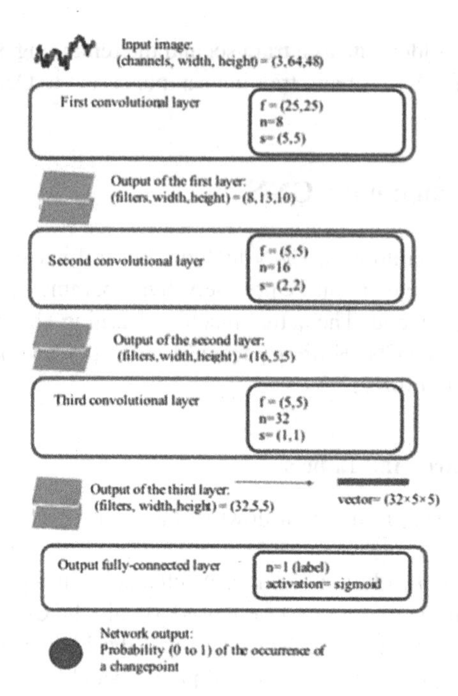

Fig. 2. The final structure of the ChP-c model.

3.3 CNN Training

The choice of model hyperparameters can have a decisive influence on the simulation result, however, the problem of finding the optimal combination of hyperparameters is nontrivial and requires a series of experiments. For simple models, automated procedures are suitable - full or randomized search on a grid composed of all possible combinations of hyperparameters. For more complex models, gridsearch is usually associated with technical difficulties, since running one cycle can take several hours or even days. In such situations, heuristics obtained as a result of similar studies and logical analysis of pitfalls of each experiment play an important role. Within the framework of this study, it was revealed that the choice of the type of the loss function and the learning rate α had a decisive influence on the learning outcome. Variations of other parameters either did not have any significant effect or influenced only the duration and resource intensity of the learning process.

3.4 CNN Validation

The CNN validation procedure is similar to other machine learning models and is determined by the type of problem. Nevertheless, it should be kept in mind, that the final prediction of the model is formed as a result of the interaction of all three submodels (ChP-c, ChP-r and TF). Another reason why standard quality metrics are not quite applicable is the time factor: even if the data is taken with a skip = 5 days, a shift in predictions means only a slight delay in the model, and, therefore, a slight decrease in the financial result. This leads to the conclusion that in our case, from the practical point of view, the main quality criterion is the returns obtained as a result simulation.

4 The Results

4.1 Submodel Results

ChangePoints_classifier. The train and test data set for the ChP-c models have around 90 thousand and 40 thousand of observations accordingly (the exact number depends on n_days and skip parameters). In total 24 more or less successful ChP-c models were trained (we do not consider experiments with the absolutely unfortunate choice of hyperparameters). They differ in the n_days parameter (25 and 75), dataset choice (labelled by both expert or only one to exclude contradictions), loss function modification and the number of iterations. The quality metrics of the two best models, calculated on the test set, are summarized in Table 1. The F-score in parentheses contains the values of the F-score metric for the minority class - F-score$_{y=1}$.

Table 1. The best ChP-c models.

Model id	ChP-c_4 (n_days = 25)	ChP-c_6 (n_days = 75)
Accuracy	81.92%	66.42%
AUC	54.90%	56.43%
F-score	83% (24%)	64% (28%)
False negatives	85%	76%
False positives	11%	17%

Despite all the efforts made to deal with the imbalanced dataset, both models have the same pitfall: they often miss real changepoints (give false negative predictions) - that is they ignore the minority class. Let us keep in mind, however, that because of the contradictions issue, the metrics based on ground truth may be inaccurate.

ChangePoints_regression. In total 19 ChP-r models were trained. They differ in the n_days parameter, skip parameter, dataset choice, loss function modification, number of iterations, number of layers and other hyperparameters. The values of the n_days parameter were chosen in compatibility with the corresponding ChP-c models. The sample size can be adjusted by reducing the skip to 2 or 5 days to give around 100 thousand and 50 thousand observations on train and test accordingly. The quality metrics of the two best models on the test set are summarized in Table 2.

Despite the fact that the distribution of labels was uniform in the range from 0 to 1 and numerous experiments with different loss functions, the predictions of ChP-r models are inevitably concentrated in the middle of the time segment (0.5), which gives the average absolute error around 22–25% of the window slice width. This corresponds to approximately 17 business days for n_days = 75 or 6 business days for n_days = 25.

Table 2. The best ChP-r models.

Model id	ChP-r_13 (n_days = 25)	ChP-r_15 (n_days = 75)
R^2	0.133	0.112
MAE	21.66%	22.59%

Same as for ChP-c, additional research is required. A change in one hyperparameter may sufficiently change the result as it happened to the TF model, but this is the process of trial and error.

TF model. The dataset for the TF model uses the markup windows initially selected by the experts. The train and test data sets for TF models have around 10 thousand and 5 thousand of observations accordingly. In total, more than three dozen models were trained. However, the first twenty models consistently predicted the majority class (0 ("No trend, flat")), which automatically ensured accuracy around 60%. Only after switching to the lower resolution and changing the learning rate $\alpha = 0.001$, a leap in quality was obtained. The results for dpi = 20 and dpi = 10 of the last 12 models were almost equal but the TF model with dpi = 10 as less resource-intensive.

The selected model best determines flats (89% of correctly recognized observations of this class) and the uptrend (recall 95%). The situation is relatively worse with the downward trend: in 25% of cases the TF_19 (dpi = 10) mistakenly recognizes it as flat. For reference, we note that the TF model with dpi = 20 recognizes flats and upwards a little better (93% and 92%, respectively), but in 37% of cases it mistakenly took a downward trend for a flat.

4.2 Simulation Results

The final stage of the proposed framework is a real-time simulation. This requires the development of quality metrics focused on financial results.

Specific Metrics. Recalling that the direction of the trend can be determined by the slope of the regression line, we can calculate the profit earned during the time in position (i.e., during trends) and a couple of other metrics (see Table 3).

To compare models, the last two indicators - YearProfit and YearProfit_avg - are the most informative, because they are independent of the length of the time period and the number of stocks in the pipeline. The main pitfall of YearProfit and YearProfit_avg metrics is that they are not normalized, but we can compare our results with the results of alternative trading simulations.

Results. The trading simulation was performed for n_days = 25 (ChP-c_4, ChP-r_15, TF_19 submodels) and n_days = 75 (ChP-c_6, ChP-r_13, TF_19 submodels). In both cases, the experiment was run for the entire test dataset (that is, data for dates between October 17, 2014, and May 13, 2017) for all 700 stocks with skip = 5.

Table 3. The proposed quality metrics.

Indicator	Description
Profit	The sum of all profits earned while in position (for all the stocks)
Days_in	The total number of business days in position (for all the stocks)
Times_in	The number of times the position was opened (for all the stocks)
DayProfit	The profit per one day in position, %: DayProfit = Profit/ Days_in
YearProfit	dayProfit scaled per annum, %: YearProfit = DayProfit*250, where 250 is the average number of business days in a year
YearProfit_avg	The average annual profit, including the days not in position, %: YearProfit_avg = Profit/number of data points *250

The comparison with the results of the "average" expert indicate the need to refine the models. Even though positions are opened about 6 times more often (increases transaction costs), the final profit on them is several times less. In the case of n_days = 75 the final results are slightly higher. 85% of downwards are still recognized as flats, but the share of unrecognized upwards is 49%, which allows one to earn a little more on this trend in comparison with the case n_days = 25. Because most of the trends marked by experts are long-term and medium-term, we can conclude that using a wider data slice (n_days = 75) is more preferable to correctly identify the changepoints in market trends.

If we drill down into the P&L report of the trading simulation for n_days = 75, we can notice, that it is short positions that cause losses. While, on average, the model lets us earn when we buy on the upward trend, it generates negative returns when we enter the market and sell at the beginning of the downward trend. This fact encourages us to refuse from short positions at all as non-efficient and concentrate only on traditional deals when one buys expecting the price will rise. Excluding short deals results in an increase in YearProfit_avg from 1.05% to 2.2% for n_days = 75 case (see Table 4).

To decrease the number of unnecessary deals we can opt for a slightly different interpretation of the model signals. We will ignore flat signals which often are incorrect, that is, we will open the position at the first signal of the upward trend and close it only when we receive the first downward signal. This modification decreases the number of deals from 3365 to 1120, while YearProfit_avg reaches 4.76% and brings us closer to the "average expert" baseline - 5,83% (short positions excluded). Finally, we can compare our result with the buy-and-hold strategy, which assumes that we buy stock on the very first day and stay in position all the time. Surprisingly, it earns more than the "average expert" in terms of YearProfit_avg indicator - 9.22%, but uses money less effectively: the "average expert" stays in position 5 times less and therefore has the higher YearProfit indicator -24,8%. In conclusion we may say, that while the suggested model definitely allows to earn on increasing market it still does not outperform the alternatitives, though some additional benefit may be gained from effectively using the time not in position, e.g. switching between the stocks in portfolio.

Table 4. The simulation results (short position excluded)

Metrics (short positions excluded)	n_days = 75			
	Model	Model (flat ignored)	Expert	Buy and hold
Profit (%)	3928	8414	10324	16302
Days_in	117723	369128	104056	442115
Times_in	3365	1120	655	699
YearProfit (%)	8.34	5.7	24.8	9.22
YearProfit_avg (%)	2.22	4.75	5.83	9.22

5 Conclusions

The paper illustrates the application of CNNs, traditionally used for image detection and recognition, to the problem of long-term market trend prediction. Unlike in the traditional approaches, the labels (trend or flat) are not derived from prices but filled manually by experts who worked with stock data as with images. The task is quite challenging since we actually try to 'digitalize' successful traders' skills and we can only compare the performance of the model with the performance of the experts themselves. The comparison with the results of other researchers would be inadequate since we use a completely different source for ground truth.

The main reason for the unsatisfactory accuracy of the predictions is the shortcomings of the ChP-c and ChP-r submodels described in the sections above. They, in turn, can be caused by inconsistencies in the original data markup. We should also consider the imbalance in the class of downward trends, which led to a relatively lower accuracy of their recognition by the TF model. In general, however, the proposed CNN framework for the prediction of changepoints in long-term market trends allows us to learn from successful traders' decisions and evaluate the model performance, although the submodels require additional exploring of errors and calibration.

Improving the prediction quality of the ChP-c, ChP-r and TF models is the main direction of further research. One of the options is to use a pretrained network (for example, Alexnet [16] or YOLO [17]) and adapt it for ChP-c and ChP-r labels by changing the fully connected output layer. We also consider a radically different model design with gradient boosting algorithm XGBoost, which currently performs comparatively better, but too, requires additional research.

References

1. Henrique, B.M., Sobreiro, V.A., Kimura, H.: Literature review: Machine learning techniques applied to financial market prediction. Expert Syst. Appl. **124**, 226–251 (2019). https://doi.org/10.1016/j.eswa.2019.01.012. ISSN 0957-4174
2. Long, J., Chen, Z., He, W., Wu, T., Ren, J.: An integrated framework of deep learning and knowledge graph for prediction of stock price trend: An application in Chinese stock exchange market. Appl. Soft Comput. J. **91**, 106205 (2020). https://doi.org/10.1016/j.asoc.2020.106205

3. Moews, B., Ibikunle, G.: Predictive intraday correlations in stable and volatile market environments: Evidence from deep learning. Phys. A Stat. Mech. its Appl. **547**, 124392 (2020). https://doi.org/10.1016/j.physa.2020.124392
4. Vijh, M., Chandola, D., Tikkiwal, V.A., Kumar, A.: Stock closing price prediction using machine learning techniques. In: Procedia Computer Science (2020)
5. Zhou, F., Zhou, H.M., Yang, Z., Yang, L.: EMD2FNN: A strategy combining empirical mode decomposition and factorization machine based neural network for stock market trend prediction. Expert Syst. Appl. **115**, 136–151 (2019). https://doi.org/10.1016/j.eswa.2018.07.065
6. Hoseinzade, E., Haratizadeh, S.: CNNpred: CNN-based stock market prediction using a diverse set of variables. Expert Syst. Appl. **129**, 273–285 (2019). https://doi.org/10.1016/j.eswa.2019.03.029
7. Picasso, A., Merello, S., Ma, Y., Oneto, L., Cambria, E.: Technical analysis and sentiment embeddings for market trend prediction. Expert Syst. Appl. **135**, 60–70 (2019). https://doi.org/10.1016/j.eswa.2019.06.014
8. Chandrinos, S.K., Sakkas, G., Lagaros, N.D.: AIRMS: A risk management tool using machine learning. Expert Syst. Appl. **105**, 34–48 (2018). https://doi.org/10.1016/j.eswa.2018.03.044
9. Moews, B., Herrmann, J.M., Ibikunle, G.: Lagged correlation-based deep learning for directional trend change prediction in financial time series. Expert Syst. Appl. **120**, 197–206 (2019). https://doi.org/10.1016/j.eswa.2018.11.027
10. Jiang, M., Liu, J., Zhang, L., Liu, C.: An improved Stacking framework for stock index prediction by leveraging tree-based ensemble models and deep learning algorithms. Phys. A Stat. Mech. its Appl. **541**, 122272 (2020). https://doi.org/10.1016/j.physa.2019.122272
11. Weng, B., Lu, L., Wang, X., Megahed, F.M., Martinez, W.: Predicting short-term stock prices using ensemble methods and online data sources. Expert Syst. Appl. **112**, 258–273 (2018). https://doi.org/10.1016/j.eswa.2018.06.016
12. Ismail, M.S., Md Noorani, M.S., Ismail, M., Abdul Razak, F., Alias, M.A.: Predicting next day direction of stock price movement using machine learning methods with persistent homology: Evidence from Kuala Lumpur Stock Exchange. Appl. Soft Comput. J. **93**, 106422 (2020). https://doi.org/10.1016/j.asoc.2020.106422
13. Ma, Y., Han, R., Wang, W.: Portfolio optimization with return prediction using deep learning and machine learning. Expert Syst. Appl. **165**, 113973 (2021). https://doi.org/10.1016/j.eswa.2020.113973
14. Bisoi, R., Dash, P.K., Parida, A.K.: Hybrid Variational Mode Decomposition and evolutionary robust kernel extreme learning machine for stock price and movement prediction on daily basis. Appl. Soft Comput. J. **74**, 652–678 (2019). https://doi.org/10.1016/j.asoc.2018.11.008
15. Zhou, F., Zhang, Q., Sornette, D., Jiang, L.: Cascading logistic regression onto gradient boosted decision trees for forecasting and trading stock indices. Appl. Soft Comput. J. **84**, 105747 (2019). https://doi.org/10.1016/j.asoc.2019.105747
16. Krizhevsky, A., Sutskever, I., Hinton, G.E.: ImageNet classification with deep convolutional neural networks. Commun. ACM **60**(6), 84–90 (2017). https://doi.org/10.1145/3065386
17. Redmon, J., Divvala, S., Girshick, R., Farhadi, A.: You only look once: unified, real-time object detection. In: Proceedings of the IEEE Computer Society Conference on Computer Vision and Pattern Recognition (2016)
18. Ren, S., He, K., Girshick, R., Sun, J.: Faster R-CNN: towards real-time object detection with region proposal networks. IEEE Trans. Pattern Anal. Mach. Intell. **39**(6), 1137–1149 (2017). https://doi.org/10.1109/TPAMI.2016.2577031
19. Fama, E.F.: Efficient capital markets: a review of theory and empirical Work. J. Finan. **25**(2), 383–417 (1970). https://doi.org/10.2307/2325486. Papers and Proceedings of the Twenty-Eighth Annual Meeting of the American Finance Association, New York, N.Y., 28–30 December 1969 (May 1970)

20. Zolotareva, E.: Aiding long-term investment decisions with XGBoost machine learning model. arXiv preprint arXiv:2104.09341 (2021)
21. Seide, F.: CNTK : Microsoft ' s open-source deep-learning toolkit. In: Proceedings of the 22nd ACM SIGKDD International Conference on Knowledge Discovery and Data Mining (2016)

Fuzzy Systems and Their Applications

The Extreme Value Evolving Predictor in Multiple Time Series Learning

Amanda O. C. Ayres(✉) and Fernando J. Von Zuben(✉)

School of Electrical and Computer Engineering, University of Campinas, Campinas, Brazil
{amanda,vonzuben}@dca.fee.unicamp.br

Abstract. This paper extends the evolving fuzzy-rule-based algorithm denoted Extreme Value evolving Predictor (EVeP) to deal with multivariate time series. EVeP offers a statistically well-founded approach to the online definition of the fuzzy granules at the antecedent and consequent parts of evolving fuzzy rules. The interplay established by these granules is used to formulate a regularized multitask learning problem which employs a sparse graph of the structural relationship promoted by the rules. With this multitask strategy, the Takagi-Sugeno consequent terms of the rules are then properly determined. In this extended version, called Extreme Value evolving Predictor in Multiple Time Series Learning (EVeP_MTSL), we propose an approach that resorts to the similarity degree among the time series. The similarity is calculated by the distance correlation statistical measure extracted from a sliding window of data points belonging to the multiple time series. Noticing that each fuzzy rule is part of a specific time series predictor, the new unified model called EVeP_MTSL updates the sparse graph by composing the relationship established by each pair of fuzzy rules (already provided by EVeP) with the similarity degree of their corresponding time series. We are then exploring not only the current interplay of the multiple rules that compose each evolving predictor, but also the current correlation of the multiple time series being simultaneously predicted. Two computational experiments reveal the superior performance of EVeP_MTSL when compared with other contenders devoted to online multivariate time series prediction.

Keywords: Evolving fuzzy-rule-based systems · Online learning · Extreme Value Theory · Multitask learning · Multivariate time series prediction

1 Introduction

In recent years, many methods for time series prediction have emerged in both the statistical and computational intelligence communities. However, the

This work has been supported by grants from CNPq - Brazilian National Research Council, proc. #143455/2017-6; #307228/2018-5, and Fapesp proc. #2013/07559-3.

L. Rutkowski et al. (Eds.): ICAISC 2021, LNAI 12854, pp. 285–295, 2021.
https://doi.org/10.1007/978-3-030-87986-0_25

problem of multivariate time series prediction, which aims to simultaneously forecast the values of multiple time series in dynamical interaction along time, has received less attention. One of the few attempts in the evolving systems literature is the integrated multi-model framework (IMMF) [12]. IMMF is implemented by training a neural network to assign relative weights to predictions from models at three different data granularity levels: global, local, and transductive. The ensemble of the fuzzy set-based evolving modeling (E-FBeM) [4] is another example, where an extension of [11] is developed to deal with multivariate time series employing an ensemble approach.

Besides these solutions guided by evolving fuzzy-rule-based (eFRB) systems, others are based on different architectures. An online echo state network based on square root cubature Kalman filters, for example, is proposed in [9] and is referred to as SCKF-γESN. SCKF-γESN can learn the training data one-by-one or chunk-by-chunk and incorporate an outlier detection feature to improve the forecasting accuracy and robustness. Based on the extreme learning machine principle, which presents a simple structure and good performance, an improved Levenberg–Marquardt algorithm is introduced in [14]. A kernel recursive least squares (KRLS) algorithm designed to multivariate chaotic time series is proposed in [10], by combining approximate linear dependency, dynamic adjustment, coherence criterion and quantization. Founded on the kernel extreme learning machine, in [6] it is proposed an improved version with adaptive forgetting factor introduced into the objective function, which can be adjusted iteratively and adaptively according to the system changes.

In this work, we propose an extension of the eFRB algorithm denoted Extreme Value evolving Predictor (EVeP) [1] to multivariate time series prediction. By jointly employing the similarity degree among the time series (calculated online taking a sliding window of time series data points) and the degree of intersection among the information granules belonging to the multiple time series, a more general formulation is conceived to promote information sharing not only among the existing rules, but also among the prediction models for the time series involved. The resulting regularization structure is then employed to obtain the Takagi-Sugeno (TS) consequent parameters of the fuzzy rules that compose the multiple evolving predictors.

The remaining sections of the paper are organized as follows: Sect. 2 reviews the main aspects of EVeP, necessary to properly introduce the formal definition of the EVeP_MTSL in Sect. 3. In Sect. 4, two computational experiments are considered in order to compare the performance of EVeP_MTSL with its original version for single time series prediction and also with several state-of-the-art algorithms. Section 5 is devoted to concluding remarks and further steps of the research.

2 The Extreme Value Evolving Predictor (EVeP)

Founded on the Extreme Value Theory [5], the Extreme Value evolving Predictor (EVeP) [1] resorts to a statistically consistent approach for the definition of the fuzzy rules in eFRB systems. It employs fuzzy granules that are statistically guaranteed to be the limiting distributions of the relative proximity among the data points that support the rules. The expression for the membership function $\mu^i(\boldsymbol{z}^{[t]})$ of a data point $\boldsymbol{z}^{[t]}$, available at time instant t, for the rule R^i with center $\boldsymbol{z}_0^i$, is given by the Weibull distribution $\Psi^i(||\boldsymbol{z}_0^i - \boldsymbol{z}^{[t]}||, \kappa^i, \lambda^i)$, defined by Eq. (1):

$$\mu^i(\boldsymbol{z}^{[t]}) = \Psi^i(||\boldsymbol{z}_0^i - \boldsymbol{z}^{[t]}||, \kappa^i, \lambda^i) = \exp\left[-\left(\frac{||\boldsymbol{z}_0^i - \boldsymbol{z}^{[t]}||}{\lambda^i}\right)^{\kappa^i}\right], \qquad (1)$$

where $||\boldsymbol{z}_0^i - \boldsymbol{z}^{[t]}||$ is the distance from $\boldsymbol{z}^{[t]}$ to center $\boldsymbol{z}_0^i$, and κ^i, λ^i are, respectively, the Weibull shape and scale parameters obtained automatically by EVeP by fitting the distribution to the smallest pairwise distance $m^i = ||\boldsymbol{z}_0^i - \boldsymbol{z}_s^j||$, $j = 1, \ldots, c$, $j \neq i$, $s = 1, \ldots, N^*$, thus taken into account the relative proximity to all the data points of the other rules. N^* is the sliding window size, i.e., the maximum number of samples kept by each rule.

To define the structural relationship among the rules, [1] introduced a low-cost approach reflecting the pairwise intersection of the fuzzy granules. As illustrated by Fig. 1 for a two-dimensional case, the connection degree of a rule R^{i_2} to a rule R^{i_1} is defined according to the maximum firing degree of R^{i_1} calculated at two distance levels: the distance $d(\boldsymbol{z}_0^{i_1}, \boldsymbol{z}_0^{i_2})$ to the center of fuzzy rule R^{i_2}, and the distance to the closest point belonging to the curve of Eq. (2), calculated for $i = i_2$, which is given by $d(\boldsymbol{z}_0^{i_1}, \boldsymbol{z}_0^{i_2}) - d^{i_2}$. In Eq. (2), σ is a user-defined parameter representing the granularity of the model.

$$\Psi^i(d^i, \kappa^i, \lambda^i) = \sigma \qquad (2)$$

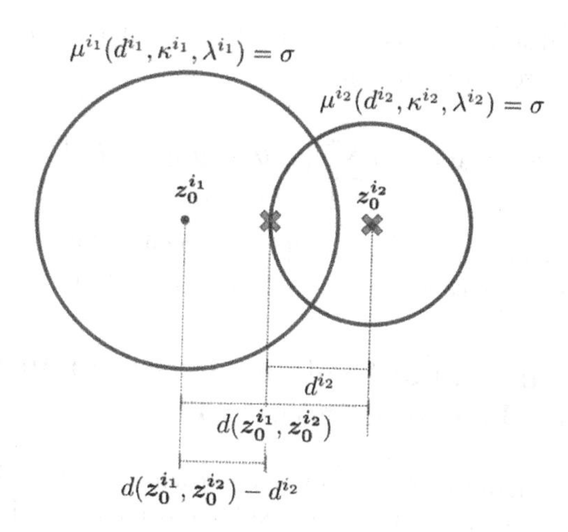

Fig. 1. Illustration of how to obtain the degree of relationship between a rule R^{i_2} and a rule R^{i_1} (adapted from [1])

The final value of the connection degree of rule R^{i_2} to rule R^{i_1} is calculated by Eq. (3), which is a composition of the degree of relationship for the input and output variables:

$$q(R^{i_1}, R^{i_2}) = \max(\max(\Psi_x^{i_1}(d(\boldsymbol{x}_0^{i_1}, \boldsymbol{x}_0^{i_2})), \Psi_x^{i_1}(d(\boldsymbol{x}_0^{i_1}, \boldsymbol{x}_0^{i_2}) - d_x^{i_2})), \\ \max(\Psi_y^{i_1}(d(\boldsymbol{x}_0^{i_1}, \boldsymbol{x}_0^{i_2})), \Psi_y^{i_1}(d(y_0^{i_1}, y_0^{i_2}) - d_y^{i_2}))) \tag{3}$$

The similarity measure $s(R^{i_1}, R^{i_2})$ between two rules R^{i_1} and R^{i_2} is then calculated as the maximum value of the connection degrees $q(R^{i_1}, R^{i_2})$ and $q(R^{i_2}, R^{i_1})$, according to Eq. (4):

$$s(R^{i_1}, R^{i_2}) = \max(q(R^{i_1}, R^{i_2}), q(R^{i_2}, R^{i_1})). \tag{4}$$

To calculate the parameters $\boldsymbol{\theta}^i$, $i = 1, \ldots, c$, of the TS consequent part of the rules, EVeP resorts to the benefits of the information sharing among the rules by means of a multitask learning (MTL) approach, employing a generalization of the Sparse Structure-Regularized Learning with Least Squares Loss (Least SRMTL) [3,15] to represent the structural dependencies among the rules. Considering the training data set $\{\boldsymbol{x}^{i^{[t]}}, y^{i^{[t]}}\}_{t=1}^{N_i}$, where N_i is the quantity of data points assigned to the ith rule R^i, $\boldsymbol{x}^{i^{[t]}} \in \mathbb{R}^n$, $y^{i^{[t]}} \in \mathbb{R}$, then the matricial form of the input-output dataset associated with the ith rule is expressed by:

$$X^i = \begin{bmatrix} 1 & \boldsymbol{x}^{i^{[1]}T} \\ 1 & \boldsymbol{x}^{i^{[2]}T} \\ \vdots & \vdots \\ 1 & \boldsymbol{x}^{i^{[N_i]}T} \end{bmatrix}, \quad \boldsymbol{y}^i = \begin{bmatrix} y^{i^{[1]}} \\ y^{i^{[2]}} \\ \vdots \\ y^{i^{[N_i]}} \end{bmatrix}. \tag{5}$$

The final optimization problem to calculate the matrix of parameters $\Theta = [\boldsymbol{\theta}^1, \boldsymbol{\theta}^i, \ldots, \boldsymbol{\theta}^c]$ is represented by Eq. (6):

$$\Theta^* = \arg \min_{\Theta} \sum_{i=1}^{c} ||X^i \boldsymbol{\theta}^i - \boldsymbol{y}^i||_2^2 + \Omega(\Theta), \tag{6}$$

where $||\cdot||_2^2$ is the squared l_2-norm and $\Omega(\Theta)$ is a regularization term that encodes the structural dependencies among the learning tasks, to be presented in Sect. 3.

3 The Extreme Value Evolving Predictor in Multiple Time Series Learning (EVeP_MTSL)

The fundamental idea of the Extreme Value Evolving Predictor in Multiple Time Series Learning (EVeP_MTSL) is to extend the original EVeP [1] to deal with multivariate time series by taking advantage of their joint learning at the calculation of the TS consequent parameters. Instead of training a separated model for each time series, as conducted in EVeP, EVeP_MTSL employs the degree of similarity among these time series to train a unique model that generates individual predictions for each time series.

The general operation of EVeP_MTSL follows the same principles of its predecessor EVeP [1]. The creation and removal of rules are executed independently for each time series, as well as the incremental updates in their antecedents. The prediction mechanism is also kept individualized for each series. What changes in this extended version is the training strategy of the parameters of the consequent part of the rules. We introduce a unified and regularized model that, by reflecting the interconnections among all the time series' rules, benefits from the joint learning in a generalized approach.

To calculate the similarity among the time series, we use the distance correlation measure [13], which is dynamically calculated based on a sliding window of data points and taking all pairs of time series. Let Y_k^w and Y_l^w, $w = t - N^*, \ldots, t$ be the last N^* values of the time series Y_k and Y_l. The distance correlation measure $\gamma(Y_k^w, Y_l^w)$ between Y_k^w and Y_l^w is calculated by Eq. (7):

$$\gamma(Y_k^w, Y_l^w) = \frac{dCov(Y_k^w, Y_l^w)}{\sqrt{dVar(Y_k^w)\; dVar(Y_l^w)}}, \quad w = t - N^*, \ldots, t \tag{7}$$

where $dCov$ refers to the distance covariance and $dVar$ refers to the distance variance. Details about these computations can be found on [13]. The distance covariance measure was chosen to calculate the similarity degree among the time series because it measures both linear and nonlinear dependencies between two random variables. It comes in contrast to Pearson's correlation, for example, which can only detect linear dependencies.

The similarity between two rules R_k^i and R_l^j associated with the prediction models for time series Y_k and Y_l, for any k and l, is calculated according to Eq. (8), which considers the similarity measure between the rules, given by Eq. (4), weighted by the distance correlation between their respective time series, given by Eq. (7).

$$s'(R_k^i, R_l^j) = s(R_k^i, R_l^j)\; \gamma(Y_k^w, Y_l^w), \quad w = t - N^*, \ldots, t \tag{8}$$

Let X_k^i be the input variables of rule R_k^i of time series k, $\boldsymbol{y}_k^i$ be the corresponding output variables and $\boldsymbol{\theta}_k^i$ be the parameters of the consequent part of the ith rule R_k^i of the prediction model for time series k. The final optimization problem to obtain the consequent parameters Θ^* is given by Eq. (9):

$$\Theta^* = \arg \min_{\Theta} \sum_{k=1}^{p} \sum_{i=1}^{c^k} (||X_k^i \boldsymbol{\theta}_k^i - \boldsymbol{y}_k^i||_2^2 + \rho \sum_{l=k+1}^{p} \sum_{\substack{j=1 \\ s'(R_k^i, R_l^j) > \eta}}^{c^l} s'(R_k^i, R_l^j) ||\boldsymbol{\theta}_k^i - \boldsymbol{\theta}_l^j||_2^2), \tag{9}$$

where p is the number of time series and c^k is the number of rules that compose the prediction model for time series k.

The first term, $||X_k^i \boldsymbol{\theta}_k^i - \boldsymbol{y}_k^i||_2^2$, seeks to minimize the joint error produced by the prediction models for every time series. The second term, $s'(R_k^i, R_l^j)||\boldsymbol{\theta}_k^i - \boldsymbol{\theta}_l^j||_2^2$, which corresponds to the regularization term $\Omega(\Theta)$ of Eq. (6), forces the corresponding models at the consequent part of rules R_k^i and R_l^j to exhibit a similar behavior [1] whenever the similarity measure $s'(R_k^i, R_l^j)$ exceeds the connection threshold parameter η. The user-defined parameter ρ controls the influence of the regularization term in the calculation of the matrix of TS parameters Θ. When compared to what would happen in the single-task learning approach ($\rho = 0$), the larger the relation, the more intense the reduction in the Euclidean distance involving θ_k^i and θ_l^j [1].

3.1 Numerical Example

Suppose the existence of three ($p = 3$) time series with $c^1 = c^2 = 2$ and $c^3 = 1$, and distance correlation measures currently estimated as $\gamma(Y_1^w, Y_2^w) = 0.86$, $\gamma(Y_1^w, Y_3^w) = 0.67$ and $\gamma(Y_2^w, Y_3^w) = 0.49$, $w = t - N^*, \ldots, t$. Let the similarity matrix S, calculated by obtaining $s(R_k^i, R_l^j)$ for each pair of rules of the three time series, be given by Eq. (11):

$$S = \begin{bmatrix} s(R_1^1,R_1^1) & s(R_1^1,R_1^2) & s(R_1^1,R_2^1) & s(R_1^1,R_2^2) & s(R_1^1,R_3^1) \\ s(R_1^2,R_1^1) & s(R_1^2,R_1^2) & s(R_1^2,R_2^1) & s(R_1^2,R_2^2) & s(R_1^2,R_3^1) \\ s(R_2^1,R_1^1) & s(R_2^1,R_1^2) & s(R_2^1,R_2^1) & s(R_2^1,R_2^2) & s(R_2^1,R_3^1) \\ s(R_2^2,R_1^1) & s(R_2^2,R_1^2) & s(R_2^2,R_2^1) & s(R_2^2,R_2^2) & s(R_2^2,R_3^1) \\ s(R_3^1,R_1^1) & s(R_3^1,R_1^2) & s(R_3^1,R_2^1) & s(R_3^1,R_2^2) & s(R_3^1,R_3^1) \end{bmatrix} = \begin{bmatrix} 1 & 0.38 & 0.17 & 0.48 & 0.17 \\ 0.38 & 1 & 0.13 & 0.24 & 0.36 \\ 0.17 & 0.13 & 1 & 0.86 & 0.85 \\ 0.48 & 0.24 & 0.86 & 1 & 0.47 \\ 0.17 & 0.36 & 0.85 & 0.47 & 1 \end{bmatrix} \quad (10)$$

The final similarity matrix S', which takes into account the similarity $s'(R_k^i, R_l^j)$ calculated among all pairs of rules and also the distance correlation among their corresponding time series, is given by Eq. (12):

$$S' = \begin{bmatrix} s'(R_1^1,R_1^1) & s'(R_1^1,R_1^2) & s'(R_1^1,R_2^1) & s'(R_1^1,R_2^2) & s'(R_1^1,R_3^1) \\ s'(R_1^2,R_1^1) & s'(R_1^2,R_1^2) & s'(R_1^2,R_2^1) & s'(R_1^2,R_2^2) & s'(R_1^2,R_3^1) \\ s'(R_2^1,R_1^1) & s'(R_2^1,R_1^2) & s'(R_2^1,R_2^1) & s'(R_2^1,R_2^2) & s'(R_2^1,R_3^1) \\ s'(R_2^2,R_1^1) & s'(R_2^2,R_1^2) & s'(R_2^2,R_2^1) & s'(R_2^2,R_2^2) & s'(R_2^2,R_3^1) \\ s'(R_3^1,R_1^1) & s'(R_3^1,R_1^2) & s'(R_3^1,R_2^1) & s'(R_3^1,R_2^2) & s'(R_3^1,R_3^1) \end{bmatrix} = \begin{bmatrix} 1 & 0.38 & 0.15 & 0.41 & 0.11 \\ 0.38 & 1 & 0.11 & 0.21 & 0.24 \\ 0.15 & 0.11 & 1 & 0.86 & 0.42 \\ 0.41 & 0.21 & 0.86 & 1 & 0.23 \\ 0.11 & 0.24 & 0.42 & 0.23 & 1 \end{bmatrix} \quad (11)$$

Table 1. Wind speed prediction for eolian farms

Model	*Rank*	$\overline{RMSE}$	#<	#>	$\overline{Rules}$
EVeP_MTSL	1.733	0.0438	0	6	2.80
EVeP [1]	3.267	0.0447	0	4	7.67
ePL-KRLS [2]	3.800	0.0459	0	3	1.78
FBeM_MTL [2]	4.067	0.0457	1	3	2.96
ePL [2]	4.533	0.0507	1	1	2.66
eTS [2]	5.733	0.0616	2	1	4.33
eTS-KRLS [2]	6.467	0.0507	4	1	3.47
eTS+ [2]	6.600	0.0509	4	1	4.50
eTS-LS-SVM [2]	8.800	0.0769	8	0	6.20

Considering a threshold $\eta = 0.25$, the resulting regularization term of Eq. (9) is given by Eq. (12):

$$\sum_{k=1}^{p}\sum_{i=1}^{c^k}\sum_{l=k+1}^{p}\sum_{\substack{j=1\\ s'(R_k^i,R_l^j)>\eta}}^{c^l} s'(R_k^i,R_l^j)||\boldsymbol{\theta}_k^i-\boldsymbol{\theta}_l^j||_2^2 = 0.38||\boldsymbol{\theta}_1^1-\boldsymbol{\theta}_1^2||_2^2 + 0.41||\boldsymbol{\theta}_1^1-\boldsymbol{\theta}_2^2||_2^2 + 0.86||\boldsymbol{\theta}_2^1-\boldsymbol{\theta}_2^2||_2^2 + 0.42||\boldsymbol{\theta}_2^1-\boldsymbol{\theta}_3^1||_2^2. \quad (12)$$

4 Computational Experiments

4.1 Wind Speed Prediction for Eolian Farms

This experiment [1,2] consists of predicting the wind speed at the three largest wind farms in the United States. For each wind farm, five well-distributed turbines were evaluated on an hourly time window basis during 2012. The data were taken from maps.nrel.gov/wind-prospector.

We compared statistically the *RMSE* obtained by the evolving algorithms employing the Friedman test [8], with $p = 0.05$ as the threshold. Whenever the null hypothesis is rejected, the Finner *posthoc* test is applied [7] with the same threshold to statistically support the advantage of an algorithm over the other. Table 1 presents the resulting statistical comparison. The table provides information on the rank of each algorithm, the average *RMSE* ($\overline{RMSE}$), the number of algorithms statistically better than the evaluated algorithm (#<), the number of algorithms statistically worse than the evaluated algorithm (#>) and also the average number of rules along the time series ($\overline{Rules}$). After the hyperparameter optimization applied taking historical data of the previous year (2011) in the interval $[0, 0.1]$ for σ, $[1, 100]$ for δ, $[1, 24]$ for N^*, $[10^{-2}, 10^3]$ for ρ and $[0, 0.9]$ for η, the user-defined parameters were set to $\sigma = 0.04$, $\delta = 56$,

$N^* = 23$, $\rho = 0.6467$ and $\eta = 0.3162$. The user-defined parameters for the other contenders were the same as reported in [1,2].

The rows of Table 1 are sorted by the rank. EVeP_MTSL was able to obtain the best rank, being statistically superior to the last six algorithms (column #>). There is no algorithm statistically better than EVeP_MTSL (column #<). One may note that EVeP_MTSL was able to obtain a better performance than its predecessor EVeP using a significantly reduced number of rules in average, which evidences the compactness of EVeP_MTSL. As there is a joint-learning involving rules belonging to the prediction models for all time series, a reduction in the number of rules seems to be promoted.

Figure 2 presents the performance of EVeP_MTSL for site 9773 of Roscoe Wind Farm according to the number of turbines—represented by their corresponding time series—considered into the model. Two cases were simulated: first, new turbines were incrementally included in the model considering its similarity to turbine 9773, from the highest to the lowest values; in the second case, the turbines were added from the lowest to the highest similarities.

The left graph of Fig. 2 shows the evolution of the RMSE according to the number of turbines. For the two scenarios, one may note a significant drop in the prediction error when the first turbines were added, until reaching 4 to 6 turbines, when the performance stabilizes. The RMSE retakes its downward trajectory only after the more similar turbines were added at the end of the simulation for the second case study.

When comparing the two strategies (adding the turbines incrementally in ascending and descending degrees of similarity), one can conclude that including the more similar turbines first tends to anticipate the achievement of maximum performance.

From the right graph of Fig. 2, one can see that the execution time per turbine dropped until reaching the number of six turbines. This initial drop results from predicting all the time series in a unique regularization framework rather than multiple separated formulations. However, after six turbines, the cost of considering several connections among the rules of all the time series

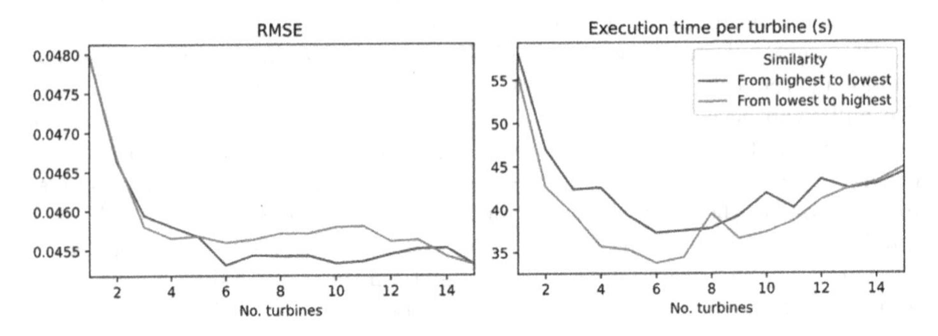

Fig. 2. Performance of EVeP_MTSL for site 9773 according to the number of turbines considered

influences the total execution time, worsening the performance as more turbines were added. The fact that the execution time is slightly shorter for the second scenario results from fewer connections among the rules of time series that are not as similar as the ones considered in the first scenario. In the end, both execution times converged to a similar value, as expected.

4.2 PM2.5 Time Series of Beijing

PM2.5 refers to the particles in the atmosphere with an aerodynamic equivalent diameter of less than 2.5 mm [6]. Their presence in the air is directly related to pollution. The concentration of PM2.5 was found to be inversely proportional to the change of wind speed. This experiment, extended from [6], aims to forecast PM2.5 values according to Beijing's historic PM2.5 time series and wind speed.

We used 960 samples of PM2.5 ($\mu g/m^3$) and wind speed (m/s) of 40 days in Beijing from November 22 to December 31, 2014. The time delay was set to $\tau_1 = \tau_2 = 1h$ and the embedding dimension as $m_1 = m_2 = 5$. The first 720 data points were used for training and the last 240 for testing, as prescribed in [6].

Table 2 presents the results. After the hyperparameter optimization applied for the training dataset in the interval $[0, 0.5]$ for σ, $[1, 200]$ for δ, $[1, 30]$ for N^*, $[10^{-2}, 10^3]$ for ρ and $[0, 0.9]$ for η, the user-defined parameters of EVeP were set to $\sigma = 0.3787$, $\delta = 166$, $N^* = 28$ and $\rho = 1.6322$. The user-defined parameters of EVeP_MTSL considering the same intervals were set to $\sigma = 0.4056$, $\delta = 94$, $N^* = 23$, $\rho = 4.0371$, and $\eta = 0.069$. The user-defined parameters for the other contenders were the same as reported in [6].

Table 2. PM2.5 Beijing

Evolving system	No. of rules (AVG.)	*RMSE*
KB-IELM	–	30.5964
NOS-KELM	–	30.9982
ALD-KOS-ELM	–	29.9865
FF-OSKELM	–	29.4089
AFF-OSKELM	–	29.4013
EVeP	27.41	30.4535
EVeP_MTSL	19.67	29.4765

As one may note, EVeP_MTSL outperformed once again its predecessor EVeP using a significantly reduced number of rules. Besides, it could obtain a very competitive performance considering contenders of robust architectures beyond the eFRB systems (for more details, see [6]). Figure 3 presents the prediction provided by EVeP_MTSL.

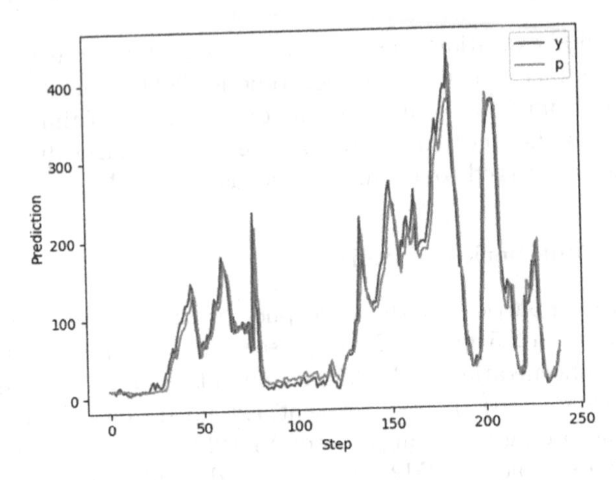

Fig. 3. Prediction of EVeP_MTSL for the PM2.5 Beijing time series

5 Concluding Remarks

This paper presents an extension of the eFRB algorithm denoted Extreme Value evolving Predictor (EVeP) for multivariate time series, named Extreme Value evolving Predictor in Multiple Time Series Learning (EVeP_MTSL). In this approach, we employed the similarity degree among the time series, calculated by the distance correlation statistical measure and updated at each new data point, and the degree of intersection among the rules that compose the multiple time series prediction models. The joint incorporation of time series correlation and rules similarity characterizes a generalized multitask learning formulation to calculate the parameters at the consequent part of the rules.

The two computational experiments conducted showed that EVeP_MTSL obtained better performance than its predecessor using fewer rules. Besides, EVeP_MTSL outperformed most of its contenders belonging to both the eFRB class of algorithms and other online and robust architectures.

As future work, we intend to adapt the proposed technique to perform multiple steps ahead prediction.

References

1. Ayres, A.O.C., Von Zuben, F.J.: The extreme value evolving predictor. IEEE Trans. Fuzzy Syst., 1–14 (2020). https://doi.org/10.1109/TFUZZ.2020.3044236
2. Ayres, A.O.C., Von Zuben, F.J.: Multitask learning applied to evolving fuzzy-rule-based predictors. Evol. Syst. **12**(2), 407–422 (2021). https://doi.org/10.1007/s12530-019-09300-w
3. Ayres, A.O.C., Von Zuben, F.J.: An improved version of the fuzzy set based evolving modeling with multitask learning. In: 2020 IEEE International Conference on Fuzzy Systems (FUZZ-IEEE), pp. 1–8. IEEE (2020)

4. Bueno, L., Costa, P., Mendes, I., Cruz, E., Leite, D.: Evolving ensemble of fuzzy models for multivariate time series prediction. In: 2015 IEEE International Conference on Fuzzy Systems (FUZZ-IEEE), pp. 1–6. IEEE (2015)
5. Coles, S.: An introduction to statistical modeling of extreme values, vol. 208. Springer (2001)
6. Dai, J., Xu, A., Liu, X., Yu, C., Wu, Y.: Online sequential model for multivariate time series prediction with adaptive forgetting factor. IEEE Access **8**, 175958–175971 (2020)
7. Finner, H.: On a monotonicity problem in step-down multiple test procedures. J. Am. Stat. Assoc. **88**(423), 920–923 (1993)
8. Friedman, M.: The use of ranks to avoid the assumption of normality implicit in the analysis of variance. J. Am. Stat. Assoc. **32**(200), 675–701 (1937)
9. Han, M., Xu, M., Liu, X., Wang, X.: Online multivariate time series prediction using SCKF-γESN model. Neurocomputing **147**, 315–323 (2015)
10. Han, M., Zhang, S., Xu, M., Qiu, T., Wang, N.: Multivariate chaotic time series online prediction based on improved kernel recursive least squares algorithm. IEEE Trans. Cybern. **49**(4), 1160–1172 (2018)
11. Leite, D., Ballini, R., Costa, P., Gomide, F.: Evolving fuzzy granular modeling from nonstationary fuzzy data streams. Evol. Syst. **3**(2), 65–79 (2012)
12. Pears, R., Widiputra, H., Kasabov, N.: Evolving integrated multi-model framework for on line multiple time series prediction. Evol. Syst. **4**(2), 99–117 (2013)
13. Székely, G.J., Rizzo, M.L., Bakirov, N.K., et al.: Measuring and testing dependence by correlation of distances. Ann. Stat. **35**(6), 2769–2794 (2007)
14. Wang, X., Han, M.: Improved extreme learning machine for multivariate time series online sequential prediction. Eng. Appl. Artif. Intell. **40**, 28–36 (2015)
15. Zhou, J., Chen, J., Ye, J.: User's Manual MALSAR: Multi-tAsk Learning via StructurAl Regularization (2012). www.MALSAR.org

Towards Synthetic Multivariate Time Series Generation for Flare Forecasting

Yang Chen(✉), Dustin J. Kempton, Azim Ahmadzadeh, and Rafal A. Angryk

Georgia State University, Atlanta, GA 30302, USA
ychen113@student.gsu.edu, {dkempton1,aahmadzadeh1,angryk}@cs.gsu.edu

Abstract. One of the limiting factors in training data-driven, rare-event prediction algorithms is the scarcity of the events of interest resulting in an extreme imbalance in the data. There have been many methods introduced in the literature for overcoming this issue; simple data manipulation through undersampling and oversampling, utilizing cost-sensitive learning algorithms, or by generating synthetic data points following the distribution of the existing data. While synthetic data generation has recently received a great deal of attention, there are real challenges involved in doing so for high-dimensional data such as multivariate time series. In this study, we explore the usefulness of the conditional generative adversarial network (CGAN) as a means to perform data-informed oversampling in order to balance a large dataset of multivariate time series. We utilize a flare forecasting benchmark dataset, named SWAN-SF, and design two verification methods to both quantitatively and qualitatively evaluate the similarity between the generated minority and the ground-truth samples. We further assess the quality of the generated samples by training a classical, supervised machine learning algorithm on synthetic data, and testing the trained model on the unseen, real data. The results show that the classifier trained on the data augmented with the synthetic multivariate time series achieves a significant improvement compared with the case where no augmentation is used. The popular flare forecasting evaluation metrics, TSS and HSS, report 20-fold and 5-fold improvements, respectively, indicating the remarkable statistical similarities, and the usefulness of CGAN-based data generation for complicated tasks such as flare forecasting.

Keywords: Multivariate time series · Class imbalance · Generative adversarial network · Flare forecasting

1 Introduction

In February 2010, NASA launched the Solar Dynamics Observatory, the first mission of NASA's Living with a Star program, which is a long term project dedicated to the study of the Sun and its impacts on human life [1]. The SDO mission is an invaluable instrument for researching solar activity, which can produce damaging space weather. This space weather activity can have drastic

L. Rutkowski et al. (Eds.): ICAISC 2021, LNAI 12854, pp. 296–307, 2021.
https://doi.org/10.1007/978-3-030-87986-0_26

impacts on space and air travel, power grids, GPS, and communications satellites [2]. For example, in March of 1989, geomagnetically induced currents, produced when charged particles from a coronal mass ejection impacted the earth's atmosphere, caused power blackouts and direct costs of tens of millions of dollars to the electric utility "Hydro-Qubec" [3]. If a similar event would have happened during the summer months, it is estimated that it would likely have produced widespread blackouts in the northeastern United States, causing an economic impact in the billions of dollars [3].

A solar flare is an event occurring in the solar corona that is characterized by a sudden orders-of-magnitude brightening in Extreme Ultra-Violet (EUV) and X-ray, and for large events, gamma-ray emissions, from a small area on the Sun, lasting from minutes to a few hours [4,5]. The classification system for solar flares are on a logarithmic scale and uses the letters A, B, C, M or X, according to the peak X-ray flux. In a typical binary classification strategy, M and X classes are identified as the positive class while no flare occurrence and flares of A, B and C classes are identified as the negative class.

The goal of this project is to generate synthetic data, especially for multivariate time series of magnetic filed parameters leading up to solar flares. As discussed in [6,7], the extreme class-imbalance between positive and negative classes in the solar flare data, and the improper treatment of said extreme imbalance between classes, can result in unrealistic and unreliable analyses, with little practical value in flare forecasting, flare classification, and flare clustering. Therefore, solving the issue of insufficient positive class of flare data is an important problem for current research in this domain. As such, this project is dedicating to generate realistic flare data based on real data, in order to provide a balanced training dataset for use in such problems.

2 Related Work

Although remedies such as oversampling, undersampling, and cost-sensitive learning, have been considered to tackle this problem [8,9], these methods can only provide limited improvements since they do not introduce or utilize any new data. The development of generative modeling provides an attractive alternative and potentially more domain-specific approach for data augmentation. For example, the Generative Adversarial Network can be trained to learn data distributions of the minority classes, thereby generating synthetic data for constructing a balanced and larger dataset to train more unbiased and powerful classifiers.

2.1 Generative Adversarial Network (GAN)

First proposed in [10], the Generative Adversarial Network (GAN) tries to learn an implicit density of real samples. The GAN trains the two components in an adversarial way. First, the generator is used to sample initial inputs from a latent space, which is used to produce data similar to real data. Next step, both generated samples and real data are used as inputs of a discriminator, and the discriminator assigns the label of samples after processing inputs with a neural network. Eventually, the predicted labels are used to calculate errors with a defined objective function, and the result is used for adjusting the whole model. This mechanism can help a generator gradually generate better realistic samples under the supervision of real samples, and this process keeps running until the discriminator cannot distinguish real data and synthetic samples.

There have been various types of GANs proposed as an extension of the vanilla GAN to deal with different demands. For instance, in the computer vision domain, the Deep Convolutional GAN [11] has been applied to learn reusable feature representations and generate synthetic images by utilizing convolutional neural networks as the generator and discriminator. The Wasserstein GAN [12] commits to improve the stability of learning and provide a meaningful learning curve. The Info GAN [13] incorporates the representation learning by encoding features into the latent vector. The Conditional GAN (CGAN) [14] is dedicated to improving the quality of generated samples and controlling the classes of synthetic samples by utilizing conditional information. Finally, the advantage of controlling the mode of generated samples makes CGAN become the most appropriate framework in our study since our goal is to generate synthetic samples of multiple minority classes for tackling the class imbalance issue.

2.2 Time Series Generation

There are already several projects in different domains that have worked on generating time series data utilizing the Generative Adversarial Network. In [15], the RGAN was utilized to generate medical time series data implemented with Long Short-Term Memory (LSTM) network. The motivation of their work was to develop a privacy-preserving method of generating synthetic medical data for machine learning modeling since actual patient data is sensitive to privacy issues. In [16], the use of a C-RNN-GAN was proposed as a method to generate musical data. This method differs from [15], in that it applied a unidirectional LSTM for the generator and a bidirectional LSTM for the discriminator. Then, in [17], the TimeGAN was proposed that combines the versatility of the unsupervised GAN approach with the control over conditional temporal dynamics. This method has two more autoencoding components, including an embedding function and a recovery function trained jointly with generator and discriminator components. This structure enables the model can learn to encode features, generate representations and iterate across time simultaneously.

2.3 Multivariate Time Series Dataset

The data primarily used in this project is a benchmark dataset, named as Space Weather ANalytics for Solar Flares (SWAN-SF), recently released by [18]. SWAN-SF is a comprehensive, multivariate time series (MVTS) dataset extracted from solar photospheric vector magnetograms in HMI Active Region Patch (HARP) data made available as the Spaceweather HMI Active Region

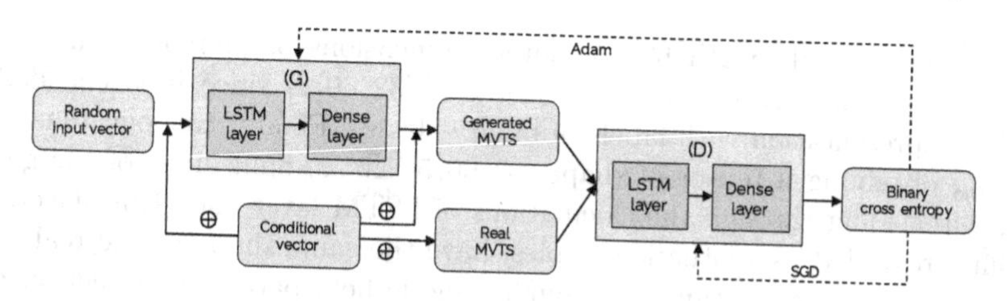

Fig. 1. This is the framework of the CGAN model, including components of the generator (G) and the discriminator (D). Each component is processed by the combination of the LSTM layer and the Dense layer. The inputs of the generator are random input vectors concatenated with conditional vectors. The inputs of the discriminator are either synthetic or real multivariate time series with conditional vectors. The binary cross-entropy is the criterion for optimizing the model.

Patch (SHARP) series [19,20]. The SWAN-SF is made up of five temporally non-overlapping partitions covering the period from May 2010 through August 2018 [21]. Each partition contains approximately an equal number of X- and M-class flares, and there is a total of 6,234 flare records and 324,952 no-flare records. Comparing the amount of two kinds of data, we can find an extremely imbalanced issue in this dataset. As mentioned previously, [7] showed that the extreme class imbalance between positive and negative classes in the solar flare data, and the improper treatment of said extreme imbalance can result in unrealistic and unreliable analyses. Furthermore, each flare record is a multivariate time series with 60 time steps, and each time step has 51 magnetic field parameters. This work will focus on four parameters, including TOTUSJH, ABSNJZH, SAVNCPP, and TOTBSQ, as a representative subset of the full 51 field parameters (for the definition of parameters see Table 1 in [21]). This is because many of the parameters are highly correlated, leading most studying flare forecasting to utilize some subset of the full set.

3 Methodology

We decide to use the Conditional Generative Adversarial Network (CGAN) in this study for several reasons. First, the advantage of controlling the mode of generated samples allows us to generate samples of minority classes to tackle the class imbalance issue. Second, CGAN can provide a stable and faster training compared to the vanilla GAN. Moreover, the SWAN-SF dataset is labelled, therefore it can provide conditional information for us to train CGAN models. LSTM network is utilized as the basic components in both the generator and the discriminator illustrated in Fig. 1 since we are processing time series data.

The ultimate goal of a generator G is to generate an output with similar characteristics as the real data. As seen in Fig. 1, the model takes in a random input vector Z_n, which is a tensor with the shape of [$batch\_size$, $sequence\_length$, $latent\_dim$]. In this study, the shape is $[32, 60, 3]$ for 32 multivariate time series in a batch, each of length 60 and latent dimensions 3. Moreover, the conditional vector, namely C_n, has the shape of $[32, 60, 2]$ since it is encoded in one-hot representation with labels of binary classes. Finally, we concatenate Z_n and C_n, obtaining a tensor of shape $[32, 60, 5]$ as the final input of the generator. After going through the calculations of LSTM layer and Dense layer, the outputs, regarded as synthetic samples, have the same shape as the real data, i.e., $[32, 60, 4]$ where 4 stands for four magnetic field parameters mentioned in Sect. 2.3.

The task of a discriminator D is to classify inputs as either being real or synthetic samples generated by the generator. In Fig. 1, it can be seen that the discriminator takes two forms of multivariate time series (MVTS) as the input: the real and the generated MVTS samples. To simplify the representation for our discussion about the discriminator, we denote X_n as a uniform set of inputs. Through feeding C_n into D, the discriminator not only produces judgments about whether the data is synthetic or real but also evaluates the correspondence of the synthetic sample to its conditional information. Finally, the binary cross-entropy loss calculated between the prediction and the ground truth is used to update the parameters of both the generator and the discriminator with the back-propagation algorithm.

So far, we have comprehended the structures and functionalities of the generator and discriminator. Next, we will define the objective function used for optimizing the discriminator and the generator. In our framework, the objective function is divided into two parts, including the generator loss and the discriminator loss. First, the discriminator loss which is calculated as the cross-entropy between the ground-truth and outputs of a discriminator, is defined as:

$$Loss_D(X_n|C_n, y_n) = -\mathrm{CE}\Big(D(X_n|C_n), y_n\Big) \tag{1}$$

In this equation, X_n is the set of inputs of the discriminator, and C_n is the conditional vector. $D(X_n|C_n)$ returns the probability of X_n being a real or synthetic sample by taking X_n and C_n as inputs. Note that X_n in this equation is composed of two different types of data sources:

$$X_n = \begin{cases} X_n & \textbf{if } \textit{inputs are real samples,} \\ \mathrm{G}(Z_n|C_n) & \textbf{if } \textit{inputs are synthetic samples.} \end{cases} \tag{2}$$

Correspondingly, the y_n takes two different values, dependent upon the source of the sample in X_n,

$$y_n = \begin{cases} \mathit{1} & \textbf{if } \textit{inputs are real samples,} \\ \mathit{0} & \textbf{if } \textit{inputs are synthetic samples.} \end{cases} \tag{3}$$

The generator loss is also formulated with cross entropy as below:

$$Loss_G(Z_n|C_n) = -\mathrm{CE}\bigg(D\Big(G(Z_n|C_n)|C_n\Big), \mathbf{1}\bigg) \tag{4}$$

where the input $G(Z_n|C_n)$ is the synthetic samples, and its corresponding predictions are $D\Big(G(Z_n|C_n)|C_n\Big)$. In calculating the loss of the generator, the label of a synthetic sample is held constant as **1** s, since the goal of the generator is to generate realistic-enough samples such that the discriminator can no longer distinguish them from the real samples.

4 Experiments

Despite the model that can be optimized according to the objective function defined in Sect. 3, the loss cannot objectively reflect the convergence of the training progress or the quality of generated samples [12]. To determine when to stop training models is a well-known and up in the air question in the GAN study. Unlike many image-based GAN projects, such as deepfake [22] and GauGAN [23], the visual verification of synthetic time series as outputs does not give us much evidence as to whether the generated data are realistic or not. In this section, we present three types of evaluation methods, in both qualitative and quantitative metrics, to verify the effectiveness and correctness of the CGAN model.

4.1 Experimental Settings

We have implemented our model using the TensorFlow 2.0 library [24], and the code can be found at our repository for the project[1]. After we explored various settings based on the defined objective function, we found that using the Adam Optimizer for the generator and the Gradient Descent Optimizer for the discriminator produced our best results. Moreover, we concluded our parameter settings of the learning rate to 0.1, the LSTM hidden size to 100, and the batch size to 32. The model is trained with 300 epochs, and intermediate models are saved every five epochs. In Sects. 4.2 and 4.3, we utilize 1,254 real flare samples in partition 1 of SWAN-SF and 1,254 synthetic samples generated by the CGAN model in our evaluations. In Sect. 4.4 though, we utilize the entire partition 1, including 1,254 real flares and 72,238 no-flare samples, and generate 70,984 synthetic flare samples to balance the training dataset. More details regarding the experimental setup will be presented in that section.

4.2 Evaluation Using the Distributions of Statistical Features

To provide a statistical evaluation for our model, we utilize the statistical features extracted from the time series data, including the mean, median, and standard deviation statistics. For visualization, we construct frequency distributions of these values for both the real input data and the synthetic data produced through binning the values into 20 equal-width bins. It is our assumption that if the distributions of the statistic values for both the real data and generated data are

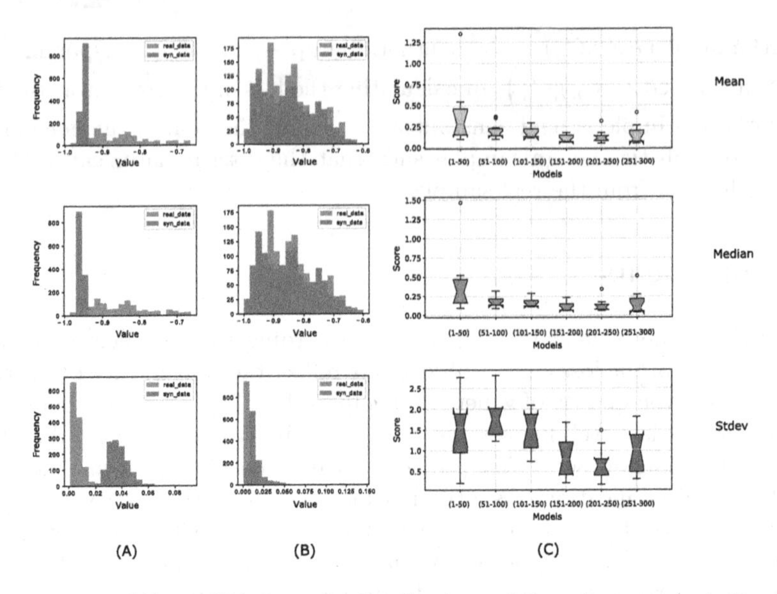

Fig. 2. Columns (A) and (B) show the distributions of three features, including mean, median, and standard deviation, of TOTUSJH at two intermediate epochs with a bin size of 20. Red bars stand for the real data, and blue bars stand for the synthetic data. Specifically, (A) is the result of the model at the 50th epoch, and (B) is the result of the model at the 250th epoch. Column (C) shows the distributions of KL divergence scores calculated by comparing distributions of synthetic samples and real samples across all intermediate models divided into six groups. (Color figure online)

similar, then the generated samples should be similar enough to perform well for our final task of producing minority class samples for model training.

We have conducted statistical feature evaluation on all the physical parameters, but for brevity, we present only the results of TOTUSJH. The columns (A) and (B) in Fig. 2 show the distributions of three features evaluated based on two intermediate models saved in the training process. (A) is resulted after training 50 epochs, which is the last model of the first group (1–50 epoch), and (B) is resulted after training 250 epochs, which is the last model of the fifth group (201–250 epoch). From (A) to (B), it can be found that the generator can gradually produce time series which have similar statistic attributes to the real data. Additionally, we calculate the Kullback–Leibler (KL) divergence between distributions of real data and generated samples with all features at different epochs. We observe, as shown in column (C) of Fig. 2, that the KL divergences of all three features are decreasing as training progresses, which means that the two distributions are getting more and more similar as the model evolves. We found that models between the 201–250 epochs achieve the best performance, with lower KL divergence for the mean, median, and standard deviation value distributions. We also see that the variance between the results produced by

[1] https://bitbucket.org/gsudmlab/mvts-gann.

intermediary models trends downward until we surpass the 250 epoch mark. This is regarded as the first criterion of the model selection in this study.

4.3 Evaluation Using Adversarial Accuracy

The Adversarial Accuracy, as formulated in Eq. 5, is put forward by Yale [25], which is used for comparing the similarity of two sets of data samples.

$$AA_{TS} = \frac{1}{2}(\frac{1}{n}\sum_{i=1}^{n}\mathbf{1}(d_{TS}(i) > d_{TT}(i)) + \frac{1}{n}\sum_{i=1}^{n}\mathbf{1}(d_{ST}(i) > d_{SS}(i))) \quad (5)$$

In the definition, the variable T stands for the set of real data, and the variable S stands for the set of synthetic data. For calculating $\mathbf{1}(d_{TS}(i) > d_{TT}(i))$, each real sample from T are compared with all synthetic data points in S to calculate the shortest distance $d_{TS}(i)$, and compared with all the other real data points in T to calculate the shortest distance $d_{TT}(i)$. The most straightforward distance metric is the Euclidean distance which is our choice in this study as well. The shortest distance generally means the highest similarity between two data points. If $d_{TS}(i) > d_{TT}(i)$, it means no synthetic data point can be found that is more similar to the current real data point than other real data points. Otherwise, a synthetic data point, which is more similar to the current real data point, can be found. The $d_{TS}(i) < d_{TT}(i)$ indicates a realistic samples is generated. The second part, $\mathbf{1}(d_{ST}(i) > d_{SS}(i))$, is implemented in a similar manner, except that here each synthetic sample will be compared with not only all the real samples but also all the other synthetic samples. Overall, the best balance result of Adversarial Accuracy should equal 0.5, which implies the generator can generate realistic samples.

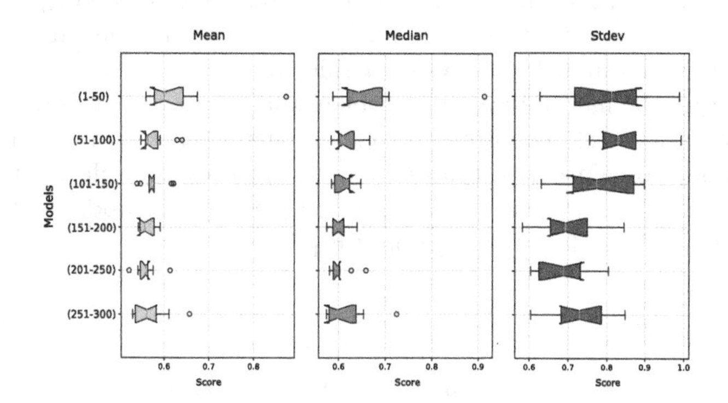

Fig. 3. The box plots show the distributions of Adversarial Accuracy of three features, including mean, median, and standard deviation, of TOTUSJH evaluated with all intermediate models by divided into six groups.

In our experiments, each time series is represented by extracting statistical attributes of the mean, median, and standard deviation. Then, we utilize each attribute to calculate the Adversarial Accuracy among all samples using Eq. 5 with all intermediate models. Figure 3 shows the results of the three statistical features of TOTUSJH. Through observing the box plots, we find that the models between 201 to 250 epochs can achieve 0.55 in mean, 0.60 in median, and 0.68 in standard deviation on average, which shows that the CGAN model can generate synthetic samples by maintaining a good balance between underfitting and overfitting with real samples. Moreover, the Adversarial Accuracy results have a consistent conclusion with the KL divergence experiment in Sect. 4.2 regarding the model selection.

4.4 Evaluation Using SVM Classifiers

In this section, we evaluate how well the generated data remedy the class-imbalance issue in classification of flaring and non-flaring instances of SWAN-SF. First, we train the generator on partition 1 of SWAN-SF, with the four magnetic field parameters mentioned in Sect. 2.3. The outputs of the generator are the synthetic multivariate time series most similar to the actual multivariate time series of flares. Then, we move to train two classifiers: one on the highly imbalanced real data without any change, and the other, on the data that is made balanced by adding synthetic samples. Of course, the synthetic samples are only used for the purpose of training, and the validation and test sets are made entirely of real data. We consider the former as the baseline. The only difference between the two classification strategies is that they are trained with different training datasets.

Both classifiers are trained on partition 1 and evaluated on partitions 2, 3, and 5. Partition 4 is used for tuning hyperparameters. We choose the SVM as the standard classifier for both classification tasks with the same settings that conclude the hyperparameters C and γ to 0.25 and 0.25. For preprocessing of SWAN-SF, we linearly transform all five partitions to the range $[-1, 1]$ for training the CGAN model and SVM classifiers.

Considering the flare forecasting problem that we approach is in fact a rare-event classification task, choosing a proper evaluation is important. From years of exploration, domain experts have come to agree on the effectiveness of two metrics, namely the *true skill statistic* (*TSS*) [26] and the updated *Heidke skill score* (*HSS*2) [27] as shown in Eq. 6 and Eq. 7.

$$TSS = \frac{tp}{tp + fn} - \frac{fp}{fp + tn} \tag{6}$$

$$HSS2 = \frac{2 \cdot ((tp \cdot tn) - (fn \cdot fp))}{(tp + fn) \cdot (fn + tn) + (fp + tn) \cdot (tp + fp)} \tag{7}$$

As the forecasting results reported in Fig. 4, we observe that the performance of the classifier trained on the training dataset balanced using our generated synthetic samples is remarkably higher than that of the baseline classifier, by both metrics, TSS and HSS2. This observation confirms that the model generally performs best when classes in the training dataset are roughly equal in

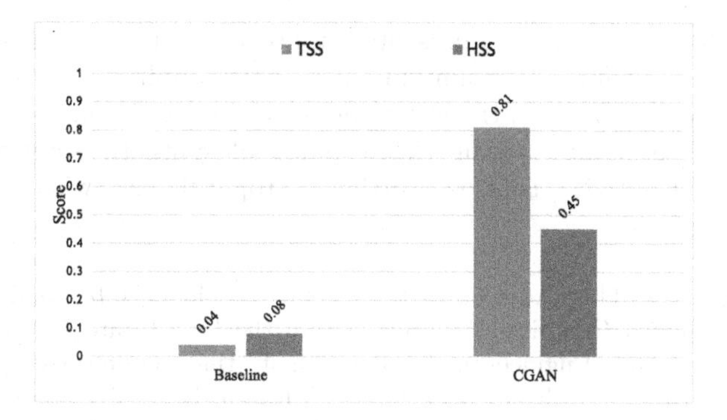

Fig. 4. This is the evaluation of flare forecasting based on SVM classifiers. The Baseline experiment (Left) is trained on partition 1 (imbalanced) of SWAN-SF, and the CGAN experiment (Right) is trained on the balanced partition 1 by adding synthetic samples generated by the CGAN model. Both experiments are evaluated on the same test dataset: partitions 2, 3, and 5 of SWAN-SF. From the results, we find notable improvements in terms of TSS and HSS. This experiment shows the CGAN can be considered as an effective remedy for tackling the class imbalance issue.

size. Specifically, the CGAN classifier results in an over 20-fold improvement compared to the baseline experiment in terms of TSS (an increase from 0.04 in average to 0.81). Moreover, the HSS2 improves an over 5-fold improvement from 0.08 to 0.45. The experiment results show that the CGAN model can successfully capture statistical features of real MVTS samples by learning the data distribution and, therefore, generating realistic MVTS samples. Overall, the CGAN model can be regarded as an effective remedy for tackling the class imbalance issue.

5 Conclusion and Future Work

In this project, we utilized the conditional generative adversarial network (CGAN) to overcome the class imbalance issue with a multivariate time series dataset. We generated synthetic samples and evaluated them by conducting two sets of experiments: First, experiments based on the data distributions of statistical features and the Adversarial Accuracy verified that synthetic samples are indeed similar to the real data. Next, our classification experiment showed that generating synthetic samples of the minority class in order to balance the training dataset can remarkably boost the performance of classification models. Therefore, we concluded that the CGAN method can be considered as an effective remedy for tackling the class imbalance issue in flare forecasting.

Of course, this study is only the first attempt towards generating a reliable synthetic dataset with meaningful physical features. There are still many aspects to be improved upon, such as incorporating an advanced loss function of Wasserstein GAN, introducing the representation learning of Info GAN, or exploring more complex structures of the generator and the discriminator.As our future work, we plan to compare the presented approach with other class-imbalance remedies including simple oversampling and undersampling strategies, in order to provide more insights into the effectiveness of the CGAN approach. Furthermore, exploring and interpreting the meaning of synthetic samples from the astrophysics point of view is also a worthwhile topic that we wish to investigate in the future.

Acknowledgment. This project has been supported in part by funding from the Division of Advanced Cyberinfrastructure within the Directorate for Computer and Information Science and Engineering, the Division of Atmospheric & Geospace Sciences within the Directorate for Geosciences, under NSF awards #193155 and # 1936361.

References

1. Withbroe, G.L.: Living With a Star. American Geophysical Union, pp. 45–51 (2013). [Online] https://doi.org/10.1029/GM125p0045
2. N.R. Council: Severe Space Weather Events-Understanding Societal and Economic Impacts: A Workshop Report. Washington, D.C., The National Academies Press, 2008. [Online] https://doi.org/10.17226/12507
3. Boteler, D.H.: Geomagnetic hazards to conducting networks. Nat. Hazards **28**(2), 537–561 (2003). [Online] https://doi.org/10.1023/A:1022902713136
4. Benz, A.O.: Flare observations. Living Rev. Sol. Phys. **5**(1) (2008). [Online] https://doi.org/10.12942/lrsp-2008-1
5. Martens, P.C., Angryk, R.A.: Data handling and assimilation for solar event prediction. In: Proceedings of the International Astronomical Union, vol. 13, no. S335, pp. 344–347 (2017). [Online] https://doi.org/10.1017/S1743921318000510
6. Hostetter, M., et al.: Understanding the impact of statistical time series features for flare prediction analysis. In: 2019 IEEE International Conference on Big Data (Big Data), 9–12 December 2019, Los Angeles, CA, USA, pp. 4960–4966. IEEE (2019). [Online] https://doi.org/10.1109/BigData47090.2019.9006116
7. Ahmadzadeh, A., et al.: How to train your flare prediction model: revisiting robust sampling of rare events. arXiv e-prints arXiv:2103.07542, March 2021
8. Ahmadzadeh, A., et al.: Challenges with extreme class-imbalance and temporal coherence: a study on solar flare data. In: 2019 IEEE International Conference on Big Data (Big Data) 2019, pp. 1423–1431 (2019). [Online] https://doi.org/10.1109/BigData47090.2019.9006505
9. Ahmadzadeh, A., et al.: Rare-event time series prediction: a case study of solar flare forecasting. In: 2019 18th IEEE International Conference On Machine Learning And Applications (ICMLA), pp. 1814–1820 (2019). [Online] https://doi.org/10.1109/ICMLA.2019.00293
10. Goodfellow, I.J., et al.: Generative adversarial nets. In: Proceedings of the 27th International Conference on Neural Information Processing Systems - Volume 2, ser. NIPS 2014. Cambridge, MA, USA, pp. 2672–2680. MIT Press (2014). [Online] https://doi.org/10.5555/2969033.2969125
11. Radford, A., et al.: Unsupervised representation learning with deep convolutional generative adversarial networks. CoRR, vol. abs/1511.06434 (2016)
12. Arjovsky, M., Chintala, S., Bottou, L.: Wasserstein generative adversarial networks. In: Proceedings of the 34th International Conference on Machine Learning - Volume 70. JMLR.org 2017, pp. 214–223 (2017). [Online] https://dl.acm.org/doi/10.5555/3305381.3305404

13. Chen, X., et al.: InfoGAN: interpretable representation learning by information maximizing generative adversarial nets. In: Proceedings of the 30th International Conference on Neural Information Processing Systems, ser. NIPS 2016. Red Hook, NY, USA: Curran Associates Inc., pp. 2180–2188 (2016). [Online] https://doi.org/10.5555/3157096.3157340
14. Mirza, M., Osindero, S.: Conditional generative adversarial nets (2014). [Online] arXiv:1411.1784
15. Esteban, C., Hyland, S.L., Rätsch, G.: Real-valued (medical) time series generation with recurrent conditional GANs (2017). arXiv:1706.02633
16. Mogren, O.: C-RNN-GAN: a continuous recurrent neural network with adversarial training. In: Constructive Machine Learning Workshop (CML) at NIPS 2016 (2016)
17. Yoon, J., et al.: Time-series generative adversarial networks. In: Advances in Neural Information Processing Systems, pp. 5508–5518 (2019)
18. Middelkamp, A.: Online. Praktische Huisartsgeneeskunde, vol. 3(4), 3 (2017). https://doi.org/10.1007/s41045-017-0040-y
19. Hoeksema, J.T., et al.: The Helioseismic and magnetic imager (HMI) vector magnetic field pipeline: overview and performance. Sol. Phys. **289**(9), 3483–3530 (2014). [Online] https://doi.org/10.1007/s11207-014-0516-8
20. Bobra, M.G., et al.: The Helioseismic and magnetic imager (HMI) vector magnetic field pipeline: sharps-space-weather HMI active region patches. Solar Phys. **289**(9), 3549–3578 (2014). [Online] https://doi.org/10.1007/s11207-014-0529-3
21. Angryk, R.A., et al.: Multivariate time series dataset for space weather data analytics. Sci. Data **7**(1) (2020). [Online] https://doi.org/10.1038/s41597-020-0548-x
22. Chan, C., et al.: Everybody dance now. In: 2019 IEEE/CVF International Conference on Computer Vision (ICCV). IEEE, October 2019. [Online] https://doi.org/10.1109/iccv.2019.00603
23. Park, T., et al.: GauGAN: semantic image synthesis with spatially adaptive normalization. In: ACM SIGGRAPH 2019 Real-Time Live!, ser. SIGGRAPH 2019. New York, NY, USA. Association for Computing Machinery (2019). [Online] https://doi.org/10.1145/3306305.3332370
24. Abadi, M., et al.: TensorFlow: large-scale machine learning on heterogeneous systems (2015). Software available from tensorflow.org
25. Yale, A., et al.: Assessing privacy and quality of synthetic health data. In: Proceedings of the Conference on Artificial Intelligence for Data Discovery and Reuse, ser. AIDR 2019. New York, NY, USA. Association for Computing Machinery (2019). [Online] https://doi.org/10.1145/3359115.3359124
26. Hanssen, A., Kuipers, W.: On the Relationship Between the Frequency of Rain and Various Meteorological Parameters: (with Reference to the Problem Ob Objective Forecasting), ser. Koninkl. Nederlands Meterologisch Institut. Mededelingen en Verhandelingen. Staatsdrukkerij- en Uitgeverijbedrijf (1965). [Online] https://books.google.com/books?id=nTZ8OgAACAAJ
27. Balch, C.C.: Updated verification of the space weather prediction center's solar energetic particle prediction model. Space Weather Int. J. Res. Appl. **6**(1) (2008). [Online] https://doi.org/10.1029/2007SW000337

The Streaming Approach to Training Restricted Boltzmann Machines

Piotr Duda(✉), Leszek Rutkowski, Piotr Woldan, and Patryk Najgebauer

Czestochowa University of Technology, Czestochowa, Poland
piotr.duda@pcz.pl

Abstract. One of the greatest challenges facing researchers of machine learning algorithms nowadays is the desire to minimize the training time of these algorithms. One of the most promising and unexplored structures of the neural network is the Restricted Boltzmann Machine. In this paper, we propose to use the BBTADD algorithm for RBM training. The performance of the algorithm has been illustrated on one of the most popular data sets.

Keywords: Restricted Boltzmann Machines · Drift detectors · Neural networks

1 Introduction

It is not hard to see that the development of machine learning techniques is hardly related to the size of analyzed data. If we take a look on a state of the art of modern neural networks we can find that there were trained for a relatively very long time. The authors of AlexNet trained their network five to six days [19], and VGG - two to three weeks [31]. Of course, nowadays computers and software allow to train those network significantly faster, however, the number of the stored data is still increasing. According to [11] over 70% of data are useless, but currently, only 2% of stored data are analyzed. It means that there is about 20% gap of data that are sored; they carry on some useful information, but still are unanalyzed. It shows that if we want to create more accurate models, we will be forced to create deeper architectures and analyzed more data. To minimize the cost of those analyses we should enhance infrastructure, using GPU, or TPU, but it is also important to propose a new algorithm that can significantly reduce the time of model training [2].

If we are aware that we have to analyze a huge amount of data, and we can assume that part of those data is redundant, it seems reasonable to try to choose only part of them, let's say the most important data elements. Most neural networks are trained with the epoch based approach. The data elements for the training set are subsequently put to the input of the neural network to obtain

This work was supported by the Polish National Science Centre under grant no. 2017/27/B/ST6/02852.

L. Rutkowski et al. (Eds.): ICAISC 2021, LNAI 12854, pp. 308–317, 2021.
https://doi.org/10.1007/978-3-030-87986-0_27

prediction, and in a back phase, the weights are updated. As a consequence, many data elements (e.g. those correctly classified) will be processed contributing only to insignificant changes in the neural network. It seems to be a lack of time. On the other hand, we cannot permanently delete those elements from the training set. During the initial epochs of training, the neural network adjusts its parameters to cover the general concept of the considered issue. When the network starts to stabilize then more detailed information is needed to obtain the best accuracy of the model. Most often this is done by decreasing a learning rate after some time of training. It is known that during the training of neural networks importance of particular data elements can be changed.

Based on the above-mentioned reasons, our goal is to propose an approach to training a neural network, that will be able to: 1) automatically select the essential data elements, 2) detect a moment when the importance of the data is changing, 3) reduces the time of training (understood as the ratio of the model accuracy to the number of processed elements).

The various approaches can be used to train various types of neural networks, as they are designed for different issues. To mention only the most vivid examples, convolutional neural networks (CNN) are most often applied to images [20], natural language processing and classification [27], recurrent neural networks [24], [34] are a good choice for speech recognition, and autoencoders to feature extraction. One of the most promising models is Restricted Boltzmann Machines (RBM). This is a type of the two-layer neural network, where each neuron in the visible layer is connected to every neuron in a hidden layer, but there are no connections between neurons in the same layer. The special feature of these networks is that the signal is sent back and forth between the layers. RBMs can be seen as a special type of autoencoders. It finds much application, e.g. in recommender systems [30], and is the base to create more complex structures like Deep Belief Networks [18].

In this paper, we are trying to apply the fast approach to neural network training (previously applied to CNN [5] and autoencoders [7]) to work with RBMs. To present the effectiveness of the proposed algorithm, the simulations were carried on the real-world dataset. The rest of the paper is organized as follows. In Sect. 2, the related work about the application of RMBs and techniques to detect changes in data, are presented. Section 3 presents details about proposed algorithm. Simulation results are depicted in Sect. 4, and the paper is concluded in Sect. 5.

2 Related Works

Since RBMs have been proposed some time ago [32], they seem to be one of the most promising structures of neural networks, which still focus the attention of many scientists [13]. They are developed in many ways. In [29] the authors concidering the application of dropout during training of RBMs. In [26] the authors investigate the importance of temperature in Boltzmann-related distribution for Deep Boltzmann Machines. On the other hand, they found many applications. In

[8] the authors used RBMs for intrusion detection in the context of smart cities. The application to detect malicious attacks can be found in [21]. Authors of [23] found it useful for simulating complex wavefunctions in quantum many-body physics. A recent overview of RBMs is presented in [33].

Data streams are possible infinite sequences of data which distributions can change over time. Researchers from many years are trying to find the best method to detect the moment of changes in a coming data. For this purpose, a few techniques have been proposed, see e.g. [9]. One of the most popular is the Drift Detection Method (DDM) [10] that monitors the correctness of classification by the current model. Treating observations as a result of Bernoulli trials, the authors propose a statistical test to inform about the warning or alarm states. Another approach is a procedure based on the ADWIN algorithm [1] and the Page-Hinkley test [25]. In [4], the authors proposed the WSTD algorithm, which applied the Wilcoxon rank-sum statistical test to improve false positive detection. The method based on a random forest algorithm, and depending decision on the measures of features importances has been proposed in [6]. The application of various diversity measures to drift detection is investigated in [22].

It should also be noted that several authors tried to merge the fields of deep learning and data stream mining. In [3] the authors combined the evolving deep neural network with the Least Squares Support Vector Machine. Deep neural networks were also successfully applied in semi-supervised learning tasks in the context of streaming data. In [28] the idea was to train the Deep Belief Network in an unsupervised manner based on the unlabeled data from the stream. Then, few available labeled elements were used to occasionally fine-tune the model to the current data concept. In [14] and [15] the authors proposed to apply the RBM as a concept drift detector. It was demonstrated that the properly learned RBM can be used to monitor possible changes in the underlying data distribution. This method was further analyzed from the resource-awareness perspective in [17] and missing data perspective in [16].

3 The BBTADD Algorithm for RBMs

The main idea of this paper is mainly based on the BBTADD algorithm proposed in [5], which is a combination of the boosting technique with drift detectors. For convenience let us introduce some notations.

Let D be a set of d-dimensional feature vectors X_i for $i = 1, \ldots, n$, i.e.

$$D = \{X_i | i = 1, \ldots, n, X_i \in \mathbf{A}\}, \tag{1}$$

where $\mathbf{A}$ is a d-dimensional feature space. To each data element, a new factor has been added to model a probability of drawing (*pod*) from the training data.

$$D' = \{(X_i, w_i) | X_i \in D, w_i \in (0, 1)\}. \tag{2}$$

Now we can create a possibly infinite stream of data S by subsequent drawing data element concerning a probability distribution given by *pod* $\{w_i\}_{i=1}^{n}$

$$S_t = (Y_1, \ldots, Y_t | Y_i = (X_{j_i}, w_{j_i}), 1 \leq i \leq t, 1 \leq j_i \leq n), \tag{3}$$

where t is the number of the elements coming from the stream.

3.1 Background of RBMs

As it was mentioned in Sect. 1, the RBM is a type of neural network consisting of two layers. The first one, a visible layer, is consists the same number of the neuron as the dimension of data elements. They are designed to process binary data. The number of neurons in a hidden layer M can be chosen by the user (see Fig. 1). The weights connecting neurons of opposite layers allow sending a signal in both ways (from visible to the hidden layer and from hidden to visible layer). Let us denote the weight between the i-th neuron of the visible layer with the j-th neuron of the hidden layer by $c_{i,j}$. Then the operation of the RBM can be partitioned into two-phase. The first phase, where the visible values $\mathbf{v} = [v^{(1)}, ..., v^{(d)}]$ are given, then values of hidden neurons $\mathbf{h} = [h^{(1)}, ..., h^{(M)}]$ are established with probability equal to

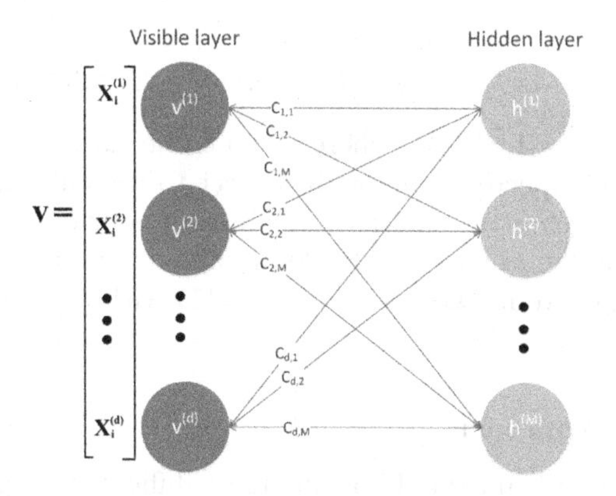

Fig. 1. The schema of the RBM

$$P(h^{(j)} = 1|\mathbf{v}) = \sigma(\sum_{i=1}^{d} c_{i,j} v^{(i)} + b_j), \tag{4}$$

and the second (reconstruction) phase, where the values of visible neurons $\hat{\mathbf{v}} = [\hat{v}^{(1)}, ..., \hat{v}^{(M)}]$ are established with probability equal to

$$P(\hat{v}^{(i)} = 1|\mathbf{h}) = \sigma(\sum_{j=1}^{M} c_{i,j} h^{(j)} + a_i), \tag{5}$$

where σ is a sigmoid function, and a_i and b_j are related to the bias in neurons of the visible and hidden layer, respectively.

The RBMs is energy-based model [13] and the energy function is given as follows.

$$E(\mathbf{v}, \mathbf{h}) = -\sum_{i=1}^{d} v^{(i)} a_i - \sum_{j=1}^{M} h^{(j)} b_j - \sum_{i=1}^{d} \sum_{j=1}^{M} v^{(i)} h^{(j)} c_{i,j} \tag{6}$$

Then the probability that the RBM is in a given state is given by the Boltzmann distribution

$$P(\mathbf{v}, \mathbf{h}) = \frac{\exp(-E(\mathbf{v}, \mathbf{h}))}{\sum_{v \in A} \sum_{h \in \{0,1\}^M} \exp(-E(v, h))}. \quad (7)$$

The RBM can be trained by minimizing the negative log-likelihood

$$C(\mathbf{v}) = -\log(P(\mathbf{v})), \quad (8)$$

what can be done in a mini-batch manner with the application of the stochastic gradient descent method. Then the weights are updated according to the formulas

$$c_{i,j} := c_{i,j} - \eta(E_{recon}[v^{(i)}h^{(j)}] - E_{data}[v^{(i)}h^{(j)}]), \quad (9)$$

$$a_i := a_i - \eta(E_{recon}[v^{(i)}] - E_{data}[v^{(i)}]), \quad (10)$$

$$b_j := a_j - \eta(E_{recon}[h^{(j)}] - E_{data}[h^{(j)}]), \quad (11)$$

for $i = 1, \ldots, d$ and $j = 1, \ldots, M$, where $\eta > 0$ is the learning rate and E_{recon} is an estimator of the expected value with respect to the current model obtained by contrastive divergence [12] and is E_{data} an estimator of expected value with respect to the current mini-batch. More details about training RBM can be found in [13]. In this paper we merge Formulas (9)–(11) with the procedure proposed in [5].

3.2 Streaming Approach

For the network to be trained only on the most difficult to reproduce data (i.e. those for which RBM is capable of high energy), instead of training the model on mini-batches created from consecutive data, the elements are each time sampled from the training set. The sampling depends on the weights w_i, $i = 1, ..., n$. Let us denote the size of the mini-batch as N. To set the initial values of the weights w_i, during the first two epochs, the data were processed in the epoch-based approach. Then the values of the weights are established based on the value of cost function

$$v_i' = tanh(C(X_i))/M_i, \quad (12)$$

where M_i indicates the number of times the i-th data element was drawn. Big values of the cost function indicate a higher value of weight. To ensure that weights define a probability distribution they have to be normalized.

$$v_i = \begin{cases} v_i'/Z, & \text{for } x_i \in B \\ v_i/Z, & \text{for } x_i \in D \backslash B \end{cases} \quad (13)$$

where Z is a normalization factor, given as

$$Z = \sum_{\{v_i' | X_i \in D\}} v_i'. \quad (14)$$

The mini-batches are subsequently processed until a stopping criterion is fulfilled. As this approach may result in a constant sampling of the same data, there should be used a method that will be able to detect when it needs to re-initiate the sampling.

One of the best tools for this is the CUSUM algorithm, given by the following formula

$$\begin{cases} Cus_0 = 0, \\ Cus_i = max(0, Cus_{i-1} + C(B_{i-1}) - C(B_i) - \alpha), \end{cases} \tag{15}$$

for $i = 1, 2, \ldots$, where $C(B_i)$ is an average of the cost function for the data elements from the i-th mini-bath and α is a fixed parameter. In the case that Cus_i exceeds the value of the threshold λ, then the change is detected, and weights w_i have to be re-initialized to equal values.

The BBTADD (Boosting-Based Training Algorithm with Drift Detector) algorithm for RBM is presented in Algorithm 1.

```
   Input: S - data stream, N - batch size, α, λ
 1 CuSum = 0 ;
 2 Collect a new batch B from the stream S;
 3 for every data element in B do
 4     Increase counter of drawn of the current element;
 5     Train the network on current element;
 6     Compute cost function for a current data element;
 7     Update w_i according to (12)
 8 for every data element in D do
 9     Update pods according to (13)
10 Compute cost function on a validation set;
11 Update CuSum according to (15);
12 if CuSum > λ then
13     Reinitialize pod's values;
14     Return to line 1;
15 else
16     Return to line 2;
```

Algorithm 1: The BBATDD for RBM algorithm.

4 Experimental Results

To demonstrate the usefulness of the proposed method, the application to denoising images was investigating. The simulations were carried out on one of the most popular real datasets, e.g. the MNIST data set. The dataset consists of images of hand-drawn numbers. The dataset is split into train and test sets, with

60 000 and 10 000 data elements, respectively. The training of the network was compared to the RBM trained with the traditional, epoch-based approach.

The processed images have a resolution 28×28, so each image can be seen as a vector of length equal to 784. As its coefficients take values from 0 to 255, they have to be preprocessed to adjust data to RBM. In this paper, each coefficient higher than zero was set to 1. In consequence, we obtained more raw data. Examples of the original and preprocessed images are depicted in Fig. 2

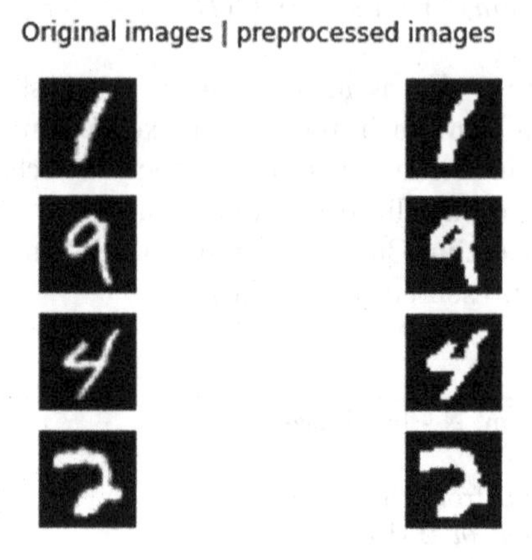

Fig. 2. Examples of original (left-hand side) and preprocessed images (right-hand side)

Then the noise was added to the images. For this purpose, the value of the coefficients for each pixel has been changed from 0 to 1 with a probability equal to 0.25. The data prepared in this way was fed to the RBM input. Next, the model was trained using the BBTADD for the RBM algorithm. The network consisted of 784 neurons in the visible layer and 256 neurons in the hidden layer. It was trained by 46 789 batches (each with 128 elements), which corresponds to 100 epochs. A sample of effects for images from the test set is depicted in Fig. 3. The presented result seems to be satisfactory.

Finally, the averaged values of the cost function calculated for successive batches for the proposed and epochal approaches were compared. As it is presented in Fig. 4, the averaged value of the cost function for the proposed approach decreases faster compared to the epochal approach. Moreover, changes between its values for subsequent mini-batches seem to be more stable.

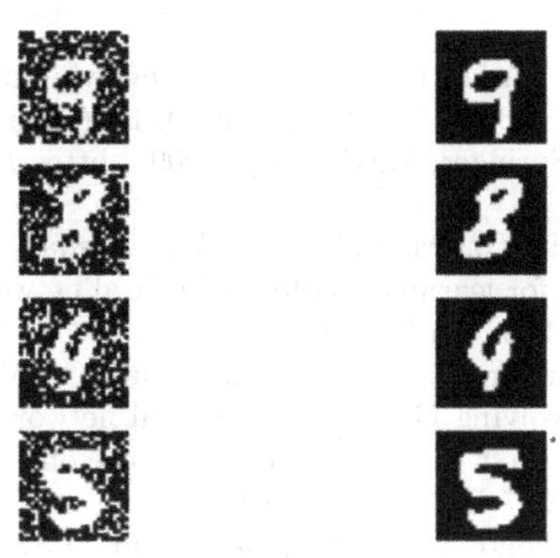

Fig. 3. Examples of input (left-hand side) and output images (right-hand side)

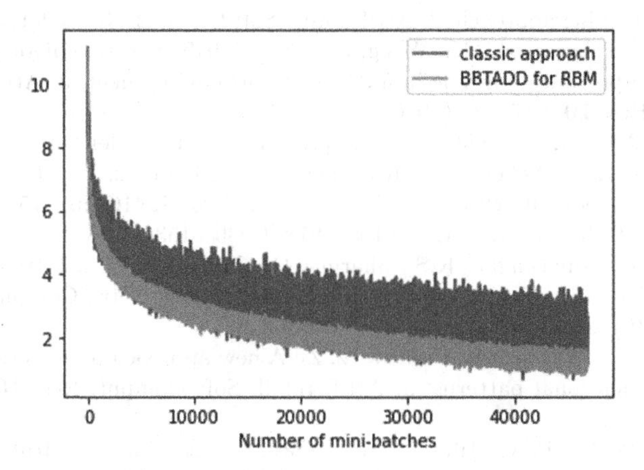

Fig. 4. The averaged cost function computed on the test set for subsequent batches

5 Conclusions

In this paper, we explored the possibility of the application of a streaming approach to training Restricted Boltzmann's Machines. For this purpose, the CUSUM drift detector, which is one of the most popular techniques in data stream analysis, was used. As a result, we obtained the algorithm that allows us to minimalize the cost function faster than the classic approach. On the other hand, this algorithm requires additional calculations related to the sampling of data from the training set. Further work can be directed to relaxing the computational burden of the presented approach.

References

1. Bifet, A., Gavaldà, R.: Adaptive learning from evolving data streams. In: Adams, N.M., Robardet, C., Siebes, A., Boulicaut, J.-F. (eds.) IDA 2009. LNCS, vol. 5772, pp. 249–260. Springer, Heidelberg (2009). https://doi.org/10.1007/978-3-642-03915-7_22
2. Bilski, J., Kowalczyk, B., Marchlewska, A., Zurada, J.M.: Local Levenberg-Marquardt algorithm for learning feedforwad neural networks. J. Artif. Intell. Soft Comput. Res. **10**(4), 299–316 (2020)
3. Bodyanskiy, Y., Vynokurova, O., Pliss, I., Setlak, G., Mulesa, P.: Fast learning algorithm for deep evolving GMDH-SVM neural network in data stream mining tasks. In: 2016 IEEE First International Conference on Data Stream Mining Processing (DSMP), pp. 257–262, August 2016
4. deBarros, R.S.M., Hidalgo, J.I.G., de Lima Cabral, D.R.: Wilcoxon rank sum test drift detector. Neurocomputing **275**, 1954–1963 (2018)
5. Duda, P., Jaworski, M., Cader, A., Wang, L.: On training deep neural networks using a streaming approach. J. Artif. Intell. Soft Comput. Res. **10**(1), 15–26 (2020)
6. Duda, P., Przybyszewski, K., Wang, L.: A novel drift detection algorithm based on features' importance analysis in a data streams environment. J. Artif. Intell. Soft Comput. Res. **10**, 287–298 (2020)
7. Duda, P., Wang, L.: On a streaming approach for training denoising auto-encoders. In: Rutkowski, L., Scherer, R., Korytkowski, M., Pedrycz, W., Tadeusiewicz, R., Zurada, J.M. (eds.) ICAISC 2020. LNCS (LNAI), vol. 12416, pp. 315–324. Springer, Cham (2020). https://doi.org/10.1007/978-3-030-61534-5_28
8. Elsaeidy, A., Munasinghe, K.S., Sharma, D., Jamalipour, A.: Intrusion detection in smart cities using restricted Boltzmann machines. J. Netw. Comput. Appl. **135**, 76–83 (2019)
9. Gałłowski, T., Krzyzak, A., Filutowicz, Z.: A new approach to detection of changes in multidimensional patterns. J. Artif. Intell. Soft Comput. Res. **10**(2), 125–136 (2020)
10. Gama, J., Medas, P., Castillo, G., Rodrigues, P.: Learning with drift detection. In: Bazzan, A.L.C., Labidi, S. (eds.) SBIA 2004. LNCS (LNAI), vol. 3171, pp. 286–295. Springer, Heidelberg (2004). https://doi.org/10.1007/978-3-540-28645-5_29
11. Gantz, J., Reinsel, D.: The digital universe in 2020: Big data, bigger digital shadows, and biggest growth in the far east. IDC iView: IDC Analyze Future **2007**(2012), 1–16 (2012)
12. Hinton, G.E.: Training products of experts by minimizing contrastive divergence. Neural Comput. **14**(8), 1771–1800 (2002)
13. Hinton, G.E.: A practical guide to training restricted Boltzmann machines. In: Montavon, G., Orr, G.B., Müller, K.-R. (eds.) Neural Networks: Tricks of the Trade. LNCS, vol. 7700, pp. 599–619. Springer, Heidelberg (2012). https://doi.org/10.1007/978-3-642-35289-8_32
14. Jaworski, M., Duda, P., Rutkowski, L.: On applying the restricted Boltzmann machine to active concept drift detection. In: 2017 IEEE Symposium Series on Computational Intelligence (SSCI), pp. 1–8. IEEE (2017)
15. Jaworski, M., Duda, P., Rutkowski, L.: Concept drift detection in streams of labelled data using the restricted Boltzmann machine. In: 2018 International Joint Conference on Neural Networks (IJCNN), pp. 1–7. IEEE (2018)
16. Jaworski, M., Rutkowski, L., Angelov, P.: Concept drift detection using autoencoders in data streams processing. In: Rutkowski, L., Scherer, R., Korytkowski, M.,

Pedrycz, W., Tadeusiewicz, R., Zurada, J.M. (eds.) ICAISC 2020. LNCS (LNAI), vol. 12415, pp. 124–133. Springer, Cham (2020). https://doi.org/10.1007/978-3-030-61401-0_12
17. Jaworski, M., Rutkowski, L., Duda, P., Cader, A.: Resource-aware data stream mining using the restricted Boltzmann machine. In: Rutkowski, L., Scherer, R., Korytkowski, M., Pedrycz, W., Tadeusiewicz, R., Zurada, J.M. (eds.) ICAISC 2019. LNCS (LNAI), vol. 11509, pp. 384–396. Springer, Cham (2019). https://doi.org/10.1007/978-3-030-20915-5_35
18. Krizhevsky, A., Hinton, G.: Convolutional deep belief networks on CIFAR-10. Unpublished Manuscript **40**(7), 1–9 (2010)
19. Krizhevsky, A., Sutskever, I., Hinton, G.E.: Imagenet classification with deep convolutional neural networks. Adv. Neural. Inf. Process. Syst. **25**, 1097–1105 (2012)
20. Kumar, R., Weill, E., Aghdasi, F., Sriram, P.: A strong and efficient baseline for vehicle re-identification using deep triplet embedding. J. Artif. Intell. Soft Comput. Res. **10**(1), 27–45 (2019)
21. Li, C., Wang, J., Ye, X.: Using a recurrent neural network and restricted Boltzmann machines for malicious traffic detection. NeuroQuantology **16**(5) (2018)
22. Mahdi, O.A., Pardede, E., Ali, N., Cao, J.: Diversity measure as a new drift detection method in data streaming. Knowl.-Based Syst. **191**, 105227 (2020)
23. Melko, R.G., Carleo, G., Carrasquilla, J., Cirac, J.I.: Restricted Boltzmann machines in quantum physics. Nat. Phys. **15**(9), 887–892 (2019)
24. Niksa-Rynkiewicz, T., Szewczuk-Krypa, N., Witkowska, A., Cpalka, K., Zalasinski, M., Cader, A.: Monitoring regenerative heat exchanger in steam power plant by making use of the recurrent neural network. J. Artif. Intell. Soft Comput. Res. **11**(2), 143–155 (2021)
25. Page, E.S.: Continuous inspection schemes. Biometrika **41**(1/2), 100–115 (1954)
26. Passos, L.A., Papa, J.P.: Temperature-based deep Boltzmann machines. Neural Process. Lett. **48**(1), 95–107 (2018)
27. Rahman, J.S., Gedeon, T., Caldwell, S., Jones, R., Jin, Z.: Towards effective music therapy for mental health care using machine learning tools: Human affective reasoning and music genres. J. Artif. Intell. Soft Comput. Res. **11**(1), 5–20 (2020)
28. Read, J., Perez-Cruz, F., Bifet, A.: Deep learning in partially-labeled data streams. In: Proceedings of the 30th Annual ACM Symposium on Applied Computing, SAC 2015, pp. 954–959. ACM, New York, NY, USA (2015)
29. Roder, M., de Rosa, G.H., de Albuquerque, V.H.C., Rossi, A.L.D., Papa, J.P.: Energy-based dropout in restricted Boltzmann machines: why not go random. IEEE Trans. Emerg. Top. Comput. Intell., 1–11 (2020). https://doi.org/10.1109/TETCI.2020.3043764
30. Salakhutdinov, R., Mnih, A., Hinton, G.: Restricted Boltzmann machines for collaborative filtering. In: Proceedings of the 24th International Conference on Machine Learning, pp. 791–798 (2007)
31. Simonyan, K., Zisserman, A.: Very deep convolutional networks for large-scale image recognition. arXiv preprint arXiv:1409.1556 (2014)
32. Smolensky, P.: Information processing in dynamical systems: Foundations of harmony theory. Technical report, Colorado University at Boulder Department of Computer Science (1986)
33. Zhang, N., Ding, S., Zhang, J., Xue, Y.: An overview on restricted Boltzmann machines. Neurocomputing **275**, 1186–1199 (2018)
34. Zini, J.E., Rizk, Y., Awad, M.: An optimized parallel implementation of non-iteratively trained recurrent neural networks. J. Artif. Intell. Soft Comput. Res. **11**(1), 33–50 (2020)

Abrupt Change Detection by the Nonparametric Approach Based on Orthogonal Series Estimates

Tomasz Gałkowski[1(✉)] and Adam Krzyżak[2,3]

[1] Institute of Computational Intelligence Czestochowa University of Technology, Częstochowa, al. Armii Krajowej 36, 42-200 Częstochowa, Poland
tomasz.galkowski@pcz.pl

[2] Department of Computer Science and Software Engineering, Concordia University, Montreal, Quebec H3G 1M8, Canada
krzyzak@cs.concordia.ca

[3] Department of Electrical Engineering, Westpomeranian University of Technology, 70-310 Szczecin, Poland

Abstract. Many algorithms have been proposed for detection possible deviations and/or narrow changes in the data. A key problem is verification whether the characteristics of information sources have changed. If the change occurred then we would like to know the essence of this change and when or where it happened. Nowadays, well-known methods of mathematical statistics have been successfully applied to address this problem. Recently, a new approach based on nonparametric regression estimation has been proposed. The idea based on Parzen kernel has been studied in depth. This article presents an alternative approach for detecting abrupt change in data based on nonparametric orthogonal series estimation. The proposed method is validated in experiments on noisy data.

Keywords: Abrupt change detection · Nonparametric regression · Orthogonal series estimation

1 Introduction

In practice we are often interested in determining whether stored or transmitted data are genuine, or reliable. It is important to be convinced that information sources did not change their characteristics in time, for instance. We are very interested in determining the essence of the changes, if they have occurred, as well as determining where and when they have arisen. Statistical tests or model-building strategies are usually caused by an incomplete mathematical description of the processes generating the data, especially observed in the presence of an additive random noise.

In our paper, we focus on abrupt, narrow changes, called jumps or edges. The aim is to detect a significant aberration in the data observed so far.

Part of this research was carried out by the second author during his visit of the Westpomeranian University of Technology while on sabbatical leave from Concordia University.

L. Rutkowski et al. (Eds.): ICAISC 2021, LNAI 12854, pp. 318–327, 2021.
https://doi.org/10.1007/978-3-030-87986-0_28

The response to the detected change depends obviously on the problem at hand. In many cases, they can indicate a problem that requires an urgent response and alert. Examples of such important problems include: abnormalities in physiological parameters of hospital patients, abrupt changes in the stock market, changes in geological phenomena, especially in seismology, and in several industrial processes. Jump increase in the network traffic, various sensor data, can precede probable hacker attack. This may indicate a general network insecurity and system risks.

But there are many tasks where finding sudden changes is needed to achieve the desired result. For instance narrow changes of light in the pictures or videos may be applied in pattern recognition systems, detecting edge curves in satellite photos can help in e.g. water resources exploration, or in cartography in preparing maps of certain areas, etc.

2 A Brief Survey on the Used Methodologies

There are known and successfully applied the several algorithms developed to detect abnormalities or deviations in the data. The brief review of the edge detection techniques in image processing can be found in, e.g., [2,66]. The authors described the methods of abrupt change detection via classical gradient-based operations involving first order derivatives such as Sobel, Prewitt, Robert's [43] and Canny [3]. The distribution of intensity values in the neighborhood of given pixels determines the probable edges. The algorithms using second order derivatives such as the Laplacian and Gaussian filtering for detecting of zero-crossings also allow edge detection in images [40].

The natural approach is to model the data via densities or distributions [9]. The significant features of the process or object could be compared using different sample sets by using mathematical statistics and the representative templates like means and/or linear regression. This comparison may result in detection of change in certain parameters.

More general criteria applying mean square error method are also often used to detect changes. Many statistical tests like the Kolmogorov-Smirnov test or Wilcoxon test have been exploited, for instance (see [6]). The main idea is to compute a scalar function of the data (so-called test statistics) and compare the values to determine whether a significant change (defined before) has occurred. The Kullback-Leibler distance [37] (also named relative entropy) is one of the most common distribution distance measures. The application of the Kulback-Leibler divergence one may find in e.g. [16]. A compromise between Hoteling (parametric detector) and non-parametric Kulback-Leibler divergence was also presented in [16] using, among others, the Mahalanobis distance and Gaussian mixture of distributions.

The enumerated methods are efficient when the data volumes are not very large. They are usable off-line. They are not applicable directly for data streams. One way of detecting change is to compare likelihood between the subsequent examples using adjacent sliding time-windows, for previous elements in the stream and the further ones. The point p could be estimated when we observe a decreasing likelihood.

Several interesting approaches concerning various regression models for stream data mining are studied in [11–13,30,32,42,56–59].

The objects and/or processes in general can be described mathematically as a function $R(.)$ of the d-dimensional vector of random variable $\mathbf{X}$. Then the methods based on regression function analysis can be applied. An abrupt change of the function $R(.)$ value at point p may be recognized as a jump discontinuity of the function. In one dimensional case ($d = 1$) it may be observed as a steep change in function value. The main problem is to determine the point p at which this occurs. In case $d > 1$ the change location (edge) takes form of a curve in d-dimensional space (across which R has jump discontinuity). It is more difficult to establish it and the calculation requires much more computational effort.

This article concerns techniques useful in the wide range of fields such as classification, computer vision, diagnostics etc. (see e.g. [4,7,25–28,36,48,49,63,64,67,69, 70]). The approach based on regression analysis is developed as an attractive tool also in classification and modelling of objects (e.g. [38,39]), forecasting of phenomena (e.g., [5,14,35,46]) and entire methodology of machine learning like neural networks, fuzzy sets, genetic algorithms (e.g. [8,31,41,60,61]). Nonparametric approach to analysis and modelling of various systems one may found e.g. in [10,29,47,50,53–55].

In this paper, we focus our attention on the challenge of abrupt change detection (also called edge detection problem) by presenting a new original approach.

3 Algorithm for Abrupt Change Detection Using Orthogonal Series

The goal of this paper is to introduce a new method of edge detection derived from the nonparametric approach based on orthogonal series. Algorithms use the estimates of unknown functions and their derivatives from the set of noisy measurements.

We consider model of the object in the form:

$$y_i = R(\mathbf{x}_i) + \varepsilon_i,\ i = 1, ..., n \tag{1}$$

where $\mathbf{x_i}$ is assumed to be the d-dimensional vectors of deterministic input, $\mathbf{x}_i \in R^d$, y_i is the scalar random output, and ε_i is a measurement noise with zero mean and bounded variance. $R(.)$ is assumed to be completely unknown function.

We investigate the fixed-design problem (see, e.g., [15]) which means that the experimenter decides on the input values set. The domain area D (the space where function R is defined) is partitioned into n disjunctive nonempty sub-spaces D_i and the measurements $\mathbf{x}_i$ are chosen from D_i, i.e.: $\mathbf{x}_i \in D_i$. For instance, in one-dimensional case let the $D = [0,1]$, then $\cup D_i = [0,1]$, $D_i \cap D_j = \emptyset$ for $i \neq j$, the points x_i are chosen from D_i, i.e.: $x_i \in D_i$.

The set of input values $\mathbf{x}_i$ (independent variable in the model (1) are chosen by the experimenter in the phase of recording data e.g., equidistant samples of ECG signal in time domain, or stock exchange information, or internet activity of the web/ftp server logs, etc. These data points should provide a balanced representation of function R in the domain D. The maximum diameter of set D_i should tend to zero if n tends to infinity (compare convergence conditions in, e.g., [17,18,22]).

We start with estimator $\hat{R}_n(\mathbf{x})$ of function $R(.)$ at point $\mathbf{x}$ based on the set of measurements y_i, $i = 1, ..., n$.

The orthogonal series based algorithm is defined in one-dimensional case by:

$$\hat{R}_n(x) = \sum_{k=0}^{N} g_k(x) \cdot \hat{a}_k \tag{2}$$

where

$$\hat{a}_k = \sum_{i=1}^{n} y_i \int_{Q_i} g_k(u)du \tag{3}$$

and $g_k, k = 0, ..., N$ is an orthogonal series (for more discussion of the orthogonal series refer to the next paragraph). Equations (2) and (3) can be rewritten in compact formula

$$\hat{R}_n(x) = \sum_{k=0}^{N} g_k(x) \sum_{i=1}^{n} y_i \int_{Q_i} g_k(u)du \tag{4}$$

Equation (4) can be presented in the form

$$\hat{R}_n(x) = \sum_{i=1}^{n} y_i \sum_{k=0}^{N} \left[\int_{Q_i} g_k(u)du \right] \cdot g_k(x) \tag{5}$$

Equation (5) describes an estimate of function $R(\cdot)$ at point x defined as the weighted sum of observations y_i.

The most known and commonly used in function recovery related problems is the trigonometric Fourier orthonormal system. The Hermite orthogonal system approach can be also applied in regression estimation (see, e.g., [24]), and/or the orthonormal systems constructed by orthonormalizing piecewise polynomials, see, [34]. For more information on orthogonal series theory and applications see e.g. [1,33,52,65,68,71]. In this work, the proposed algorithm uses the Fourier cosine orthogonal series (see, simulation example). Recently have been also studied another nonparametric algorithms applying the weights defined using the Parzen kernel, see, e.g. [20,21,44,45].

The main idea of the paper is to deduce the dynamics of changes from the course of the first derivative estimated from sample. The more rapidly the change occurs - the higher the first derivative (or speed). Using only the first derivative we need the appropriate thresholding strategy to detect jumps in function R, however, applying simultaneously the second derivative can help to determine edges directly by detecting the zero-crossing point.

The algorithm for estimating the first derivative is based on differentiation of the functions $g_k(x)$ in expansion (5):

$$\hat{R'}_n(x) = \sum_{i=1}^{n} y_i \sum_{k=0}^{N} \left[\int_{Q_i} g_k(u)du \right] \cdot g'_k(x) \tag{6}$$

Subsequently, the estimate of m–th derivative of the regression function in point x can be defined as follows:

$$\hat{R}_n^{(m)}(x) = \sum_{i=1}^{n} y_i \sum_{k=0}^{N} \left[\int_{Q_i} g_k(u)du \right] \cdot g_k^{(m)}(x) \tag{7}$$

The nonparametric approach in application to estimation of unknown functions and their derivatives by the Parzen kernel methods was previously proposed and studied in, e.g., [19,23].

4 Simulation Tests

Next we present simulation results of detection of the abrupt change points for trial one-dimensional case. The function chosen for testing has three discontinuities (similar to this proposed by Romani et al. in [51] and not detailed here). As an example the orthogonal trigonometric Fourier cosine series were applied in the estimation algorithm. The set of measurements has $n = 500$ sample points artificially generated with additive noise. Next, the function and its first derivative estimates were calculated using algorithm (6). Figure 1 presents three diagrams where different numbers N of Fourier series components were used. The red marked points, signing the local maxima of the first derivative, significantly of higher amplitude than neighboring ones, are the detected jumps in function $\dot{R}()$. The yellow marked point in upper diagram indicates the false detection because of the too small number ($N = 20$) of components in series. Next two diagrams show the proper points detected. Such result strictly relates with known Gibbs phenomenon: when the approximated function has the discontinuity at the point p in the Fourier series, the high frequency components arise, so the corresponding Fourier coefficients take larger absolute values in this region. A suitable thresholding strategy could be used to detect point p.

5 Conclusions

This paper considers the important problem of detection the sudden change occurred in the function, and where or when it arise. The proposed algorithm is derived from the orthogonal series nonparametric regression estimation techniques, with fixed-design of unknown functions. Furthermore, our algorithm can scale up and it does not require the samples to be uniformly spaced. The algorithm based on trigonometric cosine Fourier series is presented in detail. The detection algorithm uses the first derivative estimates. Simulation results showed in diagrams confirmed utility and of the proposed approach in practical cases. From the presented Figs. 1 one may observe that the potential effectiveness of the method improves when the Fourier series is longer i.e. when the number N of its components is greater. The extension of the edge detection algorithm to multivariate case is planned in future works.

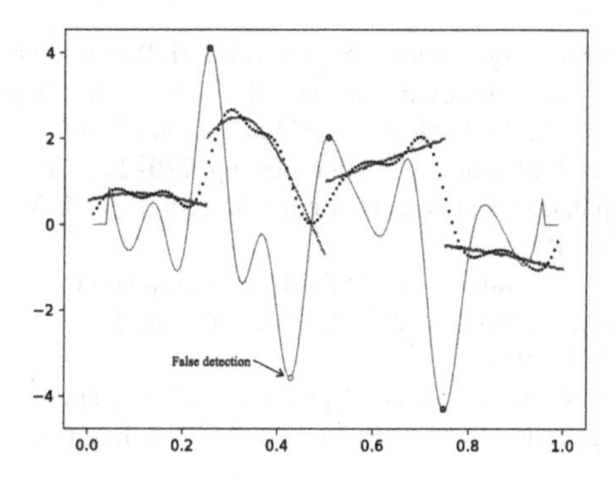

Number of measurement points n=500, number of Fourier components N=20

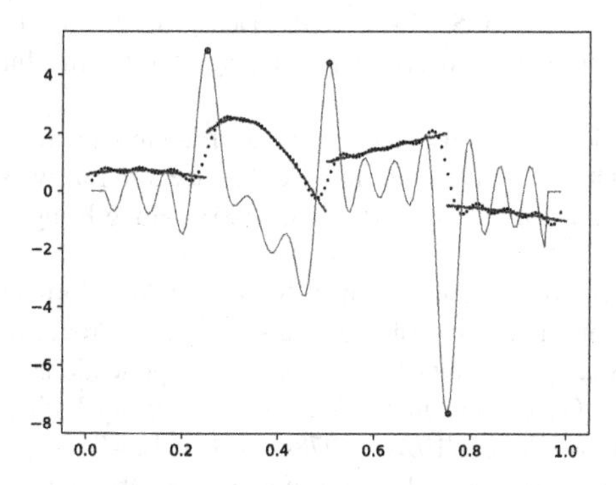

Number of measurement points n=500, number of Fourier components N=30

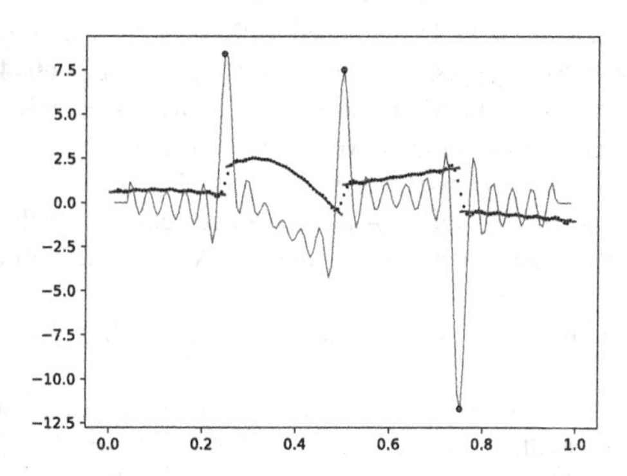

Number of measurement points n=500, number of Fourier components N=50

Fig. 1. Simulation example: trial function jumps edge detection with respect to Fourier series components number (Color figure online)

References

1. Bary, N.K.: A Treatise on Trigonometric Series, vol. I. II. Pergamon Press, New York (1964)
2. Bhardwaj, S., Mittal, A.: A survey on various edge detector techniques, Elseiver, SciVerse ScienceDirect. In: Procedia Technology 4, 2nd International Conference on Computer, Communication, Control and Information Technology, pp. 220–226 (2012)
3. Canny, J.F.: A computational approach to edge detection. IEEE Trans. Patt. Anal. Mach. Intell. **8**(6), 679–698 (1986)
4. Cao, Y., Samidurai, R., Sriraman, R.: Stability and dissipativity analysis for neutral type stochastic Markovian jump static neural networks with time delays. J. Artif. Intell. Soft Comput. Res. **9**(3), 189–204 (2019)
5. Cierniak, R., Pluta, P., Kaźmierczak, A.: A practical statistical approach to the reconstruction problem using a single slice Rebinning method. J. Artif. Intell. Soft Comput. Res. **10**(2), 137–149 (2020)
6. Corder, G.W., Foreman, D.I.: Nonparametric Statistics: A Step-by-Step Approach. Wiley, New York (2014)
7. Costa, M., Oliveira, D., Pinto, S., Tavares, A.: Detecting driver's fatigue. Distraction and activity using a non-intrusive AI-based monitoring system. J. Artif. Intell. Soft Comput. Res. **9**(4), 247–266 (2019)
8. Cpałka, K., Rutkowski, L., Evolutionary learning of flexible neuro-fuzzy systems. In: Proceedings of the 2008 IEEE International Conference on Fuzzy Systems (IEEE World Congress on Computational Intelligence, WCCI 2008), Hong Kong, 1–6 June CD, pp. 969–975 (2008)
9. Dasu, T., Krishnan, S., Venkatasubramanian, S., Yi, K.: An information-theoretic approach to detecting changes in multi-dimensional data streams. In: Proceedings of the Symposium on the Interface of Statistics, Computing Science, and Applications, pp. n/a, (2006)
10. Devroye, L., Lugosi, G.: Combinatorial Methods in Density Estimation. Springer-Verlag, New York (2001). https://doi.org/10.1007/978-1-4613-0125-7
11. Duda, P., Jaworski, M., Rutkowski, L.: Convergent time-varying regression models for data streams: tracking concept drift by the recursive Parzen-based generalized regression neural networks. Int. J. Neural Syst. **28**(2), 1750048 (2018)
12. Duda, P., Jaworski, M., Rutkowski, L.: Knowledge discovery in data streams with the orthogonal series-based generalized regression neural networks. Inf. Sci. **460–461**, 497–518 (2018)
13. Duda, P., Rutkowski, L., Jaworski, M., Rutkowska, D.: On the Parzen Kernel-based probability density function learning procedures over time-varying streaming data with applications to pattern classification. IEEE Trans. Cybern. **50**, 1–14 (2018)
14. Duda, P., Przybyszewski, K., Wang, L.: A novel drift detection algorithm based on features? Importance analysis in a data streams environment. J. Artif. Intell. Soft Comput. Res. **10**(4), 287–298 (2020)
15. Eubank, R.L.: Nonparametric Regression and Spline Smoothing, 2nd edn. Marcel Dekker, New York (1999)
16. Faithfull, W.J., Rodriguez, J.J., Kuncheva, L.I.: Combining univariate approaches for ensemble change detection in multivariate data. Elseiver, Inf. Fusion **45**, 202–214 (2019)
17. Gałkowski, T., Rutkowski, L.: Nonparametric recovery of multivariate functions with applications to system identification. Proc. IEEE **73**, 942–943 (1985)
18. Gałkowski, T., Rutkowski, L.: Nonparametric fitting of multivariable functions. IEEE Trans. Autom. Control **AC-31**, 785–787 (1986)
19. Gałkowski, T., On nonparametric fitting of higher order functions derivatives by the kernel method - a simulation study. In: Proceedings of the 5-th International Symposium on Applied Stochastic Models and data Analysis, Granada, Spain, pp. 230–242 (1991)

20. Gałkowski, T., Krzyżak, A., Filutowicz, Z.: A new approach to detection of changes in multidimensional patterns. J. Artif. Intell. Soft Comput. Res. **10**(2), 125–136 (2020)
21. Gałkowski, T., Krzyżak, A.: Edge curve estimation by the nonparametric Parzen Kernel method. In: Yang, H., et al. (eds.) ICONIP 2020, CCIS 1332, pp. 377–385 (2020)
22. Gasser, T., Müller, H.-G.: Kernel estimation of regression functions. Lecture Notes in Mathematics, vol. 757, pp. 23–68. Springer-Verlag, Heidelberg (1979). https://doi.org/10.1007/BFb0098489
23. Gasser, T., Müller, H.-G.: Estimating regression functions and their derivatives by the kernel method. Scandinavian J. Stat. **11**(3), 171–185 (1984)
24. Greblicki, W., Pawlak, M.: Fourier and Hermite series estimates of regression functions. Annals Inst. Stat. Math. **37**(3), 443–454 (1985)
25. Grycuk, R., Scherer, R., Gabryel, M.: New image descriptor from edge detector and blob extractor. J. Appl. Math. Comput. Mech. **14**(4), 31–39 (2015)
26. Grycuk, R., Knop, M., Mandal, S.: Video key frame detection based on SURF algorithm. In: Rutkowski, L., Korytkowski, M., Scherer, R., Tadeusiewicz, R., Zadeh, L.A., Zurada, J.M. (eds.) ICAISC 2015. LNCS (LNAI), vol. 9119, pp. 566–576. Springer, Cham (2015). https://doi.org/10.1007/978-3-319-19324-3_50
27. Grycuk, R., Gabryel, M., Scherer, M., Voloshynovskiy, S.: Image descriptor based on edge detection and Crawler algorithm. In: Rutkowski, L., Korytkowski, M., Scherer, R., Tadeusiewicz, R., Zadeh, L.A., Zurada, J.M. (eds.) ICAISC 2016. LNCS (LNAI), vol. 9693, pp. 647–659. Springer, Cham (2016). https://doi.org/10.1007/978-3-319-39384-1_57
28. Grycuk, R., Wojciechowski, A., Wei, W., Siwocha, A.: Detecting visual objects by edge crawling. J. Artif. Intell. Soft Comput. Res. **10**(3), 223–237 (2020)
29. Györfi, L., Kohler, M., Krzyżak, A., Walk, H.: A Distribution-Free Theory of Nonparametric Regression. Springer (2002). https://doi.org/10.1007/b97848
30. Härdle, W.: Applied Nonparametric Regression. No. 19. Cambridge University Press, Cambridge (1990)
31. Homenda, W., Jastrzębska, A., Pedrycz, W., Yu, F.: Combining classifiers for foreign pattern rejection. J. Artif. Intell. Soft Comput. Res. **10**(2), 75–94 (2020)
32. Jaworski, M., Duda, P., Rutkowski, L.: New splitting criteria for decision trees in stationary data streams. IEEE Trans. Neural Netw. Learn. Syst. **29**(6), 2516–2529 (2018)
33. Katznelson, Y.: An Introduction to Harmonic Analysis, 3rd edn. Cambridge University Press, Cambridge (2004)
34. Kohler, M.: Nonlinear orthogonal series estimates for random design regression. J. Stat. Plann. Inference **115**(2), 491–520 (2003)
35. Krell, E., Sheta, A., Balasubramanian, A.P.R., King, S.A.: Collision-free autonomous robot navigation in unknown environments utilizing PSO for path planning. J. Artif. Intell. Soft Comput. Res. **9**(4), 267–282 (2019)
36. Korytkowski, M., Senkerik, R., Scherer, M.M., Angryk, R.A., Kordos, M., Siwocha, A.: Efficient image retrieval by fuzzy rules from boosting and metaheuristic. J. Artif. Intell. Soft Comput. Res. **10**(1), 57–69 (2020)
37. Kullback, S., Leibler, R.A.: On information and sufficiency. Ann. Math. Stat. **22**(1), 79–86 (1951)
38. Łapa, K., Cpałka, K., Przybył, A., Grzanek, K.: Negative space-based population initialization algorithm (NSPIA). In: Rutkowski, L., Scherer, R., Korytkowski, M., Pedrycz, W., Tadeusiewicz, R., Zurada, J.M. (eds.) ICAISC 2018. LNCS (LNAI), vol. 10841, pp. 449–461. Springer, Cham (2018). https://doi.org/10.1007/978-3-319-91253-0_42
39. Łapa, K., Cpałka, K., Przybył, A.: Genetic programming algorithm for designing of control systems. Inf. Technol. Control **47**(5), 668–683 (2018)
40. Marr, D., Hildreth, E.: Theory of edge detection, Proc. R. Soc. London, B-207, pp. 187–217 (1980)

41. Oded, K., Hallin, C.A., Perel, N., Bendet, D.: Decision-making enhancement in a big data environment: application of the k-means algorithm to mixed data. J. Artif. Intell. Soft Comput. Res. **9**(4), 293–302 (2019)
42. Pietruczuk, L., Rutkowski, L., Jaworski, M., Duda, P.: How to adjust an ensemble size in stream data mining? Information Sciences, Elsevier Science Inc., vol. 381, No. C, pp. 46–54 (2017)
43. Pratt, W.K.: Digital Image Processing, 4th edn. John Wiley Inc., New York (2007)
44. Qiu, P.: Nonparametric estimation of jump surface. Indian J. Stat. Ser. A **59**(2), 268–294 (1997)
45. Qiu, P.: Jump surface estimation, edge detection, and image restoration. J. Am. Stat. Assoc. **102**, 745–756 (2007)
46. Rahman, M.W., Zohra, F.T., Gavrilova, M.L.: Score level and rank level fusion for Kinect-based multi-modal biometric system. J. Artif. Intell. Soft Comput. Res. **9**(3), 167–176 (2019)
47. Rafajłowicz, E., Schwabe, R.: Halton and Hammersley sequences in multivariate nonparametric regression. Stat. Probab. Lett. **76**(8), 803–812 (2006)
48. Rafajłowicz, E., Wnuk, M., Rafajłowicz, W.: Local detection of defects from image sequences. Int. J. Appl. Math. Comput. Sci. **18**(4), 581–592 (2008)
49. Rafajłowicz, E., Rafajłowicz, W.: Testing (non-)linearity of distributed-parameter systems from a video sequence. Asian J. Control **12**(2), Special Issue, 146–158 (2010)
50. Rafajłowicz, W.: Nonparametric estimation of continuously parametrized families of probability density functions - Computational aspects. Wrocław University of Science and Technology, Wrocław, Preprint of the Department of Engineering Informatics (2020)
51. Romani, L., Rossini, M., Schenone, D.: Edge detection methods based on RBF interpolation. J. Comput. Appl. Math. **349**, 532–547 (2019)
52. Rutkowski, L.: Orthogonal series estimates of a regression function with applications in system identification. In: Grossmann, W., Pflug, G.C., Wertz, W. (eds.) Probability and Statistical Inference. Springer, Dordrecht, p. n/a, (1982). https://doi.org/10.1007/978-94-009-7840-9_32
53. Rutkowski, L.: Application of multiple Fourier-series to identification of multivariable non-stationary systems. Int. J. Syst. Sci. **20**(10), 1993–2002 (1989)
54. Rutkowski, L., Rafajłowicz, E.: On optimal global rate of convergence of some nonparametric identification procedures. IEEE Trans. Autom. Control **34**(10), 1089–1091 (1989)
55. Rutkowski, L.: Identification of MISO nonlinear regressions in the presence of a wide class of disturbances. IEEE Trans. Inf. Theory **37**(1), 214–216 (1991)
56. Rutkowski, L., Pietruczuk, L., Duda, P., Jaworski, M.: Decision trees for mining data streams based on the McDiarmid's bound. IEEE Trans. Knowl. Data Eng. **25**(6), 1272–1279 (2013)
57. Rutkowski, L., Jaworski, M., Pietruczuk, L., Duda, P.: Decision trees for mining data streams based on the Gaussian approximation. IEEE Trans. Knowl. Data Eng. **26**(1), 108–119 (2014)
58. Rutkowski, L., Jaworski, M., Pietruczuk, L., Duda, P.: The CART decision tree for mining data streams. Inf. Sci. **266**, 1–15 (2014)
59. Rutkowski, L., Jaworski, M., Pietruczuk, L., Duda, P.: A new method for data stream mining based on the misclassification error. IEEE Trans. Neural Net. Learn. Syst. **26**(5), 1048–1059 (2015)
60. Rutkowski, T., Romanowski, J., Woldan, P., Staszewski, P., Nielek, R., Rutkowski, L.: A content-based recommendation system using neuro-fuzzy approach. In: International Conference on Fuzzy Systems: FUZZ-IEEE, pp. 1–8 (2018)
61. Rutkowski, T., Romanowski, J., Woldan, P., Staszewski, P., Nielek, R.: Towards interpretability of the movie recommender based on a neuro-fuzzy approach. In: Rutkowski, L., Scherer, R., Korytkowski, M., Pedrycz, W., Tadeusiewicz, R., Zurada, J.M. (eds.) ICAISC 2018. LNCS (LNAI), vol. 10842, pp. 752–762. Springer, Cham (2018). https://doi.org/10.1007/978-3-319-91262-2_66

62. Rutkowski, L., Jaworski, M., Duda, P.: Stream Data Mining: Algorithms and Their Probabilistic Properties. Springer, Heidelberg (2019). https://doi.org/10.1007/978-3-030-13962-9
63. Rutkowski, T., Łapa, K., Nielek, R.: On explainable fuzzy recommenders and their performance evaluation. Int. J. Appl. Math. Comput. Sci. **29**(3), 595–610 (2019)
64. Rutkowski, T., Łapa, K., Jaworski, M., Nielek, R., Rutkowska, D.: On explainable flexible fuzzy recommender and its performance evaluation using the Akaike information criterion. In: Gedeon, T., Wong, K.W., Lee, M. (eds.) ICONIP 2019. CCIS, vol. 1142, pp. 717–724. Springer, Cham (2019). https://doi.org/10.1007/978-3-030-36808-1_78
65. Sansone, G.: Orthogonal Functions. Interscience (1959)
66. Singh, S., Singh, R.: Comparison of various edge detection techniques. In: 2nd International Conference on Computing for Sustainable Global Development, pp. 393–396 (2015)
67. Szczypta, J., Przybył, A., Cpałka, K.: Some aspects of evolutionary designing optimal controllers. In: Rutkowski, L., Korytkowski, M., Scherer, R., Tadeusiewicz, R., Zadeh, L.A., Zurada, J.M. (eds.) ICAISC 2013. LNCS (LNAI), vol. 7895, pp. 91–100. Springer, Heidelberg (2013). https://doi.org/10.1007/978-3-642-38610-7_9
68. Szegö, G.: Orthogonal Polynomials. American Mathematical Society (1975)
69. Zalasiński, M., Cpałka, K., Hayashi, Y.: New fast algorithm for the dynamic signature verification using global features values. In: Rutkowski, L., Korytkowski, M., Scherer, R., Tadeusiewicz, R., Zadeh, L.A., Zurada, J.M. (eds.) ICAISC 2015. LNCS (LNAI), vol. 9120, pp. 175–188. Springer, Cham (2015). https://doi.org/10.1007/978-3-319-19369-4_17
70. Zhao, X., Song, M., Liu, A., Wang, Y., Wang, T., Cao, J.: Data-driven temporal-spatial model for the prediction of AQI in Nanjing. J. Artif. Intell. Soft Comput. Res. **10**(4), 255–270 (2020)
71. Zygmund, A.: Trigonometric series, vol. I. II. Third edition. Cambridge University Press, Cambridge (2002)

Recommendation System for Signal Processing in SHM

Jakub Gorski[1], Michal Dziendzikowski[2], and Ziemowit Dworakowski[1](✉)

[1] AGH University of Science and Technology, Mickiewicza 30, Cracow, Poland
zdw@agh.edu.pl

[2] Air Force Institute of Technology, Ksiecia Boleslawa 6, Warsaw, Poland

Abstract. In this article, the recommendation system for processing signals was presented. It contains a database and two fuzzy modules composed within the system. Based on the contextual knowledge provided by the user, collected database, and fuzzy rules, the system suggests processing methods and features. The article presents an evaluation of the proposed system on a two-stage gearbox dataset. The system results are a list of recommended processing methods and extracted features, which allow for a more accurate data classification.

Keywords: Recommendation · Fuzzy-logic · SHM · Condition monitoring

1 Introduction

In the age of information, making decisions is a difficult task. The number of choices is overwhelming, and it would take a long time to analyze all the possible options. To facilitate decision-making, engineers developed recommendation algorithms based on soft computing methods. Examples include recommendation of films [1–3] or products in e-commerce [4,5].
A similar problem often arises in Structural Health Monitoring (SHM), in which engineering structures are subjected to continuous monitoring over extended periods of time. The common approach requires the acquisition of large numbers of time-domain signals (e.g., vibration signatures, waveforms, etc.), then perform a processing chain including preprocessing with various filtration or domain-changing techniques, feature extraction, and finally, setup of a classifier that can provide autonomous decisions. Expert knowledge is required to choose and configure methods correctly at each stage of this processing chain: This research aims to automate the method-selection procedures based on a set of historical processing chains.

Supported by the National Centre for Research and Development in Poland.

L. Rutkowski et al. (Eds.): ICAISC 2021, LNAI 12854, pp. 328–337, 2021.
https://doi.org/10.1007/978-3-030-87986-0_29

2 System Concept

The basic idea of the system involves decision-making based on an extensive database of historical information containing similar objects and results of their monitoring. In other words: the system proposes solutions that are likely to work because they worked before on similar objects.

The schematics of the recommendation system is presented in Fig. 1, where it is composed of 5 main blocks: the category and sub-category of monitored structure surveys, the list of context information surveys, structure selector blocks and signal processing, feature selection blocks, and the database. The brown arrows indicate the flow of information from surveys to specific blocks. The numbers in Fig. 1 indicate a sequence of actions performed by the system.

A series of context information parameters define each monitoring object. Utilizing this information, a measure of similarity is calculated between the analyzed structure and those stored in the database. As a result, the list of structures with the highest similarity is selected. The system returns final results as a list of recommended signal processing methods and features based on the signal processing and features information from stored structures.

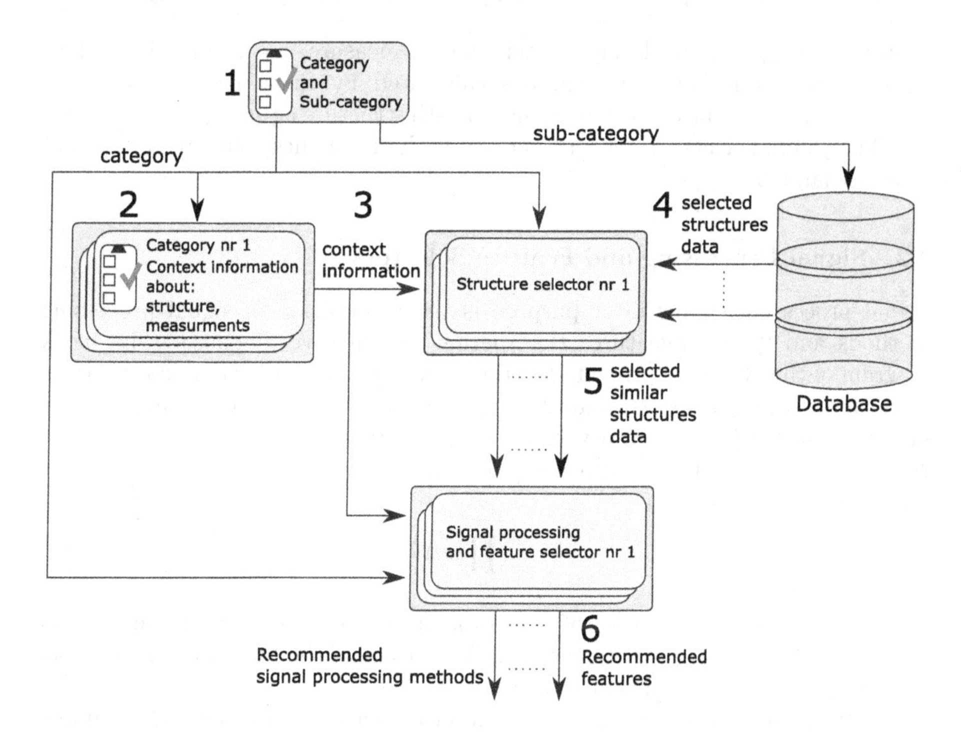

Fig. 1. Block diagram of system

2.1 Structure Selectors

The block diagram of the structure selector is shown in Fig. 2. The completed survey provided the context information in a vector form. Based on the category, a single fuzzy system is selected with three categorized embedding dimensions:

- Embedding 1 - Variation of operational parameters,
- Embedding 2 - Method of conducting the experiment.
- Embedding 3 - Type of problem (Condition classification, Novelty detection, Prediction),

The structures' data for the selected category and sub-category was acquired from the database as presented in Fig. 1. For each object from the database, the Euclidean similarity measure is calculated. The described operation allows to narrow down the number of considered structures. Afterward, the sub-system assigns an embedding category. The smallest distance in one of the three embedding dimensions is calculated. For this purpose, the modified Manhattan similarity measure was applied:

$$E_{man1D} = max(exp(-|e_1 - e_{1si}|), exp(-|e_2 - e_{2si}|), exp(-|e_3 - e_{3si}|)) \quad (1)$$

Where E_{man1D} is Manhattan similarity measure for single dimention, e_1, e_2, e_3 are values of 3 embeddings calculated by fuzzy logic system and $e_{1si}, e_{2si}, e_{3si}$ are values of 3 embeddings for i-th structure aquired from database.

The system selects up to 5 structures with the highest similarity measure greater than 0.5.

2.2 Signal Processing and Feature Selectors

Signal processing chains filter purpose is the proposition of signal processing methods and types of features that permit an effective diagnosis. The block diagram of the system is presented in Fig. 2. As a result of the completed survey, the context information is quantitatively encoded in a vector. A single fuzzy system is selected to assign a weight for each signal processing chain based on the category. Thus the utility value is calculated, which can be express by the equation:

$$U = (\prod_{i=1}^{n} w_i)^{\frac{1}{n}} \quad (2)$$

where w_i is the weight factor of a method from the filter, n is the number of methods in the signal processing chain. As a result of the filtration, the number of signal processing chains is reduced.

In the next part of the algorithm, the probability of selecting a chain under the condition of success diagnose is calculated.

$$P(c_i) = \frac{E_i}{N_c} \quad (3)$$

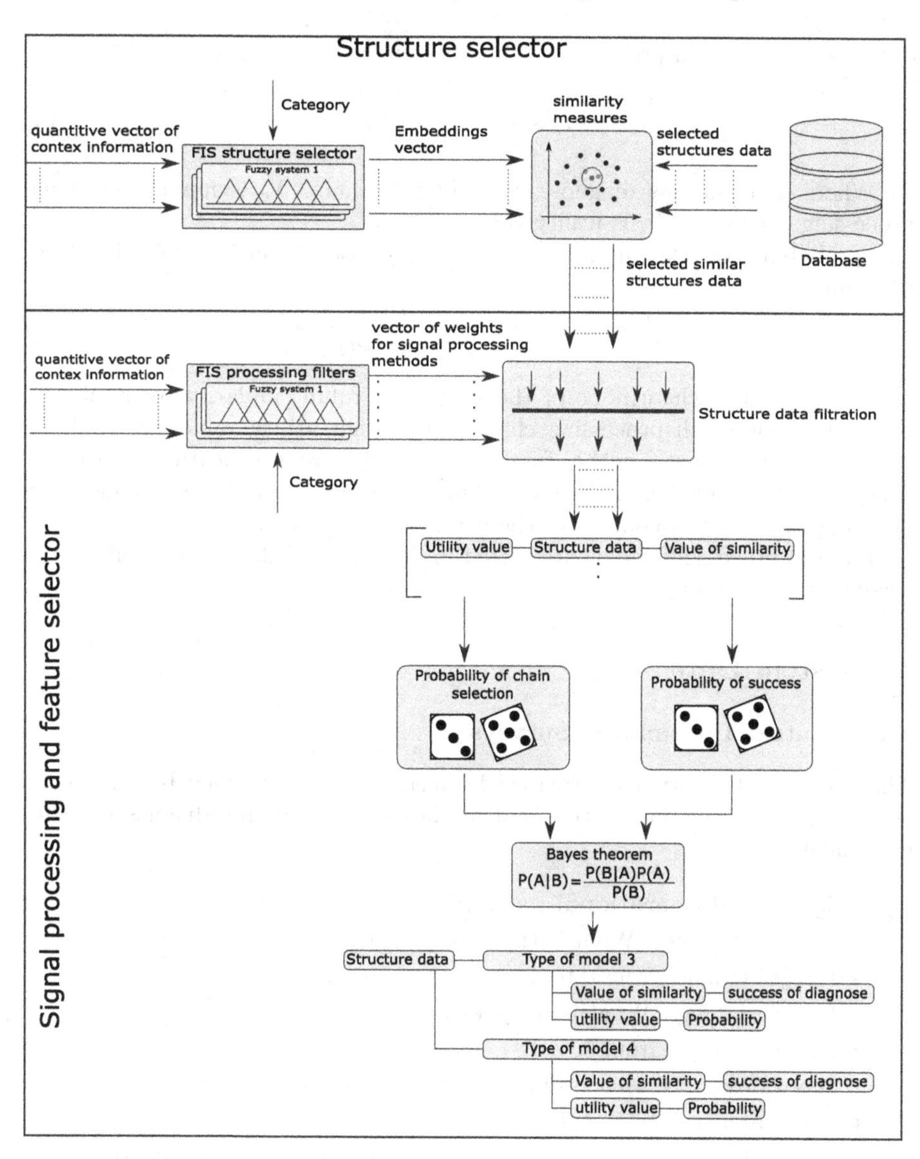

Fig. 2. Block diagram of system selector and signal processing methods filtering.

where c_i is i-th signal processing chain, E_i is the similarity value for i-th signal processing chain, and N_c is the number of signal processing chains. Each signal processing chain contains an attribute called the success of diagnosing, which quantitatively describes chain effectiveness in diagnose. By utilizing this information and the utility value, the probability of success (s_i) under the condition of the selection signal processing chain is given by

$$P(s_i|c_i) = \frac{d_i}{N_c} * U_i \tag{4}$$

Where d_i is success diagnose value of i-th chain, N_c is number of signal processing chains and U_i is utility value.
The probability of selecting a chain can be expressed by introducing the Bayes theorem.

$$P(c_i|s_i) = \frac{P(c_i) * P(s_i|c_i)}{\sum_{j=1}^{N_c} P(c_j) * P(s_j|c_j)} \tag{5}$$

For situations where none of the structures fulfill similarity requirements, the system filters all processing chains and assign utility values by applying the Formula (2). Next, another fuzzy-based model proposes features to extract from each processing chain. The probability of selecting a chain is calculated by applying expressions (4) and (5) where $P(c_i) = P(c_j) = 0.5$.

Filtered processing chains are sorted in decreasing order of probability, and presented to the user.

3 System Data

3.1 Context Information Surveys

The context information is provided by means of surveys filled by an experienced operator. The diagnostic problem classes used for vibrodiagnostics category include:

1. **Variability of operational parameters**
   - speed_variability: What is the speed variation?
   - Integer number from 0 to 4
   - load_variability: What is the load variation?
   - Integer number from 0 to 4
   - speed_order: By what order of magnitude the speed changes?
   - Real number from 0 to 10.
   - load_order: By what order of magnitude the load change? Answer:
   - Real number from 0 to 3.
2. **Method of experimenting**
   - A vector including determination of sensor bandwidth and flags for the presence of speed information, type, direction, and location of sensors.
3. **Type of problem**
   - struct_labels: Are there labels for structure states?
   - Integer number from 0 to 4
   - problem_type: What is the type of problem?
   - Integer number from 0 to 4

3.2 Fuzzy Logic System for Structure Selector

The fuzzy system configuration for structure selector is presented in Fig. 3a. The architecture of each system is a fuzzy tree. The single branch is designated to evaluate the value of a single embedding.

Figure 3b presents detailed block diagram of the developed embedding branches.

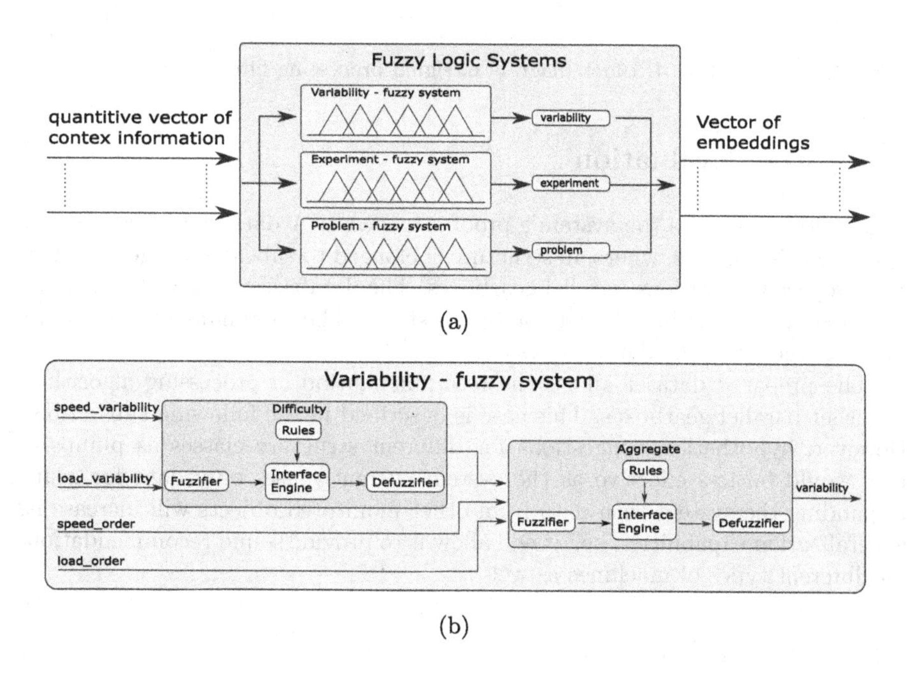

Fig. 3. Block diagram of: (a) structure selector model; (b) one of its branches.

The branch labeled **Variability** is a fuzzy tree composed of 2 Fuzzy subsystems. The first combines information about load with the speed variation to form a hidden variable called difficulty. The second combines the calculated hidden variable with load and speed order to determine embedding value labeled as **variability**. The branches for **Experiment** and **Problem** work in a similar way.

3.3 Fuzzy Logic System for Signal Processing Filter

The fuzzy system configuration for signal processing and feature selectors are presented in Fig. 4. Each system is a parallel fuzzy tree where a single branch is designated to recommend a single processing method or feature. Each branch is Mamdani based fuzzy system.

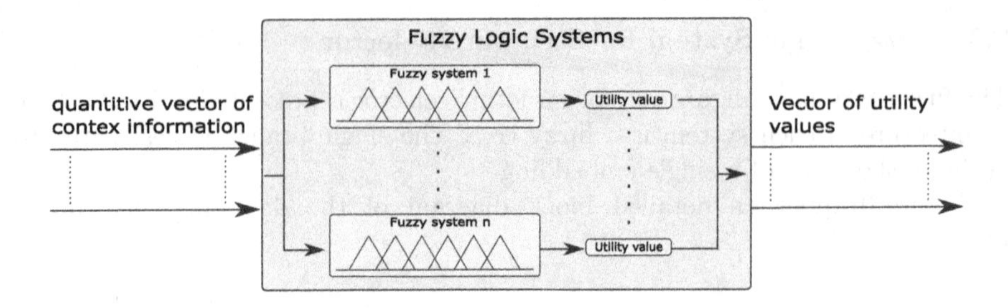

Fig. 4. Block diagram of signal processing filter.

4 System Evaluation

This section contains the system's proof of concept. A dataset has been developed for this purpose, where all structures belonged to subcategory **gears** which was prepared for various parallel gearboxes. The database contains 28 one-stage gearboxes operating in different conditions states. The total number of measurements gathered in the database was 11680.

The prepared dataset allows the recommendation of processing algorithms for other parallel gearboxes. This case is described in the following subsections. However, hypothetical suggestions for different structure classes as pumps or fans would be less effective as the system currently uses only data for gears. Expanding the system with data from other monitored objects will increase its generalization capabilities, i.e., it will allow it to provide sound recommendations for different types of machines as well.

4.1 Test Structures

For the system evaluation, the dataset was obtained for a two-stage gearbox, which schematics is presented in Fig. 5(a). The vibration signals were acquired for multiple faults as missing tooth, root crack, spalling, and chipping tip with five different levels of severity. The dataset was first presented in work [6].

The inputs vector for analyzed gearbox and 3 vectors for similar structures are stored in Table 1.

4.2 Two Stage Gearbox Recommendation Results

The recommendation system result is a list of signal processing chains and corresponding features. The list is presented in the Table 2. The system recommended three different processing chains. To validate obtained results, a processing chain was selected at random from the database.

The first chain is dedicated to the calculation of 3 features from the envelope spectrum. In literature [7], the envelope analysis is convenient for detecting cyclostationary components generated by faulty rolling-element bearings and gears meshing.

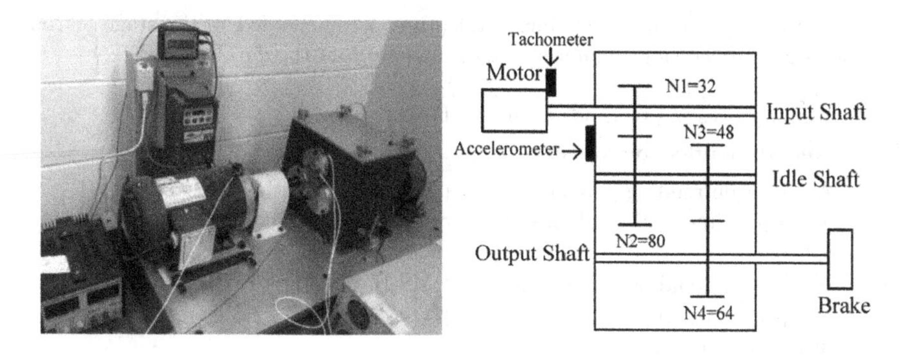

Fig. 5. Two-stage gearbox picture (left) and kinematic diagram (right) [6].

Table 1. Context information.

Question tag	Gearbox	Similar database objects			Randomly selected object
speed_variability	1	1	0	1	2
load_variability	0	0	0	0	0
speed_order	0	0.0071	0	0.02	0.4
load_order	0	0	0	0	0
speed_information	0	0	0	0	1
sensors_bandwidth	20	25	25	25	40
sensors_location	1	0	0	0	0
sensors_direction	0	1	1	1	1
sensors_type	0	0	0	0	0
struct_labels	1	1	1	3	3
problem_type	1	0	0	1	1
Similarity value	–	0.6775	0.6769	0.6527	0.4698

The second one is the frequency spectrum analysis, which is fundamental in terms of time signals.

The last of the proposed chains was developed for the velocity signal extraction. Such an analysis allows the detection of shaft damage, manifested in low frequencies [8].

The features were utilized for a similar structure, where they revealed good class separability.

The two processing chains presented in Table 2 were applied to data acquired from the structure under investigation. The normalized results are shown in Fig. 6. Figure 6(a) contains results for the recommended chains. Figure 6(b) provides comparison for a non-recommended processing. Every point corresponds to a single measurement. The attached legend presents the labels containing information about the structure state.

Table 2. Signal processing algorithms and features recommended for two stage gearbox and one signal processing chain selected at random from the database.

Signal processing methods chain	Features
Recommended by system	
A: linear detrending - signal envelope - linear detrending - spectrum	peak-to-RMS-A, Kurtosis-A, Skewness-A
B: linear detrending - spectrum	RMS-B, Skewness-B
C: linear detrending - highpass filtration (10 Hz cutoff) - integration - spectrum	RMS-C
Non-recommended by system	
D: linear detrending - demodulation - spectrum	peak-to-RMS-D, Kurtosis-D, Shannon entropy-D

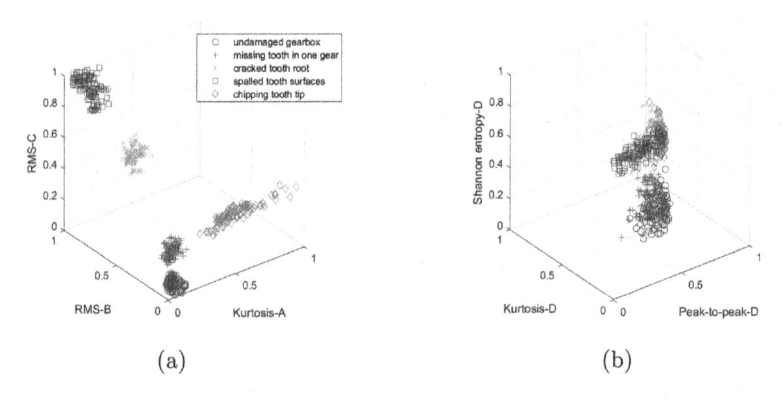

Fig. 6. (a) Three selected features obtained from recommended chains; (b) Three features obtained from a not-recommended chain.

The data processing routines recommended by the system allowed for a good separability of all the classes. In contrast, the non-recommended processing algorithms resulted in significant overlap of the state classes, leading to a higher misclassification rate. It appears that, while the system is not yet tested in a blind trial and comparison with the capabilities of the human expert, its current results are already far better than random guessing. They thus can be used as a starting point for further optimization.

The results presented by the system are selected based on the previous results for historical data. The proposed processing chains and features gained the highest probability value for the most similar gearboxes stored in the database. The global minimum of the optimization criterion, i.e., the best possible separation of classes, is unlikely because the system is incapable of proposing an entirely new processing chain and is restricted to working based only on historical capabilities of the processing methods.

5 Summary and Conclusions

In this paper, the fuzzy-based recommendation system for context-aware signal processing method choice was presented. The system structure was implemented for testing purposes with a database containing information about gearbox monitoring. The system was evaluated on a dataset obtained from a two-stage gearbox, for which it proposed the processing recommendations. The analysis of the results reveals that the recommendation is useful for damage classification algorithms development because it allows building a decision space with clearly separable clusters representing different faults. The system was tested on a condition monitoring data, parallel gearbox in particular, but its general principle allow it to be used for any typical condition monitoring or SHM scenario. However, the database must contain many similar structures like the one under monitoring. It is thus rather a population-based approach than a system for solving novel problems.

Acknowledgement. The work presented in this paper was supported by the National Centre for Research and Development in Poland under the research project no. LIDER/3/0005/L-9/17/NCBR/2018.

References

1. Xiaowei, S.: An intelligent recommendation system based on fuzzy logic. Inform. Control Autom. Robot. **I**, 105–109 (2006). https://doi.org/10.1007/1-4020-4543-3_12
2. Davidson, J., Liebald, B., Liu, J., Nandy, P., Van Vleet, T.: The YouTube video recommendation system. In: RecSys'10 - Proceedings of the 4th ACM Conference on Recommender Systems, pp. 293–296 (2010). https://doi.org/10.1145/1864708.1864770
3. Siddiquee, M.M.R., Haider, N., Rahman, R.M.: A fuzzy based recommendation system with collaborative filtering. In: SKIMA 2014–8th International Conference on Software, Knowledge, Information Management and Applications (2014). https://doi.org/10.1109/SKIMA.2014.7083524
4. Cao, Y., Li, Y.: An intelligent fuzzy-based recommendation system for consumer electronic products. Expert Syst. Appl. **33**(1), 230–240 (2007). https://doi.org/10.1016/j.eswa.2006.04.012
5. Gipson, A.: [CF] Amazon Recommendations Item-to-Item Collaborative Filtering (Amazon 2003). Mississippi Legislature (February), 2 (2018). https://doi.org/10.1038/sj.onc.1203797
6. Cao, P., Zhang, S., Tang, J.: Preprocessing-free gear fault diagnosis using small datasets with deep convolutional neural network-based transfer learning. IEEE Access **6**, 26241–26253 (2018). https://doi.org/10.1109/ACCESS.2018.2837621
7. Condition Monitoring Algorithms in MATLAB®. STME, Springer, Cham (2021). https://doi.org/10.1007/978-3-030-62749-2
8. Randall, R.B.: Vibration-based condition monitoring (2013). https://doi.org/10.1007/978-94-007-6422-4_11

Monitoring of Changes in Data Stream Distribution Using Convolutional Restricted Boltzmann Machines

Maciej Jaworski[1(✉)], Leszek Rutkowski[2,3], Paweł Staszewski[2], and Patryk Najgebauer[2]

[1] Department of Computer Science, Cracow University of Technology, Cracow, Poland
maciej.jaworski@pk.edu.pl

[2] Department of Computational Intelligence, Czestochowa University of Technology, Czestochowa, Poland

[3] Information Technology Institute, University of Social Sciences, Łódź, Poland
{leszek.rutkowski,pawel.staszewski,patryk.najgebauer}@pcz.pl

Abstract. In this paper, we propose the Convolutional Restricted Boltzmann Machine (CRBM) as a tool for detecting concept drift in time-varying data streams. Recently, it was demonstrated that the Restricted Boltzmann Machine (RBM) can be successfully applied to this task. A properly learned RBM contains information about the data probability distribution. Trained on one part of the stream it can be used to detect possible changes in the incoming data. In this work we replace the fully-connected layer in the standard RBM with the convolutional layer, composing the CRBM. We show that it is more suitable for the drift detection task regarding the image data. Preliminary experimental results demonstrate the usefulness of the CRBM as a tool for drift detection in data streams with such type of data.

Keywords: Convolutional Restricted Boltzmann Machine · Data stream mining · Concept drift detection

1 Introduction

Data stream mining attracts attention of many machine learning researchers around the world [2,4,7,10,18,24,25,27–32]. Data streams are potentially infinite sequences of data, which often are fed into the system with very high rates. Therefore, the properly designed data stream mining algorithms should take into account the limited amounts of resources like memory or computational power [3,11,20,37]. In the literature, there can be found many valuable data stream mining algorithms based on some standard machine learning approaches, e.g. the decision trees [5,18,33], neural networks [25,26,34] or ensemble methods [21,27].

The data stream can be denoted as an ordered sequence of data elements

$$S = (\mathbf{s}_1, \mathbf{s}_2, \dots), \tag{1}$$

L. Rutkowski et al. (Eds.): ICAISC 2021, LNAI 12854, pp. 338–346, 2021.
https://doi.org/10.1007/978-3-030-87986-0_30

In this paper, each data element is a D-dimensional binary vector

$$\mathbf{s}_n = [s_{n,1}, \ldots, s_{n,D}] \in \{0; 1\}^D \tag{2}$$

If we consider image data, then this vector is formed into a 2-dimensional array. It is very common that the distribution of data elements in the stream change over time [6,8,9,12,35,36,38]. It is commonly known in the literature under the name 'concept drift'. Then, the algorithm should be either equipped with a special mechanism allowing it to deal with possible changes or an external drift detector should be applied. In this paper, we consider the latter approach and we focus on applying the Restricted Boltzmann Machine as a concept drift detector. The idea was presented in [16]. It was extended in [19] to make the method more resource-aware. In [17] the mechanisms for dealing with missing and incomplete data were introduced. In this paper, we continue this research direction by replacing the RBM with the Convolutional Restricted Boltzmann Machine (CRBM), which is better suited for image data.

The rest of the paper is organized as follows. The RBM is briefly recalled in Sect. 2. In Sect. 3 the CRBM is described. In Sect. 4 the idea of data stream monitoring using properly trained RBM or CRBM is presented. Preliminary results obtained for the CRBM as a drift detector as well as its comparison with the standard RBM are demonstrated in Sect. 5. Conclusions are drawn in Sect. 6.

2 The Restricted Boltzmann Machine

The RBM [15] is a neural network consisting of two layers of neurons. The visible layer contains D neurons $\mathbf{v} = [v_1, \ldots, v_D]$ and the hidden layer is formed by H neurons $\mathbf{h} = [h_1, \ldots, h_H]$. There are no inter-layer connections in the RBM. The probability distribution of states $(\mathbf{v}, \mathbf{h})$ is defined using the energy function

$$P(\mathbf{v}, \mathbf{h}) = \frac{\exp\left(-E(\mathbf{v}, \mathbf{h})\right)}{Z}, \tag{3}$$

where Z is a normalization constant called partition function. The energy for a state of neurons $(\mathbf{v}, \mathbf{h})$ takes the following value

$$E(\mathbf{v}, \mathbf{h}) = -\sum_{i=1}^{D} v_i a_i - \sum_{j=1}^{H} h_j b_j - \sum_{i=1}^{D} \sum_{j=1}^{H} v_i h_j w_{ij}. \tag{4}$$

Here w_{ij}, a_i and b_j denote RBM weights, visible layer biases, and hidden layer biases, respectively. The RBM is usually trained using minibatches of data. Let S_t denote the t-th minibatch of size B

$$S_t = \left(\mathbf{s}_{Bt+1}, \ldots, \mathbf{s}_{Bt+B}\right), \ t = 0, 1, \ldots. \tag{5}$$

Then, the cost function, which is the negative log-likelihood in the case of the RBM, for minibatch S_t can be calculated as follows

$$C(S_t) = -\log P(S_t) = -\frac{1}{B} \sum_{n=1}^{B} \sum_{\mathbf{h}} \log P(\mathbf{s}_{Bt+n}, \mathbf{h}). \tag{6}$$

The gradient of the cost function with respect to weight w_{ij} is given by (the derivations can be found in e.g. [1,13])

$$\frac{\partial C(S_t)}{\partial w_{ij}} = \sum_{\mathbf{v},\mathbf{h}} P(\mathbf{v},\mathbf{h}) v_i h_j - \frac{1}{B} \sum_{n=1}^{B} \sum_{\mathbf{h}} P(\mathbf{h}|\mathbf{v} = \mathbf{s}_{Bt+n}) v_i h_j. \quad (7)$$

Because the first term on the right-hand side is intractable, the gradient can be only approximated, using for example the Contrastive Divergence (CD) algorithm [14]. In this paper, we propose to replace the RBM with the Convolutional RBM [23] to create a concept drift detector more suitable for image data.

3 The Convolutional Restricted Boltzmann Machine

In this paper, we attempt to apply the same idea of using the RBM as a drift detector to image data. To this end, we replaced the RBM with the Convolutional RBM [23], as the convolutional layers turned out to be more relevant for processing images. The CRBM differs from the standard RBM in several details. First of all, the input units are formed into a 2-dimensional array. Therefore, the visible units $v_{i,j}$ have two indices, $i, jin\{1, \ldots, d\}$ (without using the generality we assume that the images have the shape of square). In this case, the total number of visible neurons is $D = d^2$. The hidden layer consists of K groups, where each group is a 2-dimensional layer of $(h \times h)$ neurons. Summarizing, there are totally $H = Kh^2$ hidden neurons $h^k_{m,n}$, $m, n \in \{1, \ldots, h\}$. The number K corresponds to the number of filters. Each filter is of size $W \times W$ and the relation between the filter, visible layer, and hidden layer sizes are given as follows

$$W = d - h + 1. \quad (8)$$

Let $w^k_{r,s}$ denote the (r, s)-th element of the k-th filter. The weights of each filter are shared across all the hidden neurons belonging to the same group. Moreover, each group of hidden neurons has its own bias b_k. There is also one bias value a shared among all the visible units. Having all the above notations introduced, we can now write down the formula for the energy of the CRBM

$$E(\mathbf{v},\mathbf{h}) = -a \sum_{i,j=1}^{d} v_{i,j} - \sum_{k=1}^{K} b_k \sum_{i,j=1}^{h} h^k_{i,j} - \sum_{k=1}^{K} \sum_{i,j=1}^{h} \sum_{r,s=1}^{W} v_{i+r-1,j+s-1} h^k_{i,j} w^k_{r,s}. \quad (9)$$

Obviously, Formula (3) also applied in the case of the CRBM to calculate the probability of state $(\mathbf{v}, \mathbf{h})$. Analogously to the case of the standard RBM, the energy function allows us to analytically derive the conditional probabilities of visible and hidden units states

$$P(h^k_{i,j} = 1|\mathbf{v}) = \sigma \left(\sum_{r,s=1}^{W} v_{i+r-1,j+s-1} w^k_{r,s} + b_k \right) \quad (10)$$

$$P(h^k_{i,j} = 1|\mathbf{v}) = \sigma \left(\sum_{k=1}^{K} \sum_{r,s=1}^{W} h^k_{i-r+1,j-r+1} w^k_{r,s} + a \right) \quad (11)$$

4 Changes Detection with the CRBM

The idea of applying the CRBM as a tool for concept drift detection is analogous to the one presented for the standard RBM [18]. First, we have to choose the evaluation measure. The most natural is the log-likelihood itself. However, because of the partition function Z the exact computation of it would be intractable. Therefore, we decided to apply a reconstruction error instead, which is defined as follows. Let v_{ij}, $i, j, \in \{1, \ldots, d\}$ be initial data element passed to the visible layer of the CRBM. Then we apply Formula (10) to compute the probabilities for all hidden units and sample the states of these units according to it

$$h_{i,j}^k \sim P(h_{i,j}^k = 1|\mathbf{v}),\ i, j \in \{1, \ldots, h\}, k \in \{1, \ldots, K\}. \tag{12}$$

Then we calculate the reconstructed version $\tilde{v}_{i,j}$, $i, j \in \{1, \ldots, d\}$ of the input data element using the conditional probabilities given by (11)

$$\tilde{v}_{i,j} = P(v_{i,j} = 1|\mathbf{h}),\ i, j, \in \{1, \ldots, d\} \tag{13}$$

Then the reconstruction error for data element $\mathbf{v}$ is calculated as a square root of squared error between the original and reconstructed values

$$R(\mathbf{v}) = \sqrt{\sum_{i,j=1}^{d} (v_{i,j} - \tilde{v}_{i,j})^2}. \tag{14}$$

It is worth averaging the reconstruction error over the batch of data elements. Then the obtained value has a lower variance. For the t-th minibatch of data given S_t by (6) the average reconstruction error is given by

$$R(\mathbf{S}_t) = \frac{1}{B} \sum_{m=1}^{B} R(\mathbf{s}_{t*B+m}) \tag{15}$$

To monitor the possible changes of data distribution in the data stream first we have to train the CRBM on a part of the stream. The learning should be long enough to ensure that the CRBM is trained properly. Then the learning is stopped and the CRBM is used to check to values of reconstruction error of the incoming data. If it is kept on the same level, then it means that the distribution of data does not change with a high probability. When the drift occurs, the monitored value of reconstruction error increases. It may increase suddenly or gradually, depending on the type of the detected drift.

5 Experimental Results

In this section, the preliminary results of applying the CRBM as a drift detector are presented. The numerical simulations were carried out on the MNIST dataset [22], which consists of 60000 gray-scale images of handwritten digits. Each image is of size 28×28 pixels (hence $d = 28$). Since we need data streams with concept drifts in our experiments, we prepared synthetically two datasets for our purposes based on the original MNIST data. First, we created two sets corresponding to different digits: $S1$, consisting of digits $0, 2, 4, 7$, and $S2$, consisting of digits $1, 3, 5, 6, 8, 9$. Then, based on sets $S1$ and $S2$ two datasets with artificially imputed sudden and gradual concept drifts were created

$$S_s = (S_{s,1}, S_{s,2}, \dots), \tag{16}$$

$$S_g = (S_{g,1}, S_{g,2}, \dots) \tag{17}$$

The dataset S_s contains a sudden concept drift and is constructed as follows: the data element $S_{s,i}$ is taken from $S1$ if $i < 30000$ and from $S2$ if $i \geq 30000$. In the case of the dataset S_g, in which the gradual drift is introduced, the procedure of selecting elements goes as follows: $S_{g,i}$ is taken from $S1$ if $i < 30000$ and from $S2$ if $i \geq 40000$. In the interval $i \in [30000; 40000)$ the data element $S_{g,i}$ is drawn from $S1$ with probability $\frac{40000-i}{10000}$ and from $S2$ with probability $\frac{i-30000}{10000}$.

The CRBM is created with the following parameters: $d = 28$, $h = 18$, $W = 11$ and $K = 5$. The training process is conducted with minibatches of size $B = 20$. The learning rate was set to $\eta = 0.05$. The stochastic gradient descent procedure with momentum was applied with the friction parameter $\gamma = 0.8$. For comparison, we run also the simulations using the standard RBM, with $D = 784$ and $H = 200$. The same learning parameters and procedure were applied. In both experiments learning of the RBM and the RBM took place for the first 25000 data elements. Then, the neural model was used only to monitor incoming data. The results obtained for data streams with the sudden and gradual drift are demonstrated in Fig. 1 and Fig. 2, respectively. Both networks provide similar order of values of the reconstruction error, although the CRBM is slightly better. However, the most important observation is that in the case of the CRBM the changes of values for data before and after the drift are more vivid. Therefore, the CRBM can potentially provide a stronger signal to trigger a data stream mining algorithm to rebuild the current model.

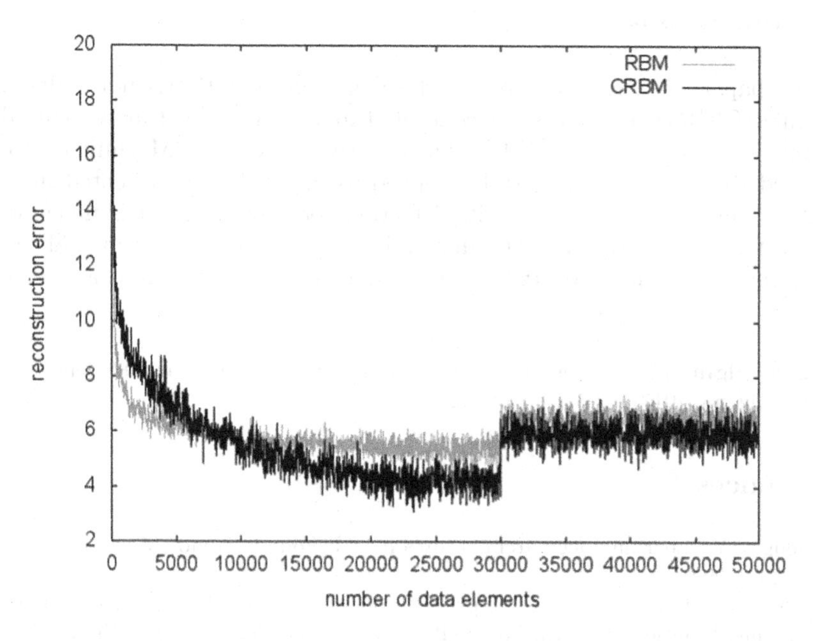

Fig. 1. Reconstruction error of the RBM and the CRBM for the data stream S_s with sudden concept drift at the 30000-th data element .

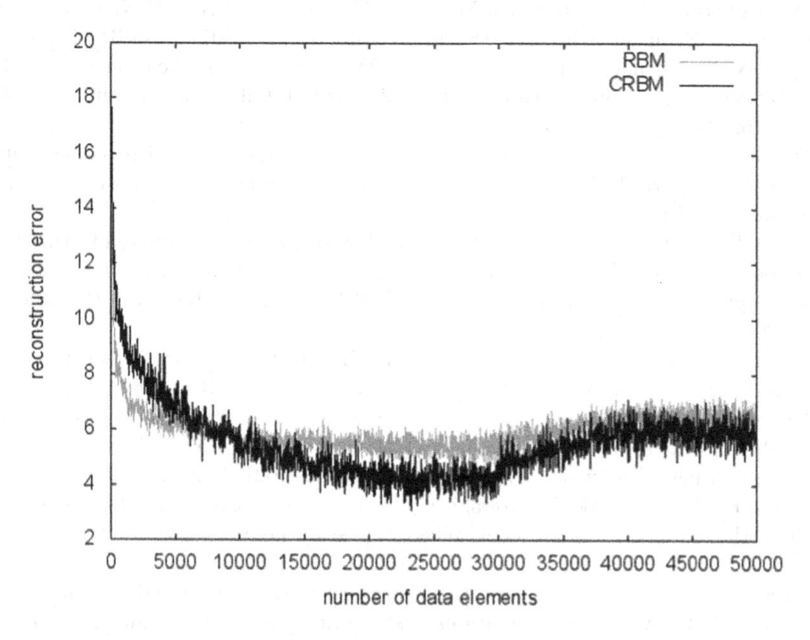

Fig. 2. Reconstruction error of the RBM and the CRBM for the data stream S_g with gradual concept drift for data elements from 30000 to 40000.

6 Conclusions

In this paper, we proposed to use the Convolutional Restricted Boltzmann Machine (CRBM) in the task of concept drift detection for time-varying data streams. It was demonstrated that the properly trained CRBM, similarly to the standard RBM, can be successfully applied to this task. We presented that the CRBM is more valuable as a drift detection tool concerning the image data. This statement was verified experimentally in preliminary experimental results conducted on a simple MNIST dataset, artificially modified into a streaming version.

Acknowledgments. This work was supported by the Polish National Science Centre under grant no. 2017/27/B/ST6/02852.

References

1. Bengio, Y.: Learning deep architectures for AI. Found. Trends Mach. Learn. **2**(1), 1–127 (2009)
2. Bifet, A.: Adaptive stream mining : pattern learning and mining from evolving data streams. Frontiers in Artificial Intelligence and Applications, IOS Press, Amsterdam, Berlin (2010)
3. Bilski, J., Kowalczyk, B., Grzanek, K.: The parallel modification to the Levenberg-Marquardt algorithm. In: Rutkowski, L., Scherer, R., Korytkowski, M., Pedrycz, W., Tadeusiewicz, R., Zurada, J.M. (eds.) ICAISC 2018. LNCS (LNAI), vol. 10841, pp. 15–24. Springer, Cham (2018). https://doi.org/10.1007/978-3-319-91253-0_2
4. Devi, V.S., Meena, L.: Parallel MCNN (PMCNN) with application to prototype selection on large and streaming data. J. Artif. Intell. Soft Comput. Res. **7**(3), 155–169 (2017)
5. Domingos, P., Hulten, G.: Mining high-speed data streams. In: Proceedings of the 6th ACM SIGKDD International Conference on Knowledge Discovery and Data Mining, pp. 71–80 (2000)
6. Duda, P., Rutkowski, L., Jaworski, M., Rutkowska, D.: On the Parzen Kernel-based probability density function learning procedures over time-varying streaming data with applications to pattern classification. IEEE Trans. Cybern. **50**(4), 1683–1696 (2020)
7. Duda, P., Jaworski, M., Cader, A., Wang, L.: On training deep neural networks using a streaming approach. J. Artif. Intell. Soft Comput. Res. **10**(1), 15–26 (2020)
8. Duda, P., Jaworski, M., Rutkowski, L.: Convergent time-varying regression models for data streams: tracking concept drift by the recursive Parzen-based generalized regression neural networks. Int. J. Neural Syst. **28**(02), 1750048 (2018)
9. Duda, P., Jaworski, M., Rutkowski, L.: Knowledge discovery in data streams with the orthogonal series-based generalized regression neural networks. Inf. Sci. **460–461**, 497–518 (2018)
10. Dyer, K.B., Capo, R., Polikar, R.: Compose: a semisupervised learning framework for initially labeled nonstationary streaming data. IEEE Trans. Neural Netw. Learn. Syst. **25**(1), 12–26 (2014)
11. Gaber, M., Zaslavsky, A., Krishnaswamy, S.: Mining data streams: a review. SIGMOD Record **34**(2), 18–26 (2005)

12. Gałkowski, T., Krzyżak, A., Filutowicz, Z.: A new approach to detection of changes in multidimensional patterns (2020)
13. Goodfellow, I., Bengio, Y., Courville, A.: Deep Learning. MIT Press, Cambridge (2016). http://www.deeplearningbook.org
14. Hinton, G.E.: Training products of experts by minimizing contrastive divergence. Neural Comput. **14**(8), 1771–1800 (2002)
15. Hinton, G.E., Sejnowski, T.J., Ackley, D.H.: Boltzmann machines: constraint satisfaction networks that learn. Technical Report CMU-CS-84-119, Computer Science Department, Carnegie Mellon University, Pittsburgh, PA (1984)
16. Jaworski, M., Duda, P., Rutkowski, L.: On applying the restricted Boltzmann machine to active concept drift detection. In: Proceedings of the 2017 IEEE Symposium Series on Computational Intelligence, pp. 3512–3519. Honolulu, USA (2017)
17. Jaworski, M., Duda, P., Rutkowska, D., Rutkowski, L.: On handling missing values in data stream mining algorithms based on the restricted Boltzmann machine. In: Gedeon, T., Wong, K.W., Lee, M. (eds.) ICONIP 2019. CCIS, vol. 1143, pp. 347–354. Springer, Cham (2019). https://doi.org/10.1007/978-3-030-36802-9_37
18. Jaworski, M., Duda, P., Rutkowski, L.: New splitting criteria for decision trees in stationary data streams. IEEE Trans. Neural Netw. Learn. Syst. **29**(6), 2516–2529 (2018)
19. Jaworski, M., Rutkowski, L., Duda, P., Cader, A.: Resource-aware data stream mining using the restricted Boltzmann machine. In: Rutkowski, L., Scherer, R., Korytkowski, M., Pedrycz, W., Tadeusiewicz, R., Zurada, J.M. (eds.) ICAISC 2019. LNCS (LNAI), vol. 11509, pp. 384–396. Springer, Cham (2019). https://doi.org/10.1007/978-3-030-20915-5_35
20. Kopczynski, M., Grzes, T.: Hardware rough set processor parallel architecture in FPGA for finding core in big datasets. J. Artif. Intell. Soft Comput. Res. **11**(2), 99–110 (2021). https://doi.org/10.2478/jaiscr-2021-0007
21. Krawczyk, B., Cano, A.: Online ensemble learning with abstaining classifiers for drifting and noisy data streams. Appl. Soft Comput. **68**, 677–692 (2018)
22. LeCun, Y., Cortes, C.: MNIST handwritten digit database (2010). http://yann.lecun.com/exdb/mnist/
23. Lee, H., Grosse, R., Ranganath, R., Ng, A.Y.: Convolutional deep belief networks for scalable unsupervised learning of hierarchical representations. In: Proceedings of the 26th Annual International Conference on Machine Learning, pp. 609–616. ICML 2009. Association for Computing Machinery, New York, NY, USA (2009)
24. Lemaire, V., Salperwyck, C., Bondu, A.: A survey on supervised classification on data streams. In: Zimányi, E., Kutsche, R.-D. (eds.) eBISS 2014. LNBIP, vol. 205, pp. 88–125. Springer, Cham (2015). https://doi.org/10.1007/978-3-319-17551-5_4
25. Ludwig, S.A.: Applying a neural network ensemble to intrusion detection. J. Artif. Intel. Soft Comput. Res. **9**(3), 177–188 (2019)
26. Niksa-Rynkiewicz, T., Szewczuk-Krypa, N., Witkowska, A., Cpałka, K., Zalasinski, M., Cader, A.: Monitoring regenerative heat exchanger in steam power plant by making use of the recurrent neural network. J. Artif. Intell. Soft Comput. Res. **11**(2), 143–155 (2021). https://doi.org/10.2478/jaiscr-2021-0009
27. Pietruczuk, L., Rutkowski, L., Jaworski, M., Duda, P.: How to adjust an ensemble size in stream data mining? Inf. Sci. **381**(C), 46–54 (2017)
28. Rafajłowicz, E., Rafajłowicz, W.: Iterative learning in repetitive optimal control of linear dynamic processes. In: Rutkowski, L., Korytkowski, M., Scherer, R., Tadeusiewicz, R., Zadeh, L.A., Zurada, J.M. (eds.) ICAISC 2016. LNCS (LNAI), vol. 9692, pp. 705–717. Springer, Cham (2016). https://doi.org/10.1007/978-3-319-39378-0_60

29. Rafajłowicz, E., Rafajłowicz, W.: Iterative learning in optimal control of linear dynamic processes. Int. J. Control **91**(7), 1522–1540 (2018)
30. Ramirez-Gallego, S., Krawczyk, B., García, S., Woźniak, M., Herrera, F.: A survey on data preprocessing for data stream mining: current status and future directions. Neurocomputing **239**, 39–57 (2017)
31. Rutkowski, L., Jaworski, M., Duda, P.: Stream Data Mining: Algorithms and Their Probabilistic Properties. SBD, vol. 56. Springer, Cham (2020). https://doi.org/10.1007/978-3-030-13962-9
32. Rutkowski, L., Jaworski, M., Pietruczuk, L., Duda, P.: A new method for data stream mining based on the misclassification error. IEEE Trans. Neural Netw. Learn. Syst. **26**(5), 1048–1059 (2015)
33. Rutkowski, L., Pietruczuk, L., Duda, P., Jaworski, M.: Decision trees for mining data streams based on the McDiarmid's bound. IEEE Trans. Knowl. Data Eng. **25**(6), 1272–1279 (2013)
34. Simões, D., Lau, N., Reis, L.P.: Multi agent deep learning with cooperative communication. J. Artif. Intell. Soft Comput. Res. **10**(3), 189–207 (2020). https://doi.org/10.2478/jaiscr-2020-0013
35. Tsymbal, A.: The problem of concept drift: definitions and related work. Technical report TCD-CS-2004-15, Computer Science Department, Trinity College Dublin, Ireland, April 2004
36. Zhao, X., Song, M., Liu, A., Wang, Y., Wang, T., Cao, J.: Data-driven temporal-spatial model for the prediction of AQI in Nanjing. J. Artif. Intell. Soft Comput. Res. **10**(4), 255–270 (2020). https://doi.org/10.2478/jaiscr-2020-0017
37. Zini, J.E., Rizk, Y., Awad, M.: An optimized parallel implementation of non-iteratively trained recurrent neural networks. J. Artif. Intell. Soft Comput. Res. **11**(1), 33–50 (2021). https://doi.org/10.2478/jaiscr-2021-0003
38. Zliobaite, I., Bifet, A., Pfahringer, B., Holmes, G.: Active learning with drifting streaming data. IEEE Trans. Neural Netw. Learn. Syst. **25**(1), 27–39 (2014)

A Proposal for Hybrid Memories Management Exploring Fuzzy-Based Page Migration Policy

Lizandro de Souza Oliveira(✉), Rodrigo Costa de Moura, Guilherme Bayer Schneider, Adenauer Correa Yamin, and Renata Hax Sander Reiser

Federal University of Pelotas (UFPEL), Gomes Carneiro, 1, Pelotas 96010-610, Brazil
{lsoliveira,rcmoura,gbschneider,adenauer,reiser}@inf.ufpel.edu.br
https://wp.ufpel.edu.br/computacao/ppgc/

Abstract. This work presents the *Intf-HybridMem* architecture, a proposal for page migration in hybrid memories using fuzzy systems to support decision making. The fuzzy approach is explored to model the uncertainties of the data access profile and the characteristics of the memory modules. Additionally, the *Intf-HybridMem* evaluation was carried out, identifying the limit of its accuracy when comparing with the Oracle mechanism.

Keywords: Fuzzy decision making · Decision support · Hybrid memory

1 Introduction

Main limitations concerned with static energy consumption and restriction on reducing the size of memory cells are related to the traditional memory applied on computational systems which are based on DRAM technology. In such context, several research efforts have been applied to the development of new memory technologies.

The non-volatile memories (NVM) emerge as an alternative to DRAM technology. NVM would be a promise to overcome DRAM restriction in order to achieve high memory capacity demand. In addition, they unable to achieve a lower energy consumption in actual use conditions. However, NVM also have some limitations, such as the low endurance of the material, which makes them up and the difficulty of reaching the same access rates like DRAM.

This work was supported by Brazilian Funding Agencies CAPES and CNPq; PQ Grants (310106/ 2016-8); PqG/FAPERGS 02/2017 (17/2551-0001207-0) and FAPERGS/CNPq 12/2014 - PRONEX (16/2551-0000488-9): Green-Cloud and Sustainable Computing.

L. Rutkowski et al. (Eds.): ICAISC 2021, LNAI 12854, pp. 347–357, 2021.
https://doi.org/10.1007/978-3-030-87986-0_31

Considering this limitation, NVM can be combined with DRAM in a new hybrid configuration exploring the best characteristics of each technology. Therefore, it is necessary to choose, at each moment, which memory module should be used. This decision must consider the data profile, such as frequency of access and the type of operations. Thus, algorithms are developed to manage data in hybrid memories to explore the memory modules better.

This work aims to the evaluation of the *Intf-HybridMem* architecture, which considers the interval-valued fuzzy sets as the logical support to decision making in the management of hybrid memories. This architecture seeks to explore the uncertainty resulting from the acquisition and storage of information about memory operations.

2 Preliminaries

In this section, studies of main characterization of updated memory technologies and foundations underlying interval-valued inference fuzzy systems are reported.

2.1 Memory Technologies

NVMs are classified according to their functional properties concerning to programming and erasing operations [14]. New techniques have considered the NVM architecture because of its advantages over traditional DRAM and SRAM memories. Moreover, there are emerging memories in the literature, such as PCM (Phase Change Memory Random-Access Memory), MRAM (Magnetoresistive RAM), STT-MRAM (Spin-Transfer Torque Magnetic RAM), RRAM (Resistive RAM), FRAM (Ferromagnetic RAM) and DWM (Domain Wall Memory).

Several types of memory optimizations have been focused on hardware or software optimizations, both also can be combined in new memory arquitectural alternatives. Several works take advantage of new memory technologies to replace DRAM as main memory and improve life time [23], performance [11], energy consumption and performance [29], also including works which turn their attention to energy optimization [26].

NVM technologies have been proposed as an alternative to mitigate the disadvantages of traditional DRAM and SRAM memories. In this scenario, hybrid architectures have been used to combine desirable characteristics of volatile and NVM. However, according to results on [3], there are significant performance issues caused by the interference data migration between DRAM and NVM and a the lack of effective migration policies. Researchers have proposed the use of NVM to address DRAM limitations, becoming relevant the data management in hybrid memories to deal with multiple parameters in volatile and NVM. In this case, it is meaningful to investigate related work of data management strategies in hybrid memories.

Several approaches are considering the hybrid memories see, e.g., DRAM and PCM [1], DRAM and a generic NVM [12] or combining STT-RAM and PCM [13]. Other works take advantage of cache approach to deal with the high cost of read/write operation in NVM and consequently to deal with this cost in hybrid memories that uses NVM ([7,16]). Moreover, Fuzzy Logic for decision support in cache memories is presented in ([5,18]).

2.2 Fuzzy Logic Foundations

The theory of Interval-valued fuzzy logic (IT2FL) arise in 1975 with different names and motivations, introduced by Zadeh in [28] as Interval-valued fuzzy sets, a particular case of type-2 fuzzy sets (T2FS) as well as other ones, which provided simultaneous but independent researches, as in [22] and [9] works. For further details see [4,19,24].

The inherent uncertainties related to the antecedent and consequent membership functions in the logical approach of IT2FL enable the manipulation of imprecise terms throughout its fuzzy inference system [10] and the imprecision (non-specificity) reflecting by the length of the interval membership degree. Such intervals can model uncertainty about the degrees, forms, or parameters of the membership functions. It also provides a potential strategy for the treatment of uncertainties in information models based on multiple-criteria obtained from distinct specialists and/or extracted from simulators.

Let $L([0,1])$ be the set of all closed real intervals in the unitary interval $[0,1]$ i.e., $L([0,1]) = \{[a,b] : 0 \leq a \leq b \leq 1\}$ and χ be a nonempty universe set. An interval-valued fuzzy set (IT2FS) A on χ is a function $A : \chi \rightarrow L([0,1])$ [19,21,24] such that, for each $X = [x_1, x_2] \in L([0,1]$, the projections $l_{L([0,1])}, r_{L([0,1])} : L([0,1]) \rightarrow U$ are defined as $l_{L([0,1])}(X) = l_{L([0,1])}([x_1,x_2]) = x_1$ and $r_{L([0,1])}(X) = r_{L([0,1])}([x_1,x_2]) = x_2$, respectively. And thus, the bounds of $X \in L([0,1])$ are $\underline{X}$ and $\overline{X}$, respectively. Among different relations on IT2FS, we consider the component wise *Kulisch-Miranker order*, also called the *product order* given as follows: $X \leq_{L([0,1])} Y \Leftrightarrow \underline{X} \leq \underline{Y} \wedge \overline{X} \leq \overline{Y}$, $\forall X, Y \in L([0,1])$.

Thus, $\mathbf{0} = [0,0] \leq_{L([0,1])} X \leq_{L([0,1])} \mathbf{1} = [1,1], \forall X \in L([0,1])$, and the set $L([0,1])$ equipped with the product order is characterized as a bounded lattice, taking $\mathbf{0}$ as its bottom, $\mathbf{1}$ as the top, and the infimum and supremum given as follows $X \wedge Y = [\min(\underline{X}, \underline{Y}), \min(\overline{X}, \overline{Y})]$ and $X \vee Y = [\max(\underline{X}, \underline{Y}), \max(\overline{X}, \overline{Y})]$, respectively. So,IT2FS are a particular case of T2FS [6,15]. In addition, the corresponding complement of an interval X is given as $X_C = [1 - \overline{A(x)}, 1 - \underline{A(x)}$.

In a system based on IT2FL, one can estimate input and output functions by using heuristic and interval techniques. In the following, its main blocks presented in Fig. 1 are described:

1 *Fuzzification Interface*, associating an input value with an interval function and not simply with a single value of χ. By inserting into the mechanism of inference, we model the uncertainty regarding the input membership function and, for each IT2FS $A(x)$, an input vector $\mathbf{x} = (x_1, x_2, \ldots, x_n) \in \chi^n$ when $n \in \mathbb{N}^*$ is related to a pair of vectors in $(L_n([0,1])^n$ given as follows:

$$\langle(\overline{A(x_1)}, \overline{A(x_2)}, \ldots, \overline{A(x_n)}), (\underline{A(x_1)}, \underline{A(x_2)}, \ldots, \underline{A(x_n)})\rangle.$$

2 *Rule Base (RB)*, composing rules to classify linguistic variables (LVs) according to the IT2FS;
3 *Logic Decision Unity*, executing inference operations between the input data and the rules defined in the RB to obtain performance by the system action;
4 *Defuzzification*: Considering two main stages of IT2FS:
(i) Type Reducer, selecting the best fuzzy set representing the IT2FS which satisfy the following premise: When all uncertainties disappear, the result of System Based on interval-valued fuzzy rules (SBFR2) is reduced to a System Based on Fuzzy Rules (SBFR1) [27];
(ii) CoA-Method, obtaining a crisp output as the average of limit points of IT2FS B, given as $1/2 \cdot \left(\underline{B(x)} + \overline{B(x)}\right), \forall x \in \chi$, applying KM algorithm.

3 *Intf-HybridMem*: Architecture Proposal

The *Intf-HybridMem* is an architecture proposed for the management of hybrid main memory systems using a fuzzy-based approach. This proposal is organized into two main components: (i) Hardware-Based Data Acquisition and a (ii) Fuzzy Based Migration Policy.

This proposal considers a hybrid memory composed of two memory modules: a DRAM and an NVM module. This is a general approach that can admit different sizes of memories and any type of NVM. Based on NVM constraints related to endurance and energy consumption, executions dealing with multiples write operations are not recommended. Thus, it is desirable to store on NVM only pages with a high read rate, and keep all other pages on DRAM.

So, the strategy of *Intf-HybridMem* for managing hybrid memories is by migrating pages between memory modules. Based on the access patterns of each page, the *Intf-HybridMem* can perform two types of page migration: (i) *Promotion*, as page migration from NVM to DRAM, and (ii) *Demotion*, as page migration from DRAM to NVM. For deciding which pages migrate each time, a fuzzy system is applied.

3.1 Hardware Based Data Acquisition

The *Intf-HybridMem* architecture (Fig. 1) integrates *Access Updater*, as a component (Hardware Core) for storing the page accesses in a buffer and also, periodically sending this information to *Intf-HybridMem*. Also, this module assigns to DMA the selected pages to be migrated.

Access Updater Module. This module is composed of three components: (i) a Buffer, (ii) a mechanism to update the Buffer, and an (iii) interface to periodically call *Intf-HybridMem* Migration Policy module. The Buffer preserves values from data access, meaning the volume of data read or write and the recency of each page access. Once memory access occurs, the Access Updater intercepts it updating the Buffer. This module stores the page address whenever

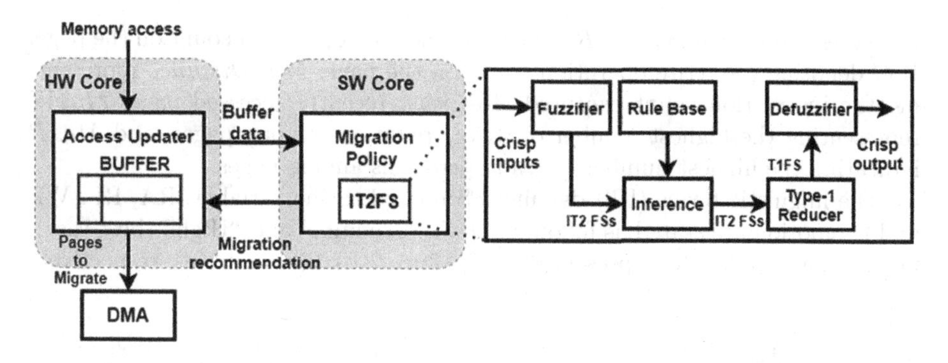

Fig. 1. The *Intf-HybridMem* architecture.

it is not already stored, increasing its read or write counters according to the type of memory accesses.

Access Updater requests to *Intf-HybridMem* Migration Policy (Software Core), located in the operating system, to recommend page migrations. Thus, the Access Updater sends all the information stored in the buffer to the *Intf-HybridMem* Migration Policy module and receives a list of migrations recommendations. Lastly, the Access Updater, based on these recommendations, assigns to Direct Memory Access (DMA) the migration of the pages in the memory.

3.2 Fuzzy Modeling of Migration Policy

Intf-HybridMem Migration Policy verifies the priority of each page be switched from one to another memory, taking into account a Rule Base acting on the three steps: Fuzzification, Inference, and Defuzzification. So, *Intf-HybridMem* returns as output the priority of each page, considering the Juzzy [25] module.

Membership Functions. During the study of variables considering the opinions of experts, each one of the linguistic variables was associated with four distinct FS, using the trapezoidal graphical representation to corresponding membership functions. A setting reading related to the memory environment is performed to measure attributes as Recency of Access (RA), Read Frequency (RF), and Write Frequency (WF). The reading values are applied to the standard scale, considering the interval $[0, 10]$ obtaining their membership degrees as follows:

$$RA = p_i(LA)/MaxDistance \tag{1}$$

$$RF = h_i(RC)/MaxR \tag{2}$$

$$WF = p_i(WC)/MaxW \tag{3}$$

Following parameters are considered: p_i denoting the $page(i)$ of the memory environment; LA indicating the executed memory instructions count after the last access on the same page; RC denoting the read operation count on the page; WC denoting the write operation count on the page; $MaxDistance$ representing the instruction count executed after least recently accessed page; $MaxW$ representing the highest number of write operations among pages; and $MaxR$ indicating the highest number of read operations among pages.

The linguistic terms (LT) defining FS related to the variables RA, RF, WF and Promotion are stated as follows: "Low", "Medium" and "High" (best case), in $[0, 10]$, as graphically represented in Fig. 2(a), 2(b), 2(c) and 2(d), respectively

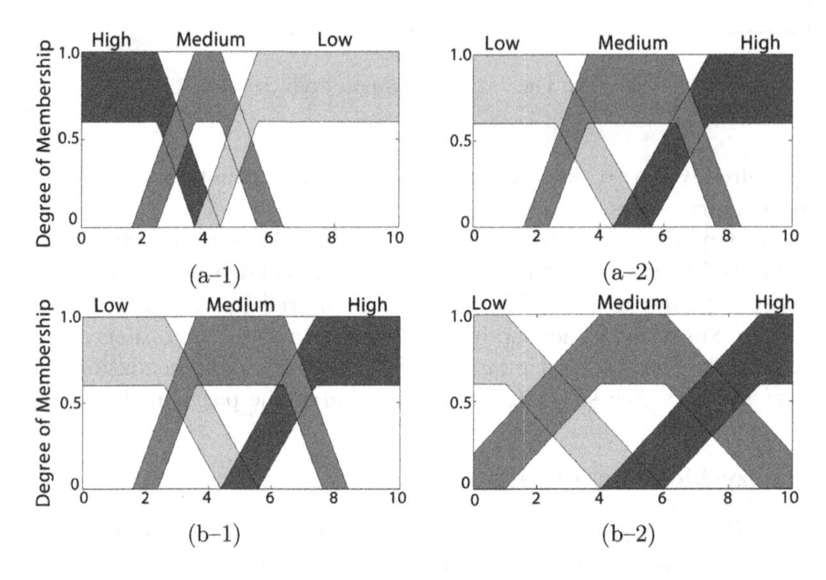

Fig. 2. Membership functions (MFs) of T2FS, representing linguistic values low, medium and high, of the (a) RA, (b) RF, (c) WF and (d) Promotion attributes.

Fuzzification. At this stage, the input values are mapped to the fuzzy domain.

Rule Base. In *Intf-HybridMem* Migration Policy component, the development of the RB, presented in Table 1, was based on the expertise of specialists. RB should be easily understandable and editable since there is no difficulty in adding new rules whether other input variables are desired to be manipulated. Three factors are considered in its construction: (i) LV as name FS, turning the modeling closer to the real-world system; (ii) Type "AND" connections are taken into account to create the relationship among the input variables; (iii) Type fuzzy implications performing the affirmative modus in inference scheme, related to the generalized modus ponens (GMP): "if X is A, then Y is B".

Table 1. Rule base of *Intf-HybridMem*.

Attr	Rule																										
	1	2	3	4	5	6	7	8	9	10	11	12	13	14	15	16	17	18	19	20	21	22	23	24	25	26	27
RA	L	L	L	L	L	L	L	L	L	M	M	M	M	M	M	M	M	M	H	H	H	H	H	H	H	H	H
RF	L	L	L	M	M	M	H	H	H	L	L	L	M	M	M	H	H	H	L	L	L	M	M	M	H	H	H
WF	L	M	H	L	M	H	L	M	H	L	M	H	L	M	H	L	M	H	L	M	H	L	M	H	L	M	H
PR	L	M	H	L	H	H	L	M	H	L	M	H	L	M	H	L	M	H	L	L	M	L	L	M	L	L	M

Inference. In the inference process, the composition operators are performed over FS relating the antecedents of rules to implications using the generalized modus ponens operator:
(i) It performs the application of fuzzy operators when input data consist of three values resulting from fuzzification, applying the fuzzy implication $I(x, y) = MAX(1 - x, y)$. The "AND" fuzzy operator aggregates the main rules and, the method MIN (minimum) returning the related fuzzification values;
(ii) Implication Fuzzy Method Application, performs a combination of the value obtained in the fuzzy operator applied and the values of FS output rule, using the method MIN (minimum) on these combinations;
(iii) Aggregation Fuzzy Method Application, resulting composition of the fuzzy output of each rule by using the method MAX (maximum), thus creating a single fuzzy region to be analyzed by the next Fuzzy process module.

Defuzzification. In this step, the region transformation results in a discrete value (related to promotion) applying the center of the area. This method calculates the centroid (u) of the area consisting of the output of the fuzzy inference system (connecting of all contribution rules) according to the following equation: $u = \sum_{i=1}^{N} u_i \mu_{OUT}(u_i) / \sum_{i=1}^{N} \mu_{OUT}(u_i)$.

4 *Intf-HybridMem*: Evaluation

In this section is discussed the accuracy evaluation of *Intf-HybridMem* architecture proposal. In this evaluation is considered: (i) the achieved number of reads and writes characteristic of each benchmark; (ii) Oracle Mechanism to define the best R/W ratio possible for a migration mechanism and (iii) the R/W ratio obtained by *Intf-HybridMem* proposed architecture.

4.1 Oracle Mechanism

Oracle is a mechanism which aims to make migrations based on the knowledge of future instructions. Thus, this mechanism can anticipate which pages will have more reads or more writes and positioning them in the appropriate memory module. Although a near-optimal approximation, the oracle can help to define the limit of the migration mechanism for hybrid architectures, mainly considering two phases: (i) Oracle Eye (OE); and (ii) Oracle Predictor (OP).

The first phase, OE, works offline, analyzing instruction intervals of memory accesses. For each interval, the oracle keeps in its memory the pages with the highest writes/reads (W/R) and reads/writes (R/W) ratio. The second part, OP, works online, performing the migrations based on the pages stored in the oracle's memory in the OE step. For each instruction interval, the OP reads the pages previously stored on the oracle's memory and performs the migration if the R/W and W/R ratios are greater than a threshold. Whether they are greater than the threshold, and if the pages are already in memory, then the pages can be migrated. Pages with the highest ratio of R/W are migrated from DRAM to NVM and those with the highest W/R ratio are migrated from NVM to DRAM. The algorithms and the complete evaluation of the Oracle were presented in [17].

4.2 Tests and Results

Tests performed to evaluate our architecture used an in-house simulator to model the hybrid main memory and the Access Updater. Based on the above migration policy, we implemented a fuzzy decision system based on interval-valued fuzzy sets, as described in Sect. 3.2. Considering input for the tests, traces of memory accesses were collected by running benchmarks from Mibench [8], over GEM5 [2] and NVMain simulators [20]. We selected a subset of evaluation benchmarks, considering their different natures: *basicmath*, *FFT*, *qsort* and *typeset*.

Since the main migration objective is to perform more read operations on NVM, the performance was measured by analyzing the R/W ratio on NVM. A high value for the R/W ratio on NVM means that the architecture was able to place the more read pages on NVM correctly. So, Fig. 3 shows the comparison between three R/W ratios: (i) The total, which represents the original proportion

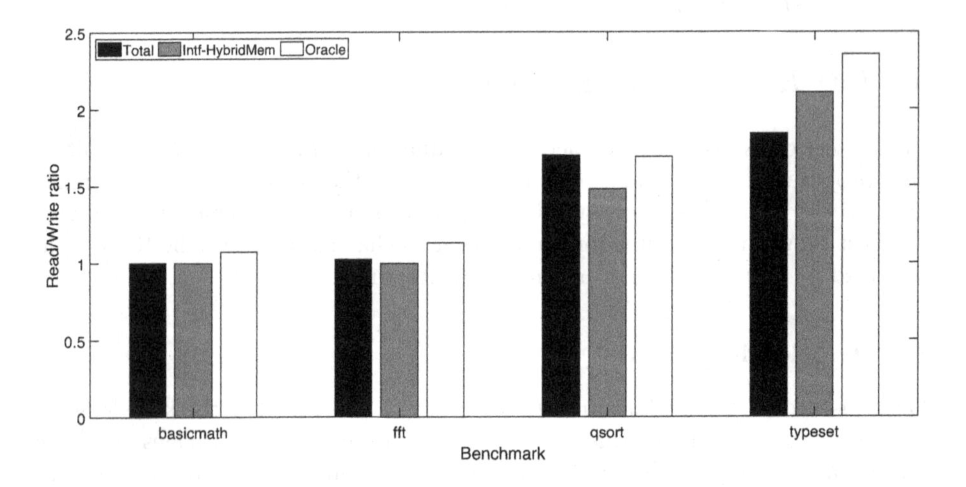

Fig. 3. Read/Write ratio comparison: Total & *Intf-HybridMem* & Oracle.

of reads and writes of each benchmark before the migration mechanism, (ii) the Fuzzy R/W ratio, which provides the result of the *Intf-HybridMem* Migration Policy, and (iii) the Oracle result.

When comparing the parameters, Total and Oracle in Fig. 3, it is possible to observe the potential of the benchmark to be explored by migration techniques. For the benchmarks *basicmath* and *FFT*, the Oracle results are slightly superior to the Total, representing a small potential for migration techniques. Regarding *qsort*, it presents no potential for migration techniques. The *typeset* benchmark was the one that presented the most potential.

Analyzing the results related to the Fuzzy and Total parameters, the *basicmath* presented no gain by comparing them. Since this benchmark has a small potential for migration techniques, this result is expected. For the benchmarks *FFT* and *qsort*, the results were inferior to Total, because of the low potential of the benchmarks and the cost added by the migration mechanism. Regarding *typeset*, due to the large potential of the benchmark, the Fuzzy approach could achieve 51.9% of its potential.

5 Conclusions

In this paper, the *Intf-HybridMem* data management proposal for hybrid memories based on fuzzy systems is presented. The proposal is conceived as two modules: the Hardware-Based Data Acquisition and the Fuzzy Based Migration Policy. Tests were conducted to evaluate the performance of the proposed migration policy, comparing it to the performance limit of the Oracle mechanism. Main results show that for the benchmark *typeset* it was possible to achieve 51.9% of the performance limit defined by the Oracle mechanism. For the benchmarks *basicmath*, *FFT* and *qsort*, which have low limit of performance, the Fuzzy approach could not achieve positive results.

It is shown that the proposed architecture for page migration in hybrid memories using fuzzy systems for decision making is an approach that still allows for improvements to the page migration technique in the *Intf-HybridMem*. As future work, we intend to make adjustments in fuzzy rules and on the membership functions to achieve better results even in the benchmarks with a low limit of performance. Ongoing work considers the fuzzy system qualification, by applying through the application of penalty function on consensus methods.

References

1. Aghaei Khouzani, H., Hosseini, F.S., Yang, C.: Segment and conflict aware page allocation and migration in dram-pcm hybrid main memory. IEEE Trans. on Comp.-Aided Des. Integr. Circuits Syst. **36**(9), 1458–1470 (2017)
2. Binkert, N., et al.: The gem5 simulator. ACM SIGARCH Comput. Arch. News **39**(2), 1–7 (2011)
3. Bock, S., Childers, B.R., Melhem, R., Mossé, D.: Characterizing the overhead of software-managed hybrid main memory. In: MASCOTS 2015, pp. 33–42 (October 2015)

4. Bustince, H., et al.: A historical account of types of fuzzy sets and their relationships. IEEE Trans. Fuzzy Syst. **24**(1), 179–194 (2016)
5. Diab, H., Kashani, A., Nasri, A.: Cache replacement engine: a fuzzy logic approach. In: International Conference on the Current Trends in Information Technology (CTIT), pp. 1–7 (December 2009)
6. Gehrke, M., Walker, C., Walker, E.: Some comments on interval valued fuzzy sets. Intl. J. Intell. Syst. **11**(10), 751–759 (1996)
7. Guo, Y., Xiao, W., Liu, Q., He, X.: A cost-effective and energy-efficient architecture for die-stacked dram/nvm memory systems. In: IEEE 37th International Performance Computing and Communications Conference (IPCCC), pp. 1–2 (November 2018)
8. Guthaus, M.R., Ringenberg, J.S., Ernst, D., Austin, T.M., Mudge, T., Brown, R.B.: Mibench: a free, commercially representative embedded benchmark suite. In: IEEE International Workshop on Workload Characterization, pp. 3–14. IEEE (2001)
9. Jahn, K.: Intervall-wertige mengen. Math. Nachr. **68**, 115–132 (1975)
10. Karnik, N.N., Mendel, J.M.: Introduction to type-2 fuzzy logic systems. In: 1998 IEEE International Conference on Fuzzy Systems Proceedings. IEEE World Congress on Computational Intelligence, vol. 2, pp. 915–920 (1998)
11. Li, B., Shan, S., Hu, Y., Li, X.: Partial-set: write speedup of pcm main memory. In: DATE 2014, pp. 1–4. IEEE (2014)
12. Liu, L., Yang, S., Peng, L., Li, X.: Hierarchical hybrid memory management in os for tiered memory systems. TPDS **30**(10), 2223–2236 (2019)
13. Maddah, R., Seyedzadeh, S.M., Melhem, R.: Cafo: cost aware flip optimization for asymmetric memories. In: HPCA 2015, pp. 320–330 (February 2015)
14. Meena, J.S., Sze, S.M., Chand, U., Tseng, T.Y.: Overview of emerging nonvolatile memory technologies. Nanoscale Res. Lett. **9**(1), 526 (2014)
15. Mendel, J.M., John, R.I., Liu, F.: Interval type-2 fuzzy logic systems made simple. IEEE Trans. Fuzzy Syst. **14**(6), 808–821 (2006)
16. Mittal, S., Vetter, J.S.: Ayush: a technique for extending lifetime of sram-nvm hybrid caches. IEEE Comp. Archit. Lett. **14**(2), 115–118 (2015)
17. de Moura, R.C., de Souza Oliveira, L., Schneider, G.B., Pilla, M.L., Yamin, A.C., Reiser, R.H.S.: Intf-hybridmem: page migration in hybrid memories considering cost efficiency. Sustain. Comput. Inf. Syst. **29**, 100466 (2021)
18. Niu, J., Xu, J., Xie, L.: Online fuzzy logic control with decision tree for improving hybrid cache performance symposia. In: 2016 12th IEEE International Conference on Control and Automation (ICCA), pp. 511–516 (June 2016)
19. Pękala, B.: Uncertainty Data in Interval-Valued Fuzzy Set Theory. SFSC, vol. 367. Springer, Cham (2019). https://doi.org/10.1007/978-3-319-93910-0
20. Poremba, M., Zhang, T., Xie, Y.: Nvmain 2.0: A user-friendly memory simulator to model (non-) volatile memory systems. IEEE Comput. Arch. Lett. **4**(2), 140–143 (2015)
21. Reiser, R.H.S., Bedregal, B.R.C., dos Reis, G.A.A.: Interval-valued fuzzy coimplications and related dual interval-valued conjugate functions. J. Comput. Syst. Sci. **80**(2), 410–425 (2014)
22. Sambuc, R.: Function ϕ-Flous, Application a l'aide au Diagnostic en Pathologie Thyroidienne. Ph.D. thesis, Univ. Marseille, Marseille, France (5 1975), thése de Doctorat en Médicine
23. Sampson, A., Nelson, J., Strauss, K., Ceze, L.: Approximate storage in solid-state memories. ACM Trans. Comput. Syst. **32**(3), 9 (2014)

24. Starczewski, J.T.: Advanced Concepts in Fuzzy Logic and Systems with Membership Uncertainty, Studies in Fuzziness and Soft Computing, vol. 284. Springer, Berlin (2013) https://doi.org/10.1007/978-3-642-29520-1
25. Wagner, C.: Juzzy - a java based toolkit for type-2 fuzzy logic. In: IEEE Symposium on Advances in Type-2 Fuzzy Logic Systems (T2FUZZ), pp. 45–52 (April 2013)
26. Wang, G., Guan, Y., Wang, Y., Shao, Z.: Energy-aware assignment and scheduling for hybrid main memory in embedded systems. Computing **98**(3), 279–301 (2016)
27. Wu, D., Nie, M.: Comparison and practical implementation of type-reduction algorithms for type-2 fuzzy sets and systems. In: FUZZ-IEEE, pp. 2131–2138. IEEE (2011)
28. Zadeh, L.: The concept of a linguistic variable and its application to approximate reasoning–i. Inf. Sci. **8**(3), 199–249 (1975)
29. Zhang, Z., Jia, Z., Liu, P., Ju, L.: Energy efficient real-time task scheduling for embedded systems with hybrid main memory. J. Sig. Process. Syst. **84**(1), 69–89 (2016)

A Novel Approach to Determining the Radius of the Neighborhood Required for the DBSCAN Algorithm

Artur Starczewski(✉)

Institute of Computational Intelligence, Częstochowa University of Technology, Al. Armii Krajowej 36, 42-200 Częstochowa, Poland
artur.starczewski@pcz.pl

Abstract. Data clustering is one of the most important methods used to discover naturally occurring structures in datasets. One of the most popular clustering algorithms is the Density-Based Spatial Clustering of Applications with Noise (DBSCAN). This algorithm can discover clusters of arbitrary shapes in datasets and thus it has been widely applied in many different applications. However, the DBSCAN requires two input parameters, i.e. the radius of the neighborhood (*eps*) and the minimum number of points required to form a dense region (*MinPts*). The right choice of the two parameters is a fundamental issue. In this paper, a new method is proposed to determine the radius parameter. In this approach the distances between each element in the dataset and its k-th nearest neighbor are used, and then in these distances abrupt changes in values are identified. The performance of the new approach has been demonstrated for several different datasets.

Keywords: Clustering algorithms · DBSCAN · Data mining

1 Introduction

Clustering refers to grouping objects into meaningful clusters so that the elements of a cluster are similar, whereas they are dissimilar in different clusters. Data clustering is a very useful technique used in many fields, such as biology, spatial data analysis, business, and others. Moreover, clustering methods can be used during the process of designing various neural networks [1,2], fuzzy and rule systems [7,9,20,29], and creating some algorithms for the identification of classes [12]. A variety of large collections of data brings a great challenge for clustering algorithms, so a lots of new different clustering algorithms and their configurations are being intensively developed, e.g. [11,13,14]. It should be noted that there is no clustering algorithm which creates the right clusters for all datasets. Moreover, the same algorithm can also produce different results depending on the input parameters applied. Therefore, cluster validation should be also used to assess the results of data clustering. So far, a number of authors have proposed different cluster validity indices or modifications of existing ones, e.g., [10,23,26,27]. Generally, clustering algorithms can be divided into four categories including partitioning, hierarchical, grid-based, and density-based clustering.

L. Rutkowski et al. (Eds.): ICAISC 2021, LNAI 12854, pp. 358–368, 2021.
https://doi.org/10.1007/978-3-030-87986-0_32

For example, the well-known partitioning algorithms are, e.g. *K-means*, *Partitioning Around Medoids* (*PAM*) [4,32] and *Expectation Maximization* (*EM*) [18], whereas the hierarchical clustering includes agglomerative and divisive approaches, e.g. the *Single-linkage*, *Complete-linkage* or *Average-linkage* or *DIvisive ANAlysis Clustering* (*DIANA*) [19,22]. Then, the grid-based approach includes methods such as e.g. the *Statistical Information Grid-based* (*STING*) or *Wavelet-based Clustering* (*WaveCluster*) [21,25,30]. The next category of clustering algorithms is the density-based approach. The *Density-Based Spatial Clustering of Application with Noise* (*DBSCAN*) is the most well-known density-based algorithm [8]. However, it is seldom used to cluster multi-dimensional data, but now the original DBSCAN has also many various extensions, e.g. [3,5,6,17,24,31]. This algorithm requires two input parameters, i.e. the *eps* and *MinPts*. Determination of these parameters is very difficult, but the right choice of those parameters is a fundamental issue. In literature, some methods have been proposed to determine these parameters, e.g. [16].

In this paper, a new approach to determining the *eps* parameter is proposed. It is based on an analysis of abrupt changes in the distances between each element of the dataset and its k-th nearest neighbor. This paper is organized as follows: Sect. 2 presents a detailed description of the *DBSCAN* clustering algorithm. In Sect. 3 the new method to determine the *eps* radius is outlined while Sect. 4 illustrates experimental results obtained on datasets. Finally, Sect. 5 presents conclusions.

2 The Description of the DBSCAN Algorithm

In this section, the basic concept of the *DBSCAN* algorithm is described. As mentioned above, it is a very popular algorithm because it can find clusters of arbitrary shapes and requires only two input parameters, i.e. the *eps* (the radius of the neighborhood) and the *MinPts* (the minimum number of points required to form a dense region). To understand the basic concept of the algorithm several terms should be explained. Let us denote a dataset by X, where point $p \in X$. The *eps* is usually determined by the user and the right choice of this parameter is a key issue for this algorithm. The *MinPts* is the minimal number of neighboring points belonging to a so-called *core point*.

Definition 1. The *eps-neighborhood* of point $p \in X$ is called $N_{eps}(p)$ and is defined as follows: $N_{eps}(p) = \{q \in X | dist(p,q) \leq eps\}$, where $dist(p,q)$ is a distance function between p and q.

When a number of points belonging to the *eps-neighborhood* of p is greater or equal to the *MinPts*, p is called the *core point*.

Definition 2. Point p is *directly density-reachable* from point q with respect to *epsilon* and the *MinPts* when q is a *core point* and p belongs to the *eps-neighborhood* of q.

When point p is *directly density-reachable* from point q and a number of points belonging to the *eps-neighborhood* of p is smaller than the *MinPts*, p is called a *border point*.

Definition 3. Point p is a *noise* if it is neither a *core point* nor a *border point*.

Definition 4. Point p is *density-reachable* from point q with respect to the *eps* and the *MinPts* when there is a chain of points $p_1, p_2, ..., p_n$, $p_1 = q$, $p_n = p$, so that p_{i+1} is *directly density-reachable* from p_i

Definition 5. Point p is *density-connected* to a point q with respect to the *eps* and the *MinPts* when there is point o, so that p and q are *density-reachable* from point o.

Definition 6. Cluster C with respect to the *eps* and the *MinPts* is a non-empty subset of X, where the following conditions are satisfied:

1. $\forall p,q$: if $p \in C$ and q is *density-reachable* from p with respect to the *eps* and the *MinPts*, then $q \in C$.
2. $\forall p,q \in C$: p is *density-connected* to q with respect to the *eps* and the *MinPts*.

The DBSCAN algorithm creates clusters according to Definition 6. At first, point p is selected randomly and if $|N_{eps}(p)| \geq MinPts$, then point p will be the *core point* and will be marked as a new cluster. Next, the new cluster is expanded by the points which are *density-reachable* from p. This process is repeated until no more cluster are found. On the other hand, if $|N_{eps}(p)| < MinPts$, then point p will be considered as a new *noise*. However, this point can be included in another cluster if it is *density-reachable* from some *core point*.

3 The New Approach to Determining the Radius of the Neighborhood

The right choice of the *eps* (the radius of the neighborhood) is a key issue for the right performance of the DBSCAN algorithm. It is a very difficult task and usually, a distance function is used to solve this problem. This distance function is denoted by the k_{dist} and it calculates distances between each element of the X dataset and its k-th nearest neighbor. The number of the nearest neighbors is the k parameter. For instance, Fig. 1 shows an example of a 2-dimensional dataset consisting of four clusters. The clusters contain 138, 119, 127, and 116 elements, respectively. First, the k_{dist} function is used to determine all distances between each element of the X dataset and its k-th

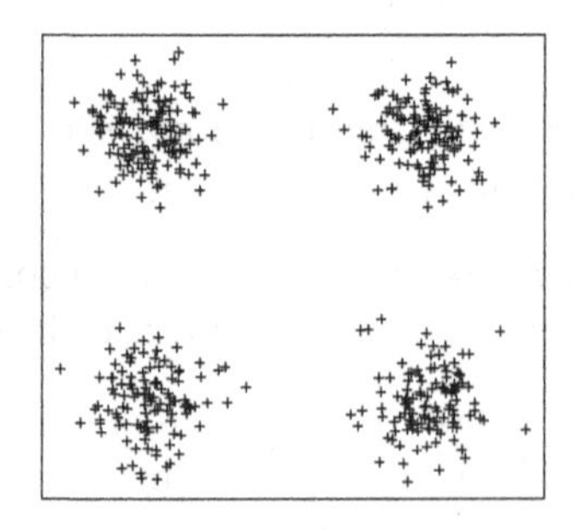

Fig. 1. An example of a 2-dimensional dataset consisting of four clusters.

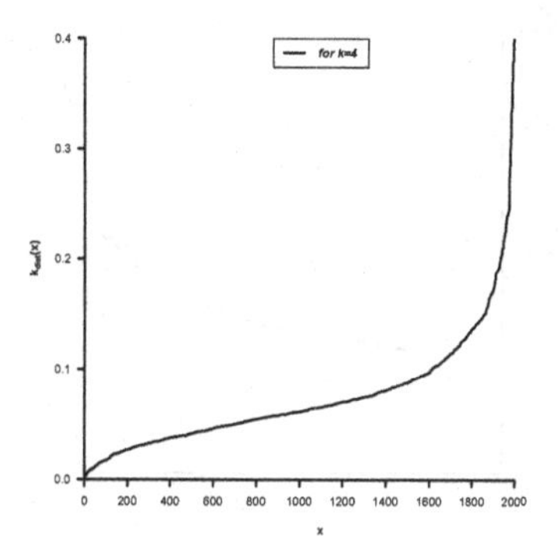

Fig. 2. Sorted values of the k_{dist} function with respect to $k = 4$ for the example dataset.

nearest neighbors. Next, the results are sorted in an ascending order. Figure 2 presents the sorted results for the k=4. It can be observed that there is a sharp change of the distances along the distance curves, i.e. values of the distances increase significantly. This place is called the "*knee*" and it can be used to determine the right value of the *eps* parameter of the DBSCAN algorithm. However, even when the "*knee*" is found correctly, the determination of the *eps* parameter is still difficult.

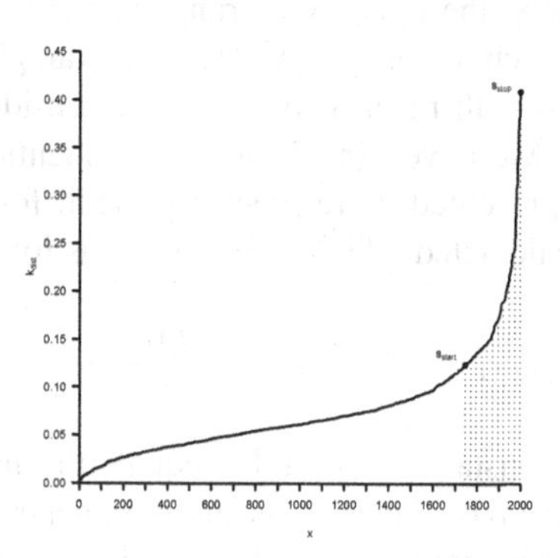

Fig. 3. Sorted values of the k_{dist} function with respect to $k = 4$ and ten identical intervals between the s_{start} and s_{stop} points.

A new approach is proposed to solve this problem. This method is a modification of the approach presented in the article [28] and it consists of a few steps. Let us denote a set of all the sorted values of k_{dist} function by S_{dist} for the X dataset. As mentioned above, the *eps* parameter depends on the "*knee*" occurring in the sorted distances (in an ascending order) for the given k parameter. It should be noted that in Fig. 2 the values of the k_{dist} function increase very abruptly when they are to the right of the "*knee*". This means that there are elements of the dataset located outside clusters and they can be interpreted as noise. Moreover, the "*knee*" usually appears at the end of the sorted values of the k_{dist} function and its size depends on the properties of the dataset.

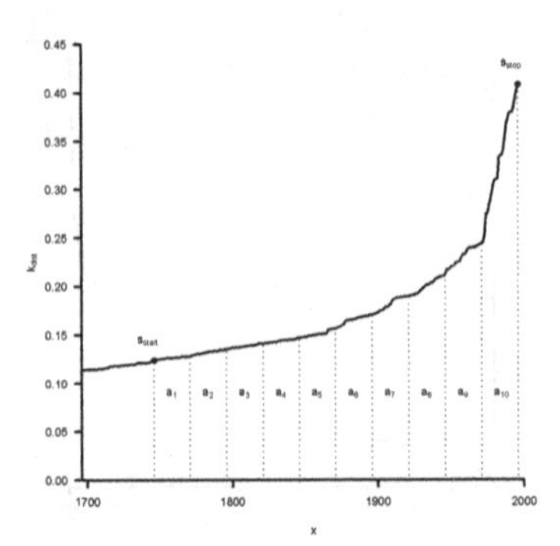

Fig. 4. Ten identical intervals between the s_{start} and s_{stop} points: $a1,...,a10$.

Among the sorted values, it is possible to define a range, which indicates the *knee* more precisely. It can be determined by the s_{start} and s_{stop} values which are calculated as follows:

$$\begin{aligned} s_{start} &= |S_{dist}| - |X| \\ s_{stop} &= |S_{dist}| \end{aligned} \tag{1}$$

where $|S_{dist}|$ is the number of the elements of S_{dist} and $|X|$ is the number of the elements of the X dataset. Furthermore, the $s_{stop} - s_{start}$ range is divided into ten equal parts, i.e.: $a1, a2, ... a10$. The size of such part is $n = |X|/10$. For example, Fig. 3 shows the sorted values of the k_{dist} function with respect to $k = 4$ and ten identical intervals between the s_{start} and s_{stop} points. Moreover, in Fig. 4 the ten identical intervals between the s_{start} and s_{stop} points are presented more precisely. Next, for the first seven intervals, the arithmetic means are calculated which is expressed as follows:

$$v_i = \frac{k_{dist}(x_i) + k_{dist}(x_i + n)}{2} \tag{2}$$

where $i = 1,...,7$, n is a constant value and it equals the number of $k_{dist}(x_i)$ values occurring in each part, i.e.: $a1,...,a10$. x_i is the parameter of $k_{dist}(x_i)$ function (see Fig. 4) and so x_1 equals the component x of the s_{start} point located at the start of the a_1 interval. Furthermore, x_2 is equal to $x_1 + n$, x_3 is equal to $x_2 + n$, and so on. However, for intervals $a8$, $a9$ and $a10$, the arithmetic means are calculated as follows:

$$v_j = \frac{k_{dist}(x_j) + k_{dist}(x_{10})}{2} \tag{3}$$

where $j = 8,9$ and 10. x_{10} equals component x of the s_{stop} point located at the end of the a_{10} interval. In Fig. 5 the average values of all the intervals are presented. These average values (see Eq. (2) and Eq. (3)) are used to calculate the *eps* parameter. It should be noted that the "*knee*" can be analyzed by these calculated average values. First, two factors $v_{7:1}$ and $v_{8:7}$ are computed and they can be expressed as follows:

$$v_{7:1} = \frac{v_7 - v_1}{2} \qquad v_{8:7} = \frac{v_8 - v_7}{2} \tag{4}$$

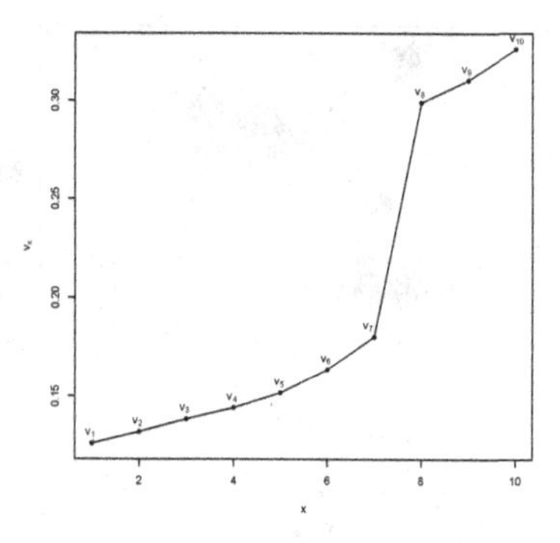

Fig. 5. Average values calculated for all intervals (i.e. $a1, ..., a10$).

These factors play a key role in the analysis of the "*knee*". For instance, when the values of k_{dist} increase very slowly in $a1, ..., a7$, average value $v_{7:1}$ does not change significantly, either. It means that the "*knee*" can be quite wide. Furthermore, if the values of the k_{dist} function increase abruptly in $a8, \ldots, a10$, then average values $v_{8:7}$ will have a large value. Thus, these factors can be used to calculate the *eps* parameter and it can be expressed as follows:

$$eps = v_{8-7} - v_{7-1} \quad (5)$$

It should be noted that v_{8-7} is close to the right value of the *eps* parameter. However, the v_{7-1} is very important because it shows how the distances increase in the "*knee*" and it is used to correct the value of v_{8-7} (see Eq. (5)).

In the next section, the results of the experimental study is presented to confirm the effectiveness of this new approach.

4 Experimental Study

In this section, several experiments have been conducted on 2-dimensional artificial datasets using the original *DBSCAN* algorithm. It should be noted that this algorithm can recognize clusters with arbitrary shapes. In these conducted experiments are used artificial datasets that include clusters of various sizes and shapes. Moreover, parameter k (i.e. $MinPts$) is equal to 4 in all the experiments and the visual inspection is used for the evaluation of the accuracy of the DBSCAN algorithm. The described new method is used to automatically determine the *eps* parameter.

It can be noted that the DBSCAN algorithm is rarely used to cluster multidimensional data due to the so-called "*curse of dimensionality*". However, different modifications of this algorithm have been proposed to solve the problem (e.g. [5]).

4.1 Datasets

Nine 2-dimensional datasets are used in the experiments and they are called *Data* 1, *Data* 2, *Data* 3, *Data* 4, *Data* 5, *Data* 6, *Data* 7, *Data* 8 and *Data* 9, respectively. Table 1

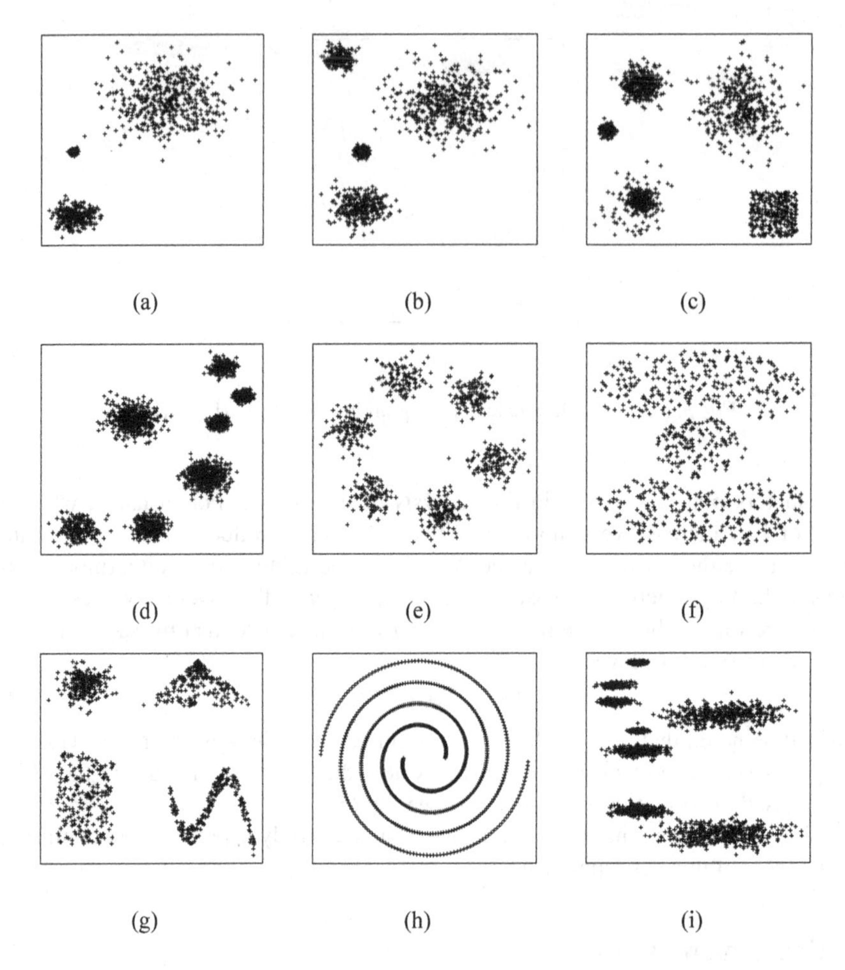

Fig. 6. Examples of 2-dimensional artificial datasets: (a) *Data* 1, (b) *Data* 2, (c) *Data* 3, (d) *Data* 4, (e) *Data* 5, (f) *Data* 6, (g) *Data* 7, (h) *Data* 8 and (i) *Data* 9.

shows a detailed description of these datasets. It should be noted that they contain varied numbers of elements and clusters (from 2 to 8 clusters). Moreover, the shapes and sizes of the clusters are also different. In Fig. 6, these datasets are presented. It can be observed that the distances between the clusters are very different, some of the clusters are very close and others quite far. For instance, in *Data* 3 the elements create five clusters with different sizes, *Data* 7 contains elements which create Gaussian, square, triangle, and wave shapes, and *Data* 8 is so-called the spirals problem, where points are on two entangled spirals.

Table 1. A detailed description of the artificial datasets

Datasets	No. of elements	Clusters
Data 1	1000	3
Data 2	1200	4
Data 3	1800	5
Data 4	2300	7
Data 5	700	6
Data 6	700	3
Data 7	900	4
Data 8	700	2
Data 9	2400	8

4.2 Experiments

In this section, the evaluation of the performance of the new method to automatically specify the eps parameter is presented. As mentioned above, the eps parameter is very important because the *DBSCAN* algorithm bases on this parameter to create the right clusters. It is usually determined by visual inspection of the sorted values of the k_{dist} function. On the other hand, the new method described in Sect. 3 allows us to determine this parameter in an automatic way. The second parameter of the *DBSCAN*, i.e. the $MinPts$ is also important but it is often chosen experimentally. In all the conducted experiments the $MinPts$ equals 4. Such a choice of the parameter guarantees the creation of various clusters with different numbers of elements. In these experiments the 2-dimensional artificial datasets are used, i.e.: *Data* 1, *Data* 2, *Data* 3, *Data* 4, *Data* 5, *Data* 6, *Data* 7, *Data* 8 and *Data* 9 sets. Thus, when the eps parameter is specified by the new method, the *DBSCAN* is used to cluster the artificial datasets. Figure 7 shows the results of the *DBSCAN* algorithm, where each cluster is marked with different signs. It should be noted that despite the fact that the differences of distances and shapes between clusters are significant, all the datasets are clustered correctly by the clustering algorithm. Moreover, the data elements classified as the *noise* are marked with a circle, and their number is small in all the datasets.

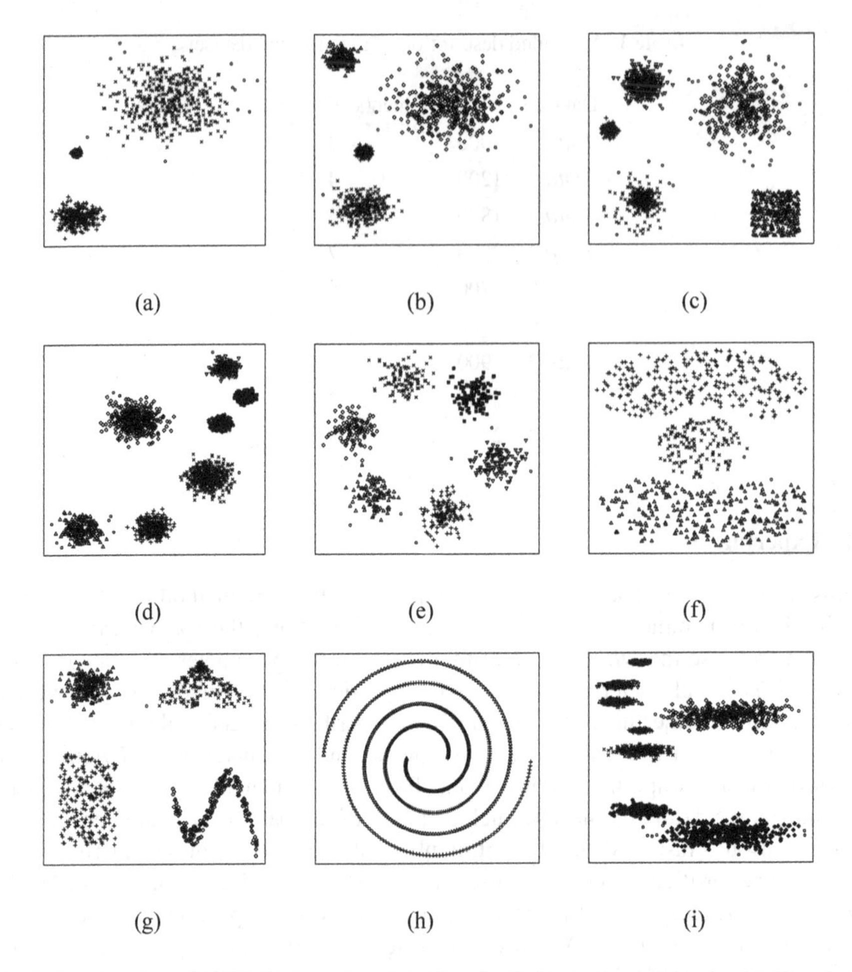

Fig. 7. Results of the *DBSCAN* clustering algorithm for 2-dimensional datasets: (a) *Data* 1, (b) *Data* 2, (c) *Data* 3, (d) *Data* 4, (e) *Data* 5, (f) *Data* 6, (g) *Data* 7, (h) *Data* 8 and (i) *Data* 9

5 Conclusions

In this paper, a new method is proposed to determine the *eps* parameter of the DBSCAN algorithm. This method bases on the k_{dist} function, which computes the distance between each element of a dataset and its *k*th nearest neighbor. Furthermore, the calculated distances are sorted in an ascending order to find out the *knee*. Next, the distances creating the *knee* are divided into several intervals and they are used to calculate the mean values (see Eq. (4)). This makes it possible to calculate the right value of the *eps* parameter. In the conducted experiments, several 2-dimensional datasets were used, where the number of clusters, sizes, and shapes varied within a wide range. From the perspective of the conducted experiments, this method for computing *eps* is very useful. All the presented results confirm a high efficiency of the newly proposed approach.

References

1. Bilski, J., Smoląg, J.: Parallel architectures for learning the RTRN and Elman dynamic neural networks. IEEE Trans. Parallel Distrib. Syst. **26**(9), 2561–2570 (2015)
2. Bilski, J., Kowalczyk, B., Marchlewska, A., Zurada, J.M.: Local levenberg-marquardt algorithm for learning feedforward neural networks. J. Artif. Intell. Soft Comput. Res. **10**(4), 299–316 (2020)
3. Boonchoo, T., Ao, X., Liu, Y., Zhao, W., He, Q.: Grid-based DBSCAN: indexing and inference. Pattern Recogn. **90**, 271–284 (2019)
4. Bradley P., Fayyad U.: Refining initial points for k-means clustering. In: Proceedings of the Fifteenth International Conference on Knowledge Discovery and Data Mining, New York, AAAI Press, pp. 9–15 (1998)
5. Chen, Y., Tang, S., Bouguila, N., Wanga, C., Du, J., Li, H.: A fast clustering algorithm based on pruning unnecessary distance computations in DBSCAN for high-dimensional data. Pattern Recogn. **83**, 375–387 (2018)
6. Darong H., Peng W.: Grid-based dbscan algorithm with referential parameters. Phys. Procedia **24**, Part B, 1166–1170 (2012)
7. Dziwiński, P., Bartczuk, Ł, Paszkowski, J.: A new auto adaptive fuzzy hybrid particle swarm optimization and genetic algorithm. J. Artif. Intell. Soft Comput. Res. **10**(2), 95–111 (2020)
8. Ester, M., Kriegel, H.P., Sander, J., Xu, X.: A density-based algorithm for discovering clusters in large spatial databases with noise. In: Proceeding of 2nd International Conference on Knowledge Discovery and Data Mining, pp. 226–231 (1996)
9. Ferdaus, M., Anavatti, S.G., Garratt, M.A., Pratama, M.: Development of C-means clustering based adaptive fuzzy controller for a flapping wing micro air vehicle. J. Artif. Intell. Soft Comput. Res. **9**(2), 99–109 (2019)
10. Fränti, P., Rezaei, M., Zhao, Q.: Centroid index: cluster level similarity measure. Pattern Recogn. **47**(9), 3034–3045 (2014)
11. Gabryel, M.: Data analysis algorithm for click fraud recognition. Commun. Comput. Inf. Sci. **920**, 437–446 (2018)
12. Gałkowski, T., Krzyak, A., Filutowicz, Z.: A new approach to detection of changes in multidimensional patterns. J. Artif. Intell. Soft Comput. Res. **10**(2), 125–136 (2020)
13. Grycuk, R., Najgebauer, P., Kordos, M., Scherer, M., Marchlewska, A.: Fast image index for database management engines. J. Artif. Intell. Soft Comput. Res. **10**(2), 113–123 (2020)
14. Hruschka E.R., de Castro L.N., Campello R.J.: Evolutionary algorithms for clustering gene-expression data, In: Fourth IEEE International Conference on Data Mining, 2004, ICDM 2004, pp. 403–406. IEEE (2004)
15. Jain, A., Dubes, R.: Algorithms for clustering data. Prentice-Hall, Englewood Cliffs (1988)
16. Karami, A., Johansson, R.: Choosing DBSCAN parameters automatically using differential evolution. Int. J. Comput. Appl. **91**, 1–11 (2014)
17. Luchi, D., Rodrigues, A.L., Varejao, F.M.: Sampling approaches for applying DBSCAN to large datasets. Pattern Recogn. Lett. **117**, 90–96 (2019)
18. Meng X., van Dyk D.: The EM algorithm - An old folk-song sung to a fast new tune. J. Royal Stat. Soc. Series B (Methodological) **59**(3), 511–567 (1997)
19. Murtagh, F.: A survey of recent advances in hierarchical clustering algorithms. Comput. J. **26**(4), 354–359 (1983)
20. Nowicki, R., Grzanek, K., Hayashi, Y.: Rough support vector machine for classification with interval and incomplete data. J. Artif. Intell. Soft Comput. Res. **10**(1), 47–56 (2020)
21. Patrikainen, A., Meila, M.: Comparing subspace clusterings. IEEE Trans. Knowl. Data Eng. **18**(7), 902–916 (2006)

22. Rohlf, F.: Single-link clustering algorithms. In: Krishnaiah, P.R., Kanal, L.N. (eds.), Handbook of Statistics, vol. 2, pp. 267–284 (1982)
23. Sameh, A.S., Asoke, K.N.: Development of assessment criteria for clustering algorithms. Pattern Anal. Appl. **12**(1), 79–98 (2009)
24. Shah, G.H.: An improved dbscan, a density based clustering algorithm with parameter selection for high dimensional data sets. In: Nirma University International Engineering, (NUiCONE), pp. 1–6 (2012)
25. Sheikholeslam, G., Chatterjee, S., Zhang, A.: WaveCluster: a wavelet-based clustering approach for spatial data in very large databases. Int. J. Very Large Data Bases **8**(3–4), 289–304 (2000)
26. Shieh, H.-L.: Robust validity index for a modified subtractive clustering algorithm. Appl. Soft Comput. **22**, 47–59 (2014)
27. Starczewski, A.: A new validity index for crisp clusters. Pattern Anal. Appl. **20**(3), 687–700 (2017)
28. Starczewski, A., Cader, A.: Determining the Eps parameter of the DBSCAN algorithm. In: Rutkowski, L., Scherer, R., Korytkowski, M., Pedrycz, W., Tadeusiewicz, R., Zurada, J.M. (eds.) ICAISC 2019. LNCS (LNAI), vol. 11509, pp. 420–430. Springer, Cham (2019). https://doi.org/10.1007/978-3-030-20915-5_38
29. Starczewski, J., Goetzen, P., Napoli, C.: Triangular fuzzy-rough set based fuzzification of fuzzy rule-based systems. J. Artif. Intell. Soft Comput. Res. **10**(4), 271–285 (2020)
30. Wang, W., Yang, J., Muntz, R.: STING: a statistical information grid approach to spatial data mining. In: Proceedings of the 23rd International Conference on Very Large Data Bases, VLDB 1997, pp. 186–195 (1997)
31. Viswanath, P., Suresh Babu, V.S.: Rough-dbscan: a fast hybrid density based clustering method for large data sets. Pattern Recogn. Lett. **30**(16), 1477–1488 (2009)
32. Zalik, K.R.: An efficient k-means clustering algorithm. Pattern Recogn. Lett. **29**(9), 1385–1391 (2008)

Evolutionary Algorithms and Their Applications

A Learning Automata-Based Approach to Lifetime Optimization in Wireless Sensor Networks

Jakub Gąsior(✉) and Franciszek Seredyński

Department of Mathematics and Natural Sciences,
Cardinal Stefan Wyszyński University, Warsaw, Poland
{j.gasior,f.seredynski}@uksw.edu.pl

Abstract. The paper examines the problem of lifetime optimization in Wireless Sensor Networks with an application of a distributed ϵ-Learning Automaton. The scheme aims to find a global activity schedule maximizing the network's lifetime while monitoring some target areas with a given measure of requested coverage ratio. The proposed algorithm possesses all the advantages of a localized algorithm, i.e., using only limited knowledge about neighbors, the ability to self-organize in such a way as to prolong the lifetime, and, at the same time, preserving the required coverage ratio of the target field. We present the preliminary results of an experimental study comparing the proposed solution with two centralized algorithms providing an exact (Integer Linear Programming (ILP)) and approximated solution (Genetic Algorithm (GA)) of the studied problem.

Keywords: Learning automata · Self-organization · Wireless Sensor Networks · Maximum lifetime coverage problem

1 Introduction

Wireless Sensor Networks (WSNs) combine a large number of tiny computer-communication devices called sensors deployed in some areas, which study a local environment, collect data depending on an application and send them via a particular node (called a sink) to an external world for further processing.

In many applications, e.g., monitoring remote and challenging access targets, sensors are equipped with single-use batteries that can not be recharged. From the point of view of Quality of Service (QoS) of such a WSNs, its operational lifetime is one of the most critical issues. There are two ways of deploying sensors to cover a target area completely, i.e., controlled deployment and random deployment. Several papers are studying optimal design patterns to assign the smallest number of sensors under the cost limitation in an area and satisfy the coverage and connectivity requirements.

However, in most cases, sensors are randomly distributed over the monitoring area in environments where human access is limited or impossible. Therefore,

L. Rutkowski et al. (Eds.): ICAISC 2021, LNAI 12854, pp. 371–380, 2021.
https://doi.org/10.1007/978-3-030-87986-0_33

batteries of sensors cannot be usually rechargeable or renewable. Such scenarios can be executed in deserts, forests, wilderness, mountain terrains, etc. Exhaustion of battery charge implies the change in the topology of WSN, quality of its work, and reduction of its lifetime. Therefore, energy-efficient management is an intrinsically important task in WSN.

With a fully connected WSN, the information about events sensed by each sensor node will be transferred to their destination (sink) in an energy-efficient multi-hop manner. Typically, sensors have four types of radio states: transmit, receive, idle, and sleep. We denote transmit, receive, and idle states as active states because each of these states consumes more energy than the sleep state.

When in an active mode, a sensor can carry out its entire operations, such as sensing, computation, and communication. To maintain these operations, sensors need to consume a relatively large amount of energy. In contrast, a sensor in a sleep mode uses only a tiny amount of energy and can be awoken in a scheduled working interval for entire operations. Since a subset of sensors in the area can already cover the target area completely, the other sensors can be scheduled to be in the sleep mode to save energy.

A group of sensors monitoring some areas is usually redundant, i.e., more than one sensor can cover monitored targets, and forms of redundancy can be different. Exploiting this redundancy in a WSN and finding out the possible scheduling sequence of sensors is crucial to maximizing the network's lifetime. By solving this coverage problem, one can indirectly also solve the maximization of the WSN lifetime. Therefore, scheduling schemes to properly alternate between active and sleep states, i.e., node wake-up scheduling protocols, are a promising method of maximizing the network lifetime.

This paper proposes an approach to the coverage/lifetime optimization based on a multi-agent interpretation of the problem. In the proposed scheduling mechanism, each sensor node is equipped with a learning automaton, which helps the node select its proper state (active or sleep) during the operation of the network.

The structure of the paper is as follows. Section 2 presents a review of the related work in the literature. Section 3 describes the problem of coverage/lifetime optimization in WSNs. In Sect. 4, we present the proposed approach of ϵ-LA optimization, and in Sect. 5, we detail the results of conducted experiments. The last section concludes the paper.

2 State of the Art

Several algorithms exist to solve the problem of coverage/lifetime maximization, in which sensing coverage and network connectivity are two fundamental issues. Centralized schemes assume the availability of complete information. The solution is usually delivered in the form of a schedule of activities of all sensors during the entire lifetime. Alternatively, in distributed schemes, a solution is found based on only partial information about the network. Because these problems are known to be NP-complete, centralized algorithms are oriented either on delivery of exact solutions for specific cases or applying heuristics and metaheuristics to find approximate solutions.

For example, in [5] authors proposed a solution for the problem of target coverage in a directional sensor network, where the sensor nodes can turn around their center. In the proposed algorithm, the sensor nodes are grouped in cover sets to avoid the redundant sensing direction, the obtained cover sets are joint, and they are actives alternatively for monitoring all targets.

An offline and centralized algorithm that allows the division of a set of sensor nodes into several subsets that are activated successively was proposed in [13]. Each subset monitors all targets. To evaluate the efficiency of the proposed algorithm, the authors compared the network lifetime obtained with the proposed solution to the maximum lifetime target coverage (*MLTC*) algorithm. Simulation results indicated that the network lifetime obtained by executing the proposed converges to *MLTC*.

Several nature-inspired algorithms applied to optimization problems in WSN have appeared in the literature in recent years. The two major groups of nature-inspired methods are swarm intelligence techniques, e.g., particle-swarm optimization [3,4,6], ant colony optimization [11]; and evolutionary computing including genetic algorithms [1,16] and memetic algorithms [7]. Another large class of scheduling methods is based on the concepts of learning [2,8,12] and cellular automata [14].

For example, in [9] authors applied a GA-based meta-heuristic to solve the target coverage problem presented as a maximum network lifetime problem (MLP). The proposed scheme groups the sensor nodes of the network into subsets covering all targets. The subsets are formed by selecting the minimum number of sensor nodes with the maximum remaining energy.

Closer to our work, in [10] authors proposed a scheduling method based on integrating a learning automaton in each sensor node, which helps to decide its state (sleep or active mode). They employed the Variable Structure Stochastic Automata (VSSA) in which the state transition probabilities are not fixed. The state transitions or the action probabilities themselves are updated at every time instant using a suitable scheme. Authors compared their solution with *MC-MIP* and *MCMCC* schemes achieving better results and fewer cover redundancies.

3 Sensor Networks and Coverage and Lifetime Problems

Let us consider a sensor network $S = \{s_1, s_2, ..., s_N\}$ consisting of N sensors randomly deployed over a two-dimensional rectangular area of size W $\times$ H m^2, where M targets (or Points of Interests (POIs)) are uniformly distributed with a step g. A sensor s_i can be defined as a point of coordinates (x_j, y_j) in two-dimensional area, with a given sensing range R_s^i and a non-rechargeable battery of capacity b_i. We assume that all sensor nodes possess the same sensing range and battery capacity.

Each sensor can work in one of two modes: an *active* mode when a battery is turned on, and a unit of its energy is consumed, and POIs in its sensing range are monitored; a *sleep* mode when a battery is turned off, and POIs in its sensing range are not monitored. Let us denote the mode of the i-th sensor during the

j-th time interval as α_i^j, where $\alpha_i^j \in \{0, 1\}$. The value of α_i^j equal to 1 means that the i-th sensor α_i during the j-th time interval is in active mode. Otherwise, it is in sleep mode.

It is further assumed that turning on/off batteries is taken in discrete moments t. It is also assumed that there exists some QoS measure evaluating the performance of WSN. As such, a measure of coverage can be defined as a ratio of POIs covered by active sensors to the whole number M of POIs.

In this paper, we focus on the problem of point coverage. A common definition of this problem is to cover (monitor) some stationary or moving target points in the sensor network area using as few sensor nodes as possible. The coverage requirement is given by a coverage degree k and a coverage ratio q, which means that at least k sensors cover at least a qth part of all targets. A point coverage problem with a k coverage degree is called a k-coverage problem in the literature. Further, we assume $k = 1$. Coverage of a target area at j-th period t_j can be denoted as $Coverage_j$:

$$Coverage_j = \frac{|M|_{obs_j}}{|M|}. \tag{1}$$

At a given moment, this ratio should not be lower than some predefined value of q $(0 < q \leq 1)$. Preserving complete area coverage is a desirable objective, but, sometimes, to achieve just a high coverage ratio may be of more practical interest.

A lifetime of WSN can be defined as a number of subsequent time intervals t_j in the schedule during which the coverage of the target area is within δ range from a given coverage ratio q, as follows:

$$Lifetime_q = \sum_{j=1}^{T_{max}} j|abs(Coverage_j - q) \leq \delta. \tag{2}$$

Naturally, a specific point within the target area may be concurrently sensed by several sensors. While this type of deployment can be beneficial in improving the quality or reliability of the data observed, this also introduces data redundancy, which results in wasted energy [15].

In this paper, we consider the Maximum Lifetime Coverage Problem (MLCP) as a scheduling problem applied to WSN, solving the point coverage problem regarding prolongation of the lifetimes of WSNs. Our objective is to prolong the lifetime of WSN by minimizing the number of redundant sensors during each time interval to minimize energy consumption.

4 Automata-Based Approach to the WSN Lifetime Optimization

In this section, firstly, we describe the concept of the Learning Automata (LA). Further, we propose an adapted ϵ-LA approach to solve the MLCP in WSNs.

An automaton is a self-operating mechanism that responds to a sequence of instructions in a certain way to achieve a particular goal. The automaton either responds to a predetermined set of rules or adapts to the environmental dynamics in which it operates.

The learning process is based on a learning loop involving the LA and random environment. A typical LA has a finite set of d actions and acts in a deterministic environment receiving reward $c = (c_1, c_2, ..., c_d)$, where c_k stands for a reward obtained for its action α_k. Whenever an automaton generates an action, the environment sends it a payoff in a deterministic way. The automaton remembers its last H actions and corresponding payoffs. The payoff (corresponding to the action) provided to the LA helps it choose the subsequent action.

The proposed scheme involving the LA mentioned above consists of two phases. During the learning phase, each node s_i randomly selects one of its actions (0 - *sleep* or 1 - *active*) and shares this information with n_i immediate neighbors (sensors sharing the same subset of POIs). Once action selection is completed, each node will receive some reward $rev_i()$, which depends on its decision and the decisions of its neighbors (see Eq. (3)) from the environment. This process repeats for H rounds, allowing us to fill the automatons' memory slots with corresponding action-payoff information. By repeating the above process, through a series of interactions, the LA finally attempts to learn the environment's optimal action. As the next action each sensor chooses the best (most profitable) action from the last H games (rounds) with the probability $1 - \epsilon$ $(0 < \epsilon \leq 1)$, and with probability $1/d$ any of its d actions. In our case $d = 2$ (0 - *sleep* or 1 - *active*). In the proposed method, the network operation phase lasts until all nodes with an active state run out of battery power.

In order to minimize the number of redundant active nodes covering the same subset of POIs, the following reward function was designed (Eq. (3)):

$$rev_i(\alpha_i, \alpha_{n_1}, \alpha_{n_2}, ..., \alpha_{n_i}) = \begin{cases} rev_i^{off+} & \text{if } (q^i_{curr} - q) \geq 0 \text{ and } \alpha_i = 0 \\ rev_i^{off-} & \text{if } (q^i_{curr} - q) < 0 \text{ and } \alpha_i = 0 \\ rev_i^{on} & \text{if } \alpha_i = 1 \end{cases} \tag{3}$$

where:

- $rev_i^{off+} = C_{rev}^{off+}$,
- $rev_i^{off-} = C_{rev}^{off-} - m^i(q - q^i_{curr})$,
- $rev_i^{on} = C_{rev}^{on}\left(C \times \frac{m^i}{m^i + \sum_i x_{n_i} \times m^i_{n_i}} + (1 - C)(1 - \frac{\sum_i x_{n_i}}{n_i + 1})\right)$,

where:

- $s_1, s_2, ..., s_n$ – a set of sensors;
- $\alpha_i, \alpha_{n_1}, \alpha_{n_2}, ..., \alpha_{n_i}$ – decisions of agent A_i and its neighbors;
- m^i – number of POIs in the sensing range of sensor s_i;
- m^i_0 – number of POIs in the sensing range of sensor s_i, not shared with neighboring sensors;
- $m^i_1, m^i_2, ..., m^i_{n_i}$ – number of POIs in the sensing range of sensor s_i, shared with the n_i-th active neighbor sensor;
- n_i – number of neighbors (sensors) of sensor s_i;

- x_{n_i} – number of sensor's s_i active neighbors (sensors sharing the same subset of POIs);
- q^i_{curr} – coverage of POIs in sensor's s_i local neighborhood;
- q – requested level of coverage;
- $C^{off-}_{rev} = 1.5, C^{off+}_{rev} = 1.25, C^{on}_{rev} = 1.0, C = 0.5$ – model constants.

We can see from Eq. (3) that an agent (automaton) A_i can receive a reward even if it is inactive state ($\alpha_i = 0$) and saves its own battery. It happens when some neighbor sensors are active Moreover, shared POIs are covered by them, and when some are not covered POIs does not exceed a specific threshold value related to a predefined coverage parameter q.

On the other hand, agent A_i receives a reward when it spends its battery energy, but this reward can be lowered when other neighbor sensors are also active and cover a shared subset of POIs. Model constants of the model were experimentally selected to ensure that sensors stay in their inactive (sleep) state whenever neighboring sensors cover a shared subset of POIs.

The purpose of each agent is to maximize its total reward, which corresponds to finding a local trade-off between the requested level of the coverage and expending battery power. This way of behavior of agents is in line with this work's primary goal: finding a global trade-off between the requested level of QoS and minimization of battery expenditure to maximize the lifetime of the WSN.

This process is visualized in Fig. 1. After the initial learning phase, each automaton selects the best action guaranteeing the highest payoff from the environment. Due to the formulation of the reward function, sensors self-organize in a way to ensure the requested level of coverage and minimal redundancy in target coverage. Once a group of sensors completely drain the battery, the network needs to reorganize in order to restore the requested level of coverage.

5 Experimental Study

In this section, we present the results of an experimental study of the proposed algorithm. The study was conducted in two steps. Firstly, several experiments were conducted in order to estimate the best values for the parameters of the ϵ-LA algorithm.

All the results in this section are based on averaging of 20 runs of different initial network states for five random sensor deployments. Sets of sensors composed of $N = \{10, 20, 30, 40\}$ were deployed over the target field containing a total of $M = \{5, 10, 20, 30\}$ targets. The requested level of POI coverage was set to a value of $q = 0.9$ with $\delta = 0.05$.

Figure 2(a) shows the relation between the averaged network lifetime and a length of ϵ-LA memory slot (H), while Fig. 2(b) show the effect of the ϵ parameter on the network's time of operation. As can be seen, the network lifetime increases with the length of automata memory up to a value of $H = 7$ and then starts to decrease slowly. In the case of the ϵ parameter, responsible for introducing

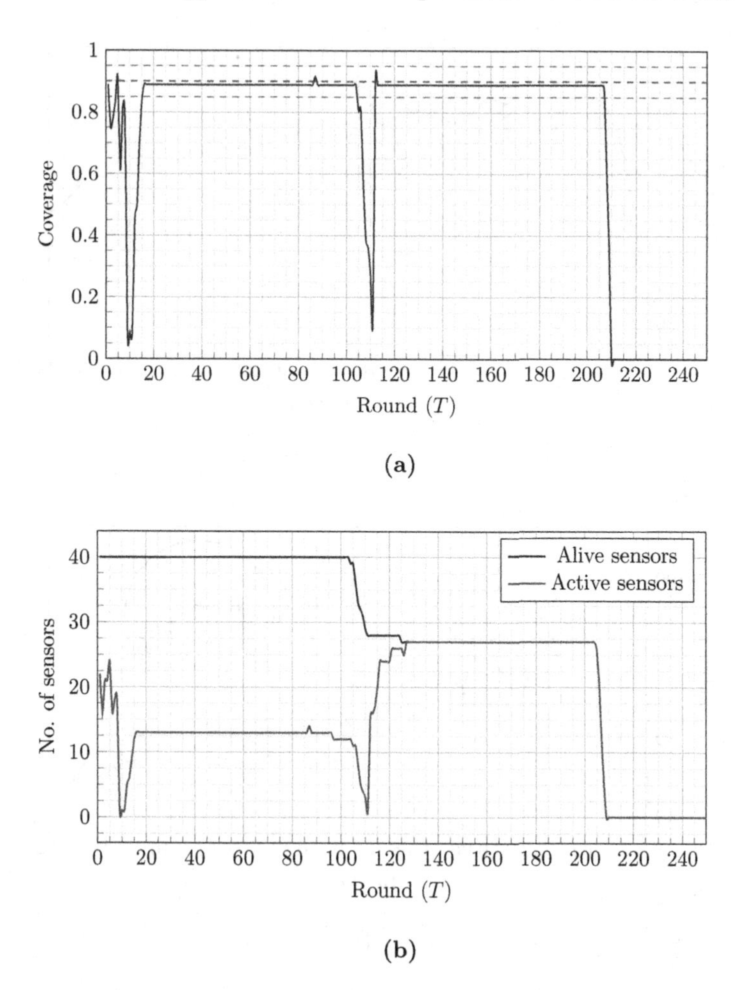

Fig. 1. Illustration of the LA-based WSN scheduling scheme consisting of a learning phase and an operation phase. Sample run for a total of $N = 40$ sensors deployed over the field containing $M = 25$ targets. The requested level of POI coverage was set to a value of $q = 0.9$ with $\delta = 0.05$: a) Target coverage vs. round, b) Number of active and alive sensors vs. round.

an element of randomness to the model, the best results can be achieved for relatively small values of ϵ. Based on these results, in the following experiments we used values of $H = 7$ and $\epsilon = 0.002$.

The next step was a comparison of the proposed decentralized, self-organizing approach with two centralized solutions employed in order to find a reference solution for the studied problem: the ILP algorithm to seek an exact solution; and the GA to reach an approximated solution in a reasonable time formulated as Non-Disjoint Set Covers (NDSC) maximization problem. The results of that comparison are presented in Table 1.

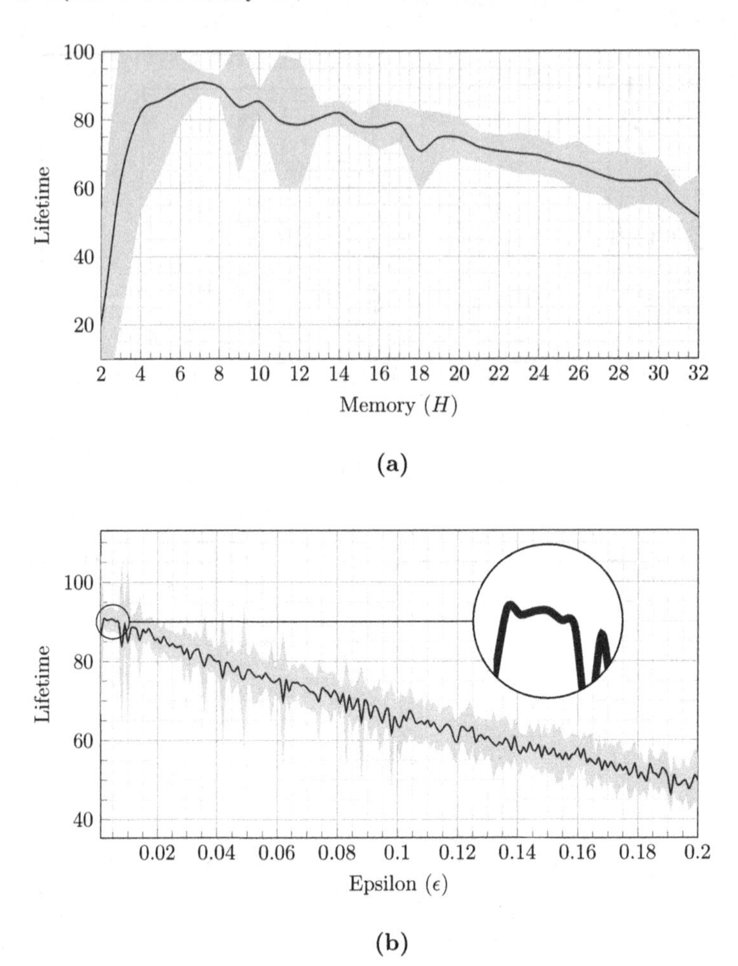

Fig. 2. Averaged results of the network lifetime vs H (a) and ϵ (b) for $N = \{10, 20, 30, 40\}$ sensors and $M = \{5, 10, 20, 30\}$ targets.

The main drawback of centralized algorithms is that a schedule of sensors' activities must be found outside the network and delivered before starting operation. It requires gathering information about the platform at a single location, which may be unrealistic for large-scale distributed systems, significantly when system parameters may continuously change.

As a result, it requires an arbitrarily large number of buffers and may induce huge latencies; it suffers from scalability issues, a single point of failure, and a lack of robustness. The latter is particularly relevant in sensor networks since sensor nodes are often deployed in hostile environments and are liable to failures.

Therefore, distributed algorithms are becoming more and more popular because they assume the reactivity of sensors in real–time and are scalable in contrast to centralized solutions. A distributed algorithmic framework enables

Table 1. Averaged results for network lifetime and solution convergence time for multiple WSNs computed with the ILP, GA and ϵ-LA schemes.

Problem instance	WSN lifetime (Convergence Time [s])		
	ILP	GA	ϵ-LA
$\{N = 10, M = 5\}$	22.5 [1.21]	22.2 [0.84]	21.4 [-]
$\{N = 10, M = 10\}$	19.3 [2.11]	19.1 [1.11]	18.5 [-]
$\{N = 20, M = 10\}$	34.2 [4.23]	33.7 [2.54]	31.2 [-]
$\{N = 20, M = 20\}$	36.4 [14.71]	35.2 [3.27]	32.8 [-]
$\{N = 30, M = 20\}$	40.8 [34.79]	39.7 [4.11]	37.5 [-]
$\{N = 30, M = 30\}$	46.8 [381.54]	45.8 [4.83]	42.6 [-]
$\{N = 40, M = 30\}$	56.7 [1453.38]	54.1 [5.75]	49.8 [-]

sensors to manage their sleep/activity cycles based on specific coverage goals. Thus, a localized algorithm solves the lifetime optimization problem by combining multiple local solutions directly, yielding a globally feasible working schedule. In our experiment, within a small WSN, the overall performance decrease was no worse than 8% in comparison with the exact solution found by the ILP method.

6 Conclusion

A localized algorithm based on the ϵ-LA concept to solve MLCP in WSN was proposed. It possesses all the advantages of a localized algorithm, i.e., it uses only limited knowledge about the neighboring sensors of the WSN, and it can self-reorganize in such a way as to preserve a required coverage ratio and prolong the lifetime of the WSN.

The experimental study results show that despite its simplicity and limited local information, it can achieve similar results (an average 8 % gap) as centralized algorithms in terms of lifetime metric without a need to compute a solution before starting the operation of the network.

Future work will include developing and studying additional localized functions based on local knowledge of neighboring nodes to find better solutions to the problem. Additional study of the relation between the experimental parameters (density of sensors and targets, ϵ-LA settings, variable range of sensors, and battery levels) and the achieved coverage and lifetime results will follow.

References

1. Charr, J., Deschinkel, K., Mansour, R.H., Hakem, M.: Optimizing the lifetime of heterogeneous sensor networks under coverage constraint : Milp and genetic based approaches. In: 2019 International Conference on Wireless and Mobile Computing, Networking and Communications (WiMob), pp. 1–6 (October 2019)

2. Gąsior, J., Seredyński, F., Hoffmann, R.: Towards self-organizing sensor networks: game-theoretic ϵ-learning automata-based approach. In: Mauri, G., El Yacoubi, S., Dennunzio, A., Nishinari, K., Manzoni, L. (eds.) ACRI 2018. LNCS, vol. 11115, pp. 125–136. Springer, Cham (2018). https://doi.org/10.1007/978-3-319-99813-8_11
3. He, X., Fu, X., Yang, Y.: Energy-efficient trajectory planning algorithm based on multi-objective pso for the mobile sink in wireless sensor networks. IEEE Access **7**, 176204–176217 (2019)
4. Jawad, H.M., et al.: Accurate empirical path-loss model based on particle swarm optimization for wireless sensor networks in smart agriculture. IEEE Sens. J. **20**(1), 552–561 (2020)
5. Jia, J., Dong, C., He, X., Li, D., Yu, Y.: Sensor scheduling for target coverage in directional sensor networks. Int. J. Distrib. Sens. Netw. **13**, 155014771771364 (2017)
6. Jiao, Z., Zhang, L., Xu, M., Cai, C., Xiong, J.: Coverage control algorithm-based adaptive particle swarm optimization and node sleeping in wireless multimedia sensor networks. IEEE Access **7**, 170096–170105 (2019)
7. Liao, C., Ting, C.: A novel integer-coded memetic algorithm for the set$k$-cover problem in wireless sensor networks. IEEE Trans. Cybern. **48**(8), 2245–2258 (2018)
8. Lin, Y., Wang, X., Hao, F., Wang, L., Zhang, L., Zhao, R.: An on-demand coverage based self-deployment algorithm for big data perception in mobile sensing networks. Future Gener. Comput. Syst. **82**, 220–234 (2018)
9. Manju, M., Chand, S., Kumar, B.: Genetic algorithm based meta-heuristic for target coverage problem. IET Wirel. Sens. Syst. **8**, 03 (2018)
10. Mostafaei, H., Meybodi, M.: Maximizing lifetime of target coverage in wireless sensor networks using learning automata. Wirel. Pers. Commun. **71**, 461–1477 (2013)
11. Rathee, M., Kumar, S., Gandomi, A.H., Dilip, K., Balusamy, B., Patan, R:. Ant colony optimization based quality of service aware energy balancing secure routing algorithm for wireless sensor networks. IEEE Trans. Eng. Manage. 1–13 (2019)
12. Razi, A., Hua, K.A., Majidi, A.: Nq-gpls: n-queen inspired gateway placement and learning automata-based gateway selection in wireless mesh network. In: Proceedings of the 15th ACM International Symposium MobiWaC 2017, pp. 41–44 (November 2017)
13. Saadi, N.: Maximum lifetime target coverage in wireless sensor networks. Wirel. Pers. Commun. 1–19 (2019)
14. Tretyakova, A., Seredynski, F., Bouvry, P.: Cellular automata approach to maximum lifetime coverage problem in wireless sensor networks. In: Wąs, J., Sirakoulis, G.C., Bandini, S. (eds.) ACRI 2014. LNCS, vol. 8751, pp. 437–446. Springer, Cham (2014). https://doi.org/10.1007/978-3-319-11520-7_45
15. Yetgin, H., Cheung, K.T.K., El-Hajjar, M., Hanzo, L.H.: A survey of network lifetime maximization techniques in wireless sensor networks. IEEE Commun. Surv. Tutorials **19**(2), 828–854 (2017)
16. Zhong, J., Huang, Z., Feng, L., Du, W., Li, Y.: A hyper-heuristic framework for lifetime maximization in wireless sensor networks with a mobile sink. IEEE/CAA J. Automatica Sinica **7**(1), 223–236 (2020)

Genetic Algorithms in Data Masking: Towards Privacy as a Service

Noel Hendrick[1] and Vicenç Torra[2,3](✉)

[1] Maynooth University, Maynooth, Ireland
Noel.hendrick.2017@mumail.ie
[2] Hamilton Institute, Department of Computer Science, Maynooth University, Maynooth, Ireland
[3] Department of Computing Sciences, Umeå University, Umeå, Sweden
vtorra@ieee.org

Abstract. Today's world is one where the number of publicly stored information and private data is growing exponentially, thus so is the need for more precise and more efficient data protection methods. Data privacy is the field that studies data protection methods as well as privacy models, tools and measures to establish when data is well protected and compliant with privacy requirements. Masking methods are used to perturb a database to permit data analysis while ensuring privacy.

This work provides a tool towards privacy as a service. Selecting an appropriate masking method and an appropriate parameterisation is an heuristic process. Our work makes use of genetic algorithms to find a good combination of masking methods and parameters. To do so, a number of solutions (masking methods, parameters) are applied and evaluated, the effectiveness of each solution is measured and well performing solutions are passed on to future generations. Effectiveness of a solution is in terms of information loss and disclosure risk.

Keywords: Data privacy · Disclosure risk · Information loss · Genetic algorithms

1 Introduction

The level of stored data is continuously growing. About 2.5 quintillion bytes of data were created each day in 2018 [19], and that number is doubling every 40 months [5]. This data includes private and confidential information.

Data privacy [10,14,26,27] is the field that studies techniques and methods so that disclosure of sensitive information does not take place. Different privacy models (computational definitions of privacy) have been defined. Privacy for reidentification and k-anonymity are two of them that focus on database releases and avoiding intruders to identify someone in the database. That is, they focus on what is known as identity disclosure. Other privacy models include differential privacy (to avoid disclosure from functions computed from databases), integral

L. Rutkowski et al. (Eds.): ICAISC 2021, LNAI 12854, pp. 381–391, 2021.
https://doi.org/10.1007/978-3-030-87986-0_34

privacy [23,24] (with a similar aim), secure multiparty computation (different organizations computing together a function of their different databases).

When a business or administration has a database and needs to share it (completely or parts of it) for its analysis, avoiding identity disclosure is a must. The difficulty of protecting data is that the removal or encryption of unique traits such as social security numbers and personal ID is not enough as the identity of individuals can be determined by combination of a few variables. For example, Sweeney [25] showed that gender, 5-digit zip code, and date of birth identified 87% of the population of the United States. Masking methods have been developed to provide solutions that avoid disclosure risk.

In the last 20 years, a large number of masking methods have been proposed. There are methods for all kind of databases (numerical, categorical) including NoSQL databases (graphs and social networks, textual documents). Informally, a masking method introduces some noise into the database with the goal that the resulting database protects the data and is still useful. That is, there is low risk and low information loss. There are a large number of methods in the literature because, up to now, there is no method that beats the other ones. While there are methods that are generally better (e.g., microaggregation [6,9] and rank swapping [20]), performance depends on the data to be protected. Therefore, in real practice, different methods and different parameterisations need to be considered and evaluated. So, evaluation needs to consider three elements:

- (i) the masking method itself (which method and with which parameters),
- (ii) the information loss a method causes into the data,
- (iii) the disclosure risk of the resulting data.

A good masking method is the one that results in a good trade-off between privacy and utility. Data masking can be a difficult process, trying to mask the data in such a way that there is a low risk of revelling confidential information while maintaining usability creates a large search space of possible masking techniques to use as well as possible parameters for these techniques. This is typically done heuristically and by an experienced data protector.

In this paper we propose the use of genetic algorithms to automate this process. By applying a genetic algorithm we can search this space in an efficient manner in order to find a solution that balances information loss and disclosure risk. So, genetic algorithms can help on building Privacy as a Service (PasS).

We use genetic algorithms (GA) in two different aspects related to data masking. First, on the selection and combination of masking methods (and their parameters). That is, we consider a database X and then its transformation by means of three masking methods ρ_1, ρ_2, and ρ_3 obtaining the protected database $X' = \rho_3(\rho_2(\rho_1(X)))$. Selection and parameterisation of ρ_i is achieved by means of GA. Second, we consider genetic algorithms to give an upper bound in disclosure risk assessment. GA are used to find the best parameters for an intruder, which will give an estimation of the upper bound of the risk.

1.1 Related Work

Genetic algorithms have been used in data privacy in two main contexts. First, in relation to the masking process. They have been used to mask data themselves as in [15,28], where the authors use genetic algorithms to find a good modification of the data. That is, each chromosome in the population represents a masked database. Second, genetic algorithms have been used in relation to some of the parameters of a masking method. For example, on finding a good parameter (a transition matrix) in PRAM [16–18], or a good partition of variables in microaggregation [3]. In contrast, in this work we use genetic algorithms to combine different masking methods and find their corresponding parameters.

We also consider genetic algorithms to find the optimal parameters for record linkage and, thus, to find an upper bound of disclosure risk. Up to our knowledge, genetic algorithms have not been used for this purpose.

1.2 Contribution

As stated above, data protection is currently heuristic and needs and experienced data protector. This work contributes in the field of data privacy providing a tool towards privacy as a service. The ultimate goal is to automatise the process of masking a file by means of selecting the best configuration (masking methods/parameters). Using genetic algorithms we can explore a larger search space than the one that can be considered by a single data protector. This includes more parameters as well as combination of methods. Our approach can outperform usual approaches at the cost of higher computational power.

We also introduce GA for disclosure risk assessment in data privacy. In this case, the goal is to obtain an estimation of the risk of the best performing intruder. To that end, we use a weighted distance and by means of genetic algorithms we optimise the weights so that the attack is as effective as possible.

1.3 Structure of the Paper

In Sect. 2 we introduce our approach and the metrics used. In Sect. 3 we describe how we validated the system and some of the experiments carried out. The paper finishes with some conclusions and lines for future work.

2 Approach and Metrics

As explained in the introduction, we have developed a system that, by means of genetic algorithms, finds a good trade-off disclosure risk-information loss. Therefore, the approach consists on protecting the data and then evaluating the resulting data set in terms of disclosure risk and information loss. In the remaining part of this section we review the data masking process and the metrics used for evaluating disclosure risk and information loss.

2.1 On Masking Data: The Genetic Algorithm Approach

Our approach is to use genetic algorithms to search through possible configurations of masking techniques in order to find a satisfactory one. Firstly the algorithm creates an initial random population of configurations for the masking techniques including the order in which they are applied. The given dataset is then masked using each configuration. The resulting masked dataset is then measured for disclosure risk (DR) and information loss (IL). These indices are combined to assign an overall assessment of the masked file, the fitness in GA.

A second generation is then created using the configurations that had the highest fitness from the first generation. This second generation is created using a combination of genetic operators. *Crossover* – selecting two parents that had a high fitness and then producing a child by combing their traits. *Mutation* – a configuration with a high scoring fitness is selected and then altered randomly to introduce new traits. Finally *asexual recombination/elitism* – highest scoring configurations immediately progress to the next generation. The fitness of the second generation is then assessed and the process is repeated.

Here, for possible configurations we understand the consecutive application of three masking methods. As briefly discussed in the introduction, our approach looks for a configuration of three good masking methods ρ_i and given the original file X computes the masked file $X' = \rho_3(\rho_2(\rho_1(X)))$.

Masking Methods and Configurations. For continuous data we use:

- Noise addition, which as the name suggests is the addition of noise to the dataset in the form [4] $X' = X + e$ where X is the original data and e is a randomly generated variable following a certain distribution;
- Microaggregation [6], which consists of building small micro clusters of records and replacing the original records with an average value representative of the entire cluster. The parameter that controls this method is the size of the cluster; finally
- Rank swapping, this consists of defining a rank of similar entries around each entry and then randomly choosing an entry within the rank to swap with [20,21]. The parameter that controls this masking method is the size of the rank.

The configurations for Continuous data contained 4 traits. That is, the chromosomes are defined by 4 elements. They are the three parameter of each of the three methods above and the order in which these methods are applied to the particular dataset. More particularly, these traits are as follows: a that defines the normal distribution $N(0, a)$ to be used to produce e in Noise addition, the minimum size k of micro clusters for Microaggregation, and p the percentage of the rank that is possible to swap an entry with in Rank swapping, and finally the order in which the masking techniques are to be applied.

For Categorical data, two masking techniques were utilised. Once again Microaggregation and a masking technique known as PRAM (multivariate Post-Randomization Method). PRAM consists of changing the categories on some categorical variables for certain records to different ones according to prescribed Markov matrices [7,12]. Each matrix contains the swapping probabilities for all the possible pairs of categories of a single variable. For PRAM, the parameter P is the probability that a category is not modified in the masking process. Then, if there are nc possible categories, the probability of swapping the original category by any other one is $(1 - P)/(nc - 1)$.

Configurations used for Categorical data contained 3 traits. They are the two parameters of the methods and the order in which these methods are applied. More particularly, we have the minimum number k of entries in micro clusters for Microaggregation, the value to generate the Markov matrix which prescribes the swapping probabilities for all possible categories, and finally the order in which the masking techniques are to be applied.

Parameters of the Genetic Operators. Implementation of the system is naturally by means of chromosomes that codify each configuration and represent a possible solution. There was 30 Chromosomes created for each generation. In the initial population all 30 were created randomly. All subsequent generations were created through the three genetic operators: Asexual Recombination, Mutation and Crossover. One third of the population was formed from Asexual Combination by picking the top 10 fittest Chromosomes from the previous generation and passing them onto the next generation. A further 33.3% of the population was formed by Mutation when the same top 10 fittest chromosomes were selected and randomly mutated (altered) in order to introduce new traits into the new population. Each chromosome had 4 traits associated with it and one was selected to be randomly changed. In this way each of the mutated chromosomes were altered by a factor of 0.25. The final 10 members (33.3%) of the generation was formed by means of Crossover. Two members of the previous population were selected according to a probability associated with their fitness (i.e. the fitter the chromosome then the higher the chance of being selected). As mentioned previously these two parent chromosomes each had four traits. For each trait there was an equal probability (0.5) it was selected from either parent and passed on to form a child chromosome.

2.2 On Computing Metrics to Evaluate Risk and Information Loss

In order to assess the masked data we need to take two measurements, disclosure risk and information loss. Disclosure risk (DR) is the probability that a record in the original file can be identified in the masked dataset. Information loss (IL) is the loss of structure and validity of the data from the original dataset. We measure slightly differently Continuous and Categorical data.

Disclosure Risk Assessment. In both continuous and categorical data disclosure risk is measured using *Distance based record linkage*. Distance is computed between every record in the original dataset and each record in the masked dataset using the Euclidean distance (for Continuous data) and common elements (for Categorical data). Given a masked record (i.e., a record in file X'), if the record (among those in the original file X) with the shortest distance is the original record then the record is said to be reidentified. The number of reidentified records over the total number of records gives a probability of disclosure risk between 0–1 [8,11,22]. In this work we use a weighted version of the Euclidean distance. That is,

$$d(x, x') = \sqrt{\sum_{i=1}^{m} w_i (x_i - x'_i)^2}$$

where x and x' are records in the original and masked files (i.e., $x \in X$ and $x' \in X'$) both in a m-dimensional space (where each dimension is one of the variable, attribute or feature) and w_i is the weight of the ith variable.

Our system attempts to adjust the weights attached to each variable in the Euclidean distance formula in order to find an optimum weighting that detects the maximum number of records. Selecting the optimum weights in this way, we try to find an upper bound of the risk. That is, the risk of intruders that are optimal in their attacks. There are different alternatives to find this optimum [1, 2]. In this project, we use genetic algorithms for this purpose. Using genetic algorithms does not ensure to find a global optimum of the solution but rather a local optimum. That is, we will find an estimation of the risk of the worst-case scenario but it is not necessarily the worst case because GA are suboptimal.

Information Loss Assessment. We consider two alternative approaches for assessing information loss. One is based on statistics. Different statistics are used for categorical and continuous data.

Information loss is measured in Categorical data through the comparison of contingency tables. A sample is taken from the original dataset as well as the masked dataset, contingency tables are then computed and a distance is computed using the number of differences between them. This is then denoted CTBIL (Contingency Table Based Information Loss measure) [8].

Information loss is measured for Continuous Data by calculating the Mean square error between the original dataset and the masked dataset. The Mean Square error measures the average squared difference between the values in original dataset and the masked dataset [8]. That is, $(1/n) \sum_{i=1}^{n} (x_i - x'_i)$ where n is the number of records, $x \in X$ and $x' \in X'$.

In addition to assessing information loss by means of statistics, we can also use machine learning techniques for the same purpose (the user of the system selects the appropriate measure). Machine learning-based information loss is computed by means of creating a model that predicts the values of a given dataset. More particularly, a model is created for the original dataset and then another model is created for the masked dataset. Both these models are compared to values in the original dataset to see how accurate there predictions are. The difference in accuracy between the Original model and the masked model is the measure of information loss [13]. This way of computing information loss is illustrated in Fig. 1.

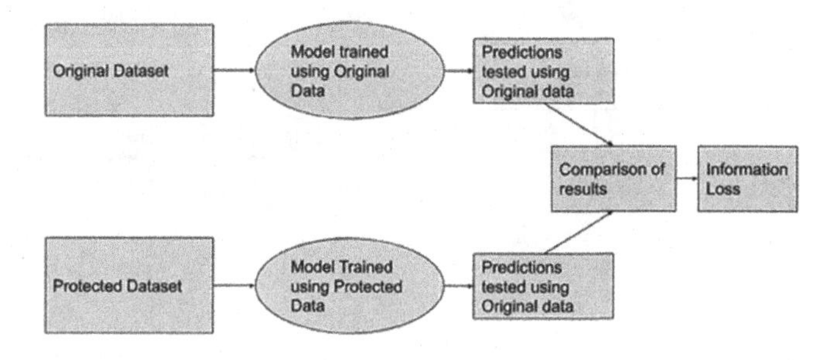

Fig. 1. Computation of information loss based on machine learning.

Quality Assessment and Fitness Function. In order to Evaluate the quality of each configuration the measurements are combined by adding them together and dividing by two. That is, we compute $(DL+IL)/2$ that gives a single fitness value that can be utilised by the genetic algorithm. Because there is a random factor to masking techniques each configuration masks the dataset three times and an average score of information loss and disclosure risk is calculated.

In order to then evaluate a population and understand how the genetic algorithm is progressing the program runs three tasks. It measures average population fitness, it tracks the fittest configuration in the population and graphs the information loss and disclosure risk values for every configuration in the population. By displaying this information it is possible to track progress and see if a satisfactory solution has been found.

Figure 2 represents the results of a generation. Each square represents a chromosome, that is, a protected database obtained applying some masking techniques. Each chromosome (each protected database) is evaluated against disclosure risk and information loss as explained above. In Fig. 2, the optimum solution is at (0,0) – i.e., no disclosure risk and no information loss – and we can see a good performing configuration by its position relative to (0,0) as the generations progress all the points move closer to (0,0) as their fitness improves.

3 Validation and Experiments

In this section we describe the experiments done to validate the system. We describe the datasets used, the experiments and some of the results obtained. The goal of the experiments was to validate the system and provide, given an original file to be masked, a good masked file. Good in terms of good trade-off between information loss and disclosure risk.

3.1 Datasets

We have used two major datasets in this project. One consisting of continuous data and another consisting of categorical data. Firstly the continuous Census dataset created in the European CASC project [29]. The dataset was obtained

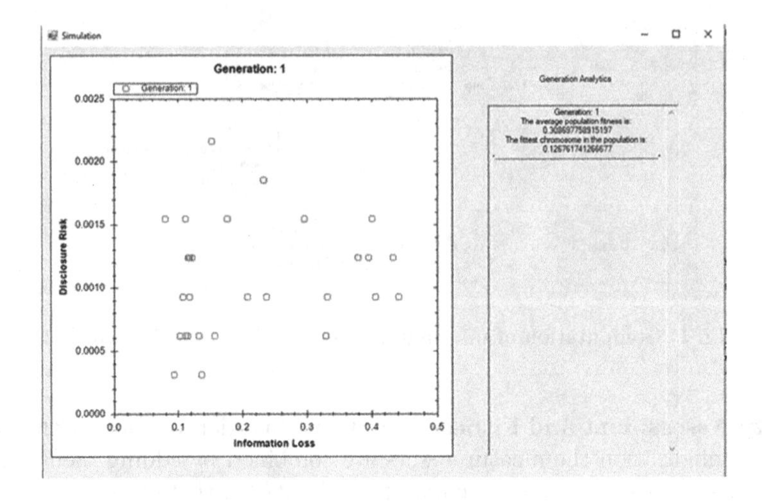

Fig. 2. Example of the result of a simulation.

using the Data Extraction System of the U.S. Census Bureau, it contains 149,642 records composed by 13 variables. A small section of it was used in these experiments consisting of 1081 rows and 13 variables for runtime purposes. For the categorical experiments a dataset based upon Car Evaluations was obtained from the UCI machine learning repository [30]. It contains 1728 rows each contains 6 variables and all rows were used in these experiments.

3.2 Experiments

As discussed previously the measures of success that are monitored are information loss, disclosure risk and the overall fitness value produced by the two.

Figure 3 (left) is an example of four generations running for numerical data (i.e., Census dataset) using the Mean Square Error as a measure of information loss. We can observe a quick drop from 0.2675 to 0.23126 in one generation and then to 0.18036 in Generation 3, after there is a decrease in population fitness by a small margin from Generation 3 to Generation 4. This highlights two key factors worth discussing. Firstly, the initial very large jump in fitness because the initial population is random while subsequent generations are selected from previous good solutions. Secondly, there is no guarantee that the average population fitness will improve from one generation to another (both stagnation and decrements are possible). What is important is the trend across multiple generations. This is the usual behaviour in genetic algorithms. Another key factor to note from the generation analytics is the progression of the fittest chromosome. Note how it takes a number of generations to incrementally increase. The fittest chromosome will always be much slower to increase than the average population fitness. However because of Asexual Recombination, where the top ten fittest chromosomes automatically progress to the next generation, the fittest

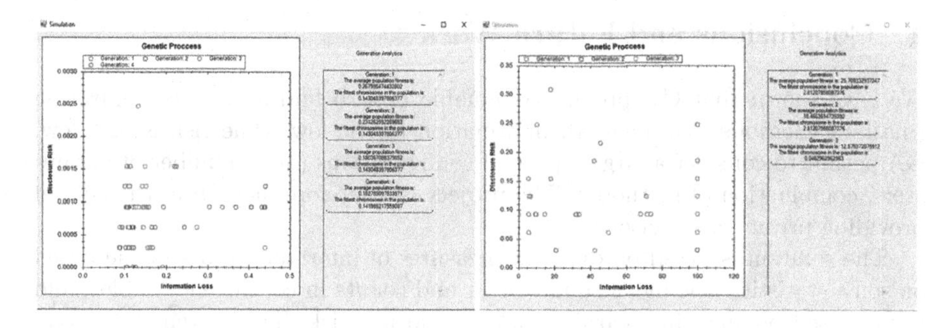

Fig. 3. Masking numerical data (left) using Mean Square Error as a measure of information loss, and (right) using machine learning as a measure of information loss.

chromosome is always guaranteed to be just as fit and never lower than the previous generation.

There are three more key factors worth discussing. Firstly dots in the graph seems to be arranged into rows with significant gaps on the Y–Axis. The Y-axis is the percentage of identified rows (and, thus, not real values are possible). Secondly, it seems that points are clumping together. Due to the use of Asexual Recombination 10 dots of the new generation are the same as previous generations and as such will appear directly on top of the dot from the previous generation. In addition, the Genetic Operator Mutation alters slightly the best performing 10 solutions in an attempt to produce a fitter chromosome. These new solutions are usually near to the one being modified. Finally there are a few dots that are complete outliers and are not close to any other dots. This is due to the Genetic Operator Sexual Recombination: two previous solutions are combined generating a completely different solution. In some cases these outliers provide a jump in fitness and become one of the fittest chromosomes. In this case, a cluster will then form around this dot as the generations progress. Nevertheless, often, these outliers are one of the least fit chromosomes and simply stay as stray dots in the graph.

In the case of categorical data, our system provides solutions that have a similar behaviour to the ones described above for numerical data. E.g., an initial big jump, slower improvement after a few iterations, clustered solutions, etc.

Figure 3 (right) is an example of Numerical data (i.e., Census dataset) using the Machine Learning as a measure of information loss. Here we use the weighted Euclidean distance, and information loss measures based on machine learning algorithms. It is relevant to underline that the use of GA to determine the weights of the Euclidean distance permits to improve in some cases the number of reidentifications. That is, we establish a larger disclosure risk than the one we could establish with just the standard distance. More particularly, in three datasets of 30 protected ones, more reidentifications were obtained than with the standard distance.

4 Conclusions and Future Work

We have shown that GA provide a valuable tool to find a good configuration (masking methods, parameters): information loss is low while risk is also low. GA permit to consider a larger space of configurations (large number of parameters, combination of methods). The project is, therefore, helpful in the goal of providing privacy as a service.

The solution is based on different measures of information loss (some based on some statistics, the mean square error and counts in contingency tables, and other based on machine learning) and disclosure risk. The system is general enough so that it is possible to incorporate other information loss measures (other types of statistics, other machine learning algorithms).

Acknowledgements. This work was designed and implemented as a final year project of the first author at Maynooth University. This work was partially supported by the Wallenberg AI, Autonomous Systems and Software Program (WASP) funded by the Knut and Alice Wallenberg Foundation.

References

1. Abril, D., Navarro-Arribas, G., Torra, V.: Improving record linkage with supervised learning for disclosure risk assessment. Inf. Fusion **13**(4), 274–284 (2012)
2. Abril, D., Navarro-Arribas, G., Torra, V.: Supervised learning using a symmetric bilinear form for record linkage. Inf. Fusion **26**, 144–153 (2015)
3. Balasch-Masoliver, J., Muntés-Mulero, V., Nin, J.: Using genetic algorithms for attribute grouping in multivariate microaggregation. Int. Data Anal. **18**, 819–836 (2014)
4. Brand, R.: Microdata protection through noise addition. In: Domingo-Ferrer, J. (ed.) Inference Control in Statistical Databases. LNCS, vol. 2316, pp. 97–116. Springer, Heidelberg (2002). https://doi.org/10.1007/3-540-47804-3_8
5. Brynjolfsson, A.M.: Big Data: The Management Revolution, The Harvard Business Review 4, (2012). Accessed March 12, 2020 https://wiki.uib.no/info310/images/4/4c/McAfeeBrynjolfsson2012-BigData-TheManagementRevolution-HBR.pdf
6. Defays, D., Nanopoulos, P.: Panels of enterprises and confidentiality: the small aggregates method. In: Proceedings of the 1992 Symposium on Design and Analysis of Longitudinal Surveys, Statistics Canada, pp. 195–204 (1993)
7. De Wolf, P.P., Van Gelder, I.: An empirical evaluation of PRAM. Discussion paper 04012. Statistics Netherlands, Voorburg/Heerlen (2004)
8. Domingo-Ferrer, J., Mateo-Sanz, J.M., Torra, V.: Comparing SDC methods for microdata on the basis of information loss and disclosure risk. In: Pre-proceedings of ETK-NTTS 2001, vol. 2, pp. 807–826 (2001) Eurostat
9. Domingo-Ferrer, J., Torra, V.: Ordinal, Continuous and heterogeneous k-anonymity through microaggregation. Data Min. Knowl. Disc. **11**(2), 195–212 (2005)
10. Duncan G.T., Elliot M., Salazar-González, J.J.: Why statistical confidentiality?. In: Statistical Confidentiality. Statistics for Social and Behavioral Sciences, pp. 1–26. Springer, New York (2011). https://doi.org/10.1007/978-1-4419-7802-8_1

11. Fellegi, I.P., Sunter, A.B.: A theory for record linkage. J. Am. Stat. Assoc. **64**(328), 1183–1210 (1969)
12. Gouweleeuw, J.M., Kooiman, P., Willenborg, L.C.R.J., De Wolf, P.-P.: Post randomisation for statistical disclosure control: theory and implementation. J. Official Stat. **14**(4), 463–478 (1998) Also as Research Paper No. 9731, Voorburg: Statistics Netherlands (1997)
13. Herranz, J., Matwin, S., Nin, J., Torra, V.: Classifying data from protected statistical datasets. Comput. Secur. **29**, 875–890 (2010)
14. Hundepool, A., et al.: Statistical Disclosure Control. Wiley, Hoboken (2012)
15. Jiménez, J., Marés, J., Torra, V.: An evolutionary approach to enhance data privacy. Soft Comput. **15**, 1301–1311 (2011)
16. Marés, J., Shlomo, N.: Data privacy using an evolutionary algorithm for invariant PRAM matrices. Comput. Stat. Data Anal. **79**, 1–13 (2014)
17. Marés, J., Torra, V., Shlomo, N.: Optimisation-based study of data privacy by using PRAM. In: Navarro-Arribas, G., Torra, V. (eds.) Advanced Research in Data Privacy. SCI, vol. 567, pp. 83–108. Springer, Cham (2015). https://doi.org/10.1007/978-3-319-09885-2_6
18. Marés, J., Torra, V.: An evolutionary algorithm to enhance multivariate post-randomization method (PRAM) protections. Inf. Sci. **278**, 344–356 (2014)
19. Marr, B.: How Much Data Do We Create Every Day? The Mind-Blowing Stats Everyone Should Read, Forbes (2018)
20. Moore, R.: Controlled data swapping techniques for masking public use microdata sets, U. S. Bureau of the Census (unpublished manuscript) (1996)
21. Nin, J., Herranz, J., Torra, V.: Rethinking rank swapping to decrease disclosure risk. Data Knowl. Eng. **64**(1), 346–364 (2007)
22. Pagliuca, D., Seri, G.: Some results of individual ranking method on the system of enterprise accounts annual survey, Esprit SDC Project, MI-3/D2 (1999)
23. Senavirathne, N., Torra, V.: Integral privacy compliant statistics computation. In: Pérez-Solà, C., Navarro-Arribas, G., Biryukov, A., Garcia-Alfaro, J. (eds.) DPM/CBT -2019. LNCS, vol. 11737, pp. 22–38. Springer, Cham (2019). https://doi.org/10.1007/978-3-030-31500-9_2
24. Senavirathne, N., Torra, V.: Integrally private model selection for decision trees. Comput. Secur. **83**, 167–181 (2019)
25. Sweeney, L.: Simple Demographics Often Identify People Uniquely, Carnegie Mellon University, Data Privacy Working Paper 3. Pittsburgh 2000 (1997)
26. Torra, V.: Data Privacy: Foundations, New Developments and the Big Data Challenge. SBD, vol. 28. Springer, Cham (2017). https://doi.org/10.1007/978-3-319-57358-8
27. Vaidya, J., Clifton, C.W., Zhu, Y.M.: Privacy Preserving Data Mining, Springer, Berlin (2006)
28. Vijayarani, S., Janakiram, M.: Genetic algorithm based confidential data protection in privacy preserving data mining. Int. J. Adv. Res. Comput. Commun. Eng. **5**(4), 158–164 (2016)
29. CASC Project. http://neon.vb.cbs.nl/casc. Accessed Jun 2020
30. UCI Machine Learning Repository. https://archive.ics.uci.edu/ml/datasets/Car+Evaluation Accessed Jan 2021

Transmission of Genetic Properties in Permutation Problems: Study of Lehmer Code and Inversion Table Encoding

Carine Khalil(✉) and Wahabou Abdou

LIB, Université de Bourgogne Franche-Comté, B.P. 47 870, 21078 Dijon, France
{carine.khalil,wahabou.abdou}@u-bourgogne.fr

Abstract. Solution encoding describes the way decision variables are represented. In the case of permutation problems, the classical encoding should ensure that there are no duplicates. During crossover operations, repairs may be carried out to correct or avoid repetitions. The use of indirect encoding aims to define bijections between the classical permutation and a different representation of the decision variables. These encodings are not sensitive to duplicates. However, they lead to a loss of genetic properties during crossbreeding. This paper proposes a study of the impact of this loss both in the space of decision variables and in that of fitness values. We consider two indirect encoding: the Lehmer code and the Inversion table.

Keywords: Genetic algorithm · Permutation problems · TSP · Encoding · Lehmer code · Inversion table

1 Introduction

Permutation-based optimization problems are widely studied in the literature because of their hardness and the diversity of their application fields. They are particularly used in the domain of network device deployment, scheduling or transportation. Solving such problems consists of finding a permutation that minimizes/maximizes some criteria.

Many efficient methods exist for solving permutation problems. This paper focuses on Genetic Algorithms (GAs) which are powerful stochastic optimization techniques. They are inspired by Darwin's theory of evolution and natural selection. GAs help with the exploration of a search space in order to find an optimal or a near optimal solution for a given problem. In GAs, a possible solution to the optimization problem is referred to as an individual. Generally, the algorithm starts with a randomly generated set of individuals (population). This population evolves throughout generations towards good solutions. At each generation of the genetic process, each individual in the population is evaluated based on objective function(s). This leads to the computation of a fitness value

L. Rutkowski et al. (Eds.): ICAISC 2021, LNAI 12854, pp. 392–401, 2021.
https://doi.org/10.1007/978-3-030-87986-0_35

which represents how good the solution is. Thereafter, individuals are selected to become parents according to their fitness values. Parents generate new individuals (offspring) using recombination operators, especially crossover and mutation operators. Offspring inherit some features from their parents. This process is iteratively repeated until a stop criterion is met.

When dealing with permutation problems, one should ensure that there are no duplicates in the permutation. This can be easily verified when generating the initial solutions. However, over generations, genetic operators such as crossover and mutation can duplicate alleles (values). There are mainly two approaches to avoid these repetitions. The first is to use mutation and crossover operators (e.g. OX, UX, or PMX [6]) which repair individuals containing duplicates. Another method consists of using a solution encoding (indirect encoding) which tolerates duplicates and defines a bijection between this indirect encoding and the permutation. The Lehmer code and the Inversion table are examples of indirect encoding. However, it should be noted that part of parents' genetic properties can be lost when generating offspring if the indirect encoding is used [9]. This paper studies the transmission of genetic properties. This study focuses on both decision variables and objective function domains. We consider a single objective traveling salesman problem (TSP) as an example of a permutation optimization problem. The TSP is a classical combinatorial optimization problem, where the goal is to find the shortest possible tour through a set of n vertices such that each vertex is visited exactly once except for the starting vertex.

The remainder of this paper is organized as follows. A brief overview of related works is introduced in Sect. 2. Section 3 describes direct and indirect encodings as well as their associated operators. Section 4 presents our experiments and results regarding the transmissions from parents to offspring and the fitness improvement. Finally, conclusions are given in Sect. 5.

2 Related Work

Initially proposed by Holland and Goldberg, genetic algorithms (GAs) have quickly evolved to solve multi-objective problems. Since Holland introduced the main mechanisms of GAs [7], many authors explained that choosing the most suitable representation/encoding is one of the major issues [1–5,8,9,11,12]. Among these authors, Goldberg [4] studied the behavior of representations and crossover-based GAs. He explained, in the schemata principle, how this representation - crossover combination should allow the transmission of meaningful building blocks from parents to offspring. Recently, Mohammed Ali et al.[9] show the impact of the choice of the couple encoding - crossover operator on the resolution of the permutation problems. They compare the characteristics related to the transmission of the properties between the parents and the offspring.

Djerid et al. [3] considered permutation encoding and crossover in a way similar to the classical schema theory [4]. They explain that encoding and crossover should be adapted according to which properties of the parents should be inherited by offspring. Pesko [12] presents an evolution algorithm for solving small (up to 32 nodes) constrained TSP. This new differential evolution algorithm with only two parameters (the population size and the number of the generations) uses the Lehmer code to encode solutions. In [11], Üçoluk proposes an inversion sequence as the representation of a permutation. This method is used for solving TSP and is compared to the well-known PMX crossover method. It is observed that Üçoluk's method outperforms PMX in convergence rate by a factor which can be as high as 11.1, on a cost of obtaining slightly worse solutions on average. Bekiroğlu [1] uses new alternatives of encoding types such as Quaternary encoding and octal encoding. He examined how they contribute to the efficiency and robustness of the genetic algorithm. He concluded that it is not possible to claim that one of the encoding types is exactly dominant over the others in all aspects such as convergence, finding the optimum solution, and the number of iteration. In [5], a GA is proposed to optimize the weight of steel truss structures. The obtained results proved the effectiveness of the genetic algorithm in relation to the classical genetic algorithm. In this case, the set of design variables consists of the collection of profiles manufactured in steel mills. Obviously, this set of profiles is discrete. The most effective type of encoding in such case is value encoding.

In [14], Rosa et al. discuss the teachers' placement (in elementary school) problem based on genetic algorithms by finding a chromosome that represents the possibility of teachers placement solution, composing a population, and finding the recommended combination of two selected mutations operators and two selected crossover operators to achieve optimal results. Another work that studies the choice in selection, crossover, and mutation operators and their impact on the performance of a genetic algorithm is [10]. The authors present a novel framework for an adaptive and modular genetic algorithm (AMGA) to discover the optimal combination of the operators in each stage of the GA to avoid premature convergence.

This paper completes the previous works. It focuses on the impact of encoding solutions thanks to Lehmer code and inversion table. It studies how genetic characteristics are inherited from parents.

3 Encoding and Recombination Operators

Encoding refers to the way decision variables are represented. When using genetic algorithms, it allows to define the genotype in the optimization process. One can use a traditional encoding (which is also referred to as direct encoding) or an indirect representation of the decision variables (indirect encoding). The choice of the encoding should take into consideration the problem and the operators to use (crossover and mutation). This mapping between the set of permutations and the set of their encoding could be used to translate any problem represented by permutations into an equivalent problem represented by another code. This may simplify some of those problems.

3.1 Classical Permutation

A permutation $\pi = \pi_1\pi_2\ldots\pi_n$ is an arrangement of the numbers $1, 2, \ldots, n$ for some positive n. Each permutation is represented by a unique number which represents its order among the $n!$ possible permutations enumerated following a lexicographic order. The classical permutation allows an easy representation of solutions. However, it can generate duplicates during the crossover and mutation steps. To deal with this issue, specific operators are defined in the literature. We can list PMX, OX, UX for the crossover, and Swap, Scramble, and Inversion for the mutation [6]. Quite often, these operators repair to non-viable solutions (those with duplicates). To avoid these repair phases, it is possible to use indirect encodings which are not sensitive to duplicates. In the following, we define two indirect encoding methods: the Lehmer code and the Inversion table.

3.2 Lehmer Code

The Lehmer code associates a unique code $L(\pi)$ to each permutation $\pi = \pi_1\pi_2\ldots\pi_n$. $L(\pi) = l_1l_2\ldots l_n$ with l_i the number of elements that are smaller than π_i and that appear to the right of π_i in the permutation.

$$l_i = Card\{j|j > i \quad \& \quad \pi_j < \pi_i\} \tag{1}$$

For example, the Lehmer code of the permutation "5 2 1 4 3" is "4 1 0 1 0". The elements l_i in the Lehmer code satisfies the condition $0 \leq l_i \leq n - i$, $\forall i$.

3.3 Inversion Table

The inversion table of a permutation $\pi = \pi_1\pi_2\ldots\pi_n$ is $T(\pi) = t_1t_2\ldots t_n$ with t_i the number of elements that are greater than i and appearing to the left of i in the permutation.

$$t_i = Card\{j|i < j \quad \& \quad j < \pi_i\} \tag{2}$$

For example the inversion table of the permutation "5 2 1 4 3" is "2 1 2 1 0". Here $t_3 = 2$ because there are two elements (5 and 2) that are greater than 1 and placed at lower positions in the permutation (to the left of 1). By definition, t_i in the Inversion table always satisfy the condition $0 \leq t_i \leq n - i$, $\forall i$.

The use of an indirect representation requires coding and decoding operations. Figure 1 illustrates these steps. First, the parents are represented as a classic permutation (Fig. 1(a)). Then, they are coded using the Lehmer code (Fig. 1(b)) before applying the crossover operator and generating offspring (Fig. 1(c)). These offspring are then decoded to be presented as a classical permutation (Fig. 1(d)). The process is the same for the Inversion table (see Fig. 2(a) to 2(d)). It is important to note here that the offspring illustrated in Fig. 1(d) (or Fig. 2(d)) have alleles that are not inherited from their parents. For example, for the offspring of Fig. 1(d), the sequences "4 1 3" and "1 3 2" are not inherited from any parent. This problem can be rephrased as: the sequences

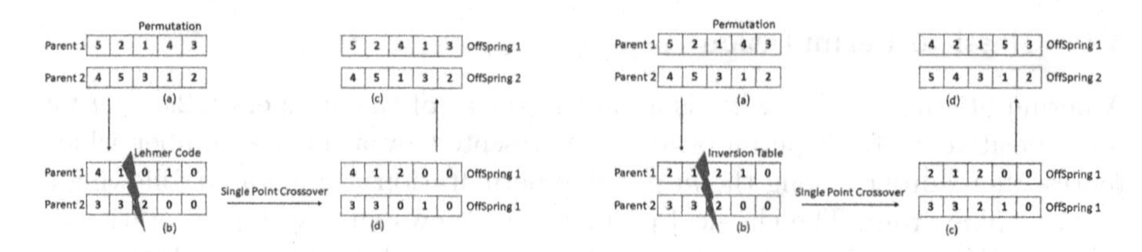

Fig. 1. Lehmer code

Fig. 2. Inversion table

"1 4 3" and "3 1 2" of the parents (Fig. 1(a)) are lost during the recombination process (without a mutation having been carried out). This paper focuses on studying the impact of the loss of these genetic characteristics over generations.

4 Experiments and Results

Our goal is, first, to study the transmission of characteristics from parents to offspring. For this, we use metrics that focus on the similarities between solutions (Sect. 4.1). Then, we evaluate the quality of the results of each encoding with respect to the objective function (Sect. 4.2). For this study, we considered the `eil51` problem (a traveling salesman problem instance of 51 cities proposed in the TSPLIB [15]). The experiment parameters used are summarized in Table 1.

The results presented in this section correspond to the averages of 10 independent runs.

Table 1. Experiment parameters

Encoding	Crossover	Mutation	Population size	Nb. of generations
Classical Permutation	PMX	Swap	200	300
Lehmer code	2-point	BitFlip		
Inversion table	2-point	BitFlip		

4.1 Assessment of Transmissions from Parents to Offspring

The goal is to evaluate the transmission of the genetic characteristics from parents to offspring in order to observe the impact of the encoding. For this purpose, several indicators are used: the Hamming distance, the edge based indicator and the position based indicator.

Hamming Distance. Measures the similarity between two solutions with respect to the decision variables. Given two permutations say π and π', the Hamming distance $HD(\pi, \pi')$ is equal to the number of positions in which π differs from π' (Eq. 3). It can be used to assess diversity.

$$HD(\pi, \pi') = \sum_{i=1}^{n} x_i \qquad where \qquad x_i = \begin{cases} 1\, if \quad \pi_i \neq \pi'_i \\ 0\, otherwise \end{cases} \tag{3}$$

Figure 3 shows the performance of the Hamming distance between parents and offspring over generations. It considers the maximum value of the Hamming distance between each solution and its two parents. We observe that this metric gradually decreases over the generations because the genetic algorithm will tend to exploit the neighborhood of the best solutions. However, due to the appearance of new alleles (as shown in Fig. 1 and 2), the Lehmer code and the Inversion table show higher HD values than the classical permutation.

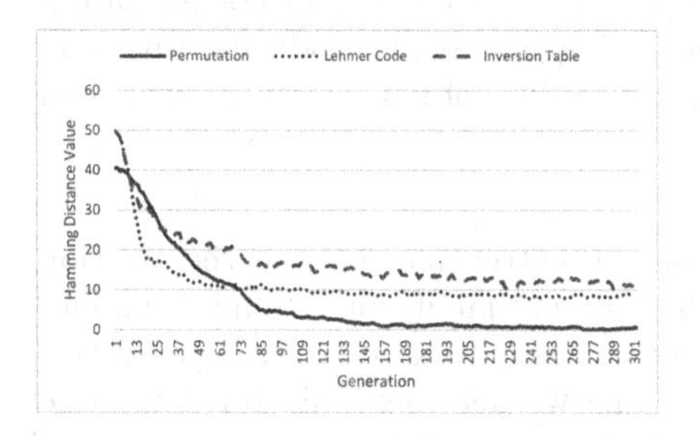

Fig. 3. Hamming distance

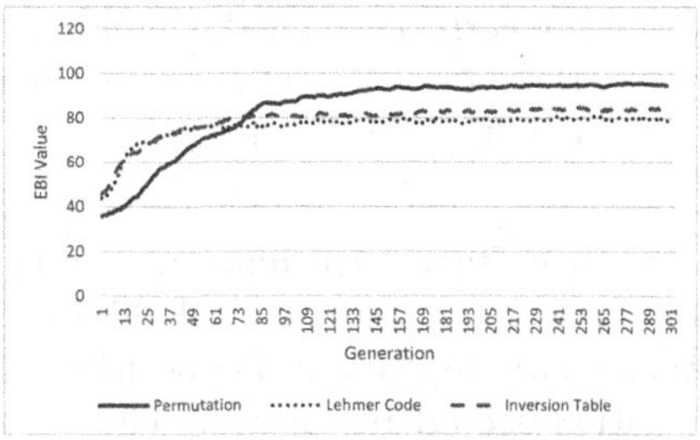

Fig. 4. Edge based indicator

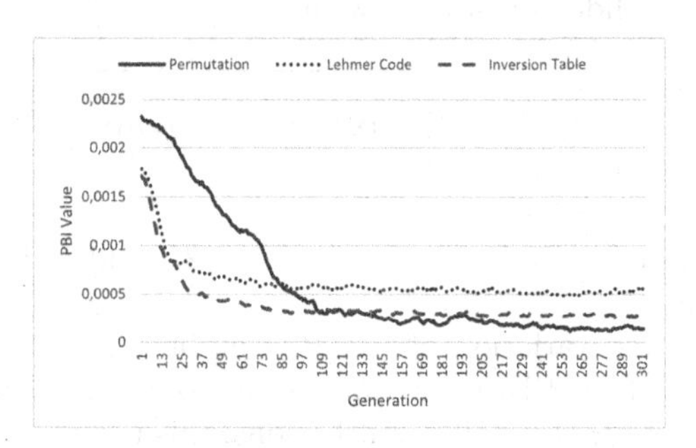

Fig. 5. Position based indicator

Edge Based Indicator. (EBI) [3] counts the edges which are present both in parent and in offspring. It is defined by Eq. 4.

$$EBI = \frac{NET \times 100}{2n} \tag{4}$$

Where $NET = \sum_{i=1}^{n} \sum_{j=1, \neq i}^{n} NE(v_i, v_j)$, such that v_i and v_j are two consecutive values inside a given chromosome.

$$NE(v_i, v_j) = \begin{cases} 2 \text{ if } v_i \text{ and } v_j \text{ are consecutive in both parents and offspring} \\ 1 \text{ if } v_i \text{ and } v_j \text{ are consecutive in one parent and offspring} \\ 0 \text{ Otherwise} \end{cases} \tag{5}$$

In the case of a vehicle routing problem (aka TSP), the EBI estimates the proportion of consecutive cities that will be traveled in the same order for parents and children. The greater this value, the better the hereditary transmission occurs between parents and children. Figure 4 shows that, even though not all parent genes are correctly copied in children when using the Lehmer code and Inversion table, these two indirect encodings obtain high values of EBI (slightly lower than the classic permutation).

Position Based Indicator. The PBI [3] is a metric that looks at the position of cities among parents and children. The idea is that if, for example, the fact that a city appears in 3rd position in the tour leads to a good result, then this position should be found in children too. The weaker this indicator, the more the positions are respected. The PBI is calculated using Eq. 6.

$$PBI = \frac{\sum_{i=1}^{2} \sum_{j=1}^{2} PB_{ij}}{4} \times \frac{6}{n(n+1)(2n+1)} \tag{6}$$

where PB_{ij} is the euclidean distance, with

$$PB_{ij} = \sqrt{\sum_{k=1}^{n} (PO_j(k) - PP_i(k))^2}$$

$PP_i(k)$ is the position of k in the parent i, $i \in \{1, 2\}$, and $PO_j(k)$ is the position of k in the Offspring j, $j \in \{1, 2\}$.
Figure 5 illustrates the variation of the PBI according to the three proposed encodings. In terms of the PBI values for the case of permutation, it increases to reach of the value zero with small perturbations, this means that, at some point, starting from around the generation 100, there is no distance between the permutations representing the parents and the offspring. The other two encodings have much faster growth but to a limit different than null. This can be explained by the diversity of individuals in these cases, since the distance from the parent is low but not zero.

4.2 Assessment of Fitness Improvement

In this section, we study the evolution of the fitness value over generations and compare the performance of the different types of encoding. This is done using a method, inspired by the set of experimentation introduced by Portman and Vignier in [13]. They classify solutions, for each generation, in five classes based on the fitness values. The classes represent five equal intervals of values such that the overall interval is bounded by the minimal and maximal values of the objective function at each generation so that all the values are represented. This means, at each generation, the intervals that define the classes depend on the values in this specific generation. Solutions for the first class are better than the ones in the second, and so on, so the last class is the worst.

At each generation, the solutions will be distributed in each quintile called class 1 to class 5. For a good execution of the optimization process, the number of solutions in class 1 should increase over the generations. This class would therefore become predominant over the others. However, if almost all the solutions were found in a single class, this could show a premature convergence and a possible trap due to a local optimum.

Figure 6 shows that for the classical permutation, around the 100th generation, almost all the solutions are in class 1. The other classes are barely represented. There is therefore a problem of diversity within the population. This is corroborated by the Hamming distance indicator of the classical permutation (see Fig. 3) where it was observed that parents and children were almost identical after the 100th generation.

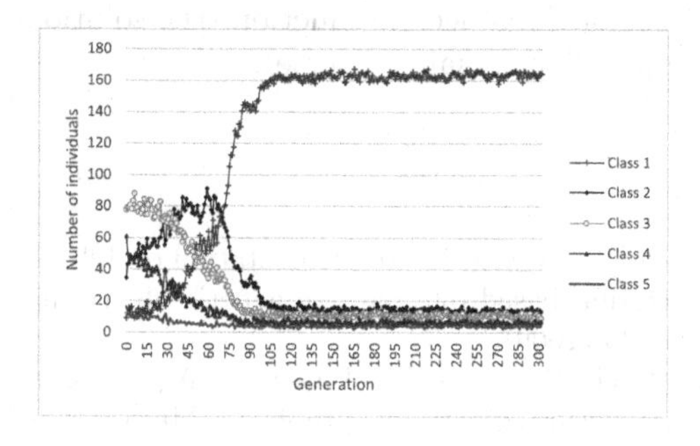

Fig. 6. Classification using the classical permutation encoding

Figure 7 and Fig. 8 show that for the indirect encoding (Lehmer code and Inversion table), greater diversity is observed because the solutions of class 1 correspond to the majority, but the proportions of the solutions in the other classes are not negligible. This helps alleviate the pressure from elitism.

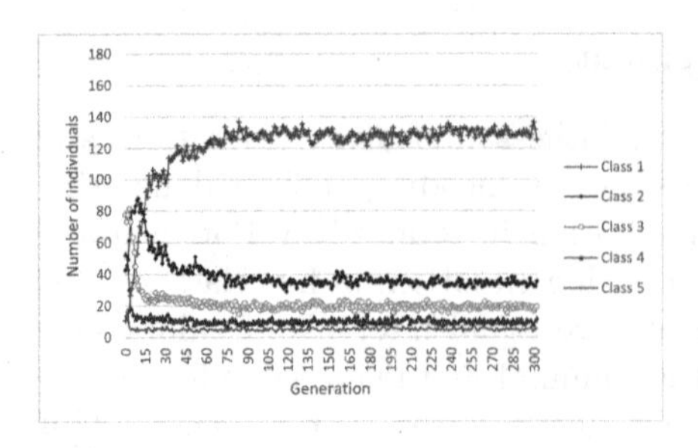

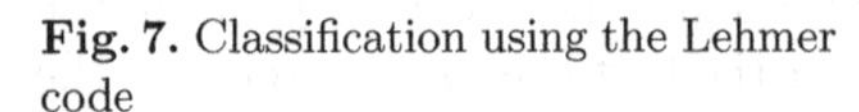
Fig. 7. Classification using the Lehmer code

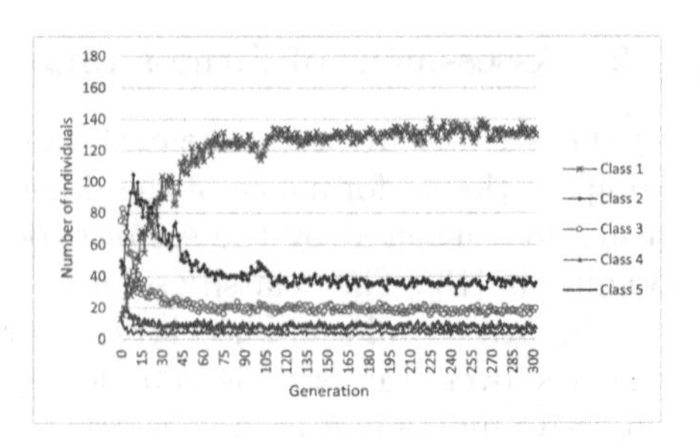

Fig. 8. Classification using Inversion table

5 Conclusions

This paper studies the transmission of genetic characters in an optimization process. It presents three encodings. The first, called direct encoding is the classical permutation. The other two are referred to as indirect encodings: the Lehmer code and the Inversion tables. Indirect coding has the advantage of not being sensitive to the appearance of duplicates during crossover operations. However, there is a partial loss of the genetic properties between parents and children. The study of the impact of these encodings was made both in the space of decision variables (thanks to the Hamming distance, EBI, PBI) and in the space of fitness values (distribution of solutions in quintiles). The results show that indirect encoding makes it possible to preserve diversity within populations without losing quality in terms of the objective function (the solutions in class 1 are the majority and the population remains diverse).

References

1. Bekiroğlu, S., Dede, T., Ayvaz, Y.: Implementation of different encoding types on structural optimization based on adaptive genetic algorithm. Finite Elem. Anal. Des. **45**(11), 826–835 (2009)
2. Contreras, J.P., Bosch, P., Varas, M., Basso, F.: A new genetic algorithm encoding for coalition structure generation problems. Math. Prob. Eng.. 1–13 (2020). https://doi.org/10.1155/2020/1203248
3. Djerid, L., Portmann, M.C., Villon, P.: Performance analysis of permutation crossover genetic operators. J. Decis. Syst. **5**(1–2), 157–177 (1996)
4. Goldberg, David, E.: Genetic Algorithms in Search, Optimization and Machine Learning. Addison-Wesley, Reading (1989)
5. Grygierek, K.: Optimization of trusses with self-adaptive approach in genetic algorithms. Archit. Civ. Eng. Environ. **9**(4), 67–78 (2016)
6. Haj-Rachid, M., Ramdane-Cherif, W., Chatonnay, P., Bloch, C.: Comparing the performance of genetic operators for the vehicle routing problem. IFAC Proc. Volumes **43**(17), 313–319 (2010)
7. Holland, J.H.: Outline for a logical theory of adaptive systems. J. ACM **9**(3), 297–314 (1962)

8. Liu, Q., Li, X., Gao, L., Li, Y.: A modified genetic algorithm with new encoding and decoding methods for integrated process planning and scheduling problem. IEEE Trans. Cybern. 1–10 (2020). https://doi.org/10.1109/TCYB.2020.3026651
9. Mohammed Ali, H., Abdou, W., Chatonnay, P., Bloch, C., Spies, F.: Behaviour study of an evolutionary design for permutation problems. International Congress on Information and Communication Technology, London, UK (2018)
10. Ohira, R., Islam, M.S., Jo, J., Stantic, B.: AMGA: an adaptive and modular genetic algorithm for the traveling salesman problem. In: Abraham, A., Cherukuri, A.K., Melin, P., Gandhi, N. (eds.) ISDA 2018 2018. AISC, vol. 941, pp. 1096–1109. Springer, Cham (2020). https://doi.org/10.1007/978-3-030-16660-1_107
11. Üçoluk, G.: Genetic algorithm solution of the tsp avoiding special crossover and mutation. Intell. Autom. Soft Comput. **8**(3), 265–272 (2002)
12. Pesko, S.: Differential evolution for small tsps with constraints. In: Fourth International Scientific Conference : Challenges in Transport and Communications (2006)
13. Portmann, M.C., Vignier, A.: Performances study on crossover operators keeping good schemata for some scheduling problems. In: Proceedings of the 2nd Annual Conference on Genetic and Evolutionary Computation, GECCO 2000, pp. 331–338. Morgan Kaufmann Publishers Inc., San Francisco, CA, USA (2000)
14. Rosa, P., Sriwindono, H., Nugroho, A., Polina, A., Pinaryanto, K.: Comparison of crossover and mutation operators to solve teachers placement problem by using genetic algorithm. In: Journal of Physics Conference Series (2020)
15. TSPData: Tsplib. http://elib.zib.de/pub/mp-testdata/tsp/tsplib/tsp

Population Management Approaches in the OPn Algorithm

Krystian Łapa[1(✉)], Krzysztof Cpałka[1], and Adam Słowik[2]

[1] Czestochowa University of Technology, 42-201 Czestochowa, Poland
{krystian.lapa,krzysztof.cpalka}@pcz.pl

[2] Koszalin University of Technology, 75-453 Koszalin, Poland
aslowik@ie.tu.koszalin.pl

Abstract. This paper deals with the problem of selecting the population size for the population-based algorithm with dynamic selection of operators (OPn). This research was undertaken to check how population size changes affect the optimization of problems in which both the parameters of the solution and its structure should be selected. Moreover, variants in which the size of the population changes dynamically were considered. The simulations were performed for a small selection/variety of examples of control problems in which the structures and parameters of controllers based on PID systems had to be selected.

Keywords: Population-based algorithms · Evolutionary algorithms · Operators selection · Population size

1 Introduction

Choosing a population-based algorithm for a given problem can be very difficult [39]. This is due to the fact that there are currently over 200 different population-based algorithms [5], each of which has its own variations and hybrid versions (see e.g. [10,11]). It also has to do with the fact that more and more problems are considered to be solved by applying population algorithms (see e.g. [24,25,34,43]) which include among others regression, control or classification problems (see e.g. [4,13,29,30,41]). Of course, this is related to the entire field of artificial intelligence, which, by expanding its applicability and implementation (see e.g. [2,3,17,33]), creates more and more applications for the optimization algorithms under consideration. Moreover, each of population-based algorithms may work differently for a given simulation problem or a given group of problems, which is due to the more No Free Lunch Theorem [40]. Therefore, it is not known which mechanisms (operators) of the given algorithms work best for a problem under consideration. This is especially important because the area of application of population-based algorithms is constantly expanding (see e.g. [15,16,37,42]).

Notwithstanding, there are solutions in the literature in which mechanisms from different algorithms may operate simultaneously, e.g. the previously mentioned hybrid algorithms (see e.g. [1,9,22]). Moreover, there are papers

L. Rutkowski et al. (Eds.): ICAISC 2021, LNAI 12854, pp. 402–414, 2021.
https://doi.org/10.1007/978-3-030-87986-0_36

presenting approaches in which the operation of operators can be dynamically controlled. In such approaches, the selection of the optimal operator adapts to the needs in a given iteration or given state of search of the algorithm and may change over time (see e.g. [20,21]). As a result, algorithms of this type work more universally and can be applied to numerous types of problems. Of course, there is a great likelihood of standard population algorithms that give better results; however, it is not possible to test all algorithms for a given problem, thus choosing universal methods seems to be justified.

The above mentioned universal approaches do not remove all issues related to population-based algorithms. It is also important to choose the size of the population (see e.g. [6,27]). On the one hand, more complex algorithms with larger population sizes are developed, reducing the chances of getting stuck in a local minimum (see e.g. [12]). On the other hand, less complex micro-algorithms with a small population size are created, which allows for obtaining similar, and even better results when the number of algorithm's iterations is increased (see e.g. [36]). Reducing the computational cost of the algorithm is important as it increases the applicability of population-based algorithms. This is also confirmed by the use of surrogate models of simulation problems (see e.g. [18]). A solution to the above issue may be algorithms in which the size of the population or sub-population is dynamically changed during the operation of the algorithm (see e.g. [8,26,28]).

In this paper, a decision was made to investigate the impact of using different population sizes on the population-based algorithm with dynamic selection of operators (OPn [20]). Moreover, the mechanism in which the population size increases and decreases, and vice versa: decreases and increases, is also investigated. This has not yet been taken up in the literature. In addition, different variants related to the population size were considered, including variants where the population size changes according to the iteration of the algorithm. Finding the optimal approach to the population size for an algorithm with dynamic operator selection could bring benefits in both reducing computational costs and improving the accuracy of the solutions found. The developed approach was tested for the control problems where both the parameters and the structure of PID controllers have to be selected.

The structure of this paper is as follows: in Sect. 2 the proposed solution is presented, in Sect. 3 the simulations are described, and in Sect. 4 the conclusions are drawn.

2 Proposed Solution

This section describes an algorithm that allows for the automatic selection of operators, parameters, and the structure of solutions obtained by using the algorithm and coded in the population's individuals. The method allows us to apply mechanisms for managing the population size which are considered in this paper.

2.1 Operator Population-Based Algorithm with n Operators (OPn)

The OPn is based on a mechanism similar to that derived from the PSO [14], which means that beside parameters vector x_d (where $d = 1, ..., D$ and D stands for the number of parameters), it uses a velocity vector v_d, and a vector that stores the best set of parameters x'_d. The main difference, however, is the introduction of an additional binary operators vector which decides which operators should be used to modify the x_d parameters - o_j (where $j = 1, ..., J$ and J stands for the number of operators). In the proposed solution, control problems were considered as example simulation problems for which it would be possible not only to select parameters, but also to select the PID controller structure. To make it possible, each individual has been extended with an additional binary structure vector that determines which controller element is reduced - b_c (where $c = 1, ..., C$ and C stands for the number of structure elements that can be reduced - the structure of the controller may depend on the simulation problem, hence the value of C may vary). The complete structure of an OPn's individual is shown in Fig. 1.

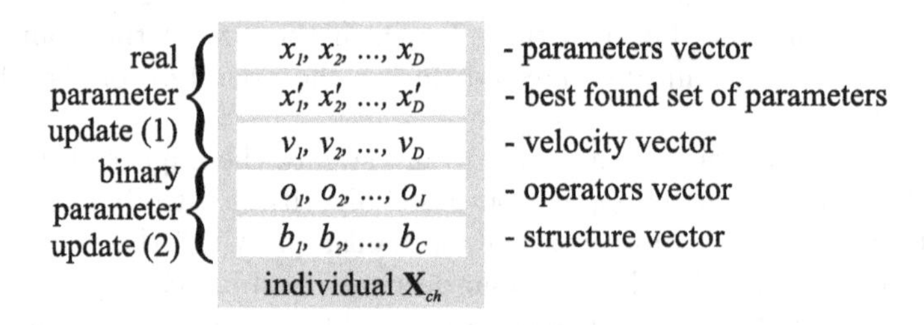

Fig. 1. Individual structure for an OPn with its parameters and structure selection.

In the first step of the algorithm, N individuals are initialized and evaluated. The vectors $\mathbf{x}$ and $\mathbf{v}$ are initialized randomly according to the ranges resulting from simulation problems, the vector $\mathbf{x}'$ is set to $\mathbf{x}$, the vectors $\mathbf{o}$ and $\mathbf{b}$ are randomly initialized from the set $\{0, 1\}$. After the evaluation, the best parameters found are stored as vector $\mathbf{x}^*$. Then, the main iterative loop of the algorithm begins, in which for each individual $\mathbf{X}_{ch}$ ($ch = 1, ..., N$) the following steps are performed:

– Vectors $\mathbf{v}$ and $\mathbf{x}$ are updated as follows:

$$\begin{cases} v_d := w \cdot v_d + \sum_{j=1}^{J} o_j \cdot op_j \left(x_d, x^*_d, x'_d, \mathbf{x}^p_d \right) \\ x_d := x_d + v_d, \end{cases} \tag{1}$$

where w is the inertia weight, op_j are the functions related to operators from different algorithms (for the details see [20]), $\mathbf{x}^*$ is the vector of the best found-so-far parameters, $\mathbf{x}^p$ are the vectors of the parameters from the individuals selected in relation to the used operators (for the details see [20]).

- Vector $\mathbf{x}$ values are narrowed to the search space according to the problem under consideration.
- Vectors $\mathbf{o}$ and $\mathbf{b}$ are updated as follows:

$$\begin{cases} o_j := (o_j + 1)\,\%2 \text{ if } rnd < m_o \\ b_c := (b_c + 1)\,\%2 \text{ if } rnd < m_b, \end{cases} \tag{2}$$

  where there is a random number from the range $rnd \in \langle 0, 1\rangle$, and where $m_o \in \langle 0, 1\rangle$, and $m_b \in \langle 0, 1\rangle$ are mutation probabilities.
- The individual is evaluated according to a fitness function related to the simulation problem under consideration.
- If the individual fitness function value is superior to the value of the fitness function for its $\mathbf{x}'$ vector, then the $\mathbf{x}'$ is updated: $\mathbf{x}' = \mathbf{x}$.
- If the individual fitness function value is superior to the value of the fitness function for best found parameter vector $\mathbf{x}^*$, then $\mathbf{x}^*$ is updated: $\mathbf{x}^* = \mathbf{x}$.

After all individuals have been updated, the iteration is complete. Then, the stop condition is checked (e.g. reaching a certain number of iterations). If the stop condition is not met, the algorithm starts another iteration. Otherwise, the best found solution, represented by the found parameters vector $\mathbf{x}^*$, is presented.

2.2 OPn and Population Management Approach

In addition to adjusting the OPn algorithm to the selection of the structure and parameters of PID controllers, the mechanisms which allowed us to dynamically change the size of the population during the operation of the algorithm were also included. After each iteration of the algorithm (see Sect. 2.1), the number of individuals for a given iteration is determined according to the following formula:

$$N = \text{int}\left(N_A + (N_B - N_A) \cdot \frac{|2 \cdot iter_{cur} - iter_{max}|}{iter_{max}}\right), \tag{3}$$

where $\text{int}(\cdot)$ is a function that returns the integer part of a floating number, N_A is an initial and final number of individuals, N_B is the number of individuals when the algorithm reaches half of its iterations, $iter_{cur}$ is the current iteration and $iter_{max}$ is the number of iterations. The above formula makes it possible for a number of individuals to change smoothly from N_A to N_B, and then back to N_A, so that the number of individuals in the population, depending on the iteration, may look as shown in Fig. 2.

A change in the number of individuals means that individuals must be added to or removed from the population during the algorithm operation. If the current number of individuals is greater than that determined with (3), the worst individuals from the population are sequentially removed. However, when new individuals must be added to the population, they are created with the use of a mechanism based on a crossover from the genetic algorithm: for two parents selected using the roulette wheel method, one new individual is added to the population, and its parameters are determined as follows:

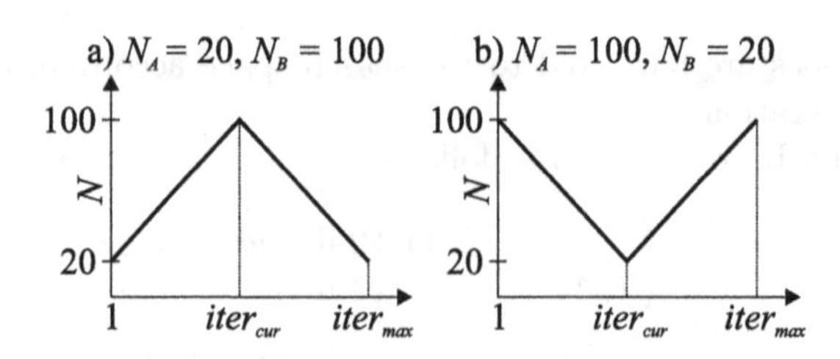

Fig. 2. Examples of changes in the number of individuals using (3).

$$\mathbf{X}^{\mathbf{new}} = \begin{cases} x_d = x_d^p + rnd \cdot (x_d^q - x_d^p), d = 1, ..., D \\ v_d = v_d^p + rnd \cdot (v_d^q - v_d^p), d = 1, ..., D \\ x'_d = x_d, d = 1, ..., D \\ o_j = o_j^p \text{ for } rnd < p_c \text{ or else } o_j = o_j^q, j = 1, ..., J \\ b_c = b_c^p \text{ for } rnd < p_c \text{ or else } b_c = b_c^q, c = 1, ..., C, \end{cases} \tag{4}$$

where p and q indexes indicate the parameters of the parents, $p_c \in \langle 0, 1 \rangle$ stands for crossover probability, and $\mathbf{X}^{\mathbf{new}}$ stands for a new individual. The notation used in formula (4) allows us to refer to all components of the created individual $\mathbf{X}^{\mathbf{new}}$ (see Fig. 1).

The described approach will allow us, for example, to use more individuals in the early and final stages of the algorithm, or vice versa. Moreover, by using the values N_A and N_B shown in the example in Fig. 2, the average number of individuals will be comparable to the cases with a fixed number of individuals.

3 Simulations

In the paper a few approaches to population management were considered: when the population size is static ($N = 100$, $N = 60$, $N = 20$), when it is variable as described in Sect. 2.2 (N : LHL, $N_A = 20$, $N_B = 100$ - see Fig. 2; N : HLH, $N_A = 100$, $N_B = 20$) and when it is constant but the number of iterations is increased five times (N : 20E) and thus the complexity is identical to the algorithm with the number of individuals $N = 100$. With fixed population sizes, it can check how different algorithms deal with the selection of structure and parameters in control systems and how the OPn operates in this comparison. The considered variable size of the population has the same complexity as the case when $N = 60$, which will allow to check whether this approach makes it possible to improve the algorithm's operation in relation to this case. On the other hand, increasing the number of iterations with a small population might verify whether the use of the micro version of the OPn algorithm has any potential for the considered simulation problems.

3.1 Simulation Problems

The three control problems described in this section were considered and a cascade PID controller with the possibility of reducing its elements was used as a control system for these problems.

PID Controller with a Dynamic Structure. In the simulations, the control problems with the use of PID controllers with a dynamic structure were considered. A typical implementation mechanism for a PID algorithm can be written as follows:

$$u(t) = K_p \cdot e(t) + K_i \cdot \int_0^t e(t)dt \cdot K_d \cdot \frac{de(t)}{dt}, \tag{5}$$

where $e(\cdot)$ is the offset, $u(\cdot)$ is the PID output, K_p, K_i and K_d are proportional, integral and derivative coefficients respectively. The considered PID controller has been expanded by four binary values $b_1, ..., b_4 \in \{0, 1\}$ allowing for the reduction of its elements:

$$u(t) = b_1 \cdot \left(b_2 \cdot K_p \cdot e(t) + b_3 \cdot K_i \cdot \int_0^t e(t)dt + b_4 \cdot K_d \cdot \frac{de(t)}{dt}\right). \tag{6}$$

In addition, a cascade connection of a number of PID controllers was used, which allows for using and processing more signals and providing a better control accuracy (see e.g. [19]). In such a case, the binary parameters **b** enabling the automatic reduction of redundant elements of the structure are marked with successive index numbers.

Water Tank Test (WTT). The first problem goal is to maintain the desired water level h^* in the tank by changing the water inflow g_{in}. The WTT model is defined as follows [31]:

$$\dot{h} = \frac{1}{A}\left(q_{in} + q_{ex} - q_{em} - s \cdot \sqrt{2gh}\right), \tag{7}$$

where A is a surface area, q_{ex} is an external water inflow, q_{em} is an additional emergency water outflow, s is a water outflow, and $g = 9.81$ m/s^2 is the gravitational acceleration. The proposed controller structure for this problem is shown in Fig. 3a.

Mass Spring Damper (MSD). In the mass spring damper problem the aim is to maintain the desired position s^* of mass m_1 by managing the control force F. The mass is connected via a spring to the mass m_2, and then by another spring to the constant point y. The MSD model is defined as follows [19]:

$$\begin{cases} s_1 = v_1 \cdot t + \frac{1}{2} \cdot a_1 \cdot t^2 \; v_1 = a_1 \cdot t \; a_1 = \frac{k \cdot (s_2 - s_1) - v_1 \cdot y}{m_1} \\ s_2 = v_2 \cdot t + \frac{1}{2} \cdot a_2 \cdot t^2 \; v_2 = a_2 \cdot t \; a_2 = \frac{k \cdot (F - s_2) - v_2 \cdot y}{m_2}, \end{cases} \tag{8}$$

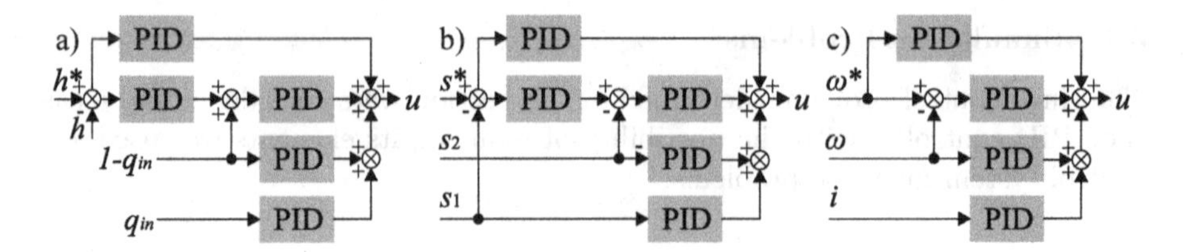

Fig. 3. Proposed cascade PID structures for control problems under consideration: a) WTT, b) MSD, and c) DCM. The structures enable automatic reduction of redundant elements of PID blocks, and thus the entire signals. The design draws on the authors' experience (see e.g. [19]).

where s_1 and s_2 are the positions of masses m_1 and m_2, k is the stiffness constant for both masses. The proposed controller structure for this problem is shown in Fig. 3b.

DC Motor (DCM). In the last problem the goal is to maintain the desired motor speed ω^* by managing the input voltage V. The DCM model is defined as follows [7]:

$$\begin{cases} \dot{\omega} = \frac{K_t \cdot i - b \cdot \omega}{J} \\ \dot{i} = \frac{-R \cdot i + V - K_e \cdot \omega}{L} \end{cases}, \tag{9}$$

where ω is the motor speed, J is the moment of inertia of the rotor, b is the viscous friction constant, L is the electric inductance, R is the electric resistance, $T = K_t \cdot i$ is the motor torque (where i is the armature current and K_t is the motor torque constant), and $e = K_e \cdot \dot{\omega}$ is the counter-electromotive force (K_e is the electromotive force constant). The proposed controller structure for this problem is shown in Fig. 3c.

Fitness Function. To evaluate the controller, three control criteria with the goal of minimization were used (see Table 1). The first criterion (ACC) is concerned with the accuracy of operation, so the difference between the setpoint and the actual signal (offset) is minimized. The second criterion (OVS) concerns the overshoot, thus minimizing the maximum difference between the actual and the set signal. The purpose of the last criterion (OSC) is to minimize the oscillation of the control system, and thus to reduce the sum of differences between successive values of the control signal (this criterion also reduces the control force). The listed criteria have been aggregated to a single fitness function as follows (a lower value is better):

$$FF = ACC \cdot w_1 + OVS \cdot w_2 + OSC \cdot w_3, \tag{10}$$

where $\mathbf{w}$ stands for the weights of components that might differ for each simulation problem (see Table 2). It is worth noting that the above function does not take into account the complexity of the controller structure. Thanks to this, the

Table 1. Fitness function criteria for the simulation problems under consideration

Problem	Criterion		
	ACC	OVS	OSC
WTT	$\sum_{i=1}^{T} \frac{\lvert h^*(t_i)-h(t_i)\rvert}{T}$	$\max_{i=1,...,T} \{h(t_i) - h^*(t_i)\}$	$\sum_{i=2}^{T} \frac{\lvert q_{in}(t_i)-q_{in}(t_{i-1})\rvert}{T-1}$
MSD	$\sum_{i=1}^{T} \frac{\lvert s^*(t_i)-s_1(t_i)\rvert}{T}$	$\max_{i=1,...,T} \{s_1(t_i) - s^*(t_i)\}$	$\sum_{i=2}^{T} \frac{\lvert F(t_i)-F(t_{i-1})\rvert}{T-1}$
DCM	$\sum_{i=1}^{T} \frac{\lvert \omega^*(t_i)-\omega(t_i)\rvert}{T}$	$\max_{i=1,...,T} \{\omega(t_i) - \omega^*(t_i)\}$	$\sum_{i=2}^{T} \frac{\lvert V(t_i)-V(t_{i-1})\rvert}{T-1}$

optimization of parameters will not strive to obtain the simplest structure, but to obtain the structure best suited to the given problem. In a case where simpler structures are needed, an additional criterion for assessing the complexity of the structure should be included.

3.2 Simulation Parameters

In the simulations, the parameters of the problems were assumed according to the literature: WTT [31], MSD [19], and DCM [7]. The other parameters related to the simulation problems are shown in Table 2. The OPn method uses 16 operators derived from various population algorithms (the details can be found in [21], where these operators were used in a different method). The OPn method was compared with known population algorithms, modified with the mechanisms that allow us to select both the parameters and the structure (an additional vector **b** was introduced in the algorithms: GA [38], DE [32], GWO [23], FWA [35], and PSO [14]). The number of iterations of all algorithms was set to 500 and the results of each simulation case were repeated 100 times and averaged. The parameters of the OPn algorithm were set as follows: $m_o = 0.1$, $m_b = 0.1$, and $p_c = 0.5$.

Table 2. Parameters related to the simulation problems.

Parameters	MSD	DCM	WTT
FF weight w_1	1.000	1.000	10.000
FF weight w_2	0.010	0.010	0.010
FF weight w_3	0.001	0.001	0.001
Individual parameters	$D = 15, C = 20$	$D = 15, C = 20$	$D = 12, C = 15$
Control signal	$u = F$	$u = V$	$u = q_{in}$

3.3 Simulation Results

The results of the comparison of the OPn with other algorithms is shown in Table 3 and Fig. 4. The comparison of population management approaches in the OPn algorithm is shown in Table 4 and Fig. 5. In addition, the examples of obtained controllers are shown in Fig. 6.

3.4 Simulation Conclusions

The conclusions from the simulations performed can be summarized as follows:

- Regardless of the simulation problem and the population size, the OPn algorithm allowed us to obtain the best results in the problem of selecting the structure and parameters of the control system (see Table 3).
- Some algorithms performed much worse with a small population size (see e.g. GA and PSO for $N = 20$ in Table 3).
- Some algorithms could not cope with the simultaneous selection of structure and parameters (see e.g. FWA and PSO for MSD, and $N = 100$ in Table 3).

Table 3. Comparison of FF values of the OPn with other algorithms and different population size N. The best values are shown in bold.

Prb. →	MSD			DCM			WTT		
Alg. ↓	$N = 100$	$N = 60$	$N = 20$	$N = 100$	$N = 60$	$N = 20$	$N = 100$	$N = 60$	$N = 20$
OPn	**0.042**	**0.044**	**0.046**	**0.020**	**0.021**	**0.024**	**0.717**	**0.729**	**0.777**
GA	0.090	0.879	3.588	0.022	0.024	0.114	0.900	1.004	1.256
DE	0.070	0.078	0.575	0.047	0.046	0.051	0.797	0.795	0.897
GWO	0.100	0.121	0.127	0.031	0.036	0.057	0.837	0.852	0.891
FWA	3.681	4.570	5.022	0.042	0.038	0.050	0.881	0.901	0.914
PSO	1.204	3.974	6.539	0.023	0.026	0.164	1.117	1.217	1.518

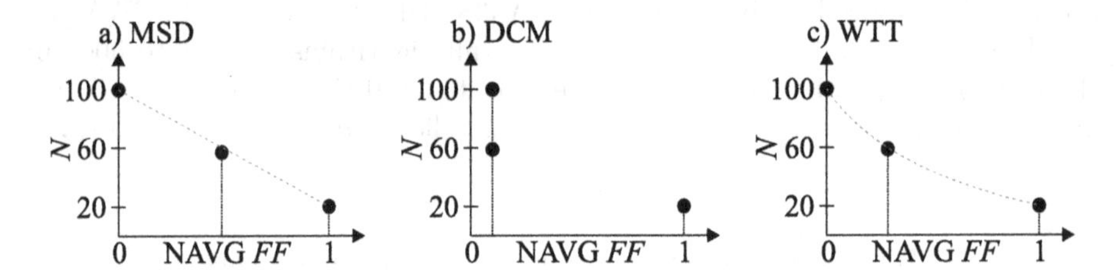

Fig. 4. Dependence of the number of individuals and normalized FF values averaged for all algorithms. NAVG stands for average fitness function values normalized individually for each problem.

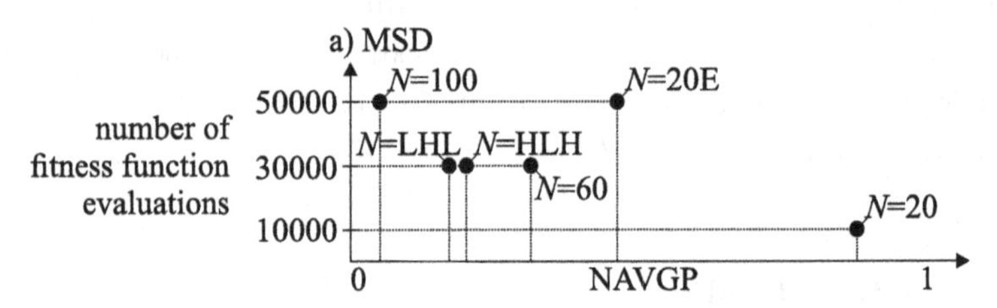

Fig. 5. Dependence of the population management and the number of evaluations for the OPn algorithm. NAVGP stands for NAVG averaged for all problems.

Table 4. Comparison of the population management approaches in the OPn algorithm. NAVG stands for the normalized for each problem average fitness function values. The best values are shown in bold.

Prb. ↓	$N = 100$	$N = 60$	$N = 20$	$N = \text{LHL}$	$N = \text{HLH}$	$N = \text{20E}$
MSD	**0.04219**	**0.04400**	0.04571	**0.04127**	0.04409	0.04858
DCM	**0.02008**	0.02131	0.02382	0.02173	**0.02058**	**0.02090**
WTT	**0.71664**	0.72941	0.77747	**0.72127**	**0.72096**	0.72618
NAVG	**0.04187**	0.30443	0.86945	**0.17276**	**0.19726**	0.45836

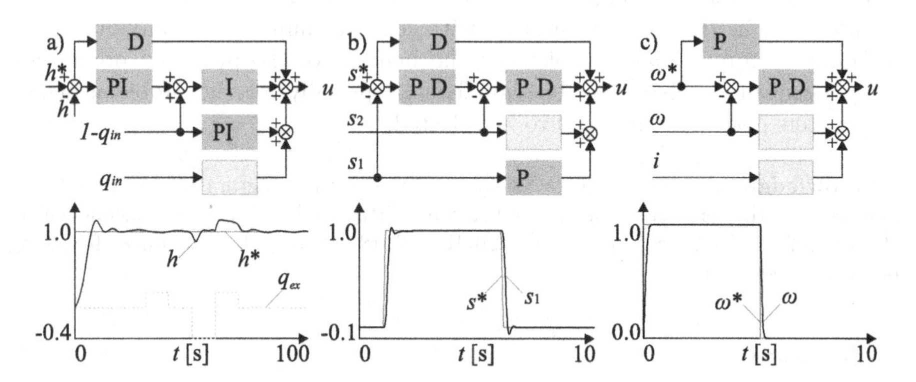

Fig. 6. Examples of obtained controllers. The reduced PID elements have been removed from the structure and grayed out in case of reducing whole PID block.

- Increasing the population benefited the MSD problem linearly and exponentially for the WTT (see Fig. 4). For the DCM problem, using a population size greater than 60 had no significant improvement.
- The approach with a small number of individuals but increasing the number of iterations of the algorithms did not bring any benefit and at the same time only worked comparably to $N = 100$ in the case of the DCM problem (see Fig. 5 and Table 4).
- In the case of a variable number of individuals, the individual nature of the problems can be seen (sometimes LHL works better and sometimes HLH works better - see Table 4).
- Comparing LHL and HLH versus $N = 60$ (a similar computational cost), LHL and HLH give much better results (see Table 4 and Fig. 5).
- For the MSD problem, the strategy of LHL with a variable number of individuals gave better results than using a more complex approach when $N = 100$ (see Table 4).
- The structures of the obtained controllers were reduced and the control error was minimized (see the differences between the set and the real signal in Fig. 6).

4 Conclusions

Selecting an appropriate number of individuals in a population has a key impact on the operation of population-based algorithms, which also translates into the results of the OPn algorithm. Increasing the population size may have a different effect on the operation of the algorithm depending on the simulation problem, and above certain limits it may not bring significant benefits if the computational complexity is being increased. In this paper it was shown that the choice of a strategy in which the number of individuals in the population for the OPn is dynamically changed brings significant benefits and allows us to obtain better results than in the case of variants with a static number of individuals and a similar computational complexity. The benefits of the proposed population size change mechanism may be significant, especially for the problems where parameters and structure need to be selected.

Acknowledgment. This paper was financed under the program of the Minister of Science and Higher Education under the name 'Regional Initiative of Excellence' in the years 2019–2022 project number 020/RID/2018/19 with the amount of financing of PLN 12 000 000.

References

1. Antonio, L.M., Coello, C.A.C.: Coevolutionary multiobjective evolutionary algorithms: survey of the state-of-the-art. IEEE Trans. Evol. Comput **22**(6), 851–865 (2017)
2. Bartczuk, Ł, Dziwiński, P., Red'ko, V.G.: The concept on nonlinear modelling of dynamic objects based on state transition algorithm and genetic programming. In: Rutkowski, L., Korytkowski, M., Scherer, R., Tadeusiewicz, R., Zadeh, L.A., Zurada, J.M. (eds.) ICAISC 2017. LNCS (LNAI), vol. 10246, pp. 209–220. Springer, Cham (2017). https://doi.org/10.1007/978-3-319-59060-8_20
3. Bartczuk, Ł, Przybył, A., Cpałka, K.: A new approach to nonlinear modelling of dynamic systems based on fuzzy rules. Int. J. Appl. Math. Comput. Sci. (AMCS) **26**(3), 603–621 (2016)
4. Bilski, J., Rutkowski, L., Smolag, J., Tao, D.: A novel method for speed training acceleration of recurrent neural networks. Inf. Sci. **553**, 266–279 (2021)
5. Campelo, F., Aranha, C.: EC Bestiary: a bestiary of evolutionary, swarm and other metaphor-based algorithms. In: Zenodo (2018)
6. Chen, T., Tang, K., Chen, G., Yao, X.: A large population size can be unhelpful in evolutionary algorithms. Theor. Comput. Sci. **436**, 54–70 (2012)
7. Cheon, K., Kim, J., Hamadache, M., Lee, D.: On replacing PID controller with deep learning controller for DC motor system. J. Autom. Control Eng **3**(6), 452–456 (2015)
8. Cui, L., et al.: A novel artificial bee colony algorithm with an adaptive population size for numerical function optimization. Inf. Sci. **414**, 53–67 (2017)
9. Dziwiński, P., Bartczuk, Ł, Paszkowski, J.: A new auto adaptive fuzzy hybrid particle swarm optimization and genetic algorithm. J. Artif. Intell. Soft Comput. Res. **10**(2), 95–111 (2020)

10. Elbes, M., Alzubi, S., Kanan, T., Al-Fuqaha, A., Hawashin, B.: A survey on particle swarm optimization with emphasis on engineering and network applications. Evol. Intell **12**(2), 113–129 (2019). https://doi.org/10.1007/s12065-019-00210-z
11. Eltaeib, T., Mahmood, A.: Differential evolution: a survey and analysis. Appl. Sci. **8**(10), 1945 (2018)
12. Fukumoto, H., Oyama, A.: Study on improving efficiency of multi-objective evolutionary algorithm with large population by M2M decomposition and elitist mate selection scheme. In: 2018 IEEE Symposium Series on Computational Intelligence (SSCI), pp. 1180–1187. IEEE (2018)
13. Galkowski, T., Pawlak, M.: Nonparametric estimation of edge values of regression functions. In: Rutkowski, L., Korytkowski, M., Scherer, R., Tadeusiewicz, R., Zadeh, L.A., Zurada, J.M. (eds.) ICAISC 2016. LNCS (LNAI), vol. 9693, pp. 49–59. Springer, Cham (2016). https://doi.org/10.1007/978-3-319-39384-1_5
14. Kennedy, J., Eberhart, R.: Particle swarm optimization. In: Proceedings of ICNN'95-International Conference on Neural Networks, vol. 4, pp. 1942–1948. IEEE (1995)
15. Korytkowski, M., Senkerik, R., Scherer, M.M., Angryk, R.A., Kordos, M., Siwocha, A.: Efficient image retrieval by fuzzy rules from boosting and metaheuristic. J. Artif. Intell. Soft Comput. Res **10**(1), 57–69 (2020)
16. Krell, E., Sheta, A., Balasubramanian, A.P.R., King, S.A.: Collision-free autonomous robot navigation in unknown environments utilizing PSO for path planning. J. Artif. Intell. Soft Comput. Res. **9**(4), 267–282 (2019)
17. Laskowski, Ł, Laskowska, M., Jelonkiewicz, J., Boullanger, A.: Molecular approach to hopfield neural network. In: Rutkowski, L., Korytkowski, M., Scherer, R., Tadeusiewicz, R., Zadeh, L.A., Zurada, J.M. (eds.) ICAISC 2015. LNCS (LNAI), vol. 9119, pp. 72–78. Springer, Cham (2015). https://doi.org/10.1007/978-3-319-19324-3_7
18. Liu, B., Yang, H., Lancaster, M.J.: Global optimization of microwave filters based on a surrogate model-assisted evolutionary algorithm. IEEE Trans. Microwave Theory Tech. **65**(6), 1976–1985 (2017)
19. Łapa, K., Szczypta, J., Venkatesan, R.: Aspects of structure and parameters selection of control systems using selected multi-population algorithms. In: Rutkowski, L., Korytkowski, M., Scherer, R., Tadeusiewicz, R., Zadeh, L.A., Zurada, J.M. (eds.) ICAISC 2015. LNCS (LNAI), vol. 9120, pp. 247–260. Springer, Cham (2015). https://doi.org/10.1007/978-3-319-19369-4_23
20. Łapa, K., Cpałka, K.: Flexible fuzzy PID controller (FFPIDC) and a nature-inspired method for its construction. IEEE Trans. Ind. Inf. **14**(3), 1078–1088 (2017)
21. Łapa, K., Cpałka, K., Laskowski, Ł, Cader, A., Zeng, Z.: Evolutionary algorithm with a configurable search mechanism. J. Artif. Intell. Soft Comput. Res **10**(3), 151–171 (2020)
22. Ma, X., et al.: A survey on cooperative co-evolutionary algorithms. IEEE Trans. Evol. Comput. **23**(3), 421–441 (2018)
23. Mirjalili, S., Mirjalili, S.M., Lewis, A.: Grey wolf optimizer. Adv. Eng. Softw **69**, 46–61 (2014)
24. Mizera, M., Nowotarski, P., Byrski, A., Kisiel-Dorohinicki, M.: Fine tuning of agent-based evolutionary computing. J. Artif. Intell. Soft Comput. Res. **9**, 81–97 (2019)
25. Ono, K., Hanada, Y., Kumano, M., Kimura, M.: Enhancing island model genetic programming by controlling frequent trees. J. Artif. Intell. Soft Comput. Res. **9**, 51–65 (2019)

26. Piotrowski, A.P.: Review of differential evolution population size. Swarm Evol. Comput. **32**, 1–24 (2017)
27. Piotrowski, A.P., Napiorkowski, J.J., Piotrowska, A.E.: Population size in particle swarm optimization. Swarm Evol. Comput. **58**, 100718 (2020)
28. Polakova, R., Tvrdik, J., Bujok, P.: Differential evolution with adaptive mechanism of population size according to current population diversity. Swarm Evol. Comput. **50**, 100519 (2019)
29. Rutkowski, L.: Multiple Fourier series procedures for extraction of nonlinear regressions from noisy data. IEEE Trans. Signal Process. **41**(10), 3062–3065 (1993)
30. Rutkowski, L.: Sequential pattern recognition procedures derived from multiple Fourier series. Pattern Recogn. Lett. **8**(4), 213–216 (1988)
31. Sabri, L.A., Al-mshat, H.A.: Implementation of fuzzy and PID controller to water level system using labview. Int. J. Comput. Appl **116**(11), 6–10 (2015)
32. Storn, R.: On the usage of differential evolution for function optimization. In Proceedings of North American Fuzzy Information Processing, pp. 519–523. IEEE (1996)
33. Szczypta, J., Przybył, A., Cpałka, K.: Some aspects of evolutionary designing optimal controllers. In: Rutkowski, L., Korytkowski, M., Scherer, R., Tadeusiewicz, R., Zadeh, L.A., Zurada, J.M. (eds.) ICAISC 2013. LNCS (LNAI), vol. 7895, pp. 91–100. Springer, Heidelberg (2013). https://doi.org/10.1007/978-3-642-38610-7_9
34. Tambouratzis, G., Vassiliou, M.: Swarm algorithms for NLP: the case of limited training data. J. Artif. Intell. Soft Comput. Res **9**, 219–234 (2019)
35. Tan, Y., Zhu, Y.: Fireworks algorithm for optimization. In: Tan, Y., Shi, Y., Tan, K.C. (eds.) ICSI 2010. LNCS, vol. 6145, pp. 355–364. Springer, Heidelberg (2010). https://doi.org/10.1007/978-3-642-13495-1_44
36. Truong, V.H., Nguyen, P.C., Kim, S.E.: An efficient method for optimizing space steel frames with semi-rigid joints using practical advanced analysis and the micro-genetic algorithm. J. Constr. Steel Res. **128**, 416–427 (2017)
37. Wei, Y., et al.: Vehicle emission computation through microscopic traffic simulation calibrated using genetic algorithm. J. Artif. Intell. Soft Comput. Res. **9**(1), 67–80 (2019)
38. Whitley, D.: A genetic algorithm tutorial. Stat. Comput. **4**(2), 65–85 (1994)
39. Wu, G., Mallipeddi, R., Suganthan, P.N.: Ensemble strategies for population-based optimization algorithms-a survey. Swarm Evol. Comput **44**, 695–711 (2019)
40. Yang, X.S.: Free lunch or no free lunch: that is not just a question? Int. J. Artificial Intelligence Tools **21**(3), 1240010 (2012). https://doi.org/10.1142/S0218213012400106
41. Zalasiński, M., Cpałka, K., Hayashi, Y.: New fast algorithm for the dynamic signature verification using global features values. In: Rutkowski, L., Korytkowski, M., Scherer, R., Tadeusiewicz, R., Zadeh, L.A., Zurada, J.M. (eds.) ICAISC 2015. LNCS (LNAI), vol. 9120, pp. 175–188. Springer, Cham (2015). https://doi.org/10.1007/978-3-319-19369-4_17
42. Zalasiński, M., Łapa, K., Cpałka, K., Przybyszewski, K., Yen, G.G.: On-line signature partitioning using a population based algorithm. J. Artif. Intell. Soft Comput. Res. **10**(1), 5–13 (2020)
43. Zhang, N., Chen, X., Kapre, N.: RapidLayout: fast hard block placement of FPGA-optimized systolic arrays using evolutionary algorithms. In: 2020 30th International Conference on Field-Programmable Logic and Applications (FPL), pp. 145–152. IEEE (2020)

Harris Hawks Optimisation: Using of an Archive

Petr Bujok(✉)

University of Ostrava, 30. dubna 22, 70200 Ostrava, Czech Republic
petr.bujok@osu.cz

Abstract. This paper proposes an enhanced variant of the novel and popular Harris Hawks Optimisation (HHO) method. The original HHO algorithm was studied in many research projects, and a lot of hybrid (cooperative) variants of HHO was proposed. In this research study, an advanced HHO algorithm with an archive of the old solutions is proposed (HHO_A). The proposed method is experimentally compared with the original HHO algorithm on a set of 22 real-world problems (CEC 2011). The results illustrate the superiority of HHO_A because it outperforms HHO significantly in 20 out of 22 problems, and it is never significantly worse. Four well-known nature-based algorithms were employed to compare the efficiency of the proposed algorithm. HHO_A achieves the best results in overall statistical comparison. A more detailed comparison shows that HHO_A achieves the best results in half real-world problems, and it is never the worst-performing method. A newly employed archive of old solutions significantly increases the performance of the original HHO algorithm.

Keywords: Swarm algorithm · Harris Hawks Optimisation · Real-world problems · Archive · Experimental comparison

1 Introduction

In this paper, a new variant of the nature-based optimisation method is proposed to tackle the global optimisation problem. There are a lot of methods to solve the global optimisation problem, where methods inspired by biological systems from nature (bio-inspired, nature-based, swarm-intelligence, etc.) are very popular and efficient in the last decades. These methods use a set of solutions called population or swarm, which is developed by typical processes to achieve the best solution of the task.

The efficiency of the various nature-based methods varies significantly, especially when various problems are solved (results of the comprehensive study illustrate the difference of these methods when solving artificial and real-world problems [3]). The main idea of newly proposed nature-inspired methods should be an application to some real-world optimisation problem. Therefore, a comparison of well-known optimisation methods (including nature-based), especially on

L. Rutkowski et al. (Eds.): ICAISC 2021, LNAI 12854, pp. 415–423, 2021.
https://doi.org/10.1007/978-3-030-87986-0_37

real-world problems, is very helpful for researchers who need to apply an existing method to their scientific optimisation task.

The rest of the paper is organised as follows. Section 2 introduces the original HHO algorithm, describes the main idea of the newly proposed HHO_A variant, and contains a brief characterisation of four well-known nature-inspired methods for comparison. In Sect. 3, settings of the experimental comparison are introduced. Section 4 illustrates numerical and graphical results from the comparison. Section 5 concludes the experimental paper.

2 Harris Hawks Optimisation

In 2019, Heidari et al. introduced an idea of a model where Harris's hawk preys rabbits to achieve the best amount of feed [5]. The Harris hawk optimisation (HHO) algorithm employs three phases - exploration, a transition from exploration to exploitation, and the exploitation phase. In the exploration phase, the solutions are updated using the random solution or the best solution from the population (controlled by control parameter q). Then, when the energy of the rabbits (prey) is low (the energy is estimated by variable E), the HHO algorithm is transferred from the exploration to the exploitation phase. The energy E is naturally dependent on the current step (generation) of the algorithm.

Algorithm 1. HHO algorithm

initialise population $P = \{\boldsymbol{x}_1, \boldsymbol{x}_2, \ldots, \boldsymbol{x}_N\}$
while stopping condition not reached **do**
  evaluate the population P
  update the best solution and position
  for $i = 1, 2, \ldots, N$ **do**
    update initial energy and jump strength
    if $|E| \geq 1$ **then**
      use exploration phase (1)
    else if $|E| < 1$ **then**
      if $r \geq 0.5$ & $|E| \geq 0.5$ **then**
        update location by soft besiege (2)
      else if $r \geq 0.5$ & $|E| < 0.5$ **then**
        update location by hard besiege (3)
      else if $r < 0.5$ & $|E| \geq 0.5$ **then**
        update location by soft besiege and progressively (4)
      else if $r < 0.5$ & $|E| < 0.5$ **then**
        update location by hard besiege progressively (5)
      end if
    end if
  end for
end while

In the exploration phase, a new position of the hawk is updated using the formula:

$$\boldsymbol{x}_{t+1} = \begin{cases} \boldsymbol{x}_{\text{rand}} - rand(\boldsymbol{x}_{\text{rand}} - 2\ rand\ \boldsymbol{x}_t) & \text{if } q \geq 0.5; \\ (\boldsymbol{x}_{\text{best}} - \boldsymbol{x}_{\text{avg}}) - rand(a + rand(b - a)) & \text{if } q < 0.5. \end{cases} \quad (1)$$

where $\boldsymbol{x}_{t+1}$ is the new position of the hawk, $\boldsymbol{x}_t$ is the current position of the hawk, $\boldsymbol{x}_{\text{rand}}$ is randomly selected hawk from the population, $\boldsymbol{x}_{\text{best}}$ is the best (rabbit's) position, $\boldsymbol{x}_{\text{avg}}$ is the average of coordinates from the whole population, and a and b are the minimal and maximal coordinates (bounds) of the search space.

When HHO is transferred to exploitation phase, it uses four different formulas to update hawk positions (based on rules from Algorithm 1):

$$\boldsymbol{x}_{t+1} = (\boldsymbol{x}_{\text{best}} - \boldsymbol{x}_t) - E\ |(2(1 - rand))\boldsymbol{x}_{\text{best}} - \boldsymbol{x}_t| \quad (2)$$

or

$$\boldsymbol{x}_{t+1} = \boldsymbol{x}_{\text{best}} - E\ |\boldsymbol{x}_{\text{best}} - \boldsymbol{x}_t| \quad (3)$$

or

$$\boldsymbol{x}_{t+1} = \begin{cases} \boldsymbol{y} = \boldsymbol{x}_{\text{best}} - E|(2(1 - rand))\boldsymbol{x}_{\text{best}} - \boldsymbol{x}_t| & \text{if } f(\boldsymbol{y}) < f(\boldsymbol{x}_t); \\ \boldsymbol{z} = \boldsymbol{y} + rand(1, D) \times LS & \text{if } f(\boldsymbol{z}) < f(\boldsymbol{x}_t). \end{cases} \quad (4)$$

or

$$\boldsymbol{x}_{t+1} = \begin{cases} \boldsymbol{y} = \boldsymbol{x}_{\text{best}} - E|(2(1 - rand))\boldsymbol{x}_{\text{best}} - \boldsymbol{x}_{\text{avg}}| & \text{if } f(\boldsymbol{y}) < f(\boldsymbol{x}_t); \\ \boldsymbol{z} = \boldsymbol{y} + rand(1, D) \times LS & \text{if } f(\boldsymbol{z}) < f(\boldsymbol{x}_t). \end{cases} \quad (5)$$

where $\boldsymbol{x}_{t+1}$ is the new position of the hawk, $\boldsymbol{x}_t$ is the current position of the hawk, $\boldsymbol{x}_{\text{best}}$ is the best (rabbit's) position, $\boldsymbol{x}_{\text{avg}}$ is the average of coordinates from the whole population, D is the dimensionality of the search space (problem), and LF represents Lèvy flight function (details are available in the original HHO paper [5]).

The behavior of HHO phases, i.e. transferring from exploration to exploitation using the control parameters illustrates Fig. 1.

The popularity of the original HHO algorithm illustrates the study, which summarises a lot of hybrid HHO variants from literature [1]. There are more than 60 research works from scientific databases published in two years what is a very promising result for the newly proposed optimisation method.

2.1 Archive of the Old Solutions - HHO$_A$

The original HHO algorithm uses five different approaches to update coordinates of the hawks in the population (Eq. (1)–(5)). In each case of update, when the new position is better than the old one, the old solution is replaced by the new

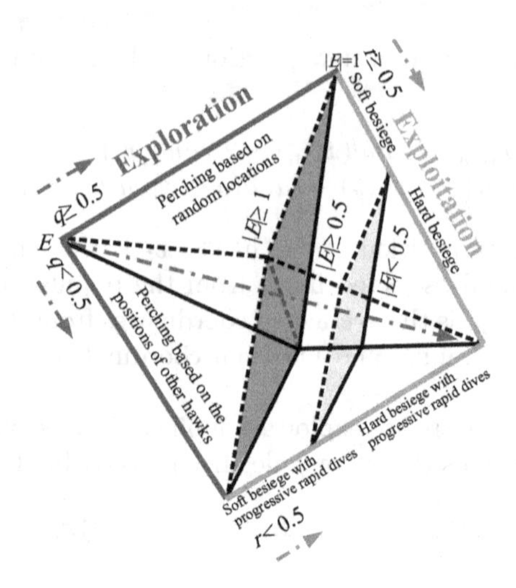

Fig. 1. An illustration of the stages of the HHO algorithm [5]

one. This idea is very efficient when the population is located in the area of the global minima of the function. In other cases, where the population occupies rather local minimum areas, the old-good (historical) positions can help to leave the population from the local minima of the function.

Based on this fact, an archive A for the old-good solutions is introduced for HHO to increase efficiency in the problems with many local minima. The idea of the archive is very simple. In the situations of HHO, when the old solution has to be replaced by a better new position, the old solution is located to archive A. The newly proposed HHO with archive A is called HHO_A. At the beginning of HHO_A, the archive is set to empty of size N and the old-good solutions are inserted gradually to the empty positions. When the archive is full, newly outperformed old-good solutions are located to randomly selected positions of A. Therefore, more current old solutions are kept for some parts of the population update.

The solutions from A are employed in selected update phases of HHO_A. At first, randomly selected individual from population $\boldsymbol{x}_{\text{rand}}$ in the exploration phase is in HHO_A selected from the union of population and archive $(P \bigcup A)$. Thus, when the archive of size N is full, it is probability 0.5 that the individual is from the population and 0.5 that it is from the archive. Next, the individual of the average coordinates from the population $\boldsymbol{x}_{\text{avg}}$ in both phases of the original HHO is computed only from coordinates of population. In HHO_A, the $\boldsymbol{x}_{\text{avg}}$ is computed as the mean of coordinates from the union of population and archive $(P \bigcup A)$. Thus, it promises more diversity when the population is trapped in the area of local minimum.

2.2 Nature-Inspired Algorithms in Comparison

Four different nature-inspired optimisation algorithms were selected for a more comprehensive comparison of the newly proposed HHO_A.

The bat algorithm (Bat hereafter) uses parameter settings that follow the original publication [8]. Maximal and minimal frequencies are set up $f_{\max} = 2$, $f_{\min} = 0$, the local-search loudness parameter is initialised $A_i = 1.2$ for each bat-individual and then reduced if a new bat position is better than the old one using coefficient $\alpha = 0.9$. The emission rate parameter is initialised for each bat-individual $r_i = 0.1$ and increased by parameter $\gamma = 0.9$ in the case of a successful offspring.

In 2014, the firefly algorithm (FFL in the experiments), was introduced as the model of real fireflies [9]. The control parameters are set to recommended values, randomisation parameter $\alpha = 0.5$, light absorption coefficient $\gamma = 1$, and attractiveness is updated using its initial and minimal values $\beta_0 = 1$, $\beta_{\min} = 0.2$.

In 2015, Kiran introduced an interesting idea of a tree-seed model (TSA) [6]. In this algorithm, a set of trees is represented by the population, which produces a bigger set of seeds. For each tree, only the best seed from the current generation is selected to compare with the tree if a better position is achieved. In 2020, a novel TSA variant using all better seeds and Eigen coordinate system was introduced, and it achieves substantially better results on real-world problems [2].

In 2015, Wang et al. proposed the elephant herd optimisation algorithm (labelled EHO) inspired by the hierarchical behaviour of elephants in elephants herd [7]. In the EHO model, elephants are structured into several clans controlled by female leaders, and some male elephants live separately from the clans. In this experiment, the recommended values of EHO control parameters are used - elitism parameter is 2, number of clans is 5, and parameters for a new clan center computation are $\alpha = 0.5$ and $\beta = 0.1$.

3 Experimental Settings

All six nature-inspired optimisation methods are applied to the set of 22 real-world problems of the CEC 2011 competition in the Special Session on Real-Parameter Numerical optimisation [4]. The true solution to these problems is unknown. The functions differ in the computational complexity and the dimensionality of the search space (from $D = 1$ to $D = 240$). For each algorithm and problem, 25 independent runs were performed. The run of the algorithm is stopped when the prescribed number of function evaluation $MaxFES = 150000$ is reached. The partial results reaching one third and two-thirds of $MaxFES$ were also mentioned. The solution of the problem is a point of the population with the smallest function value.

The population size of all algorithms is set to $N = 30$. It is clear that the population size is a crucial parameter significantly influencing the performance of evolutionary algorithms, nevertheless tuning of its value for each algorithm is computationally expensive.

Table 1. The mean ranks from the Friedman tests.

Alg.	$FES = 50000$	$FES = 100000$	$FES = 150000$
HHO_A	1.7	1.6	1.6
TSA	2.4	2.3	2.1
EHA	3.4	3.3	3.4
HHO	3.5	3.6	3.7
FFL	4.9	5.0	5.1
BAT	5.1	5.1	5.2

The remaining control parameters of all methods in comparison follow the recommendation of the authors. All the algorithms are implemented in Matlab 2020b, where statistical analysis is also assessed. All computations were carried out on a standard PC with Windows 7, Intel(R) Core(TM)i7-4790 CPU 3.6 GHz, 16 GB RAM. Source code of the algorithms in comparison are available from Mathworks.

4 Results and Discussion

The first insight into the performance of the six employed algorithms (HHO, HHO_A, EHA, TSA, FFL and BAT) on real-world problems provides the Friedman test. The test is based on the median values of the problems for all algorithms. Finally, mean ranks for each algorithm, including all real-world problems, are provided in Table 1 where lower mean rank represents a method with better overall performance. It is obvious, the order of the algorithms does not differ during the three stages, and the best performing is proposed HHO_A. On second and third position are TSA and EHA, followed by the original HHO. BAT is the worst performing method in comparison.

More details provide Table 2, where results of the Kruskal-Wallis tests are depicted. A significance level is in column *p*-level, and it is clear that the zero hypotheses are rejected for all real-world problems. In this case, algorithms are ranked from the best performing to the worst performing, where the first, second, third, and last algorithm is specified. For a better overview, the total number of these positions are in Table 3. The proposed HHO_A achieves the best results in 11 out of 22 problems, followed by TSA with seven wins. The original HHO and EHA are able to be the best in one task. On the other side, FFL and BAT are the poorest methods with the ten worst results out of 22.

Further, the proposed HHO_A is compared with each of five counterparts using the Wilcoxon rank-sum test. The count of the significantly better results of HHO_A ('+'), cases where the results are similar ($\approx$), and number of HHO_A failures ('–') are illustrated in Table 4.

HHO_A significantly outperforms HHO in 20 problems, and it is never worse (the results are similar for T03 and T11.6). It is significantly better than EHA

Table 2. The absolute ransk from the Kruskal-Wallis tests.

Fun	p-value	1st	2nd	3rd	Last
T01	1.92E–25	TSA	HHO_A	HHO	BAT
T02	1.69E–27	HHO_A	EHA	HHO	BAT
T03	1.41E–28	HHO	HHO	HHO	FFL
T04	1.30E–26	HHO_A	TSA	EHA	BAT
T05	7.25E–23	HHO_A	TSA	EHA	FFL
T06	9.60E–25	HHO_A	TSA	EHA	BAT
T07	4.64E–28	HHO_A	HHO_A	EHA	BAT
T08	4.30E–27	TSA	HHO_A	EHA	FFL
T09	5.14E–27	TSA	HHO_A	HHO	FFL
T10	9.82E–28	HHO_A	HHO	BAT	TSA
T11.1	3.12E–28	HHO_A	TSA	EHA	BAT
T11.2	1.21E–23	TSA	HHO_A	EHA	BAT
T11.3	1.14E–12	EHA	TSA	HHO_A	HHO
T11.4	4.68E–27	TSA	HHO_A	EHA	BAT
T11.5	1.14E–26	HHO	HHO_A	TSA	FFL
T11.6	2.20E–28	TSA	HHO_A	EHA	FFL
T11.7	2.00E–28	HHO_A	TSA	HHO	FFL
T11.8	5.10E–28	HHO_A	TSA	HHO	FFL
T11.9	2.37E–28	HHO_A	TSA	HHO	FFL
T11.10	4.10E–27	HHO_A	TSA	HHO	BAT
T12	5.26E–27	TSA	HHO_A	EHA	BAT
T13	2.64E–14	HHO_A	BAT	TSA	FFL

Table 3. Total count of positions from the Kruskal-Wallis tests.

Posit	HHO_A	TSA	EHA	HHO	BAT	FFL
1st	11	7	1	1	0	0
2nd	8	10	1	1	1	0
3rd	1	2	10	7	1	0
Last	0	1	0	1	10	10

in 19 cases and worse in the case of the T11.4 problem. HHO_A outperforms TSA in 12 problems, and it is worse in six (T01, T09, T10, T11.5, T11.7, T13). FFL performs worse than HHO_A in 21 problems, and it is similar in T11.4. HHO_A is better in 20 problems and performs similarly in the case of T03 and T11.4.

An ability to converge to a solution is illustrated in Fig. 2 where are the convergence plots for six selected problems. The minimum function value of the best solution in the population is recorded in 17 stages of the search process for each algorithm. It is obvious that the proposed HHO_A converges fast, and it is able to find positions with better (lower) function values even in problems where other methods get stuck (T02, T05, T11.2).

Table 4. The number of HHO_As' wins, equal results, and loses from the Wilcoxon tests.

HHO_A vs.→	HHO	EHA	TSA	FFL	BAT
$+/\approx/-$	20/2/0	19/2/1	12/4/6	21/1/0	20/2/0

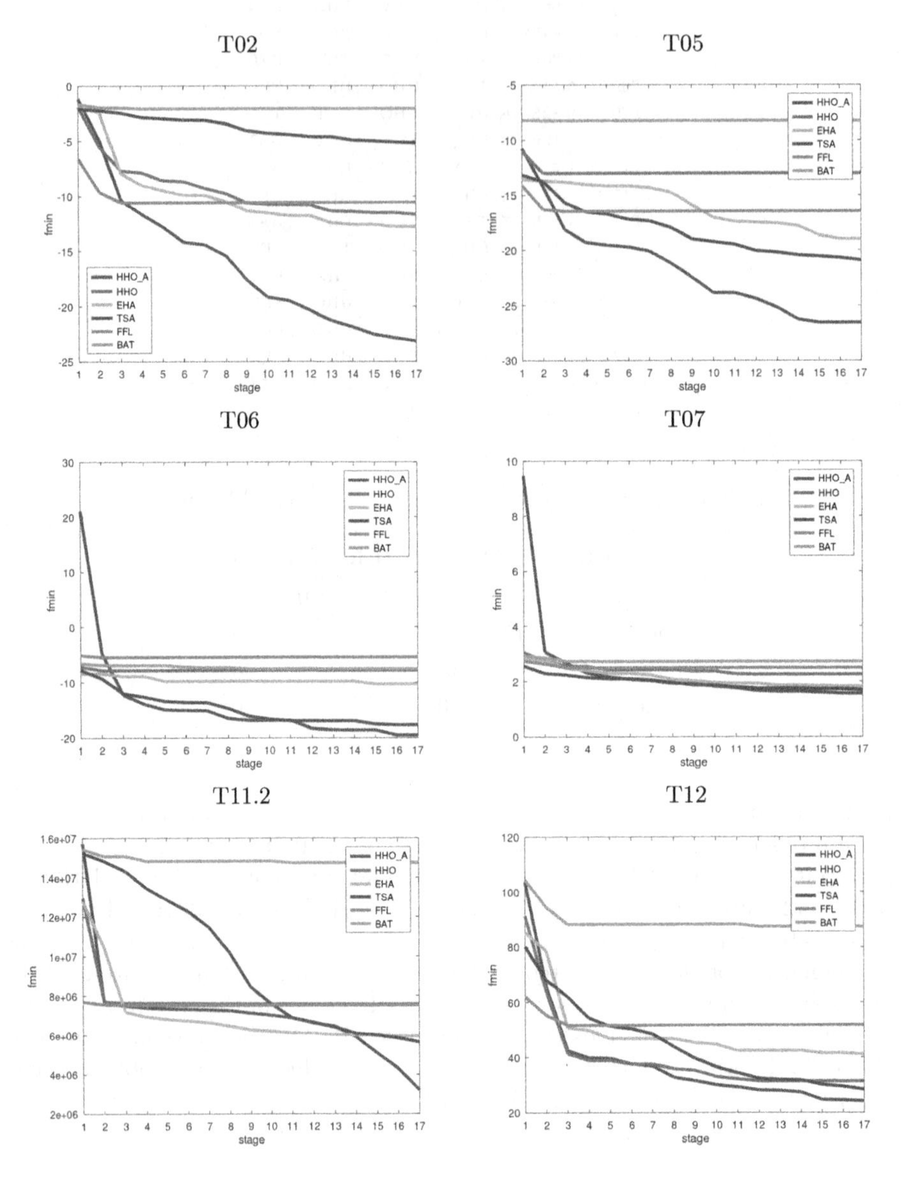

Fig. 2. Convergence of the algorithms in comparison.

5 Conclusion

Provided results from the comparison of the proposed HHO_A method with five well-known nature-inspired methods show the superiority of the new advanced algorithm. The only difference between HHO_A and the original HHO is in the archive of the old-good solutions, which is used in two rules of position update. On the other side, even the archive is substantially helpful, the HHO_A is outperformed in some problems, especially by the TSA algorithm. The archive for store the old solution to be removed is possible to employ in any of the population-based optimisation methods. Future research will be focused on studying the archive in other optimisation methods and increasing the performance of the proposed HHO_A algorithm.

References

1. Alabool, H., Al-Arabiat, D., Abualigah, L., Heidari, A.A.: Harris hawks optimization: a comprehensive review of recent variants and applications. Neural Comput. Appl. (2021). https://doi.org/10.1007/s00521-021-05720-5
2. Bujok, P.: Enhanced tree-seed algorithm solving real-world problems, pp. 12–16 (2020). https://doi.org/10.1109/ISCMI51676.2020.9311593
3. Bujok, P., Tvrdík, J., Poláková, R.: Comparison of nature-inspired population-based algorithms on continuous optimisation problems. Swarm Evol. Comput. **50**, 100490 (2019). https://doi.org/10.1016/j.swevo.2019.01.006
4. Das, S., Suganthan, P.N.: Problem definitions and evaluation criteria for CEC 2011 competition on testing evolutionary algorithms on real world optimization problems. Jadavpur University, India and Nanyang Technological University, Singapore, Technical report (2010)
5. Heidari, A.A., Mirjalili, S., Faris, H., Aljarah, I., Mafarja, M., Chen, H.: Harris hawks optimization: algorithm and applications. Future Gener. Comput. Syst. **97**, 849–872 (2019). https://doi.org/10.1016/j.future.2019.02.028
6. Kiran, M.S.: Tsa: tree-seed algorithm for continuous optimization. Expert Syst. Appl. **42**(19), 6686–6698 (2015). https://doi.org/10.1016/j.eswa.2015.04.055
7. Wang, G.G., Deb, S., Coelho, L.D.S.: Elephant herding optimization. In: 2015 3rd International Symposium on Computational and Business Intelligence (ISCBI), pp. 1–5 (2015). https://doi.org/10.1109/ISCBI.2015.8
8. Yang, X.S.: A new metaheuristic bat-inspired algorithm. In: Gonzalez, J., Pelta, D., Cruz, C., Terrazas, G., Krasnogor, N. (eds.) Nicso 2010: Nature Inspired Cooperative Strategies for Optimization. Studies in Computational Intelligence, vol. 284, pp. 65–74. Univ Laguna; Carnary Govt; Spanish Govt (2010). International Workshop on Nature Inspired Cooperative Strategies for Optimization NICSO 2008, Tenerife, Spain, 2008
9. Yang, X.S.: Nature-Inspired Optimization Algorithms. Elsevier, Amsterdam (2014)

Multiobjective Evolutionary Algorithm for Classifying Cosmic Particles

Krzysztof Pytel(✉)

Faculty of Physics and Applied Informatics, University of Lodz, Lodz, Poland
krzysztof.pytel@fis.uni.lodz.pl

Abstract. Classification is the process of predicting the class of objects. It is a type of Supervised Machine Learning, where predefined labels are assigned to objects, based on predetermined criteria. The article presents the idea of the Multiobjective Evolutionary Algorithm (MEA) that supports solving this problem. The proposed MEA uses two optimization criteria: the number of correctly assigned objects and the total distance between objects within the classes. In the process of multiobjective optimization, the algorithm minimizes the number of incorrectly assigned objects and maximizes the consistency of members within classes. The algorithm was tested on a few benchmarks and used to classify cosmic particles, based on their traces detected in Water Cherenkov Detectors (WCD). The results of the experiments suggest that the proposed algorithm takes advantage of the standard single-objective evolutionary algorithm in solving this problem. The algorithm can be also used for solving similar optimization problems.

Keywords: Classification · Evolutionary Algorithm · Multiobjective optimization

1 Introduction

Classification is a problem, where objects (observations) are grouped based on particular criteria. It is a difficult problem of Supervised Machine Learning (learning with a teacher). As an example of pattern recognition, the classification solves the identification problem of which of the set of categories the new object belongs to. Decisions are made based on a set of training data containing objects, whose category membership is known. It is a very important tool in the real world, where big sets of data are used to make decisions in government, economics, medicine, and more. While humans make classifications every day (e.g. "red" or "blue"), classification in supervised machine learning requires computers and complex algorithms. In machine learning, classification is about teaching computers to do the same. We try to build algorithms that learn how to assign class labels to all objects in a set. More information about classification can be found in publications [4,10].

L. Rutkowski et al. (Eds.): ICAISC 2021, LNAI 12854, pp. 424–433, 2021.
https://doi.org/10.1007/978-3-030-87986-0_38

Evolutionary Algorithms (EA) are optimization methods, that are inspired by the process of evolution in nature. They simulate the process of natural selection in computer software. The group of EA includes the Genetic Algorithms (GA), Evolutionary Programming (EP), and Evolutionary Strategies (ES). All these methods use the same concepts such as population, selection, reproduction, and mutation, taken from biology. They process a population of individuals in iterations called generations. The process of evolution includes fitness evaluation, reproduction, and mutation. New generations should be better fitted to the environment than their ancestors. The algorithm is terminated when a predefined criterion is met. Possible solutions to the problem (called individuals or chromosomes) are coded and represented by binary strings, real numbers, or other composite data structures. Each individual in the population has a calculated numerical value (fitness function), which describes the quality of this individual, that determines his ability to act as a parent for the next generation. Environmental pressure causes natural selection ("the survival of the fittest" rule). Individuals well-adapted to the environment are more likely to survive and pass on their genetic material to their descendants in the next generation. This increases the fitness of the entire population. Evolutionary Algorithms are usually used to solve sophisticated optimization problems in large search spaces, for which no other specialized techniques exist. They usually don't search for the global optimum but provide a near-optimum solution in an acceptable period. More information about Evolutionary Algorithms can be found in publications [5,8].

Classification can be formulated as a multi-objective optimization problem. We proposed an Evolutionary Algorithm with a two-objective function that is superior to an algorithm with a single-objective function.

Ultra-High Energy Cosmic Rays (UHECRs) are cosmic rays with an energy greater than 1 EeV (10^{18} electronvolts). It consists of sub-atomic particles traveling nearly at the speed of light. The origin of the UHECRs and their production mechanism is unknown. In Earth's atmosphere, particles interact with air particles and produce an "air shower". The surface detectors (for example in the Pierre Auger Observatory) can detect, identify and separate neutrino-origin air showers from air showers induced by regular cosmic rays for a large zenith angle. The Water Cherenkov Detector (WCD) contains $12\,\mathrm{m}^3$ of water and three photomultipliers (PMT), which receive Cherenkov's light, emitted by particles passing through the water. Each detector is equipped with its own electronics based on an FPGA chip from the Cyclone® V E family, and communications systems powered by solar energy. Signals from PMTs are digitized by 40 MHz 10-bit Analog to Digital Converters (ADCs) and sent to a Central Data Acquisition System (CDAS). CDAS combine information and identifies physical events, from the high-level trigger (T3) [1]. The limitations of software for this system are the number of programmable logic, I/O, memory resources, and power consumption. Software running in the FPGA must comply with these requirements. The ultra-high energy cosmic ray particles are very rare but they can be simulated in CORSIKA program. CORSIKA (COsmic Ray SImulations for KAscade)

[3] is a computer program that simulates the air shower, initiated by a particle of cosmic ray, i.e. protons or neutrinos. The data simulated in CORSIKA air showers are the input for the OffLine program [2]. It generates the ADC traces (signal waveforms) as a response of the Water Cherenkov Detector. Neutrinos, due to a very small cross-section for interactions, can generate showers initiated deeply into the atmosphere. Protons, with a much larger cross-section first interact usually shortly after entering the atmosphere. The simulations of air showers in CORSIKA and calculations in OffLine program showed that showers generated by protons ("old" showers) give relatively short ADC traces, while ADC traces generated by neutrinos ("young" showers) are spread in time. The timing of shower fronts directly observed as profiles of registered traces in the surface detectors is one of the fundamental criteria allowing identification of neutrino-induces showers. Figure 1 present an examples of ADC traces for protons and neutrinos.

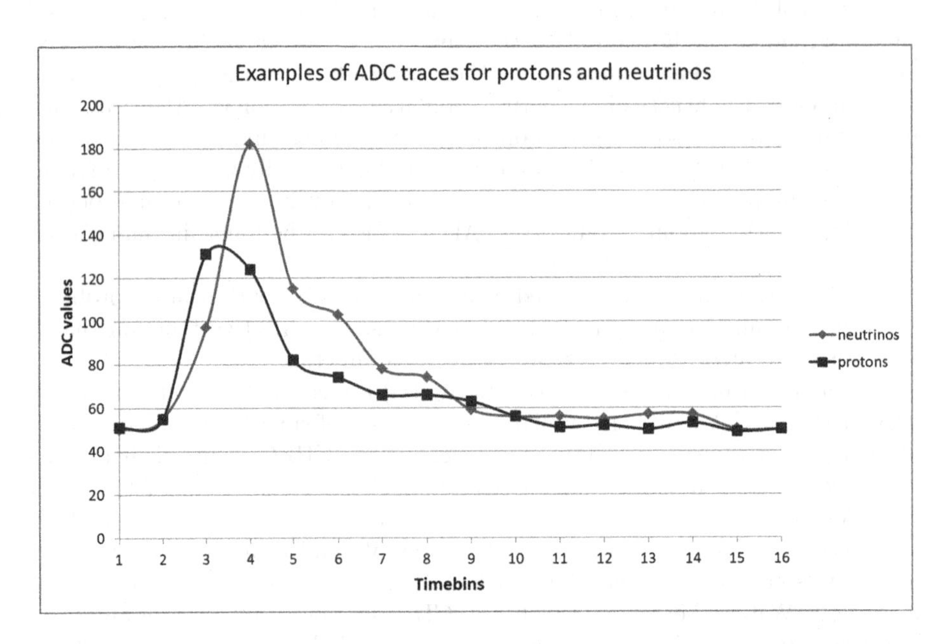

Fig. 1. Examples of ADC traces for protons and neutrinos

2 Problem Formulation

In machine learning, classification is defined as a process of supervised learning, in which the computer program learns from the training data, and then uses this knowledge to classify new objects or observations. The general aim of classification is to separate the objects into classes, using only training data. If the output label has two possible values, the problem is named binary classification. If there are more than two classes, the problem is referred to as multiclass classification.

A classification problem can be formally defined as the task of assigning the label y of a $k - dimensional$ input vector x, where $x \in X \subseteq R^k$ and $y \in Y = \{C_1, C_2, ..., C_Q\}$. This task is realized by using a classification rule or function $Y = f(X)$, able to assign the label of new objects. In the supervised learning, we are given a training set of N objects, represented by D, from which f will be adjusted, $D = \{(x_i, y_i), i = 1, ..., N\}$. The k-nearest neighbors algorithm is often used to classify objects in a variety of machine learning tasks.

The $k - nearest$ neighbors algorithm [6,7] is a supervised classification method that uses proximity as a measure for 'sameness'. The algorithm takes a bunch of labeled objects and uses them to learn how to label other objects. To label a new object, it looks at the labeled objects closest to that new object (its nearest neighbors). The geometric distance is usually used to determine which object is the nearest. After checking with k the number of the nearest neighbors, it assigns a label based on which label the most neighbors have.

The classification can be formulated as a multi-objective optimization problem, where two different functions are optimized:

- the number of correctly classified objects,
- the total distance of class members from their class centers.

$$\begin{cases} max \quad f_1(x) \\ min \quad f_2(x) \\ x \in X \end{cases} \tag{1}$$

where:

$$x = [x_1, x_2, x_3, ..., x_n] \in \Re \tag{2}$$

is an n-dimensional vector of the decision variables in search area X,
$f_1(x)$ - is a function of the number of correctly classified objects,
$f_2(x)$ - is a function of the total distance of class members from their class centers.

3 The Multobjective Evolutionary Algorithm for Solving Classification Problem

There are many publications on different methods of solving the classification problem, for example, k-means or Genetic Algorithms, but these methods do not work well for multiobjective optimization in sophisticated, multidimensional spaces. The proposed Multiobjective Evolutionary Algorithm (MEA) seeks for optimal placement of class centers, maximizes the number of correctly assigned objects, and minimizes the total distance of class members from their class centers.

In the proposed MEA the individuals' genes (potential solutions to the problem) are encoded by the means of real numbers and represent the coordinates of class centers. The algorithm uses the tournament selection method. The fitness function is computed for every individual based on its genotype. The value of the individual's fitness function is a real number, where the integer part is the number of incorrectly classified objects (inverted number of correctly classified objects) and the fractional part is the total distance between class members from their class centers divided by a constant. This constant is set depending on the task during the initial experiments.

The algorithms' parameters used in the experiments:

- the genes of individuals are coded by real numbers and represent the coordinates of class centers,
- the probability of crossover = 0.8,
- the probability of mutation = 0.1,
- the number of individuals in all populations = 25,
- the algorithms were stopped after a predefined number of generations.

4 Computational Experiments

The goal of experiments is to check the suitability of the proposed Multiobjective Evolutionary Algorithm in solving classification problems in a sophisticated, multi-dimensional environment. The experiments were divided into two stages. In the first stage, the effectiveness of the proposed algorithm was examined. The set of data from "The Fundamental Clustering Problems Suite" (FCPS) [18] has been used as a benchmark. For the experiment, two-dimensional tasks with from 400 to 4096 objects and 2 to 3 classes were selected. All tasks from selected benchmarks were solved by a k-means algorithm (we used the k-means method from the "rattle" library in R programming language [14]), proposed algorithm (MEA) and simple genetic algorithm (SGA) - an algorithm proposed in [8], and modified by me to solve a classification problem. During experiments two different fitness functions were used in SGA: the number of correctly classified objects (SGA1) and the total distance of class members from their class centers (SGA2). Each algorithm has been run a few times.

In Table 1 there are the best results obtained by each algorithm in the tests.

Table 1. The best results obtained by each algorithm in the tests.

Problem name		Lsun	TwoDiamonds	WingNut	EngyTime
Number of objects		400	800	1070	4096
Number of clusters		3	2	2	2
Number of dimensions		2	2	2	2
The number of correctly assigned objects	k-means	391	800	981	4010
	SGA1	400	800	1016	3954
	SGA2	338	800	833	2536
	MEA	400	800	1016	3953

Figure 2, 3, 4, 5 shows the distribution of objects in search space and the location of class centers obtained for each benchmark, using k-means, SGA1,

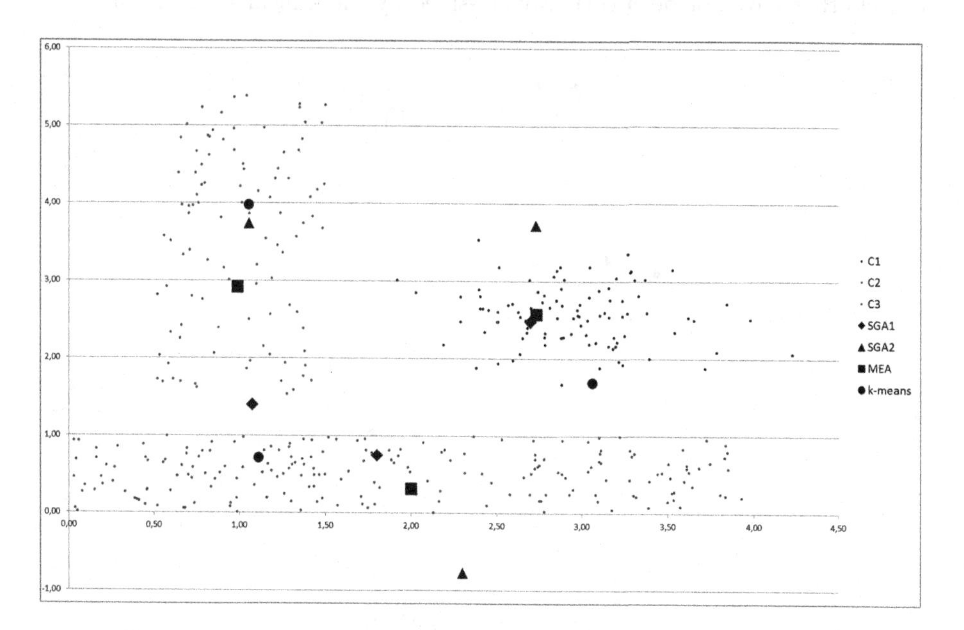

Fig. 2. The distribution of objects in search space and known a priori classifications in problem Lsun

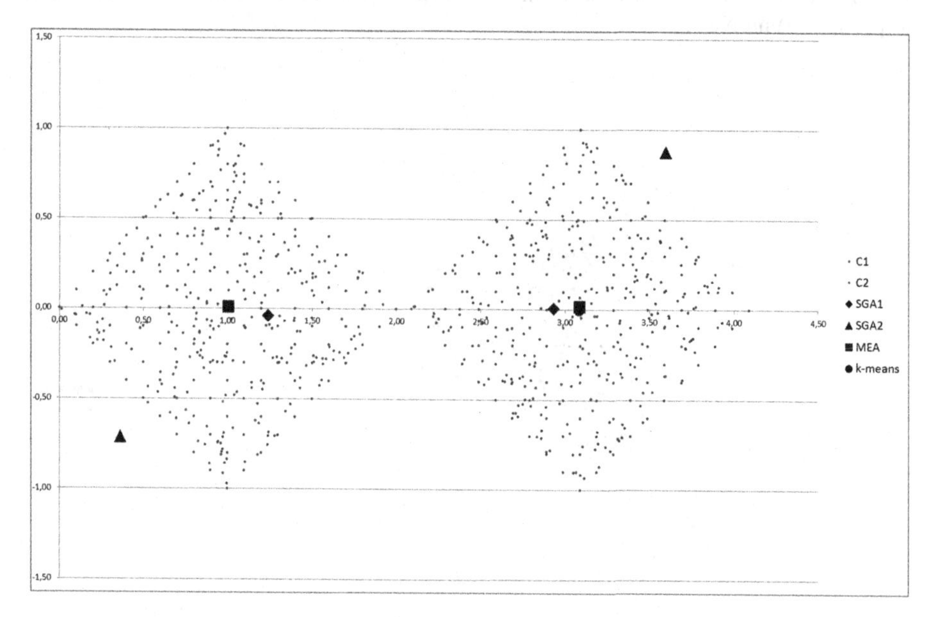

Fig. 3. The distribution of objects in search space and known a priori classifications in problem TwoDiamonds

SGA2, and MEA algorithms respectively. Classes C1, C2 and C3 represent known a priori distribution of objects. In the EngyTime task "Errors" is a class of objects that have not been correctly classified by the k-means algorithm.

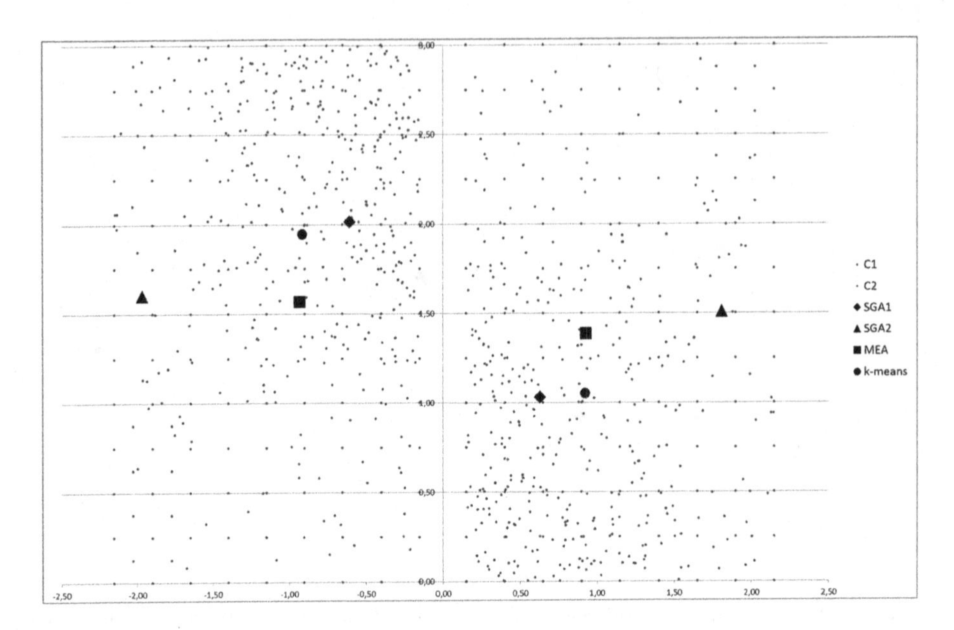

Fig. 4. The distribution of objects in search space and known a priori classifications in problem WingNut

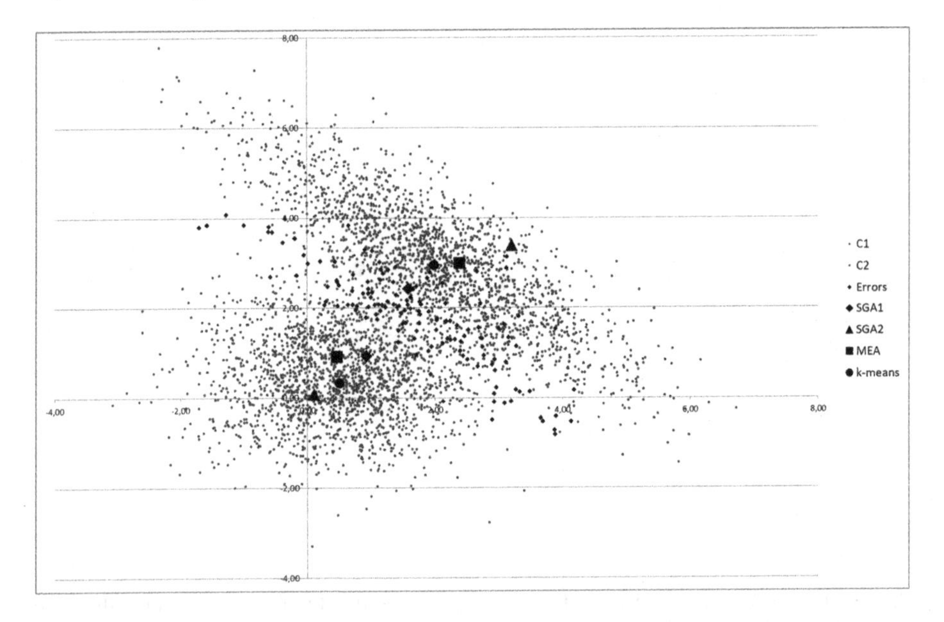

Fig. 5. The distribution of objects in search space and known a priori classifications in problem EngyTime

The construction of the fitness function causes the number of correctly classified objects a basic objective function. The total distance of class members from their class centers is the secondary objective function, that only matters if the number of correctly classified objects is equal. This construction makes it possible to find one of many equal quality solutions (the Pareto front), optimal in terms of the number of correctly classified objects, that at the same time minimizes the total distance of class members from their class centers. The SGA1 algorithm has a satisfactory result in terms of the number of correctly classified objects, but the location of the class's center is outside the area of distribution of objects from this class, e.g. in the Lsun or TwoDiamonds task. The position of class centers in MEA better represents the distribution of class objects, e.g. in the Lsun, WingNut, or EngyTime task.

The ability of the proposed system to classify cosmic particles based on its traces has been tested in the second stage of experiments. Data from simulations in CORSIKA and OffLine programs has been used as a task. The data set consists of 1242 particles including 621 neutrinos and 621 protons. The data are simulated traces of particles, that hit the Earth atmosphere in zenith angles – 80°, 85°, and 89°, and energies $3 * 10^8$, $3 * 10^9$, $3 * 10^9$, and 10^{10} GeV. The distances from the point of the first interaction of a particle with air nuclei to the detector are dependent on the type of particle. Traces, where first interactions of protons are very close to the detector, were rejected because the probability of this situation is very low. Moreover, this kind of interaction may include also the electromagnetic part of the shower and can disturbs classification. Only traces with a maximum ADC value of less than 200 were selected because the remaining particles are well detected by traditional triggers. This set of test data makes particle classification very difficult, but this range (not accessible to traditional triggers) may be very important for classification.

The task was solved by a k-means algorithm, proposed Multiobjective Evolutionary Algorithm (MEA), and simple genetic algorithm with two different fitness functions: the number of correctly classified objects (SGA1) and the total distance of class members from their class centers (SGA2). Each algorithm has been run a few times. In Table 2 there are the best results obtained by each algorithm.

Table 2. The results of cosmic particles' classification, obtained by algorithms.

	The number of correctly assigned objects	The percent of correctly assigned objects
k-means	639	51,4
SGA1	751	60.4
SGA2	621	50.0
MEA	827	66.5

5 Conclusions

The proposed Multiobjective Evolutionary Algorithm was able to find a solution for all tested problems in the first stage of experiments.

The total distance of class members from their class centers, as a fitness function (algorithm SGA2) has performed much worse than the others, except for the easy to optimize TwoDiamonds task.

In three out of four tasks from the benchmarks, the number of objects correctly classified by the proposed algorithm is equal to the algorithm with the number of correctly classified objects as a fitness function (SGA1). However, in MEA the distribution of class centers better illustrates the location of areas, where members of these classes are placed.

In Lsun and WinGnut tasks the proposed algorithm (MEA) correctly classified more objects than k-means. The class center distribution is more like a natural human classification.

Tests on data from CORSIKA and OffLine showed that MEA compared to SGA1, SGA2, and k-means classify more objects correctly. This may be due to a very large number of dimensions in the simulation data and the complexity of the solved problem.

Only traces with a low maximum signal were selected from a simulation in CORSIKA and OffLine for tests. Standard triggers used in Water Cherenkov Detectors are not capable of detecting particles with these signal values. Experiments show that the proposed algorithm can better classify cosmic particles with these signal values than other tested algorithms.

The proposed algorithm can be used to build a system consisting of fuzzy logic and an evolutionary algorithm. An example of such a system has been presented in publication [12]. The Fuzzy Logic Controller (FLC) can be used to improve the performance of the proposed algorithm [13]. In further work, these approaches will be tested. Different methods of artificial intelligence, such as Artificial Neural Networks [15,16] or Fuzzy Logic [17] were used to classify cosmic particles based on their traces. Future work will try to implement the proposed system to FPGA, test in the real-world, and compare results with traditional triggers and other methods of artificial intelligence.

The proposed algorithm can be an efficient tool for solving classification problems. It could be used for solving a very wide range of similar problems. The other way of using it is medicine, e.g. for classification of hypertension [11] or obesity [9].

References

1. Abraham, J., et al.: Trigger and aperture of the surface detector array of the pierre auger observatory. Nucl. Instrum. Methods Phys. Res. Section A **613**, 29–39 (2010) [Pierre Auger Collaboration]
2. Argiro, S., et al.: The offline software framework of the Pierre auger observatory. Nucl. Instrum. Methods Phys. Res. Section A **580**(3), 1485–1496 (2007)

3. CORSIKA an Air Shower Simulation Program [Online]. https://web.ikp.kit.edu/corsika/
4. Kaur, E.N., Kaur, E.Y.: Object classification Techniques using Machine Learning Model. Int. J. Comput. Trends Technol. (IJCTT) V **18**(4), 170–174 (2014)
5. Goldberg, D.E.: Genetic Algorithms in Search, Optimization, and Machine Learning. Addison-Wesley, Reading (1989)
6. Lloyd, S.P.: Least square quantization in PCM. Bell telephone labortories paper. Published in journal much later: Lloyd., S. P., Least squares 365 quantization in PCM. IEEE Trans. Inf. Theor. **28**(2), 129–137 (1982)
7. MacQueen, J.B.: Some Methods for classification and Analysis of Multivariate Observations. In: Proceedings of 5th Berkeley Symposium on Mathematical Statistics and Probability. University of California Press, pp. 281–370 297. MR 0214227. Zbl 0214.46201. (1967)
8. Michalewicz, Z.: Genetic Algorithms + Data Structures = Evolution Programs. Springer, Berlin (1992) https://doi.org/10.1007/978-3-662-03315-9
9. Nawarycz, T., Pytel, K., Ostrowska-Nawarycz, L.: Evaluation of health-related fitness using fuzzy inference elements. In: Rutkowski, L., Korytkowski, M., Scherer, R., Tadeusiewicz, R., Zadeh, L.A., Zurada, J.M. (eds.) ICAISC 2012. LNCS (LNAI), vol. 7267, pp. 301–309. Springer, Heidelberg (2012). https://doi.org/10.1007/978-3-642-29347-4_35
10. Osisanwo F.Y., Akinsola J.E.T., Awodele O., Hinmikaiye J. O., Olakanmi O., Akinjobi J.: Supervised Machine Learning Algorithms: Classification and Comparison. Int. J. Comput. Trends Technol. (IJCTT) V **48**(3), 128–138 (2017)
11. Pytel, K., Nawarycz, T., Ostrowska-Nawarycz, L., Drygas, W.: Anthropometric predictors and artificial neural networks in the diagnosis of hypertension. In: Proceedings of the 2015 Federated Conference on Computer Science and Information Systems, M. Ganzha, L. Maciaszek, M. pp. 2870–290 (2015). https://doi.org/10.15439/2015F246
12. Pytel, K., Nawarycz, T.: The fuzzy-genetic system for multiobjective optimization. In: Rutkowski, L., Korytkowski, M., Scherer, R., Tadeusiewicz, R., Zadeh, L.A., Zurada, J.M. (eds.) EC/SIDE -2012. LNCS, vol. 7269, pp. 325–332. Springer, Heidelberg (2012). https://doi.org/10.1007/978-3-642-29353-5_38
13. Pytel, K., Nawarycz, T.: Analysis of the distribution of individuals in modified genetic algorithms. In: Rutkowski, L., Scherer, R., Tadeusiewicz, R., Zadeh, L.A., Zurada, J.M. (eds.) ICAISC 2010. LNCS (LNAI), vol. 6114, pp. 197–204. Springer, Heidelberg (2010). https://doi.org/10.1007/978-3-642-13232-2_24
14. Rattle: A Graphical User Interface for Data Mining using R. http://rattle.togaware.com/
15. Szadkowski, Z., Pytel, K.: Artificial neural network as a FPGA trigger for a detection of very inclined air showers. IEEE Trans. Nucl. Sci. **62**, 1002–1009 (2015)
16. Szadkowski, Z., Glas, D., Pytel, K.: Optimization of an FPGA trigger based on an artificial neural network for the detection of neutrino-induced air showers. IEEE Trans. Nucl. Sci. **64**, 1271–1281 (2017)
17. Szadkowski, Z., Pytel, K.: Trigger based on a fuzzy logic for a detection of very inclined cosmic rays in the surface detector of the pierre auger observatory. In: 2019 IEEE International Instrumentation and Measurement Technology Conference (I2MTC), Auckland, New Zealand, pp. 1–5, (2019). https://doi.org/10.1109/I2MTC.2019.8827075
18. The Fundamental Clustering Problems Suite. https://www.uni-marburg.de/fb12/datenbionik/data/

A New Genetic Improvement Operator Based on Frequency Analysis for Genetic Algorithms Applied to Job Shop Scheduling Problem

Monique Simplicio Viana[1](✉), Rodrigo Colnago Contreras[2], and Orides Morandin Junior[1]

[1] Federal University of São Carlos, São Carlos, Brazil
{monique.viana,orides}@ufscar.br
[2] University of São Paulo, São Carlos, Brazil
contreras@usp.br

Abstract. Many researchers today are using meta-heuristics to treat the class of problems known in the literature as Job Shop Scheduling Problem (JSSP) due to its complexity since it consists of combinatorial problems and it is an NP-Hard computational problem. JSSPs are a resource allocation issue and, to solve its instances, meta-heuristics as Genetic Algorithm (GA) are widely used. Although the GAs present good results in the literature, it is very common for these methods that they are stagnant in solutions that are local optima during their iterations and that have difficulty in adequately exploring the search space. To circumvent these situations, we propose in this work the use of an operator specialized in conducting the GA population to a good exploration: the Genetic Improvement based on Frequency Analysis (GIFA). GIFA makes it possible to manipulate the genetic material of individuals by adding characteristics that are believed to be important, with the proposal of directing some individuals who are lost in the search space to a more favorable subspace without breaking the diversity of the population. The proposed GIFA is evaluated considering two different situations in well-established benchmarks in the specialized JSSP literature and proved to be competitive and robust compared to the methods that represent the state of the art.

Keywords: Evolutionary Algorithm · Genetic Algorithm · Genetic improvement · Job Shop Scheduling Problem · Combinatorial optimization

1 Introduction

Combinatorial optimization problems (COPs) consist of situations in which it is necessary to determine, through permutations of elements of a finite set, the configuration of parameters that is more advantageous [26]. Due to its high degree of applicability, many researchers have been using COPs in different contexts. As an example, applications in the logistics [29], vehicle routing [15], railway transport control [22], among other current problems [10]. In particular, one of the most addressed COPs in the literature is production scheduling [27], which, according to Groover [12], is part of the Production

L. Rutkowski et al. (Eds.): ICAISC 2021, LNAI 12854, pp. 434–450, 2021.
https://doi.org/10.1007/978-3-030-87986-0_39

Planning and Control activities and is responsible for determining the design of operations that will be carried out, such as: the environment in which products are processed, what resources are used and what is the start and end time for each production order.

Academic research and the development of solution methodologies have focused on a limited number of classic production scheduling problems, one of the most researched is the variation known as *Job Shop Scheduling Problem* (JSSP) [14], in which a finite set of jobs must be processed by a finite set of machines. In this category of problems, the objective is usually to determine a configuration in the order of processing of a set of jobs, or tasks, to minimize, for example, the time of using resources [39]. In this case, several performance measures are useful to evaluate how satisfactory a given configuration is for a JSSP, with *makespan* [38], which corresponds to the total time needed to finish the production of a set of jobs, one of the most used.

Belonging to the well-known class of problems *NP-Hard*, JSSP presents itself as a computational challenge, since it is not a trivial task to develop an approach to determine exact solutions that represent a configuration with an adequate performance measure, within a reasonable time, even considering small and moderate cases [35]. From this need, algorithms that present approximate results in a feasible computational time were developed and applied to JSSP. The main methods used are those composed of meta-heuristics [23], mainly by the Evolutionary Algorithm (EA) known as Genetic Algorithm (GA) [21,24,31–33]. Even so, the JSSP consists of a class of problems that remain open [6] and with many instances still unsolved in the well-known benchmarks of the area [9]. This is because the existing methods do not have the necessary efficiency to guarantee their practical use.

More specifically, it is possible to highlight some disadvantages in the use of GA in solving COPs [4,8]. In detail, it is common for this set of techniques that they become stagnant [30], during their iterations, in solutions that are local minimums, which configures the phenomenon known as premature convergence [41]. Also, GAs may require high computational time [20] to obtain good solutions to this type of problem. Therefore, for complex problems, GA needs to be assimilated to specific problem routines to make the approach effective. Hybridization can be a deeply effective way to improve the performance of these techniques. The most common form of hybridization is the addition of GAs to local search strategies and the incorporation of domain-specific knowledge in the search process [28]. In the latter, there are genetic improvement operators through manipulations in specific genes on a chromosome. These have as main objective to provide to individuals who are not able to stand out in a population the reinforcement coming from one or more individuals who have been successful in the adaptation process. In other words, these operators direct the worst individuals in a population to areas known to be good in the search space.

The authors do Amaral and Hruschka JR [2,3] present an operator in this line of reasoning, entitled transgenic operator and that simulates the process of genetic improvement. To conduct such a procedure, in one of the stages of the GA, the population of the same is replicated to four parallel sub-populations and, in each of these four populations, the best individuals transfer up to 4 genes, based on historical information, to selected individuals. Then, only the best individuals among the four sub-populations remains. Viana, Morandin Junior and Contreras [33] proposed an adaptation of the transgenic operator do Amaral and Hruschka JR [3] to solve a JSSP with GA. The authors propose the identification of relevance in the genes used in the transgenic process through a pre-processing process. However, such preprocessing is computationally time-consuming and may not be viable in large JSSPs.

In this work, we propose a new population guidance operator for GAs: the Genetic Improvement based on Frequency Analysis (GIFA) Operator. Our method consists of a new way to determine the genetic relevance based on the frequency analysis of the genes of individuals who have good fitness values in the population. We also propose the construction of a representative individual that represents this group of good individuals and that it is used in the process of genetic manipulation to guide the worst individuals towards good solutions and, possibly, that these become positive highlights in the population.

This work is divided into five sections. Specifically, in Sect. 2, we describe the JSSP basis. We present, in Sect. 3, the details about the proposed GIFA operator and the requirements that an GA needs to satisfy to use it. Experimental results on different GAs using GIFA and the advancement in the state of the art of JSSPs are presented in Sect. 4. The work is finished in Sect. 5 with conclusions about the developments carried out and future projections for improving the method and possible applications.

2 Formulation of Job Shop Scheduling Problem

We can define JSSP as a COP that has a set of N jobs that must be processed on a set of M machines. Also, each job has a script that determines the order in which it must pass through the machines for its process to be completed. Each job processing per machine represents an operation and the objective of a JSSP can be interpreted as being the challenge of determining the optimal sequencing of operations with one or more performance measures as a guide. The components of this problem follow some restrictions [39]:

- Each job can be processed on a single machine at a time;
- Each machine can process only one job at a time;
- Operations are considered non-preemptive, i.e., cannot be interrupted,
- Configuration times are included in processing times and are independent of sequencing decisions.

In this work, we adopted makespan (MKS) as a performance measure. The MKS is the total time that a JSSP instance takes to complete the processing of a set of jobs on a set of machines taking into account a given operation sequence.

Mathematically, let's assume the following components of a JSSP:

- $J = \{J_1, J_2, ..., J_N\}$ is the set of jobs;
- $\mathcal{M} = \{m_1, m_2, ..., m_M\}$ is the set of machines;
- $O = (O_1, O_2, ..., O_{N \cdot M})$ is a operation sequence that sets the priority order for processing the set of jobs in the set of machines
- $T_i(O)$ represents the time taken by the job J_i to be processed by all machines in its script according to the operation sequence defined in O.

Then, according to [7], the MKS can be defined as the total time that all jobs take to be processed according to a given operation sequence, as presented in Eq. (1).

$$\text{MKS} = \max_i T_i(O). \tag{1}$$

3 A New Genetic Improvement Operator Based on Frequency Analysis for GA Applied to JSSP

In this section, we will present in detail how the proposed method works. We will specify the ideа of determining genetic relevance by analyzing the frequency of genes that represent good characteristics in individuals with adequate fitness values in the population and, with that, we intend to obtain innovation with the following three topics:

- A new strategy for defining genetic relevance in GAs chromosomes;
- A new genetic improvement operator that is versatile and can be used in GAs variations,
- Improving the state of the art of JSSP benchmark results.

3.1 Genetic Representation

Our operator was developed to operate in all GA-like methods with minor modifications. In the meantime, we are going to conduct its fundamentation on a specific encoding. In this case, we will use the "coding by operation order" [5]. In this representation [32], the feasible space of a JSSP instance defined by N jobs and M machines is formed by chromosomes $c \in \mathbb{N}^{N \cdot M}$, such that exactly M coordinates of c are equal to i (representing the job index i), for every $i \in \{1,2,...,N\}$.

This encoding determines in chromosome the operation priority with respect to machine allocation. For example [31], let's assume $c = (2,1,2,2,1,1)$ as being a feasible solution in a JSSP instance with dimension 2×3 ($N = 2$ and $M = 3$). Thus, according to the operations defined in c, the following actions must be carried out in parallel or if the previous action has already been done.:

- 1st) Job 2 must be processed by the 1st machine of its script.
- 2nd) Job 1 must be processed by the 1st machine of its script.
- 3rd) Job 2 must be processed by the 2nd machine of its script.
- 4th) Job 2 must be processed by the 3rd machine of its script.
- 5th) Job 1 must be processed by the 2nd machine of its script.
- 6th) Job 1 must be processed by the 3rd machine of its script.

3.2 Fitness Function

The encoding used makes it natural to define the fitness function of the problem as the makespan of a JSSP instance given according to the stipulated operation sequence. That is, the fitness function [33] used is given according to Eq. (2):

$$\begin{aligned} F : \mathbb{O} &\longrightarrow \mathbb{R} \\ O &\longmapsto F(O) := \max_i T_i(O), \end{aligned} \tag{2}$$

in which $\mathbb{O}$ is the set of all possible operation sequences for the defined JSSP instance.

In this way, for this fitness function, the MKS of the JSSP instance is calculated according to a given operation sequence, then the meta-heuristic must look for an operation sequence in which the MKS is as small as possible and, consequently, the set of jobs must be processed by the set of machines taking the shortest possible time.

3.3 Proposed Genetic Improvement Based on Frequency Analysis Operator

In this work, we propose a new genetic improvement operator for evolutionary algorithms: the GIFA operator. The operator is based on a frequency analysis matrix calculated during the iterations of each GA. GIFA aims to calculate which genes on a chromosome can direct individuals with poor fitness values to better solutions and better search spaces. GIFA has two main stages: the first being defined by the making of the representative individual, that is, an individual that is determined by the configuration of the most frequent genes in the best individuals in the population; and the second stage consists of the use of the representative individual in the transgenic process, that is, the genetic manipulation through the insertion of specific genes of the representative individual in genes of the worst individuals in the population. Below, we present these steps in detail.

Stage 1: Composition of the Representative Individual. Initially, a portion of the population that presents the best fitness values is selected. Specifically, we select N_{Top} individuals who are considered good examples of solutions in the population. Then, for each index job i, a frequency vector $\vec{v}_i \in \mathbb{R}^{N \cdot M}$ is associated, in which the number of its occurrences is stored in each coordinate where the product i appears exactly at the position of this coordinate on the chromosomes selected for comparison. In Fig. 1, an example of the calculation of the frequency vectors $\vec{v}_i$ is presented when considering 4 individuals c_1, c_2, c_3, and c_4 with the best values of fitness in a 3×2 dimension JSSP instance.

Once the vector $\vec{v}_i$ has been made for every job index i, a chromosome whose coordinates are determined by the job with the highest frequency in this coordinate is defined as a representative individual. That is, each gene (coordinate) of the representative individual is defined as the job index that is most present in this coordinate in the best individuals in the population. It is also possible to establish an order of genetic relevance according to the frequency vectors $\vec{v}_i$. That is, it is possible to define which genes of the representative individual are more suitable to be transferred in the process of genetic improvement. Such relevance is also defined according to the frequency that the products present in each coordinate of the best individuals so that the genes that present the same job in many good individuals can categorize a "trend" that leads to good fitness values. Therefore, these genes must be considered to be relevant, since they describe a positive characteristic in several individuals that stand out in the population. Mathematically, the representative individual and its genetic relevance are made according to the following procedure:

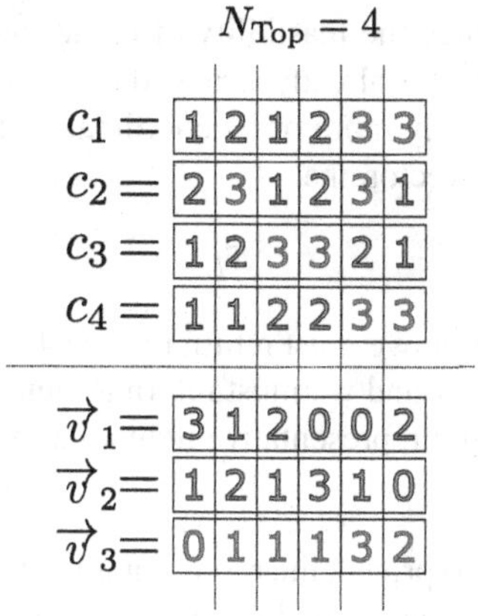

$N_{\text{Top}} = 4$

$c_1 =$	1	2	1	2	3	3
$c_2 =$	2	3	1	2	3	1
$c_3 =$	1	2	3	3	2	1
$c_4 =$	1	1	2	2	3	3
$\vec{v}_1 =$	3	1	2	0	0	2
$\vec{v}_2 =$	1	2	1	3	1	0
$\vec{v}_3 =$	0	1	1	1	3	2

Fig. 1. Calculating the frequency vectors ($\vec{v}_i$) of the three jobs in each coordinate of the four best chromosomes in the population.

1. Let **c** be the representative individual and w a vector that designates a score for each of its coordinates, initially null. In the following items, the coordinates of **c** and w are made.
2. We define I_1 as being $\arg\max_i \{\vec{v}_{i,1}\}$. That is, I_1 is the index of the job that has the highest frequency in the first coordinate of the exemplary individuals. Therefore, the first coordinate of the representative individual is defined as I_1. Mathematically,

$$\mathbf{c}_1 := I_1.$$

In addition, a w_1 score defined as the maximum frequency shown in the first coordinate of the best individuals is associated with the first coordinate of **c**. That is,

$$w_1 := \max_i \{\vec{v}_{i,1}\} = \vec{v}_{I_1,1}.$$

3. Assign the value 2 to j.
4. We define I_j as being $\arg\max_i \{\vec{v}_{i,j}\}$, that is, I_j is the most frequent product index in the j coordinate in the N_{Top} individuals. However, in order to guarantee the feasibility of the representative individual, it is necessary to establish two more restrictions:
   4.1 If the product I_j is not in M coordinates of **c**, then it is defined as I_j the j-th coordinate of the representative individual. That is,

$$\mathbf{c}_j := I_j.$$

In this case, the respective score is associated with the j -th coordinate of the representative individual as the maximum possible value presented in the j -th coordinate of the best individuals. That is,

$$w_j := \max_i \{\vec{v}_{i,j}\} = \vec{v}_{I_j,j}.$$

4.2 Otherwise, to guarantee the feasibility of **c**, the frequencies of the index job I_j are disregarded, since it is already arranged in M coordinates of **c** and, therefore, does not can occupy any more of its coordinates. To do so, we must cancel its respective frequency vector, that is,

$$\vec{v}_{I_j} := \vec{0}.$$

To make a new attempt, we must return to item 4.

5. If $j \neq N \cdot M$ then $j := j + 1$ and we must return to item 4. Otherwise, the procedure is finished and we have the representative individual pair and its respective genetic score $(\mathbf{c}, w)$.

Note that it is not necessary to project the representative individual in the feasible space of the problem since due to its construction and the item 4 above, it is already feasible. In Fig. 2, an example of the calculation of the representative individual (**c**) and the relevance of its genes (w) in a JSSP instance of dimension 4×3 is presented, taking as best individuals the $N_{\text{Top}} = 5$ individuals with the lowest fitness values available in the population.

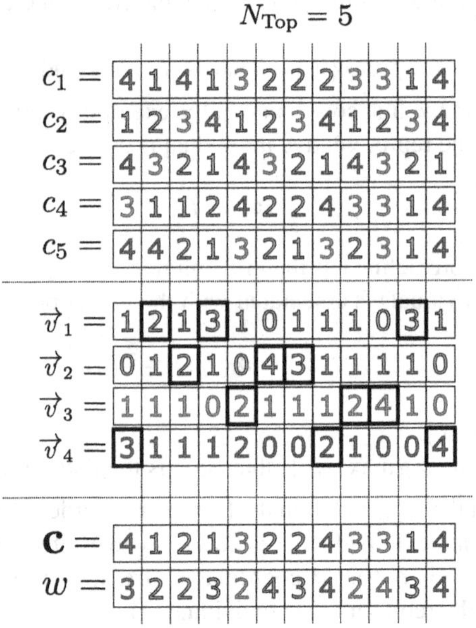

$N_{\text{Top}} = 5$

$c_1 =$	4	1	4	1	3	2	2	2	3	3	1	4
$c_2 =$	1	2	3	4	1	2	3	4	1	2	3	4
$c_3 =$	4	3	2	1	4	3	2	1	4	3	2	1
$c_4 =$	3	1	1	2	4	2	2	4	3	3	1	4
$c_5 =$	4	4	2	1	3	2	1	3	2	3	1	4
$\vec{v}_1 =$	1	2	1	3	1	0	1	1	1	0	3	1
$\vec{v}_2 =$	0	1	2	1	0	4	3	1	1	1	1	0
$\vec{v}_3 =$	1	1	1	0	2	1	1	1	2	4	1	0
$\vec{v}_4 =$	3	1	1	1	2	0	0	2	1	0	0	4
$\mathbf{c} =$	4	1	2	1	3	2	2	4	3	3	1	4
$w =$	3	2	2	3	2	4	3	4	2	4	3	4

Fig. 2. Computation of the representative individual (**c**) and its genetic relevance (w).

Stage 2: Use of the Representative Individual in Genetic Improvement. Once the representative individual and the relevance of each of its genes have been calculated, then it is proposed that its most relevant genes be transferred to the worst individuals in the population, thus simulating a mechanism for genetic improvement, or transgenics. For this, we take $P_{\text{Worst}} := \{x_1, x_2, \ldots, x_{N_{\text{Worst}}}\}$ as the set of the worst N_{Worst} individuals in a population. Subsequently, the most significant, or most relevant, N_{Genes} genes of the representative individual are transferred to all individuals in the P_{Worst} maintaining their original positions. This procedure can generate infeasible solutions. Thus, it is necessary to conduct a correction, or projection, process on the individuals resulting from this operation. For this, we carry out the projection through the Hamming distance [37] modifying only the genes that were not received from the representative individual. In this way, the individuals generated in this procedure are projected on the feasible set of the problem, giving rise to the genetically improved individuals $P_{\text{Improved}} = \{\hat{x}_1, \hat{x}_2, \ldots, \hat{x}_{N_{\text{Worst}}}\}$.

It is also necessary to establish how many genes will be transferred from the representative individual to the individuals of P_{Worst}. For this, we will follow a procedure similar to that of Viana, Morandin Junior, and Contreras [33], which empirically determine that the adequate amount of genes used in the genetic improvement process is given by the root of the number of coordinates of the chromosome. Thus, the process remains advantageous and does not cause early convergence in the population. Thus, in this work, N_{Genes} is defined as round $\left(\sqrt{N \cdot M}\right)$. In Fig. 3, an example of the determination of the most significant genes of a representative individual $\mathbf{c}$ when it is given the scores of his genes w while addressing a JSSP with dimension 4×3.

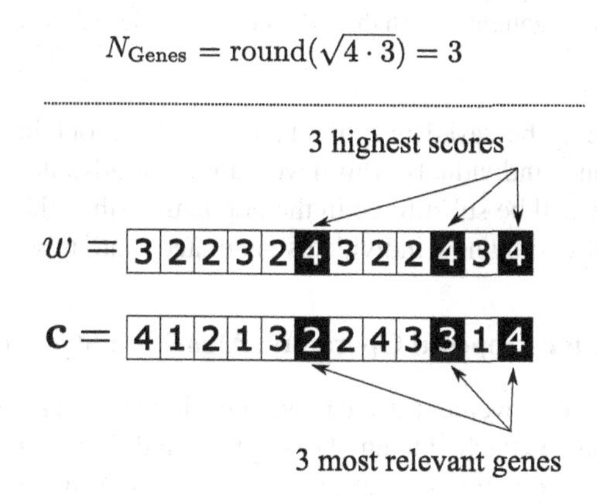

Fig. 3. Determination of the most significant genes of a representative individual.

Assuming $N_{\text{Worst}} = 3$ and $P_{\text{Worst}} = \{x_1, x_2, x_3\}$ as the set of the worst 3 individuals in a population, the improvement process is shown in Fig. 4 genetic that transfers the N_{Genes} best genes from the representative individual $\mathbf{c}$ of Fig. 4 to all individuals in the set P_{Worst}.

The genetic improvement procedure must be performed after the standard operators of the GA, or the GA-like method used, and right after the generation of a new population. Thus, the set P_{Worst} must be formed by individuals from the new population of the method. Besides, after the application of the genetic improvement, the evaluation of improvement or worsening of the affected individuals is made, so that the genetic

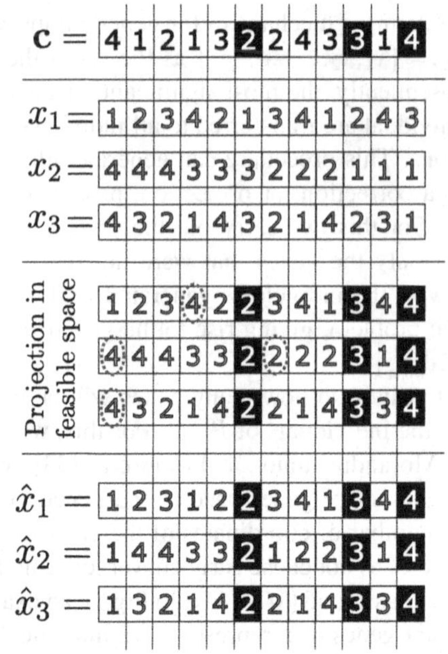

$\mathbf{c} =$ 4 1 2 1 3 2 2 4 3 3 1 4

$x_1 =$ 1 2 3 4 2 1 3 4 1 2 4 3

$x_2 =$ 4 4 4 3 3 3 2 2 2 1 1 1

$x_3 =$ 4 3 2 1 4 3 2 1 4 2 3 1

Projection in feasible space

1 2 3 4 2 2 3 4 1 3 4 4

4 4 4 3 3 2 2 2 2 3 1 4

4 3 2 1 4 2 2 1 4 3 3 4

$\hat{x}_1 =$ 1 2 3 1 2 2 3 4 1 3 4 4

$\hat{x}_2 =$ 1 4 4 3 3 2 1 2 2 3 1 4

$\hat{x}_3 =$ 1 3 2 1 4 2 2 1 4 3 3 4

Fig. 4. Genetic improvement proposed. The genes highlighted on a black background are the most relevant, while the genes highlighted with the red sectioned circle are those that need correction.

changes made will only be saved in individuals who have obtained an improvement in fitness. That is, only individuals who have gained an advantage in the process of genetic improvement will be substituted in the population; the other individuals should be discarded and replaced by new individuals generated randomly.

3.4 Scheme of Use for Proposed Operators: Algorithm Structure

The proposed genetic improvement strategy was developed to be as versatile as possible in the sense that it can be attached to any GA-like method. Thus, the proposed operator must be used after the execution of the original operators of the method considered in order to guide solutions that were not able to stand out through the traditional strategies defined in the method. In other words, to use the proposed operator in a given GA-like method, we must obey the following steps:

1. Define the initial parameters and specifics of the chosen GA-like method.
2. Execute the operators that make up the GA-like method. These being, for example, the operators of crossover, mutation, local search, creation of new population, etc.
3. At the end of an iteration involving the traditional operators of the selected GA-like method, we will make a sub-population P_{Worst} with the worst N_{Worst} individuals in the current population.
4. At the same time, we will select the best N_{Top} individuals in the population to make up the representative individual.

5. Build the representative individual using the strategy described in **Stage 1** of the Sect. 3.3.
6. Determine a relevance scale to the genes of the representative individual.
7. Conduct the genetic improvement of the P_{Worst} individuals using the most relevant N_{Genes} genes of the representative individual.
8. Replace in the current population of the method all individuals who obtained an improvement in the fitness value in the process of genetic improvement and return in the execution of the original operators of the considered GA-like method. Those who have not improved should be replaced by new individuals randomly generated according to Levy's exponential distribution, following the procedure of Al-Obaidi and Hussein [1].

In Fig. 5, we present a flowchart that illustrates the sequence of steps of the proposed genetic improvement process.

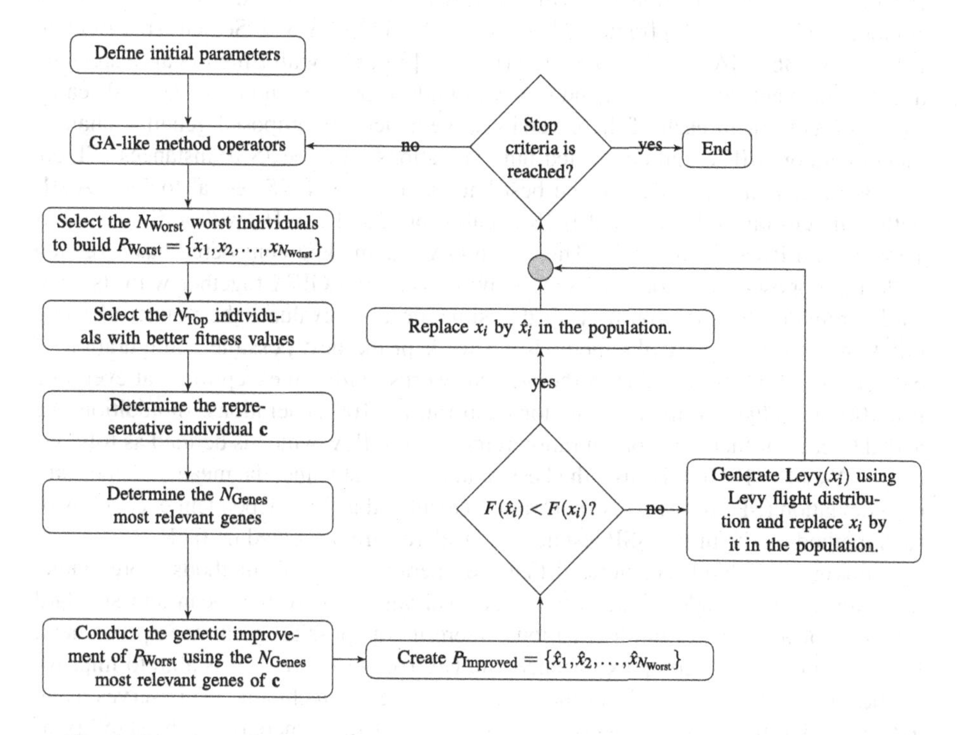

Fig. 5. Flow chart of our proposed Genetic Improvement operator for Genetic Algorithm.

4 Implementation and Experimental Results

4.1 Experimental Environment

For the conduction of the experiments, we considered two different situations: in the first, we evaluated the impact that the proposed operator causes on five GA-like methods, all of which were obtained using the framework of Viana, Morandin Junior and

Contreras [32], in three JSSP instances of varying complexity; in the second, we compare with recent methods in the literature the ability of the proposed operator to look for good solutions in 43 instances of JSSP that make up the area benchmark, with 3 from Fisher and Thompson (FT) [11] and 40 from Lawrence (LA) [19]. In detail, in this second situation, we consider relevant and recent methods which deal with the JSSP with the same specific instances and, when existing, presented in papers published in the last three years. In all, we consider for comparison the following methods: mXLSGA [32], NGPSO [40], SSS [13], GA-CPG-GT [18], DWPA [34], GWO [16], IPB-GA [17] and aLSGA [4]. The proposed algorithm is coded in MATLAB and we performed the evaluations on a computer with 2.4 GHz Intel(R) Core i7 CPU and 16 GB of RAM.

4.2 Results and Comparison with Other Algorithms

For the first testing situation, we will consider five variations of the Viana, Morandin Junior and Contreras [32] framework: a basic GA (GA), GA with Search Area Adaptation (GSA) [36], GA with Local Search (LSGA) [25], GA with Elite Local Search and agent adjustment (aLSGA) [4], and GA with multi-crossover and massive local search (mXLSGA) [32]. In each of these versions, we added the proposed genetic enhancement operator, GIFA, and conducted our evaluations on three JSSP instances: FT 06, with a dimension of 6×6, and the best-known solution (BKS) equal to 55; LA 01, with a dimension of 10×5, and BKS equal to 666; and LA 16, with a dimension of 10×10, and BKS equal to 945. Thus, each GA-like method considered has a version with the proposed operator, represented by the acronym GIFA together with its standard acronym. Our main purpose in this situation is to evaluate the impact of using GIFA in each of the GA-like methods, so we kept the best possible configuration of each of the methods available in the original works, with the exception that everyone had 100 individuals in their populations and run for 100 generations. In addition, we added to each of them the configuration referring to GIFA, which is defined as follows: $N_{\text{Top}} = N_{\text{Worst}} = 10$. In this case, the best value, the worst value, the mean, and the standard deviation (SD) of the makespan values calculated at 35 independent executions of each method on the three JSSP instances considered are presented in Table 1.

Looking at Table 1, we noticed that the operator made all methods more stable, reducing the magnitude of the worst makespan value found, the mean and standard deviation of all of them in all situations where it was possible to have improvement. However, in the most complex instance, the LA 16, our operator was able to improve the best makespan value only in the case of the aLSGA technique. This serves as an indication that the proposed operator brings a considerable increase in the stability of the method, but the ability to explore the search space still has a strong dependence on the original technique used. This is because our GIFA guides the population towards guiding individuals with bad makespan values in regions where individuals with good fitness values are known to increase local exploration and, therefore, find good solutions, but it is up to the base technique to indicate good search regions.

Thus, the second situation considered should serve as an experiment in this sense, so that we can evaluate the ability of the proposed operator to increase the search and exploration power of a given technique. For this, we will add the proposed GIFA operator in a technique already known to be effective in finding good solutions in the JSSP

Table 1. GA-like methods statistics for 35 executions of each method.

Instance	Method	Best	Worst	Average	SD
FT 06	GA	55	57	55.45	0.85
	GIFA-GA	55	56	55.14	0.21
	GSA	55	55	55	0
	GIFA-GSA	55	55	55	0
	LSGA	55	59	57.68	1.43
	GIFA-LSGA	55	56	55.84	0.73
	aLSGA	55	55	55	0
	GIFA-aLSGA	55	55	55	0
	mXLSGA	55	55	55	0
	GIFA-mXLSGA	55	55	55	0
LA 01	GA	666	712	679.02	9.98
	GIFA-GA	666	678	669.37	5.17
	GSA	666	715	677.8	13.61
	GIFA-GSA	666	687	672.34	7.43
	LSGA	666	726	697	16.65
	GIFA-LSGA	666	707	688.67	15.59
	aLSGA	666	666	666	0
	GIFA-aLSGA	666	666	666	0
	mXLSGA	666	666	666	0
	GIFA-mXLSGA	666	666	666	0
LA 16	GA	982	1100	1045.6	26.40
	GIFA-GA	982	1061	1022.89	20.51
	GSA	994	1110	1046.77	26.37
	GIFA-GSA	994	1021	1017.38	15.49
	LSGA	1016	1148	1084.25	32.27
	GIFA-LSGA	1016	1077	1037.11	26.62
	aLSGA	959	985	980.51	4.48
	GIFA-aLSGA	956	982	975.12	2.36
	mXLSGA	945	982	972.25	13.30
	GIFA-mXLSGA	945	979	959.93	6.37

instances that make up the benchmark today: the mXLSGA [32]. In this case, we will evaluate GIFA-mXLSGA at 40 instances LA and 3 FT instances. In Table 2, we presented the results derived from 10 independent executions of our method on the LA and FT instance tests. The columns indicate, respectively, the instance that was tested, the instance size (number of Jobs $\times$ number of Machines), the optimal solution of each instance, the results achieved by each method considering all the executions (best solution found and error percentage (Eq. (3)), and the mean of the error with respect to each benchmark (MErr).

$$E_{\%} = 100 \times \frac{\text{Best} - \text{BKS}}{\text{BKS}}, \quad (3)$$

in which $E_{\%}$ is the relative error, "BKS" is the best known Solution and "Best" is the best value obtained by executing the algorithm 10 times for each instance.

Analyzing Table 2, we can verify that the GIFA operator was able to improve the search capability of the mXLSGA method. Specifically, considering only the LA instances, the use of the proposed operator was able to reduce the magnitude of the average relative error by 0.12, which corresponds to a reduction of 19.67% of its value. In other words, the GIFA operator made the mXLSGA method able to find the best-known makespan in 72.5% of LA instances, obtaining an average relative error of 0.49, the lowest of all methods. Concerning FT instances, the proposed GIFA operator did not compromise the search capability of mXLSGA, causing the best-known solutions to be found in all instances. In summary, we can highlight some points when analyzing the results referring to Table 2:

- There was no worsening of the results in any instance with the use of the proposed operator;
- The GIFA-mXLSGA method obtained the lowest $E_{\%}$;
- The proposed operator made mXLSGA able to find the BKS in the LA 22 instance,
- The proposed operator improved the results of mXLSGA by 7 LA instances.

With the results of the two situations considered, we note that the proposed method is effective in increasing the stability and efficiency of finding good solutions for GA-like methods.

5 Conclusion

The objective of this work was to develop a new GA-like method operator to minimize the makespan in JSSP instances. The proposed technique was a new genetic improvement operator based on a new frequency analysis strategy to detect relevance in genes, titled GIFA operator. To evaluate the proposed approach, experiments were conducted in 43 JSSP instances of varying complexity. The instances used were FT [11] and LA [19]. The results obtained were compared with other approaches in related works: mXLSGA [32], NGPSO [40], SSS [13], GA-CPG-GT [18], DWPA [34], GWO [16], IPB-GA [17] and aLSGA [4].

We evaluate the potential of the proposed operator in two analysis situations. In the first, we involved the use of GIFA in five different GA-like methods, which were simulated by the framework of Viana, Morandin Junior and Contreras [32], and in all cases, the operator increased the stability of the technique considered improving the mean and standard deviation of the solutions found in three instances of JSSP. In the second, the performance of the GIFA used in the mXLSGA [32] was compared with methods that make up the state of the art in the specialized literature and we note that the use of the proposed operator was effective in increasing the search capacity of the

Table 2. Comparison of computational results between mXLSGA and other algorithms. The symbol "-" means "no evaluated in that instance".

Instance	Size	BKS	GIFA-mXLSGA		mXLSGA		NGPSO		SSS		GA-CPG-GT		DWPA		GWO		IPB-GA		aLSGA	
			Best	**E%**	Best	**E%**	Best	**E%**	Best	**E%**	Best	**E%**	Best	**E%**	Best	**E%**	Best	**E%**	Best	**E%**
LA01	10 × 5	666	666	0.00	666	0.00	666	0.00	666	0.00	666	0.00	666	0.00	666	0.00	666	0.00	666	0.00
LA02	10 × 5	655	655	0.00	655	0.00	655	0.00	655	0.00	655	0.00	655	0.00	655	0.00	655	0.00	655	0.00
LA03	10 × 5	597	597	0.00	597	0.00	597	0.00	597	0.00	597	0.00	614	2.84	597	0.00	599	0.33	606	1.50
LA04	10 × 5	590	590	0.00	590	0.00	590	0.00	590	0.00	590	0.00	598	1.35	590	0.00	590	0.00	593	0.50
LA05	10 × 5	593	593	0.00	593	0.00	593	0.00	593	0.00	593	0.00	593	0.00	593	0.00	593	0.00	593	0.00
LA06	15 × 5	926	926	0.00	926	0.00	926	0.00	926	0.00	926	0.00	926	0.00	926	0.00	926	0.00	926	0.00
LA07	15 × 5	890	890	0.00	890	0.00	890	0.00	890	0.00	890	0.00	890	0.00	890	0.00	890	0.00	890	0.00
LA08	15 × 5	863	863	0.00	863	0.00	863	0.00	863	0.00	863	0.00	863	0.00	863	0.00	863	0.00	863	0.00
LA09	15 × 5	951	951	0.00	951	0.00	951	0.00	951	0.00	951	0.00	951	0.00	951	0.00	951	0.00	951	0.00
LA10	15 × 5	958	958	0.00	958	0.00	958	0.00	958	0.00	958	0.00	958	0.00	958	0.00	958	0.00	958	0.00
LA11	20 × 5	1222	1222	0.00	1222	0.00	1222	0.00	1222	0.00	1222	0.00	1222	0.00	1222	0.00	1222	0.00	1222	0.00
LA12	20 × 5	1039	1039	0.00	1039	0.00	1039	0.00	-	-	1039	0.00	1039	0.00	1039	0.00	1039	0.00	1039	0.00
LA13	20 × 5	1150	1150	0.00	1150	0.00	1150	0.00	-	-	1150	0.00	1150	0.00	1150	0.00	1150	0.00	1150	0.00
LA14	20 × 5	1292	1292	0.00	1292	0.00	1292	0.00	-	-	1292	0.00	1292	0.00	1292	0.00	1292	0.00	1292	0.00
LA15	20 × 5	1207	1207	0.00	1207	0.00	1207	0.00	-	-	1207	0.00	1273	5.46	1207	0.00	1207	0.00	1207	0.00
LA16	10 × 10	945	945	0.00	945	0.00	945	0.00	947	0.21	946	0.10	993	5.07	956	1.16	946	0.10	946	0.10
LA17	10 × 10	784	784	0.00	784	0.00	794	1.27	-	-	784	0.00	793	1.14	790	0.76	784	0.00	784	0.00
LA18	10 × 10	848	848	0.00	848	0.00	848	0.00	-	-	848	0.00	861	1.53	859	1.29	853	0.58	848	0.00
LA19	10 × 10	842	842	0.00	842	0.00	842	0.00	-	-	842	0.00	888	5.46	845	0.35	866	2.85	852	1.18
LA20	10 × 10	902	902	0.00	902	0.00	908	0.66	-	-	907	0.55	934	3.54	937	3.88	913	1.21	907	0.55
LA21	15 × 10	1046	1052	0.57	1059	1.24	1183	13.09	1076	2.86	1090	4.20	1105	5.64	1090	4.20	1081	3.34	1068	2.10
LA22	15 × 10	927	927	0.00	935	0.86	927	0.00	-	-	954	2.91	989	6.68	970	4.63	970	4.63	956	3.12
LA23	15 × 10	1032	1032	0.00	1032	0.00	1032	0.00	-	-	1032	0.00	1051	1.84	1032	0.00	1032	0.00	1032	0.00
LA24	15 × 10	935	940	0.53	946	1.17	968	3.52	-	-	974	4.17	988	5.66	982	5.02	1002	7.16	966	3.31
LA25	15 × 10	977	984	0.71	986	0.92	977	0.00	-	-	999	2.25	1039	6.34	1008	3.17	1023	4.70	1002	2.55
LA26	20 × 10	1218	1218	0.00	1218	0.00	1218	0.00	-	-	1237	1.55	1303	6.97	1239	1.72	1273	4.51	1223	0.41
LA27	20 × 10	1235	1261	2.10	1269	2.75	1394	12.87	1256	1.70	1313	6.31	1346	8.98	1290	4.45	1317	6.63	1281	3.72
LA28	20 × 10	1216	1239	1.89	1239	1.89	1216	0.00	-	-	1280	5.26	1291	6.16	1263	3.86	1288	5.92	1245	2.38
LA29	20 × 10	1152	1190	3.29	1201	4.25	1280	11.11	-	-	1247	8.24	1275	10.67	1244	7.98	1233	7.03	1230	6.77
LA30	20 × 10	1355	1355	0.00	1355	0.00	1355	0.00	-	-	1367	0.88	1389	2.50	1355	0.00	1377	1.62	1355	0.00
LA31	30 × 10	1784	1784	0.00	1784	0.00	1784	0.00	1784	0.00	1784	0.00	1784	0.00	1784	0.00	1784	0.00	1784	0.00
LA32	30 × 10	1850	1850	0.00	1850	0.00	1850	0.00	-	-	1850	0.00	1850	0.00	1850	0.00	1851	0.05	1850	0.00
LA33	30 × 10	1719	1719	0.00	1719	0.00	1719	0.00	-	-	1719	0.00	1719	0.00	1719	0.00	1719	0.00	1719	0.00
LA34	30 × 10	1721	1721	0.00	1721	0.00	1721	0.00	-	-	1725	0.23	1788	3.89	1721	0.00	1749	1.62	1721	0.00
LA35	30 × 10	1888	1888	0.00	1888	0.00	1888	0.00	-	-	1888	0.00	1947	3.125	1888	0.00	1888	0.00	1888	0.00
LA36	15 × 15	1268	1295	2.12	1295	2.12	1408	11.04	1304	2.83	1308	3.15	1388	9.46	1311	3.39	1334	5.20	-	-
LA37	15 × 15	1397	1407	0.71	1415	1.28	1515	8.44	-	-	1489	6.58	1486	6.37	-	-	1467	5.01	-	-
LA38	15 × 15	1196	1246	4.18	1246	4.18	1196	0.00	-	-	1275	6.60	1339	11.95	-	-	1278	6.85	-	-
LA39	15 × 15	1233	1258	2.02	1258	2.02	1662	34.79	-	-	1290	4.62	1334	8.19	-	-	1296	5.10	-	-
LA40	15 × 15	1222	1243	1.71	1243	1.71	1222	0.00	1252	2.45	1252	2.45	1347	10.22	-	-	1284	5.07	-	-
MErr				**0.49**		**0.61**		**2.42**		**0.59**		**1.50**		**3.52**		**1.27**		**1.99**		**0.80**
FT06	6 × 6	55	55	0.00	55	0.00	55	0.00	55	0.00	55	0.00	-	-	55	0.00	55	0.00	55	0.00
FT10	10 × 10	930	930	0.00	930	0.00	930	0.00	936	0.64	935	0.53	-	-	940	1.07	960	3.22	930	0.00
FT20	20 × 5	1165	1165	0.00	1165	0.00	1210	3.86	1165	0.00	1180	1.28	-	-	1178	1.11	1192	2.31	1165	0.00
MErr				**0.00**		**0.00**		**1.28**		**0.21**		**0.60**		-		**0.73**		**1.84**		**0.00**

mXLSGA, since GIFA-mXLSGA was the method with the lowest average relative error among all the techniques considered. Thus, we conclude that the proposed operator was able to achieve the stipulated objective since it statistically directed the GA-like methods evaluated for search spaces with better solutions.

In future works, we will make use of deep learning techniques to detect relevance during GAs iterations through reinforcement learning approaches, which should make the methodology more robust and accurate. Also, we will add assessments about processing time measurements and expand the methodology to a greater number of instances and benchmarks.

Acknowledgments. This study was financed in part by the "*Coordenação de Aperfeiçoamento de Pessoal de Nível Superior - Brasil*" (CAPES) - Finance Code 001, and by the Brazilian National Council for Scientific and Technological Development, process #381991/2020-2.

References

1. Al-Obaidi, A.T.S., Hussein, S.A.: Two improved cuckoo search algorithms for solving the flexible job-shop scheduling problem. Int. J. Percept. Cogn. Comput. **2**(2), 25–31 (2016)
2. do Amaral, L.R., Hruschka, E.R.: Transgenic, an operator for evolutionary algorithms. In: 2011 IEEE Congress of Evolutionary Computation (CEC), pp. 1308–1314. IEEE (2011)
3. do Amaral, L.R., Hruschka Jr, E.R.: Transgenic: an evolutionary algorithm operator. Neurocomputing **127**, 104–113 (2014)
4. Asadzadeh, L.: A local search genetic algorithm for the job shop scheduling problem with intelligent agents. Comput. Ind. Eng **85**, 376–383 (2015)
5. Bierwirth, C., Mattfeld, D.C., Kopfer, H.: On permutation representations for scheduling problems. In: Voigt, H.-M., Ebeling, W., Rechenberg, I., Schwefel, H.-P. (eds.) PPSN 1996. LNCS, vol. 1141, pp. 310–318. Springer, Heidelberg (1996). https://doi.org/10.1007/3-540-61723-X_995
6. Çaliş, B., Bulkan, S.: A research survey: review of AI solution strategies of job shop scheduling problem. J. Intell. Manuf **26**(5), 961–973 (2015)
7. Chaudhry, I.A., Khan, A.A.: A research survey: review of flexible job shop scheduling techniques. Int. Trans. Oper. Res. **23**(3), 551–591 (2016)
8. Contreras, R.C., Morandin Junior, O., Viana, M.S.: A new local search adaptive genetic algorithm for the pseudo-coloring problem. In: Tan, Y., Shi, Y., Tuba, M. (eds.) ICSI 2020. LNCS, vol. 12145, pp. 349–361. Springer, Cham (2020). https://doi.org/10.1007/978-3-030-53956-6_31
9. Demirkol, E., Mehta, S., Uzsoy, R.: Benchmarks for shop scheduling problems. Eur. J. Oper. Res. **109**(1), 137–141 (1998)
10. Ehrgott, M., Gandibleux, X.: Multiobjective combinatorial optimization–theory, methodology, and applications. In: Multiple Criteria Optimization: State of the Art Annotated Bibliographic Surveys, pp. 369–444. Springer, Heidelberg (2003). https://doi.org/10.1007/0-306-48107-3_8
11. Fisher, C., Thompson, G.: Probabilistic learning combinations of local job-shop scheduling rules. In: Industrial Scheduling pp. 225–251 (1963)
12. Groover, M.P.: Fundamentals of Modern Manufacturing: Materials Processes, and Systems. John Wiley & Sons, Hoboken (2007)
13. Hamzadayı, A., Baykasoğlu, A., Akpınar, Ş: Solving combinatorial optimization problems with single seekers society algorithm. Knowl.-Based Syst. **201**, 106036 (2020)
14. Hart, E., Ross, P., Corne, D.: Evolutionary scheduling: a review. Genetic Program. Evol. Mach. **6**(2), 191–220 (2005)

15. James, J., Yu, W., Gu, J.: Online vehicle routing with neural combinatorial optimization and deep reinforcement learning. IEEE Trans. Intell. Transp. Syst **20**(10), 3806–3817 (2019)
16. Jiang, T., Zhang, C.: Application of grey wolf optimization for solving combinatorial problems: job shop and flexible job shop scheduling cases. IEEE Access **6**, 26231–26240 (2018)
17. Jorapur, V.S., Puranik, V.S., Deshpande, A.S., Sharma, M.: A promising initial population based genetic algorithm for job shop scheduling problem. J. Softw. Eng. Appl **9**(05), 208 (2016)
18. Kurdi, M.: An effective genetic algorithm with a critical-path-guided giffler and thompson crossover operator for job shop scheduling problem. Int. J. Intell. Syst. Appl. Eng **7**(1), 13–18 (2019)
19. Lawrence, S.: Resouce constrained project scheduling: an experimental investigation of heuristic scheduling techniques (supplement). Carnegie-Mellon University, Graduate School of Industrial Administration (1984)
20. Li, X., Gao, L.: An effective hybrid genetic algorithm and tabu search for flexible job shop scheduling problem. Int. J. Prod. Econ. **174**, 93–110 (2016)
21. Lu, Y., Huang, Z., Cao, L.: Hybrid immune genetic algorithm with neighborhood search operator for the job shop scheduling problem. In: IOP Conference Series: Earth and Environmental Science, vol. 474, 052093 (2020)
22. Matyukhin, V., Shabunin, A., Kuznetsov, N., Takmazian, A.: Rail transport control by combinatorial optimization approach. In: 2017 IEEE 11th International Conference on Application of Information and Communication Technologies (AICT), pp. 1–4. IEEE (2017)
23. Mhasawade, S., Bewoor, L.: A survey of hybrid metaheuristics to minimize makespan of job shop scheduling problem. In: 2017 International Conference on Energy, Communication, Data Analytics and Soft Computing (ICECDS), pp. 1957–1960. IEEE (2017)
24. Milovsevic, M., Lukic, D., Durdev, M., Vukman, J., Antic, A.: Genetic algorithms in integrated process planning and scheduling-a state of the art review. Proc. Manuf. Syst. **11**(2), 83–88 (2016)
25. Ombuki, B.M., Ventresca, M.: Local search genetic algorithms for the job shop scheduling problem. Appl. Intell. **21**(1), 99–109 (2004)
26. Pardalos, P.M., Du, D.-Z., Graham, R.L. (eds.): Handbook of Combinatorial Optimization. Springer, New York (2013). https://doi.org/10.1007/978-1-4419-7997-1
27. Parente, M., Figueira, G., Amorim, P., Marques, A.: Production scheduling in the context of industry 4.0: review and trends. Int. J. Prod. Res. **58**, 1–31 (2020)
28. Sastry, K., Goldberg, D., Kendall, G.: Genetic algorithms. In: Burke, E.K., Kendall, G. (eds.) Search Methodologies, pp. 97–125. Springer, Heidelberg (2005). https://doi.org/10.1007/0-387-28356-0_4
29. Sbihi, A., Eglese, R.W.: Combinatorial optimization and green logistics. Ann. Oper. Res. **175**(1), 159–175 (2010)
30. Viana, M.S., Morandin Junior, O., Contreras, R.C.: An improved local search genetic algorithm with a new mapped adaptive operator applied to pseudo-coloring problem. Symmetry **12**(10), 1684 (2020)
31. Viana, M.S., Junior, O.M., Contreras, R.C.: An improved local search genetic algorithm with multi-crossover for job shop scheduling problem. In: Rutkowski, L., Scherer, R., Korytkowski, M., Pedrycz, W., Tadeusiewicz, R., Zurada, J.M. (eds.) ICAISC 2020. LNCS (LNAI), vol. 12415, pp. 464–479. Springer, Cham (2020). https://doi.org/10.1007/978-3-030-61401-0_43
32. Viana, M.S., Morandin Junior, O., Contreras, R.C.: A modified genetic algorithm with local search strategies and multi-crossover operator for job shop scheduling problem. Sensors **20**, 5440 (2020). https://doi.org/10.3390/s20185440

33. Viana, M.S., Morandin Junior, O., Contreras, R.C.: Transgenic genetic algorithm to minimize the makespan in the job shop scheduling problem. In: Proceedings of the 12th International Conference on Agents and Artificial Intelligence, vol. 2: ICAART, pp. 463–474. INSTICC, SciTePress (2020). https://doi.org/10.5220/0008937004630474
34. Wang, F., Tian, Y., Wang, X.: A discrete wolf pack algorithm for job shop scheduling problem. In: 2019 5th International Conference on Control, Automation and Robotics (ICCAR), pp. 581–585. IEEE (2019)
35. Wang, L., Cai, J.C., Li, M.: An adaptive multi-population genetic algorithm for job-shop scheduling problem. Adv. Manuf. **4**(2), 142–149 (2016)
36. Watanabe, M., Ida, K., Gen, M.: A genetic algorithm with modified crossover operator and search area adaptation for the job-shop scheduling problem. Comput. Ind. Eng. **48**(4), 743–752 (2005)
37. Wegner, P.: A technique for counting ones in a binary computer. Commun. ACM **3**(5), 322 (1960). https://doi.org/10.1145/367236.367286
38. Wu, Z., Sun, S., Yu, S.: Optimizing makespan and stability risks in job shop scheduling. Comput. Oper. Res. **122**, 104963 (2020)
39. Xhafa, F., Abraham, A.: Metaheuristics for scheduling in industrial and manufacturing applications, vol. 128. Springer, Heidelberg (2008). https://doi.org/10.1007/978-3-540-78985-7
40. Yu, H., Gao, Y., Wang, L., Meng, J.: A hybrid particle swarm optimization algorithm enhanced with nonlinear inertial weight and gaussian mutation for job shop scheduling problems. Mathematics **8**(8) (2020). https://doi.org/10.3390/math8081355
41. Zang, W., Ren, L., Zhang, W., Liu, X.: A cloud model based DNA genetic algorithm for numerical optimization problems. Future Gener. Comput. Syst. **81**, 465–477 (2018)

Artificial Intelligence in Modeling and Simulation

Decision-Making Problems with Local Extremes: Comparative Study Case

Bartłomiej Kizielewicz, Andrii Shekhovtsov, Wojciech Sałabun(✉), and Andrzej Piegat

Research Team on Intelligent Decision Support Systems, Department of Artificial Intelligence and Applied Mathematics, Faculty of Computer Science and Information Technology, West Pomeranian University of Technology in Szczecin, ul. Żołnierska 49, 71-210 Szczecin, Poland
{bartlomiej-kizielewicz,andrii-shekhovtsov,wojciech.salabun, andrzej.piegat}@zut.edu.pl

Abstract. Many MCDA methods have been developed to support the decision-maker in solving complex decision-making problems. Most of them suppose the use of monotonic criteria, such as profit or cost. These methods do not consider the possibility of occurring local extremes in the space of the decision-making problem. Therefore, the question arises about how MCDA methods work when a decision problem consists of non-monotonic criteria.

We present a short comparative analysis for four popular MCDA methods, i.e., TOPSIS, VIKOR, PROMETHEE II and COMET. For this purpose, we have used simulations for two different decision-making models. In each case, sets of decision alternatives are generated, then evaluated by the model and selected MCDA methods. The obtained results create rankings from which rank similarity coefficients are calculated. The conducted research shows that the COMET method works better in such conditions than the others, and the VIKOR method does the least well in this task.

Keywords: MCDA · TOPSIS · VIKOR · PROMETHEE II · COMET

1 Introduction

In multi-criteria decision-making problems, we are most often faced with cost or profit criteria, which are monotonic [17]. In reality, however, these criteria may be non-monotonic [10], for example, the patient's temperature during the illness. A complete lack of fever and a too-high fever may be unfavourable to the patient, which means no or too violent an immune response from the body. Therefore, moderate fever is the most appropriate condition. This non-monotonic characteristic is also true in many technical cases, where too low or too high, a machine setting can be detrimental to its performance [5].

The consequence of non-monotonic criteria is that reference points (ideal solutions) occur inside the model space to be searched, not at the end of the

L. Rutkowski et al. (Eds.): ICAISC 2021, LNAI 12854, pp. 453–462, 2021.
https://doi.org/10.1007/978-3-030-87986-0_40

model. Thus, one or more local extremes may occur in the model being identified [9]. Multi-criteria decision-making analysis (MCDA) methods are very often used to assess a set of alternatives, also for uncertain environment [7]. In most cases, MCDA methods work linearly, so it is an interesting challenge to see how they work with such examples [3,6].

In our preliminary research, we will focus on selected popular MCDA methods. For this purpose, the following methods have been indicated: TOPSIS [1], VIKOR [16], PROMETHEE II [4] and COMET [13]. These approaches have been successfully applied to complex decision-making problems on many occasions, thus proving their usefulness [18]. However, the research question is to what extent they are suitable for problems involving non-monotonic criteria.

In this paper, an initial comparative study will be presented. Based on two simple numerical examples, the problem described in this paper will be presented, and the subsequent simulation experiments will be performed. In order to be able to illustrate the problem, the problem will be dealt with by examples with two criteria. One example will contain a single extremity and the other three extremes. In each case, we will have to deal with non-monotonic criteria. Once the problem has been presented for individual cases, we will also present comparative studies using the ranking similarity coefficients that the individual methods for our examples return.

The rest of the paper is organised as follow. Section 2 presents the methods used in the paper, i.e., the MCDA methods' parameters and the similarity coefficients. Section 3 presents two numerical examples comparing the accuracy of the methods for determining rankings. Section 4 presents the results of the conducted simulation experiments. The conclusions are presented in Sect. 5.

2 Preliminaries

2.1 The MCDA Methods

As the methods used are popular, their complete algorithms will not be presented in this paper. However, it is necessary to state the parameters in solving the problems in Sect. 3.

In this paper, the TOPSIS method was used as described in [15]. The equal criterion weights method was applied as weights, which is a very popular approach in the literature [18]. The input data were normalized using the vector method (1):

$$r_{ij} = \frac{x_{ij}}{\sqrt{\sum_{j=1}^{m} x_{ij}^2}} \tag{1}$$

where i the alternative number, j the criterion number, x_{ij} the attribute values before normalization and r_{ij} the attribute values after normalization.

For the VIKOR method, the algorithm presented in [16] was used. In this method, the parameter v is used as a weight for maximum group utility strategy, whereas $1-v$ is the weight of the individual regret. These strategies are compromised by using $v = 0.5$. Despite using the same criteria weights as in TOPSIS, we use the min-max normalization method (2) which gives better results in this method [12].

$$r_{ij} = \frac{x_{ij} - \min_j (x_{ij})}{\max_j (x_{ij}) - \min_j (x_{ij})} \tag{2}$$

From the established European decision support methods, the Promethee II method was used, where the algorithm used in [16] was applied. In this method, The min-max normalization and equal criteria weights were used. Additionally, the preference function P has to be selected, where the usual generalized criterion (3) was used in the paper [11].

$$P(d) = \begin{cases} 0 \; d \leq 0 \\ 1 \; d > 0 \end{cases} \tag{3}$$

As far as the COMET method is concerned, the most important parameter is establishing characteristic values. From these values, the characteristic objects are determined, which are then used to determine the fuzzy decision model. In both examples, the same set of characteristic values is used in the form of (4) and (4) respectively:

$$C_1 = \{0, 0.1, 0.2, 0.5, 0.65, 0.8, 0.9, 1\} \tag{4}$$

$$C_2 = \{0, 0.1, 0.2, 0.35, 0.5, 0.75, 1\} \tag{5}$$

More information on the algorithm of the COMET method used in this article can be found in [13].

2.2 Spearman's Rank Correlation Coefficient

For a sample of size N, rank values x_i and y_i are defined as (6). In this approach, the positions at the top of both rankings are more important. The weight of significance is calculated for each comparison. The element that determines the main difference to the Spearman's rank correlation coefficient examines whether the differences appeared and not where they appeared [2,8].

$$r_w = 1 - \frac{6 \sum_{i=1}^{N} (x_i - y_i)^2 \left((N - x_i + 1) + (N - y_i + 1)\right)}{N^4 + N^3 - N^2 - N} \tag{6}$$

2.3 Rank Similarity Coefficient

It is an asymmetric measure (7). The weight of a given comparison is determined based on the significance of the position in the first ranking, which is used as a reference ranking during the calculation [14].

$$WS = 1 - \sum_{i=1}^{N} 2^{-x_i} \frac{|x_i - y_i|}{max(|x_i - 1|, |x_i - N|)} \tag{7}$$

3 Comparative Study Case

This section presents a selection of two research cases, respectively, for one and three local extremes. The cases will be described step by step in order to explain the research challenge better.

3.1 Model with One Local Extreme

The first example will present an example with a single extreme for a two-criteria problem. As a reference for testing the accuracy of the obtained results, a simple function is used in the following form (8). Figure 1 shows the appearance of the function. The optimum value is located at point (0.2, 0.2).

$$f(x, y) = \frac{0.2}{0.1 + 4(x - 0.2)^2 + 2(y - 0.2)^2} \tag{8}$$

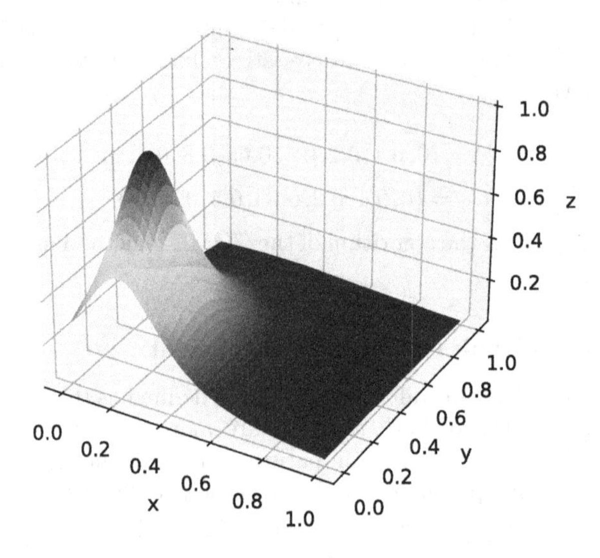

Fig. 1. Graphical representation of decision function with a single extremum.

A random selection of a set of decision alternatives is made to investigate the differences between the methods. In this way, a set of ten alternatives is determined, whose values are presented in Table 1 and visualised in Fig. 2.

Table 1. Comparison of obtained rankings for function with single extreme for a sample set of 10 alternatives.

Alternatives	Criteria		Rankings				
A_i	C_1	C_2	Reference	Comet	Vikor	Topsis	Promethee II
A_1	0.504672	0.358569	8	8	10	8	9.5
A_2	0.437299	0.090519	5	6	8	2	5.5
A_3	0.457243	0.541148	9	9	9	10	9.5
A_4	0.358303	0.185391	3	3	7	3	6
A_5	0.258697	0.632220	7	7	4	9	8
A_6	0.218545	0.559389	6	5	1	7	5.5
A_7	0.380028	0.036177	4	4	3	1	1
A_8	0.320918	0.206607	2	2	6	4	3
A_9	0.796291	0.025905	10	10	2	6	7
A_{10}	0.277296	0.298313	1	1	5	5	3

This sample set is intended to show how the simulations in Sect. 4 are run. In Fig. 2, we can see exactly how the points are distributed in space and the functions' values, which explains why linear methods may have difficulty in determining the correct rankings.

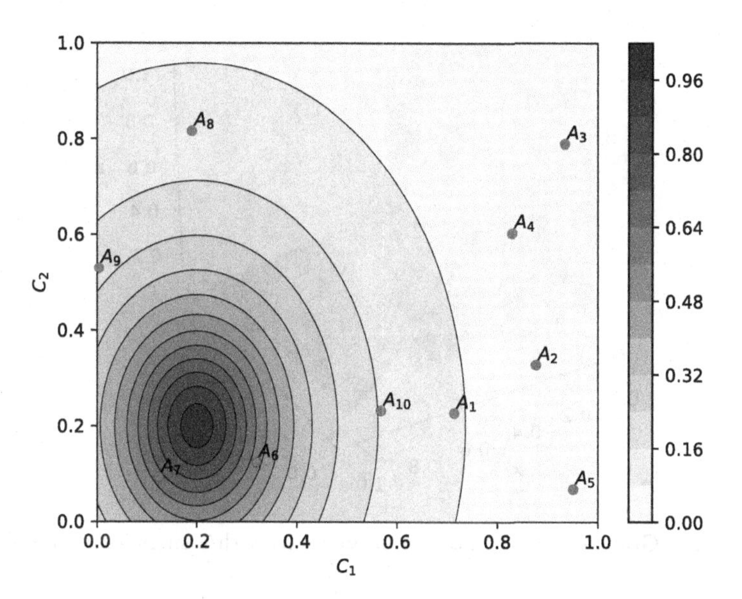

Fig. 2. Graphical representation of the sample set of ten assessed alternatives for function with single extreme.

Also, Table 1 contains the values of all the rankings together with the reference ranking determined with the help of function (8). This table is used to calculate the similarity values of these rankings, which were respectively: COMET (r_w = 0.9879, WS = 0.990625), VIKOR (r_w = −0.0457, WS = 0.5307), TOPSIS (r_w = 0.5923, WS = 0.6589), and Promethee II (r_w = 0.7645, WS = 0.7655). The best match was obtained with COMET, followed by Promethee II, TOPSIS and VIKOR; however, this example is more extensively tested in Sect. 4.

3.2 Model with Three Local Extreme

This example presents an example with three extremes. A previous function is modified to the following form (9) as a reference for testing the accuracy of the obtained results. Figure 3 shows the shape of the reference function. The extreme value are located at points (0.2, 0.2), (0.5, 0.5), and (0.8, 0.2).

$$f(x,y) = \frac{0.2}{0.1 + 4(x-0.2)^2 + 2(y-0.2)^2} + \frac{0.3}{0.1 + 6(x-0.5)^2 + 3(y-0.5)^2} + \frac{0.2}{0.1 + 20(x-0.8)^2 + 5(y-0.2)^2} \tag{9}$$

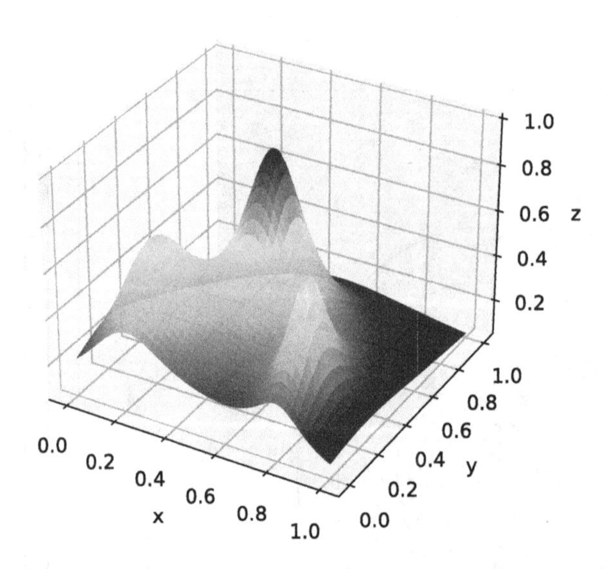

Fig. 3. Graphical representation of functions with a single extremums.

Table 2. Comparison of obtained rankings for function with three extrema for a sample set of 10 alternatives.

Alternatives	Criteria		Rankings				
A_i	C_1	C_2	Reference	Comet	Vikor	Topsis	Promethee II
A_1	0.596255	0.608955	2	2	8	9	7
A_2	0.585894	0.084457	8	5	6	5	5
A_3	0.402663	0.005038	5	7	2	3	1
A_4	0.798733	0.459945	4	3	9	8	8.5
A_5	0.013192	0.692212	10	10	3	6	6
A_6	0.882442	0.263445	3	4	7	7	8.5
A_7	0.518293	0.066079	7	6	4	4	3.5
A_8	0.682992	0.796625	9	9	10	10	10
A_9	0.130060	0.250560	1	1	5	2	3.5
A_{10}	0.007649	0.261115	6	8	1	1	2

As in the previous example, a random selection of a set of decision alternatives is made to investigate the methods' differences. Table 2 presents the selected set of alternatives with their attributes C_1 and C_2 and all available rankings. Figure 4 is visualised the set of data, where we can see why the rankings are so different.

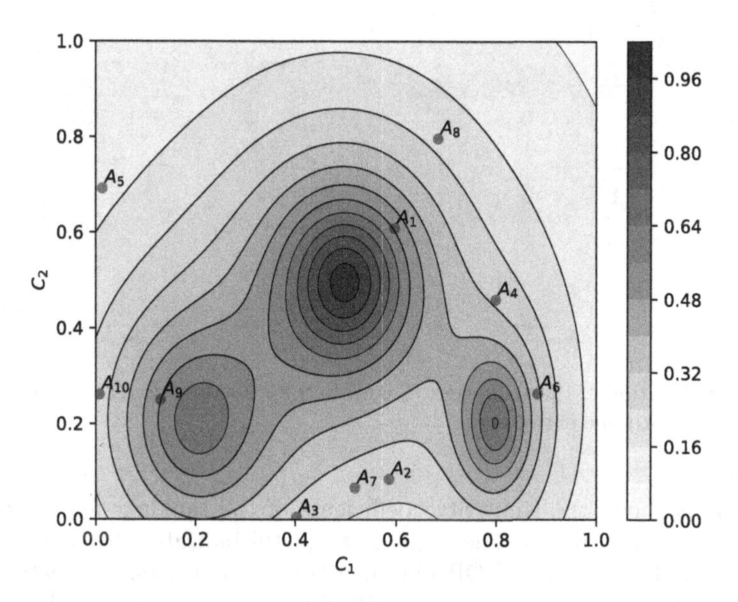

Fig. 4. Graphical representation of the sample set of ten assessed alternatives for function with three extrema.

This table is used to calculate the similarity values of these rankings, which were respectively: COMET (r_w = 0.8942, WS = 0.9500), VIKOR (r_w = −0.2066, WS = 0.4264), TOPSIS (r_w = 0.1074, WS = 0.5782), and Promethee II (r_w = 0.0384, WS = 0.5154). It is worth referring to the values we obtained in Sect. 3.1. As far as the matching order of the analysed MCDA methods itself is concerned, it is similar. The difference is because the TOPSIS method has slightly better matching in this example than Promethe II.

Without exception, all results deteriorated, but the COMET method still seems to have the best fit. However, more extensive simulation studies would need to be carried out to draw any firm conclusions, which will be presented in the next section.

4 Simulations

The short simulation section is presented to answer the research question to what extent the MCDA methods studied are suitable for analysed problems involving non-monotonic criteria. For both examples in Sect. 3, 4000 sets of alternatives are generated to show a reliable distribution of r_w and WS coefficients.

Figure 5 shows the similarity coefficients of the rankings for the example with one local extreme. The coefficient r_w indicates the best fit for the COMET method, while VIKOR obtained the worst results. The Promethee II and TOPSIS methods are very similar with one slight advantage for Promethee II. Very similar results were obtained with the WS distribution.

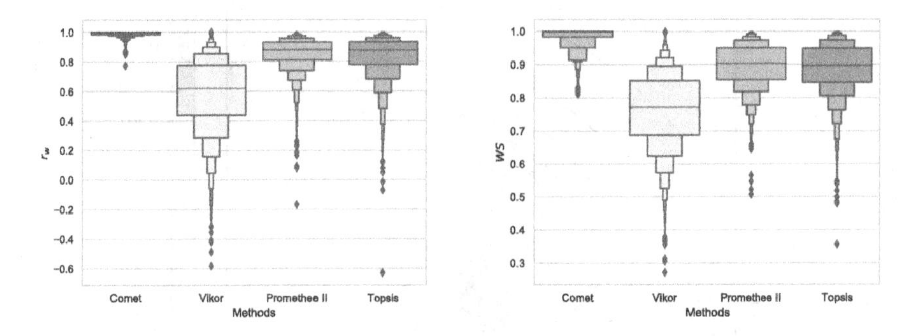

Fig. 5. Distribution of the coefficient r_w and WS for 4000 random sets of alternatives for functions with one extremum.

Figure 6 shows the similarity coefficients of the rankings for the example with three local extremes. The coefficient r_w still indicates the best fit for the COMET method, while VIKOR obtained the worst results. The deterioration of the results is significant for the TOPSIS and Promethee II methods, whose similarity distribution of rankings starts to resemble more the distribution for the VIKOR method. The results for the WS coefficient lead us to the same conclusions.

The occurrence of non-monotonic criteria and consequently, local extremes represents a significant challenge for classical MCDA methods, whose results do not correspond satisfactorily with the reference values.

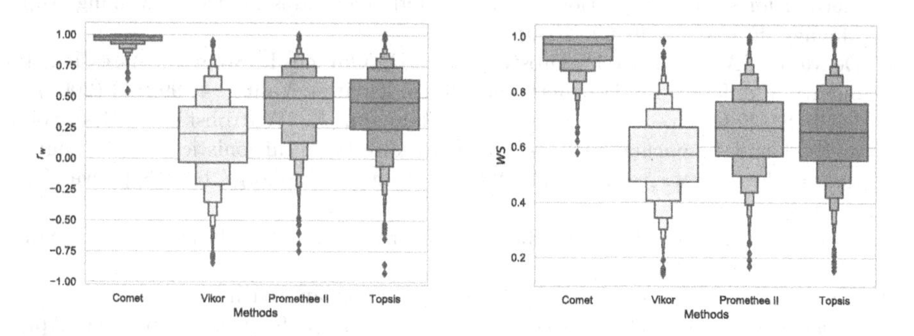

Fig. 6. Distribution of the coefficient r_w and WS for 4000 random sets of alternatives for functions with three extrema.

5 Conclusions

In this work, a problem related to decision problems taken in non-monotonic decision criteria is presented. In such cases, local extremes occur inside the identified model and not at its edge. Hence the problem with a correct representation of the decision maker's preferences by linear MCDA methods.

Our preliminary research used three linear MCDA methods, i.e., TOPSIS, PROMETHEE II, and VIKOR Besides, the COMEt method was involved, which has good properties for identifying non-linear decision models. Two examples are given to illustrate the problem, and then it is shown that linear methods perform worse in the cases studied, as indicated by a short computer simulation. More extensive simulation experiments should be carried out to confirm this problem's existence on the broader MCDA problems as a further direction of work.

Acknowledgments. The work was supported by the National Science Centre, Decision number UMO-2018/29/B/HS4/02725.

References

1. Behzadian, M., Otaghsara, S.K., Yazdani, M., Ignatius, J.: A state-of the-art survey of TOPSIS applications. Exp. Syst. Appl. **39**(17), 13051–13069 (2012)
2. Blest, D.C.: Theory & methods: rank correlation-an alternative measure. Aust. NZ J. Stat. **42**(1), 101–111 (2000)
3. Bourgeois, D., Morisseau, C., Flécheux, M.: New versions of MCDA/ICMDA algorithms applied in a nonlinear context. In: 2005 IEEE Aerospace Conference, pp. 2148–2153. IEEE (2005)

4. Brans, J.-P., Mareschal, B.: Promethee methods. In: Multiple Criteria Decision Analysis: State of the Art Surveys. ISORMS, vol. 78, pp. 163–186. Springer, New York (2005). https://doi.org/10.1007/0-387-23081-5_5
5. Chakraborty, S., Chattopadhyay, R., Chakraborty, S.: An integrated D-MARCOS method for supplier selection in an iron and steel industry. Decis. Making Appl. Manage. Eng. **3**(2), 49–69 (2020)
6. De Montis, A., De Toro, P., Droste-Franke, B., Omann, I., Stagl, S.: Assessing the quality of different MCDA methods. Altern. Environ. Valuat. **4**, 99–133 (2004)
7. Faizi, S., Sałabun, W., Nawaz, S., ur Rehman, A., Watróbski, J.: Best-worst method and hamacher aggregation operations for intuitionistic 2-tuple linguistic sets. Exp. Syst. Appl. 18, 115088 (2021). https://doi.org/10.1016/j.eswa.2021.115088
8. Genest, C., Plante, J.F.: On Blest's measure of rank correlation. Can. J. Stat. **31**(1), 35–52 (2003)
9. Guo, M., Liao, X., Liu, J.: A progressive sorting approach for multiple criteria decision aiding in the presence of non-monotonic preferences. Exp. Syst. Appl. **123**, 1–17 (2019)
10. Liu, J., Liao, X., Kadziński, M., Słowiński, R.: Preference disaggregation within the regularization framework for sorting problems with multiple potentially non-monotonic criteria. Eur. J. Oper. Res. **276**(3), 1071–1089 (2019)
11. Palczewski, K., Sałabun, W.: Influence of various normalization methods in PROMETHEE II: an empirical study on the selection of the airport location. Procedia Comput. Sci. **159**, 2051–2060 (2019)
12. Paradowski, B., Wieckowski, J., Dobryakova, L.: Why TOPSIS does not always give correct results? Proc. Comput. Sci. **176**, 3591–3600 (2020)
13. Sałabun, W., et al.: A fuzzy inference system for players evaluation in multi-player sports: the football study case. Symmetry **12**(12), 2029 (2020)
14. Sałabun, W., Urbaniak, K.: A new coefficient of rankings similarity in decision-making problems. In: Krzhizhanovskaya, V.V., et al. (eds.) ICCS 2020. LNCS, vol. 12138, pp. 632–645. Springer, Cham (2020). https://doi.org/10.1007/978-3-030-50417-5_47
15. Sałabun, W., Watróbski, J., Shekhovtsov, A.: Are MCDA methods benchmarkable? A comparative study of TOPSIS, VIKOR, COPRAS, and PROMETHEE II methods. Symmetry **12**(9), 1549 (2020)
16. Shekhovtsov, A., Sałabun, W.: A comparative case study of the VIKOR and TOPSIS rankings similarity. Procedia Computer Science **176**, 3730–3740 (2020)
17. Triantaphyllou, E., Baig, K.: The impact of aggregating benefit and cost criteria in four MCDA methods. IEEE Trans. Eng. Manage. **52**(2), 213–226 (2005)
18. Watróbski, J., Jankowski, J., Ziemba, P., Karczmarczyk, A., Zioło, M.: Generalised framework for multi-criteria method selection. Omega **86**, 107–124 (2019)

Intelligent Virtual Environments with Assessment of User Experiences

Ahmet Köse[(✉)], Aleksei Tepljakov, and Eduard Petlenkov

Department of Computer Systems, Tallinn University of Technology, Akadeemia tee 15a, 12618 Tallinn, Estonia
ahmet.kose@taltech.ee

Abstract. Virtual reality (VR) is a powerful modern medium. The advent of low-cost head-mounted display (HMD) devices made this technology accessible at large and featured VR with possibilities to monitor interactions and user's motion. However, due to lack of real-time feedback mechanism at present, the level of intelligence for virtual environments is still not sufficient to assist the experience and make group or individual assessments towards VR based applications. In this paper, we present our findings related to the problem of real-time feedback that focus on behavioral data by employing the novel feedback mechanism. Virtual-world coordinates, motions and interactions are tracked and captured in real-time while the user experiences particular application. Captured data is investigated to target the issue of complementing VR applications with features derived from real-time behavioral analysis. In our experiment, we also use collected data and provide a methodology to predict virtual-location by the nonlinear auto-regressive neural network with exogenous inputs (NARX). Results suggest employed neural network model is suitable for performing prediction which can be used to obtain a virtual environment with adaptive intelligence.

Keywords: Virtual reality · Intelligent virtual environment · Artificial neural networks · Real-time communication · Human-computer interaction

1 Introduction

Virtual Reality (VR) is a high-end human computer interface that can be defined with four essential elements: a virtual world, immersion, sensory feedback, and interactivity [1]. Furthermore, accurate information about real-world position and orientation tracking of the user has become a prominent feature of this technology. These features have become easily accessible with the advent of low-cost head-mounted display (HMD) devices in the last decade. Virtual environments (VE) formed by VR technology have proven effective in many scientific, industrial and medical applications. VR is also a powerful modern medium

Supported by the Estonian Research Council grant PRG658.

L. Rutkowski et al. (Eds.): ICAISC 2021, LNAI 12854, pp. 463–474, 2021.
https://doi.org/10.1007/978-3-030-87986-0_41

for immersive data visualization and interaction. Although VR based applications can attract even non-expert users, only a few research efforts were able to target the issue of advancing those applications with features derived from real-time human behavior analysis in VE and perform predictions accordingly. Besides that, analysis of performed activities and subjective feedback towards VR based applications are studied typically by post-experience oriented surveys. Eventually, it has become possible to alter this set of metrics with behavioral and psycho-physiological data to advance studies. However, psycho-physiological data requires additional properties and advance techniques in order to obtain fully objective measures [2]. The other affordable family of metrics is related to users' behavior inside VE such as location, motion and interactions of users, which is actually the main focus of this work [3].

Users' behavioral data-sets have also great potential for social scientists with an interest in the objective measured data in VE [4]. Human-computer interaction, namely multi-modal interactions to some degree: mouse clicks, buttons, visual and auditory signals etc. are transformed to head rotation, hand movement, speech and eye gaze in VE [5,6]. Most experiments in applied and social psychology research as well as other domain areas across a range of psychological topics face challenges to find the balance between control and ecological validity, and only few of them allow multi-modal interaction in realistic conditions [7]. Conversely, VE that are designed to study the behavior can allow users to respond in a manner that is more natural. Although recent advances in VR technology allow to trace multi-modal interactions, collecting multiple aspects of behavioral data at once remains a challenging problem. Therefore, the research in novel VR technology still requires unique considerations such as the availability of sophisticated, practical tools and protocols to measure VR outcomes with examples and data-sets [8].

In this work, we employ the novel feedback mechanism that is capable of real-time communicating in both ways between the game engine and third party software with the motivation of creating intelligent VE. As an outcome of the presented work associated with user's motions and interactions may allow real-time proactive qualities to design adaptive user interfaces and advanced recommendation systems [9]. Furthermore, applied computational intelligence methods can be used to make behavioral assessments towards other related studies. The main contribution of the present paper as follows: First, we provide a firm motivation of this work with an intelligence derived from users' behavior aspects while discussing the feedback mechanism in VR based applications. Then, we present a designed VE and the game play therein. Next, the reader is introduced to technical details including system configuration and software management to give better understanding of real-time data communication and collection. In the following section, we investigate collected data-sets with survey results towards behavioral learning. We also describe preferred computational intelligence method and the process of implementation to perform predictions. Finally, we report and address the conclusion as a result of data analysis and employed network models linked to human behavior and outline some related items for future research.

The structure of the paper is as follows. In Sect. 2, the reader is introduced to the feedback mechanism and aspects of enabled intelligent VE thereof. Our novel developments towards assessment of behavioral patterns are also explained in this section. Analysis of behavioral data in VE and primary results of performed models based on user's behavior in the application are addressed in Sect. 3. Finally, conclusions are drawn in Sect. 4.

1.1 Related Work

The idea of creating a real-time feedback mechanism to accommodate an intelligence is a recurrent interest to facilitate the broad adoption of VR technology. The advent of low-cost HMD devices has expanded the interest considerably. One of early tailor-made solutions was employed to accomplish a real-time motion tracking system. Real- time VR feedback is achieved by using a point light display while placing retro-reflective markers on the novice participants' joint centers [10]. A Few years later, an online VR based application that was introduced [11] to monitor and collect the user's behavior by built-in feedback mechanism, but that is not applicable for HMD devices. A feedback mechanism is described through a gaming-oriented VR application in [12] consists of additional hardware components to achieve a real-time haptic feedback. However, the designed architecture is not able communicate with third party software and compatible only with 32-bits applications. Finally, a recent research towards comparative study of several input technologies for VR based applications also present a solution that provides a one way communication and frequency to collect the interactions and location of user in VE [13].

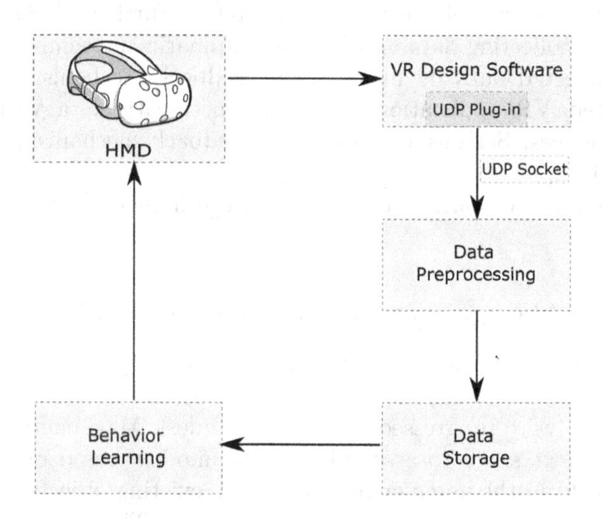

Fig. 1. Workflow of the complete feedback mechanism.

2 The Framework for Intelligent Virtual Environment

It is apparent that a complete behavioral analysis inside the VE is one of the most recurrent interests for researchers but also requires a significant amount of time and effort. In order to enable assessments towards user's behavior in VR based applications through available HMD devices, a real-time feedback mechanism that is capable of acquiring meaningful data is required. Accurate and meaningful data-sets may provide better understanding to study the behavioral aspects besides conducted traditional survey techniques. Additionally, replicability of VR based experiments in may advance behavioral studies. Consequently, understanding the behavior in VE allows to draw conclusions for creating more compelling and efficient applications.

We addressed mentioned issues by designing an architecture with the basis of the feedback mechanism implementation that can be described as follows: The game engine is employed as the visualization platform while the user experiences VE with interactive equipment. Motion capture is achieved using present HMD technology which provides access to user movement data. User Datagram Protocol (UDP) [14] serves as a communication plugin between the game engine and software environment that makes real-time data communication, simulation and feedback possible. To proceed with behavioral modeling, the data related to the user is sent out to data preprocessing layer (e.g., Python) and then to an electronic database to store and apply statistical and computational intelligence methods. Behavioral learning is occurring between the HMD device and data preprocessing layer. The workflow to illustrate the complete feedback mechanism is depicted in Fig. 1.

Appropriate access to behavioral data is a crucial requirement for better understanding of users' behavior in VR. Therefore, further efforts were devoted to demonstrate collecting data and making automatic and semi-automatic analysis of the collected data by utilizing the feedback mechanism. We used the gaming-oriented VR application to acquire necessary data with assessment of user experiences. Screenshots while the feedback mechanism is working is depicted in Fig. 2.

In what follows, we outline each stage of the implementation in a separate subsection.

2.1 Software Management and System Configuration

Efficient real-time rendering of the objects must be ensured for VR based applications. Thus, the following important considerations are in effect when modeling all objects were progressed by using Autodesk Maya software. Then, we moved to the next stage to assembly created models based on the dynamics of physical world in the game engine. Whereas, existing objects such as leaves, coins, landscape materials, etc. in VE are presented. The application is created by well-known game engine *Unreal Engine 4 (UE4)* which is a commonly used and powerful solution for VR development purposes [15]. UE4's Blueprint visual scripting system is one of the reasons to prefer UE4 to other real-time visualization engines since it allows for rapid prototyping [16].

(a) The game level is situated in the hill-top virtual environment

(b) The user experiences the application with timer and score

Fig. 2. Screenshots from VR based application.

A mix of Blueprint and C++ code allowed us to create a UDP interface to communicate between a number of third party software and the game engine in real-time. In practise, the Blueprint system with UDP socket allowed Python made script to be in charge of capturing and processing the data before accommodating into data storage. Every four bytes marked as a single input of data that is sent from VR based applications by UDP packet. A script executes the function to send captured information to the designated column of database and added time variables. The data is managed through relational database management system-SQL for the purpose; facilitate user behavior data-sets to reach. The data is transformed into numeric values to be stored in the database. To avoid problems with Blueprint multi-threading for VR based applications, the implementation uses a custom class variable to transfer data between threads to avoid a racing condition in requesting/getting new data from the UDP socket. With this approach, reliable communication via a UDP socket at sufficiently high sampling rates upwards of **fs** = **1 kHz** is achieved. Nevertheless, UDP plugin as well as exported modules can also be reused in any project.

2.2 Observation of Behavior in Immersive Environment

VR based applications are often delivered with the interaction that may also allow an evaluation of the engagement of participants [17]. According to Paliokas et Al., the observation of behavior in VE is the ability of capturing behavioral data-sets on concrete items and events under particular conditions [18]. Therefore, available sensor information (e.g. HMD devices, motion trackers, additional controller etc.), virtual-location, locomotion attempts and performance indicators (e.g. rewarding points) should be able to compose elements of metrics to make behavioral observation.

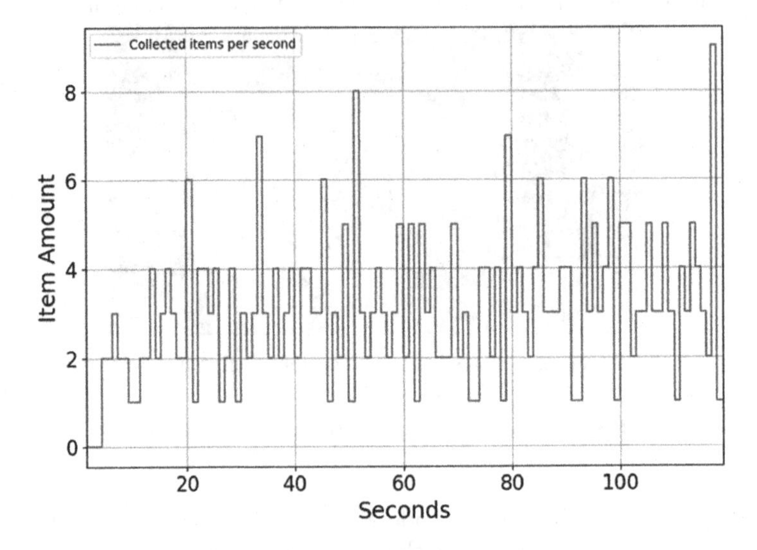

Fig. 3. Total amount of collected items per second by participants during the first try.

From the user perspective, interactions linked to the gamification based engagement (e.g. joyful learning, serious games) are of curiosity driven experimental nature [19]. Once the motivation is obtained, an accomplishment of attentional aspects should be matters of given objects and tasks. However, users may have challenges to exhibit interactions, teleportation and physical movements at once. Hence, investigation of users' experiences through traditional survey methods should be complemented by exploring all of these dynamics. The experiment to collect behavioral data-sets, demonstrate data preprocessing and storage in real-time by employing the feedback mechanism, is carried out by particular implementation-*Swedbank Experience* which is a gaming-oriented VR application that is explained with details in our earlier work [20]. Aggregated data reveals that participants were able to collect items at most in the end (9 out of 20 participants) of given time is depicted in Fig. 3.

3 System Identification and Modeling

A thorough study of the users' behavior including interactions inside VE and emotional states can compose desired intelligence. The intelligence derived by proposed feedback mechanism can be utilized at least by two major use cases at present: adaptive virtual learning environments to increase learning efficiency and user experience; user behavior modeling to reduce cost of visualization (e.g. computational power for Big Data visualization).

Featuring elements of behavioral metrics data-sets such as head motion and distances between HMD device and controllers can already be used to identify, differ and authenticate users at present. The uniqueness derived from behavioral data may advance authentication methods to customize the experience efficiently in VE [21]. Meanwhile, several studies have already been introduced through visualizations Big Data conducted with VR based applications to relief present difficulties and to engage users efficiently with provided data [22,23]. The ability to predict user's virtual-location is one of the key issue that can lead to a coherent solution. The prediction might allow to preload only the necessary part from Big Data sets. Dynamic models would also benefit particularly VR based applications to ensure a seamless immersive experience possibly with less computational power. Furthermore, dynamic user interface and mechanics can be modelled by analysing the users' visual attention in VE.

Collected data-sets referred to position degrees contain also location information. Location of the user is provided through $(\mathbf{x}, \mathbf{y}, \mathbf{z})$ coordinates in immersive environment and a sample of collected data sets in time series depicted in Fig. 4. The user behavior, $\mathbf{A}$, is translated by some $(\mathbf{x_t}, \mathbf{y_t}, \mathbf{z_t}) \in \mathbf{R}$

$$(\mathbf{x}, \mathbf{y}, \mathbf{z}) \longmapsto (\mathbf{x} + \mathbf{x_t}, \mathbf{y} + \mathbf{y_t}, \mathbf{z} + \mathbf{z_t}). \tag{1}$$

Suppose $\mathbf{A} = \mathbf{H_i}$ is translated $\mathbf{x}_t$ units in the $\mathbf{x}$ direction and $\mathbf{y}_t$ units in the $\mathbf{y}$ direction. The transformed primitive is $\mathbf{A}$ primitive of the form

$$\mathbf{H_i} = \left\{ (\mathbf{x}, \mathbf{y}, \mathbf{z}) \in \mathbf{W} | \mathbf{f_i}(\mathbf{x}, \mathbf{y}, \mathbf{z}) \leq 0 \right\} \tag{2}$$

is transformed to [24]

$$\left\{ (\mathbf{x}, \mathbf{y}, \mathbf{z}) \in \mathbf{W} | \mathbf{f_i}(\mathbf{x} + \mathbf{x_t}, \mathbf{y} + \mathbf{y_t}, \mathbf{z} + \mathbf{z_t}) \leq 0 \right\}. \tag{3}$$

Accordingly, since collected data enables investigation into position defined $(\mathbf{x}, \mathbf{y}, \mathbf{z})$ coordinates is given by described equations above, authors focus on performing predictions of user's location in the application by meaning using measured movement values to predict next values.

3.1 Dynamic Predictive Modeling of User Behavior

This section describes the process that can perform any other user behavior prediction since the approach is general to employ prediction models. The process of implementation as follows:

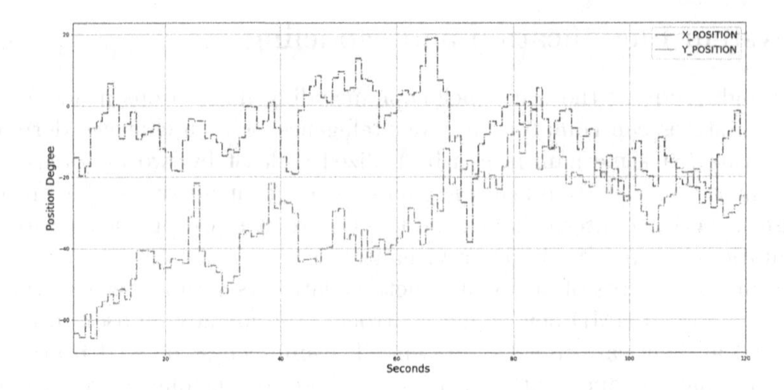

Fig. 4. An example data of position degree changes are aggregated through x and y coordinates while the user experiences VR based application in two minutes.

1. Data Communication and Collection: UDP protocol is in charge of sending the real-time data while the user experiences the application.
2. Data Processing: Besides virtual-location information, sensors of HMD device may indicate to noisy and unreliable data. Thus, rare outliers of sensor data is excluded for scaling purposes to train the model.
3. Model Training: Filtered data is divided into **70%** training data, **15%** validation and **15%** test data. The model has been trained using Levenberg-Marquardt algorithm which is often the fastest back-propagation function and commonly used [25].
4. Performance Comparison and Validation of Trained Model: Once accurate results are accomplished, test data which is not introduced to these models are set for validation purposes.

Further analysis towards the experiment in this work is devoted to investigate the location of participants in VE. The location data of users can also be traced to explore dynamic changes of location as $(\mathbf{\Delta x}, \mathbf{\Delta y}, \mathbf{\Delta z})$ coordinates. Collected data of first and second attempts indicates that motions are increased significantly through $(\mathbf{x}, \mathbf{y})$ coordinates in second time. Primary component analysis (PCA) towards virtual-coordinates of the participants in both times is depicted with the maximum two variance directions (PC1, PC2) in Fig. 5.

3.2 System Identification Using Artificial Neural Networks

ANN based prediction has been developed since it was explored because of ANN approximation and generalization property. ANN, with the ability to approximate a large class of nonlinear (NL) functions, provide a feasible uniform structure for NL system representation that is usually described with differential equations (continuous- time model) or difference equations (discrete-time model).

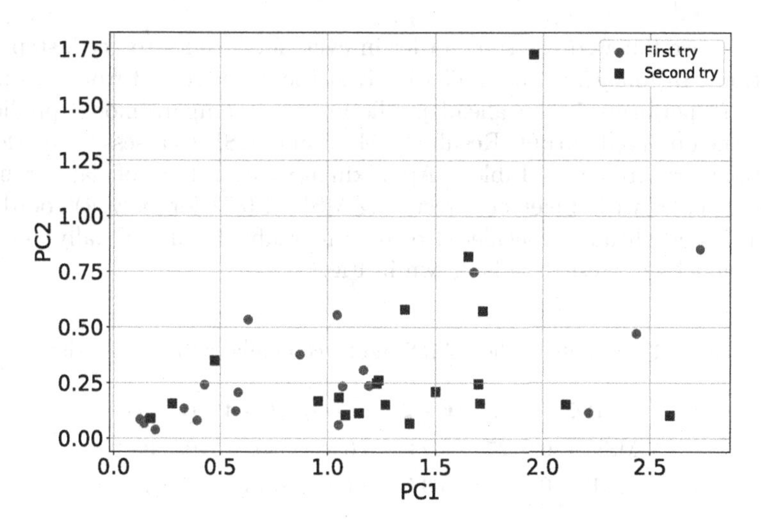

Fig. 5. PCA is applied for $(\mathbf{x}, \mathbf{y}, \mathbf{z})$ coordinates in VE while participants experience VR based application.

NARX Model. The model is built up with the wavelet network to create input and output non- linearity estimator. An important class of discrete-time nonlinear systems is NARX model is formalized by [26].

$$\mathbf{y(t) = f(u(t - n_u), \ldots, u(t - 1); u(t); y(t - n_y), \ldots, y(t - 1))} \tag{4}$$

where u(t) and y(t) represent input and output of the network at time $(\mathbf{t}, \mathbf{n_u}, \mathbf{n_y})$ are the input and output order, and the function **f** is a nonlinear function. When the function **f** can be approximated by a Multilayer Perceptron, the resulting system is called a NARX network.

3.3 Process of Model Implementation and Prediction Performances

Virtual-world position of the user based on $(\mathbf{x}, \mathbf{y}, \mathbf{z})$ coordinates is used to perform predictions by NARX neural network. The position is measured by the game engine unit (**uu**) and ($\mathbf{1uu = 1\,cm}$). Increasing the input delay by one (time-shift between inputs and outputs) corresponds to ($\mathbf{1\,s}$). The amount of regressors are 3 for NARX models respectively. Mean Absolute Error and Mean Square Error are two preferred parameters to present results of trained network models. Used data-sets consists of 18 participants virtual-coordinates with one second intervals. First, we randomly selected data of two participants for validation purposes. Next, we used rest of data to train the model. In order obtain 5-step ahead prediction for all time instances, we have created a third order NARX model of making one step ahead prediction.

The dimension of input matrix is ($\mathbf{9 \times 117}$) for each participant and by merging data from 16 participants, the dimension of input matrix becomes ($\mathbf{9 \times 1872}$), while the dimension of the output matrix is ($\mathbf{3 \times 1872}$). After data processing, we followed the same order in cycle for 5 times to get 5-step ahead predictions. In following, the model is given the input record from two participants and performs 5-step ahead predictions; we compare model predictions against the observed output. Results of MAE and MSE to assess the performed predictions are presented Table 1. Approximately over 120 s of data translates to 2160 samples with 3 elements resulting MSE of 0.71 for $(\mathbf{x}, \mathbf{y}, \mathbf{z})$ coordinates with up 5-step ahead is considered to be remarkable results. Finally, results of NARX model with test data is drawn in Fig. 6.

Table 1. Results of the NARX model compared with input delays.

NARX	$(\mathbf{t}-\mathbf{1})$	$(\mathbf{t}-\mathbf{2})$	$(\mathbf{t}-\mathbf{3})$	$(\mathbf{t}-\mathbf{4})$	$(\mathbf{t}-\mathbf{5})$
MSE	0.0127	0.023	0.039	0.051	0.709
MAE	0.068	0.097	0.129	0.141	0.454

Results of prediction performances reveal that sufficient behavioral modeling is prospective and also depends the availability of affordable and practical feedback mechanism. The mechanism is capable to measure and capture VR outcomes with examples and data-sets and is able to communicate with third party software in real-time. In other words, the desired intelligence in VE can be only obtained by assessment of user experiences towards reliable, meaningful and accurate behavioral data-sets.

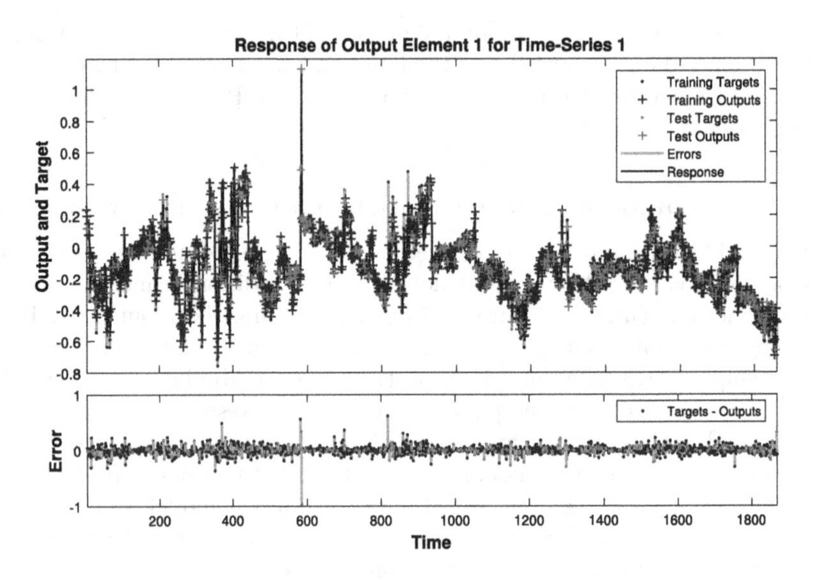

Fig. 6. Prediction performance of NARX Model.

4 Conclusions

In this paper, an approach to obtain intelligent virtual environments with assessment of user experiences is introduced. First, the novel feedback mechanism with the firm motivation to enable real -time human behavior analysis in VE is presented. Next, captured data -sets are investigated towards Virtual -world measures to perform predictions as the outcome of presented work. Prediction results with MSE of 0.039 for $\mathbf{t-3}$ and 0.79 for $\mathbf{t-5}$ between measured and predicted data imply the great potential of affordable feedback mechanism to advance the research in novel VR technology. Since the present application is envisioned to be used for real -time applications to adapt virtual environments based on assessment of user experiences, further development efforts also should be exhibited to run models simultaneously in order to feed output of collected information to employed model. In addition, presented approach may allow to immerse participants into multiple scenarios and support redirection techniques in VR based applications. In the future, the feedback mechanism may help to motion tracking method to capture whole body motion to exploit for user identification and user behavioral analysis in large population. While the location prediction may help to reduce computational power, application of similar framework for head movement data can complement efficient region -of- interest (ROI) based VR based application (e.g. VR streaming) [27].

References

1. Sherman, W., Craig, A.: Understanding Virtual Reality. Elsevier (2019)
2. Kivikangas, J.M., et al.: A review of the use of psychophysiological methods in game research. J. Gaming Virtual Worlds **3**(3), 181–199 (2011)
3. Soler-Domínguez, J.L., Contero, M., Alcañiz, M.: Workflow and tools to track and visualize behavioural data from a virtual reality environment using a lightweight GIS. SoftwareX **10**, 100269 (2019)
4. Pan, X., de C Hamilton, A.F.: Why and how to use virtual reality to study human social interaction: the challenges of exploring a new research landscape. Br. J. Psychol. **109**(3), 395–417 (2018)
5. Turk, M.: Multimodal interaction: a review. Pattern Recogn. Lett. **36**, 189–195 (2014)
6. Hansberger, J.T., Peng, C., Blakely, V., Meacham, S., Cao, L., Diliberti, N.: A multimodal interface for virtual information environments. In: Chen, J.Y.C., Fragomeni, G. (eds.) HCII 2019. LNCS, vol. 11574, pp. 59–70. Springer, Cham (2019). https://doi.org/10.1007/978-3-030-21607-8_5
7. Hunt, H.T.: Why psychology is/is not traditional science: the self-referential bases of psychological research and theory. Rev. Gen. Psychol. **9**(4), 358–374 (2005)
8. Lanier, M., et al.: Virtual reality check: statistical power, reported results, and the validity of research on the psychology of virtual reality and immersive environments. Comput. Hum. Behav. **100**, 70–78 (2019)
9. Zhang, S., Yao, L., Sun, A., Tay, Y.: Deep learning based recommender system. ACM Comput. Surv. **52**(1), 1–38 (2019)

10. Eaves, D.L., Breslin, G., van Schaik, P.: The short-term effects of real-time virtual reality feedback on motor learning in dance. Presence Teleoperators Virtual Environ. **20**(1), 62–77 (2011)
11. Johnson, A., Tang, Y., Franzwa, C.: kNN-based adaptive virtual reality game system. In: Proceedings of the 11th IEEE International Conference on Networking, Sensing and Control. IEEE (April 2014)
12. Jayaraj, L., Wood, J., Gibson, M.: Improving the immersion in virtual reality with real-time avatar and haptic feedback in a cricket simulation. In: 2017 IEEE International Symposium on Mixed and Augmented Reality. IEEE (October 2017)
13. Reski, N., Alissandrakis, A.: Open data exploration in virtual reality: a comparative study of input technology. Virtual Reality **24**(1), 1–22 (2019)
14. Serpanos, D., Wolf, T.: Architecture of Network Systems. Elsevier (2011)
15. Epic Games: Unreal Engine. https://www.unrealengine.com/en-US. Accessed 03 Jun 2020
16. Smyth, D.L., Glavin, F.G., Madden, M.G.: Using a game engine to simulate critical incidents and data collection by autonomous drones. In: 2018 IEEE Games, Entertainment, Media Conference (GEM). IEEE (August 2018)
17. Merino, L., Ghafari, M., Anslow, C., Nierstrasz, O.: CityVR: gameful software visualization. In: 2017 IEEE International Conference on Software Maintenance and Evolution (ICSME). IEEE (September 2017)
18. Paliokas, I., Kekkeris, G., Georgiadou, K.: Study of users' behaviour in virtual reality environments. Int. J. Technol. Knowl. Soc. **4**(1), 121–132 (2008)
19. Grivokostopoulou, F., Perikos, I., Hatzilygeroudis, I.: An innovative educational environment based on virtual reality and gamification for learning search algorithms. In: 2016 IEEE 8th International Conference on Technology for Education (T4E). IEEE (December 2016)
20. Kose, A., Tepljakov, A., Abel, M., Petlenkov, E.: Towards assessment of behavioral patterns in a virtual reality environment. In: De Paolis, L.T., Bourdot, P. (eds.) AVR 2019. LNCS, vol. 11613, pp. 237–253. Springer, Cham (2019). https://doi.org/10.1007/978-3-030-25965-5_18
21. Pfeuffer, K., Geiger, M.J., Prange, S., Mecke, L., Buschek, D., Alt, F.: Behavioural biometrics in VR. In: Proceedings of the 2019 CHI Conference on Human Factors in Computing Systems, CHI 2019. ACM Press (2019)
22. Millais, P., Jones, S.L., Kelly, R.: Exploring data in virtual reality. In: Extended Abstracts of the 2018 CHI Conference on Human Factors in Computing Systems - CHI. ACM Press (2018)
23. Moran, A., Gadepally, V., Hubbell, M., Kepner, J.: Improving big data visual analytics with interactive virtual reality. In: 2015 IEEE High Performance Extreme Computing Conference (HPEC). IEEE (September 2015)
24. LaValle, S.M.: Planning Algorithms. Cambridge University Press (2006)
25. Moré, J.J.: The Levenberg-Marquardt algorithm: implementation and theory. In: Watson, G.A. (ed.) Numerical Analysis. LNM, vol. 630, pp. 105–116. Springer, Heidelberg (1978). https://doi.org/10.1007/BFb0067700
26. Ibrahim, M., Jemei, S., Wimmer, G., Hissel, D.: Nonlinear autoregressive neural network in an energy management strategy for battery/ultra-capacitor hybrid electrical vehicles. Electr. Power Syst. Res. **136**, 262–269 (2016)
27. Maranes, C., Gutierrez, D., Serrano, A.: Exploring the impact of 360° movie cuts in users' attention. In: 2020 IEEE Conference on Virtual Reality and 3D User Interfaces (VR). IEEE (March 2020)

Intelligent Performance Prediction for Powerlifting

Wojciech Rafajłowicz(✉) and Joanna Marszałek

Department of Control Engineering, Wrocław University of Science and Technology, Wrocław, Poland
wojciech.rafajlowicz@pwr.edu.pl

Abstract. Artificial intelligence methods are successfully applied in many areas where a prediction or classification is needed. An example may also be the forecasting of an athlete's performance, investigated in this article. Powerlifting is a sport of widespread popularity - more and more training people are serious about the competition and prepare professionally for it. This paper aims to present and analyze the athlete's deadlift score prediction system based on the previous results and the historical data of other lifters. At first, we use the parametrics to choose an athlete with the results similar to the investigated lifter. We propose the artificial neural network application for smoothing empirical hazard and cumulative distribution functions designated for the failed deadlift attempts. We decided to involve quasi-RBF neural networks – involving the sigmoid function and nonlinear least squares learning algorithm. As a result we get the prediction whether the athlete's deadlift attempt will be valid or not.

Keywords: Artificial neural networks · Performance prediction · Artificial intelligence in sport

1 Introduction

1.1 Powerlifting

Powerlifting is a sport discipline consisting of three lifts: squat, bench press, and deadlift. The athlete's goal is to perform each of these exercises with the greatest possible external load [1]. The weight is placed on the bar with which the lifter does a squat, presses it on a flat bench and picks it from the ground (deadlift). By entering the competition, the athlete declares the weight category that she or he wants to compete in. There are also age groups and gender division to make the rivalry fair. Lifter's weight is always checked by the committee. Before a contest. During a competition, the athlete makes three attempts of each exercise. At first, the lifter declares the barbell weight to let the staff know how to load it, and – when it is her or his turn – goes to the platform. The attempt (meaning a single lift) can be valid or not – it is decided by three referees. If at least two

L. Rutkowski et al. (Eds.): ICAISC 2021, LNAI 12854, pp. 475–484, 2021.
https://doi.org/10.1007/978-3-030-87986-0_42

of them state the technical correctness of the lift, the attempt is passed. The competitor's score is determined by her or his weight and the sum of the weight from the heaviest valid attempts of each lift. Obviously, making a good decision while declaring the appropriate barbell weight before each attempt is crucial for all powerlifters taking part in various competitions.

1.2 Related Works

Nowadays, there is a growing interest in sport, including powerlifting. It is caused by the popularization of physical activities in free time, resulting from health awareness improvement. As a consequence, more and more research concerns the sports disciplines, training loads, injury risk [3] and athletes results analysis. Maximization of an athlete's performance is becoming an interesting subject for many researchers, especially for the artificial intelligence specialists. The sport results prediction problem is often solved by artificial neural networks application. Bunker *et al.* analyzed this issue in relation to both team and individual disciplines [2]. Zhao *et al.* conducted the research on wrestlers [14]. The main goal was the athlete's performance prediction and it was achieved with the use of an improved radial basis neural network. The players' performance prediction was also investigated by Saikia *et al.* [11]. The analysis subject was the athletes' classification into groups diversified in terms of their predicted results. The researchers trained a multilayer perceptron using the combined bowling rate data and successfully predicted future results. Another interesting example of artificial intelligence application to sports is swim velocity profile identification. Santos Coelho *et al.* conducted research on time series forecasting related to breaststroke and crawl style swimmers [12]. The identification problem was solved with a radial basis function neural network using the Gustafson-Kessel clustering algorithm. One of the newest proposals is the application of gravitational-double layer extreme learning machine in powerlifting analysis [4]. V. H. Chau *et al.* investigated the relationship between the weight, age and results of female powerlifters, using the performance prediction model. The research on a similar model (but concerning male lifters) was also conducted with the use of an improved artificial bee colony algorithm to optimize the kernel extreme learning machine network [5].

The modeling personalization problem can be solved with the use of neural networks. Kasabov *et al.* proposed spiking neural networks application for personalized modeling, classification and prediction of spatio-temporal patterns with a case study on stroke [8]. He noticed that this type of neural networks can be successfully used for prediction in many research ranges, including engineering, bioinformatics, neuroinformatics, medicine or economics. In fact, spiking neural networks can be also applied for stroke risk prediction [6]. In the neural network personalization process, a specialized dataset is needed [10]. It can be significantly smaller than the dataset used for pretraining. Hence, the data must represent the reality well, therefore a correctly selected dataset is crucial for proper neural network personalization.

1.3 Artificial Neural Networks

Artificial neural networks are computational tools that simulate the decision making process in the central nervous system. They consist of neurons' models, organized in layers. The input layer transfers the data to the hidden (interconnected) layers, where the input values are processing and the output layer. Each of the neurons is connected to the previous layer through the synaptic weights that can change adaptively. In other words, a weighted sum of input values is converted by a transfer function, for example, a sigmoid function can be employed:

$$f(\Sigma) = \frac{1}{1 + e^{-\Sigma}} \tag{1}$$

where $f(\Sigma)$ represents the weighted sum of input signals.

The training process consists of modifying the weight values with the use of the selected learning method. The idea of its supervised learning is to compare the model's prediction against already known target data. If some differences appear, the weights' adjustment is needed in order to decrease the error value. This process runs repeatedly until the error value is satisfactorily low, which means that the network is trained. The weights values are held as constant and used with the dataset on which the research will be conducted.

Artificial neural networks are often chosen for data mining (extracting patterns from datasets in order to get some useful knowledge) if the relationships among the input data are nonlinear, especially for classification [9]. It is a reasonable choice also when only incomplete information about the investigated phenomenon is available. The application of this flexible technique is relatively easy and it can be widely used for many statistical models approximation.

Artificial neural networks with radial basis function contain three layers: input, hidden, and output. The hidden layer is specific due to involving radial basis function – mainly Gaussian function:

$$\phi(X) = exp(-\frac{\|X - C\|^2}{\sigma}) \tag{2}$$

where X represents inputs, C – center vector, and σ – width. Therefore, the radial basis function value (ϕ) is influenced only by the distance between input and center. The hidden node's output value will increase as this distance decreases (unless the basis function is not symmetrical). Apart from the mentioned function, unipolar and bipolar sigmoid functions, hyperbolic tangent function, or conic section function are used in this type of neural networks as well. The radial basis neural network key parameters also include the number of the hidden node – often set experimentally [13]. An important advantage of radial basis neural networks (compared to the multilayer perceptron) is their ability to solve the local minima problem during the learning process.

2 Methods

2.1 Dataset

The information on 34 male powerlifters was collected from the Open Powerlifting website [15] (the access is completely open). The data includes the athlete's weight and his deadlift score in various competitions, for example, World Powerlifting Championships, World Classic Powerlifting Cup, or World Games.

For each attempt we take the following information:

- current lifter's weight,
- declared barbell weight,
- whether the attempt was valid or not.

The fragment of the dataset is shown in Table 1. 1 means that the attempt was valid and 0 stands for the failure.

Table 1. The illustrative fragment of data set

Name	Competition 1							...
	Weight	1st try		2nd try		3rd try		...
Michael Tuchscherer	125	305	1	327,5	1	332,5	1	...
Tony Cliffe	78,8	190	1	205	1	220	1	...
Viktor Samuelsson	88,8	200	1	210	1	220	0	...
Mohamed Bouafia	99,6	315	1	325	1	327,5	1	...
Richard Hozjan	80,9	220	1	230	1	240	1	...

The processing of all data and the results, as well as graphs generation, was performed using *Matlab*.

2.2 Algorithm

Before making a prediction, it is necessary to prepare the data. We have to investigate the data in order to find a lifter with results similar to the investigated athlete and designate relevant functions. The following algorithm was developed for the research:

1. Choose an athlete with a small amount of data (competition results) – lifter A.
2. Select the last competition's results for lifter A.
3. Find the most similar athlete to A using the parametrics – lifter B.
4. For lifter B designate:
   - hazard function for the valid attempts,
   - empirical distribution function for the failed attempts.

5. Smooth both designated functions using the quasi-radial basis neural network (with the sigmoid function).
6. Predict whether the attempt will be valid for the control data and compare the prediction with the actual data.

The described procedure can be repeated for another lifter or for the same one.

2.3 Selection of Similar Athlete

In order to find similarities between two lifters, each of them described by dataset, a method of measuring distance has to be proposed. Clearly in space consisting of competition results no addition can be defined. Therefore it is impossible to define strictly metrics in said space. Only parametric – often called distance – can be defined.

The best way to compare two lifters is to organize direct competition between them. Obviously it is not possible, we can only compare their results in data. This idea can be used to define distance.

The distance is loosely based on pair-wise comparison proposed in [7]. For two athletes A and B with respective n and m results, the distance (or dissimilarity) $d(A, B)$ is calculated in the following way:

1. A counter $S = 0$ is created.
2. A cartesian product of results for A and B is calculated

$$\{A_1, A_2, ..., A_n\} \times \{B_1, B_2, ..., B_m\} \tag{3}$$

3. Every pair of results $A_i\, B_j$ is treated as a mini-competition between A and B.
4. If the best valid lift for A_i is better than the best valid one for B_j then 1 is added to S, otherwise -1 is added.
5. The result is calculated by formula:

$$d(A, B) = |\frac{S}{m \cdot n}| \tag{4}$$

Let us check requirements for the parametric are fulfilled:

1. $d(A, B) \geq 0$ is obvious from the use of absolute value,
2. $d(A, A) = 0$ because of getting every pair $A_i\, A_j$ and $A_j\, A_i$ with mutually canceling of the result so $S = 0$,
3. $d(A, B) = d(B, A)$ for the same reason.

2.4 Empirical Hazard and Cumulative Distribution Functions

The data for each athlete can be divided into two separate groups: successful and unsuccessful attempts.

The hazard function is defined as:

$$H(x) = P(X > x) = 1 - F(x). \tag{5}$$

In a powerlifting setting, it gives us the probability that an athlete would lift a certain, declared weight.

The cumulative distribution function for unsuccessful attempts gives us the probability that the athlete would fail with a specified weight. In Fig. 1 the values of hazard function and cumulative distribution function are shown as small circles (hazard function of success) and crosses (cumulative distribution function for failures).

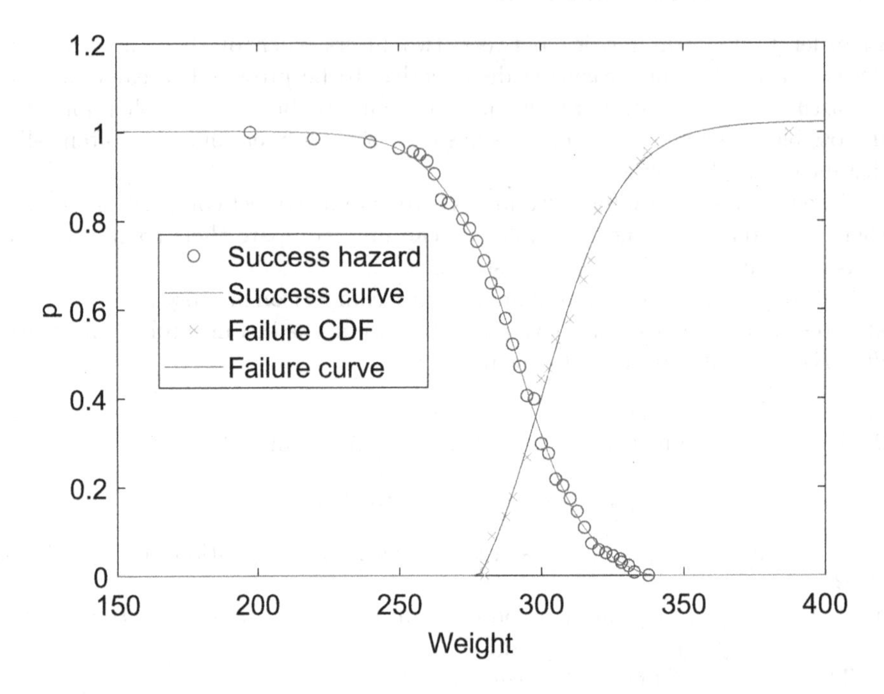

Fig. 1. Hazard and CDF functions (points) and respective curves.

2.5 Generalisation by Sigmoid RBF's

The hazard function for successes and the cumulative distribution function for failures are in the form of some number of points dependent on a number of data. It would be easier to use some sort of curve. It would give us the ability to get values for any chosen value, even when no nearby point is present.

The typical shape of the curves suggests using the sum of sigmoids with some weights to form a quasi-RBF artificial neural network with one input and functions more suitable to represent cumulative distribution function than the typical one used for distributions.

The following function was used with the additional parameters b_i and c_i:

$$s(x) = s_0 + \sum_{i=1}^{n} \frac{a_i}{1 + \exp(-(x - b_i)/c_i)} \tag{6}$$

where a_i stands for the weights. As a result, we obtain a good curve with the traditional nonlinear least squares learning algorithm. In Fig. 1 the curves represent the result of the learning process.

2.6 Prediction

In order to make a prediction, we must decide whether the probability of success given by the hazard function outweighs the probability of failure given by the cumulative distribution function. The simplest solution is to compare both probabilities and decide which one is higher. In some cases, it would be useful to require that the probability of success is higher than the probability of failure plus some additional overhead.

The model made for selected similar athlete and smoothed by quasi-RBF is used to predict results.

The examples of the results are shown in Table 2. In case of 297 kg we cannot make the prediction because the difference between the probabilities values is too small. We decide on a safe solution and we do not make a decision.

Table 2. Predictions for different weights.

Weight	p success	p failure	Prediction
250	0.9617	0	Success – 1
290	0.5247	0.1883	Success – 1
297	0.3767	0.3341	No prediction – 0
305	0.2317	0.5097	Failure – 0
355	0	0.99	Failure – 0

3 Experiment Procedure

The distance between two athletes can be calculated using the procedure outlined in Sect. 2.3.

In the Fig. 2 we can see how similar or dissimilar are all the athletes in our sample data. For each athlete, we can find the best similar data and test the prediction for each of his results. We can compare those predictions with the real results. Since we want to use larger sets of data to predict results for the athletes who do not pose such record, we need to remove cases when the record length for the athlete that we make prediction for is longer than the record used to make predictions. Due to the symmetry shown in Sect. 2.3 this would occur in half of the cases. We should also disregard the cases when we cannot find suitable data to make the prediction.

The results of the research would be discussed in the next section.

4 Results

All athletes were investigated in the way described in Sect. 3. The summary of conducted experiments is in Table 3.

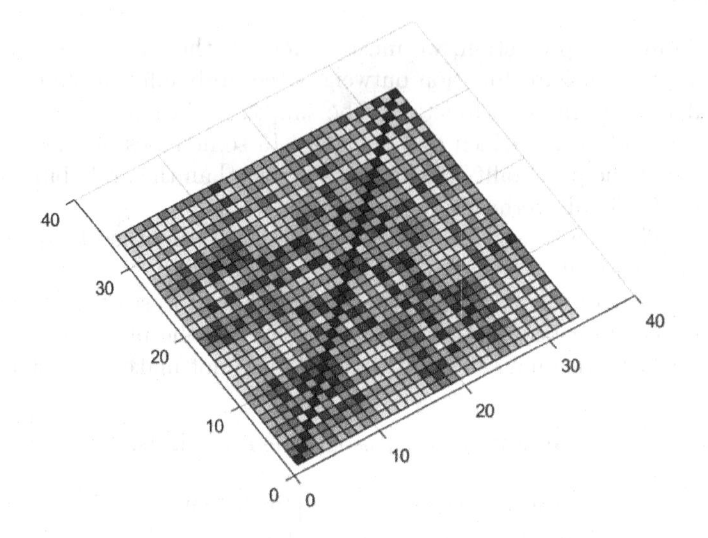

Fig. 2. Parametrics values between athletes. Darker color – smaller distance. Number on axis signify each athlete.

Table 3. Predictions for different weights.

General correct prediction rate	66.4%
Number of predictions	382
Total number of athletes	34
Number of rejected athletes	2

The distribution of prediction quality is shown in Fig. 3.

We can clearly note that – for any athlete – none of the predictions is below 50% accuracy.[1]

[1] This should be the minimal requirement in any success/failure prediction.

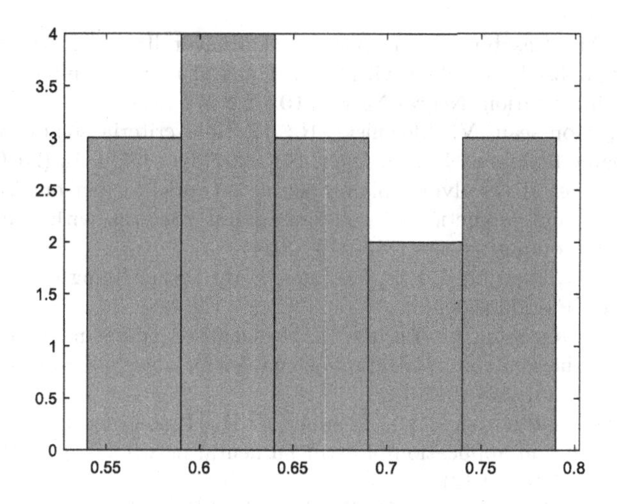

Fig. 3. The histogram of the amount of good predictions.

5 Summary

In this paper, we have presented the feasibility to make predictions for powerlifting results. The proposed system is based on quasi-RBF artificial neural networks (using a sigmoid function), using simple parametrics (distance) to find similar data among the available.

More detailed research is needed, particularly taking into account the weight of an athlete before each analyzed competition. This would require creating a better measure of distance and also a more complicated network for predictions. Also processing a larger dataset would prove beneficial.

We note that thanks to the proposed solution's simplicity, it can be successfully applied to mobile devices and provide the lifter's (legal) support in making decisions during the competition as an auxiliary tool. We can see the wide opportunities for further development in this field.

References

1. Austin, D., Mann, B.: Powerlifting, 2nd edn. Human Kinetics, Champaign (2012)
2. Bunker, R.P., Thabtah, F.: A machine learning framework for sport result prediction. Appl. Comput. Inform. **15**(1), 27–33 (2019)
3. Carey, D.L., Ong, K., Whiteley, R., Crossley, K.M., Crow, J., Morris, M.E.: Predictive modelling of training loads and injury in Australian football. Int. J. Comput. Sci. Sport **17**(1), 49–66 (2018)
4. Chau, V.H., Vo, A.T., Le, B.T.: A gravitional-double layer extreme learning machine and its application in powerlifting analysis. IEEE Access **7**, 143990–143998 (2019)
5. Chau, V.H., Vo, A.T., Le, B.T.: The effects of age and body weight on powerlifters: an analysis model of powerlifting performance based on machine learning. Int. J. Comput. Sci. Sport **18**(3), 89–99 (2019)

6. Doborjeh, M., Kasabov, N., Doborjeh, Z., Enayatollahi, R., Tu, E., Gandomi, A.H.: Personalised modelling with spiking neural networks integrating temporal and static information. Neural Netw. **119**, 162–177 (2019)
7. Greco, S., Mousseau, V., Słowiński, R.: Multiple criteria sorting with a set of additive value functions. Eur. J. Oper. Res. **207**(3), 1455–1470 (2010)
8. Kasabov, N., et al.: Evolving spiking neural networks for personalised modelling, classification and prediction of spatio-temporal patterns with a case study on stroke. Neurocomputing **134**, 269–279 (2014)
9. Mucherino, A., Papajorgji, P.J., Pardalos, P.M.: Data Mining in Agriculture, vol. 34. Springer, Florida (2009)
10. Reschke, J., Neumann, C., Berlitz, S.: Personalised neural networks for a driver intention prediction: communication as enabler for automated driving. Adv. Opt. Technol. **9**(6), 357–364 (2020)
11. Saikia, H., Bhattacharjee, D., Lemmer, H.H.: Predicting the performance of bowlers in IPL: an application of artificial neural network. Int. J. Perform. Anal. Sport **12**(1), 75–89 (2012)
12. Santos Coelho, L., da Cruz, L.F., Freire, R.Z.: Swim velocity profile identification by using a modified differential evolution method associated with RBF neural network. In: 3rd International Conference on Innovative Computing Technology, London, pp. 389–395. IEEE (2013)
13. Zhang, A., Zhang, L.: RBF neural networks for the prediction of building interference effects. Comput. Struct. **82**(27), 2333–2339 (2004)
14. Zhao, L., Liu, R., Wu, J., Gou, L.: Wrestling performance prediction based on improved RBF neural network. J. Phys. Conf. Ser. **1629**(1), 012012 (2020)
15. Open Powerlifting Homepage. http://wwww.openpowerlifting.org. Accessed 14 Dec 2020

Learning Shape Sensitive Descriptors for Classifying Functional Data

Wojciech Rafajłowicz(✉) and Ewaryst Rafajłowicz

Department of Control Systems and Mechatronics, Wroclaw University of Science and Technology, Wyb. Wyspianskiego 27, 50 370 Wrocław, Poland
{wojciech.rafajlowicz,ewaryst.rafajlowicz}@pwr.edu.pl

Abstract. We propose a new method of learning descriptors for constructing classifiers of functional data. These descriptors are moments of a curve derivative, but their learning is based solely on samples of the curve itself. Furthermore, the derivative itself is not directly estimated. This is possible due to the trick of using simultaneously two different bases of a functional space.

The advantage of extracting features from the derivative instead of from a curve itself is in raising their sensitivities to a shape of a curve. As expected, this may result in better classification accuracy. The simulation experiments that are based on an augmented real data support this claim, but it is not unconditional. Namely, noticeable improvements can be obtained when an appropriate classifier is selected.

Keywords: Functional data classification · Shape sensitive descriptors · Learning derivatives · Orthogonal expansion

1 Introduction

In recent several years, one can observe a growing interest of researchers to analyze and classify functional data (see monographs and survey papers cited in the next section). Clearly, important contributions in this direction are much earlier (see, e.g., [1,10] for reviews on classifying electrocardiogram (ECG) signals and [3,7] for electroencephalogram (EEG) signals classification and features selection [4,5] as well as [2] for the survey on the analysis of electromyography signals).

The renewed interest has its origin in growing possibilities of acquiring large number of samples from new sensors and storing them on cloud databases. An example of this kind is provided at the end of the paper, where curves from accelerometers are classified. Their distinctive feature is that they are repetitive, but in a stochastic sense, i.e., their underlying probability distributions remain the same for each class, although they are unknown. As the result, curves differ more by shape than by amplitudes. Therefore, our aim is to derive descriptors of curves that are based on curves' derivatives, without having an access to them directly.

L. Rutkowski et al. (Eds.): ICAISC 2021, LNAI 12854, pp. 485–495, 2021.
https://doi.org/10.1007/978-3-030-87986-0_43

The paper is organized as follows. In the next section, we justify using curve descriptors that are based on moments of its derivative instead of the curve itself. The main result of this section is the derivation of a relationship between these two kinds of moments. This relationship is crucial for Sect. 3 to construct an algorithm for learning the derivative descriptor, without having access to samples of the derivative curve. In Sect. 3 also an interplay between learning these descriptors and learning a classifier is described. Finally, in Sect. 4, the extensive results of testing the proposed approach on an augmented real data are summarized. They aimed to investigate an influence of a classifier on possible improvements of the classification accuracy.

2 Descriptors Based on the Derivative Moments

Our derivations are based on the notion of square root velocity (SRV) of differentiable functions $\mathbf{X}(t)$, $\mathbf{Y}(t)$, $t \in [0, T]$ that can be interpreted as signals, curves etc., defined on a finite time interval of the length $T > 0$. For those $t \in [0, T]$ for which the derivative $\mathbf{X}'(t)$ is not zero, the SRV of $\mathbf{X}$, denoted further as $q(\mathbf{X}, t)$ is defined as follows

$$q(\mathbf{X}, t) = \frac{\mathbf{X}'(t)}{\sqrt{|\mathbf{X}'(t)|}}, \quad t \in [0, T] \tag{1}$$

or, equivalently,

$$q(\mathbf{X}, t) = sgn(\mathbf{X}'(t)) \sqrt{|\mathbf{X}'(t)|}. \tag{2}$$

From (1) it is clear that the SRV description of $\mathbf{X}$ is invariant in a scale and a vertical position, i.e., for any $c > 0$ and any $\beta \in \mathbb{R}$:

$$q(c\mathbf{X}, t) = q(\mathbf{X}, t), \quad q(\beta + \mathbf{X}, t) = q(\mathbf{X}, t), \quad t \in [0, T]. \tag{3}$$

Let $\mathbf{X}'$ and $\mathbf{Y}'$ be square-integrable on $[0, T]$, $\mathbf{X}', \mathbf{Y}' \in L_2(0, T)$. Then, from (2) we immediately obtain:

$$\int_0^T q^4(\mathbf{X} - \mathbf{Y}, t)\, dt = \int_0^T [\mathbf{X}'(t) - \mathbf{Y}'(t)]^2\, dt. \tag{4}$$

Strictly speaking, the squared right hand side of (4) is not a distance measure, since $\mathbf{X} - \mathbf{Y}$ that differ by a constant yield zero in (4).

However, this expression suggests that the derivatives of $\mathbf{X}$, $\mathbf{Y}$, ... can be useful in classifying curves in a shape-sensitive way. In particular, moments of $\mathbf{X}'$, $\mathbf{Y}'$, ... with respect to a selected basis in $L_2(0, T)$ are worthwhile candidates for descriptors of $\mathbf{X}$, $\mathbf{Y}$, ... when one attempts to classify them. We shall follow this line of reasoning.

2.1 Modeling Random Curves

The main difficulty is in learning such descriptors from samples of $\mathbf{X}$, $\mathbf{Y}, \ldots$ instead of $\mathbf{X}'$, $\mathbf{Y}', \ldots$ that are frequently not directly available. In this respect we shall follow [13,14], where the approach to nonparametric estimation of derivatives from noisy observations of $\mathbf{X}(t)$ can be found. However, we emphasise that in our case $\mathbf{X}(t)$ is a random element of $L_2(0,\, T)$, which implies a different model of random errors than the one used in [13]. Furthermore, the estimation of $\mathbf{X}'$ is only an intermediate step, since our goal is to learn the moments of $\mathbf{X}'$ with respect to a selected orthonormal basis.

We refer the reader to [6,8,9,16,19] for more details on shape-sensitive description of random curves.

Let $\mathbf{v}_k(t)$, $t \in [0,\, T]$, $k = 1, 2, \ldots$ be a selected orthogonal and complete sequence in $L_2(0,\, T)$ with elements that are also normalized to 1 with respect to the standard norm $||\mathbf{v}_k||^2 =< \mathbf{v}_k,\, \mathbf{v}_k >$, where $< \mathbf{X},\, \mathbf{Y} >= \int_0^T \mathbf{X}(t)\, \mathbf{Y}(t)\, dt$. Then, $\mathbf{X} \in L_2(0,\, T)$ has the representation:

$$\mathbf{X}(t) = \sum_{k=1}^{K} a_k\, \mathbf{v}_k(t) + \mathbf{R}_K(t), \quad t \in [0,\, T], \tag{5}$$

where

$$\mathbf{R}_K(t) \stackrel{def}{=} \sum_{k=(K+1)}^{\infty} \beta_k\, \mathbf{v}_k(t) \tag{6}$$

and the coefficients are given by $a_k =< \mathbf{X},\, \mathbf{v}_k >$, $k = 1, 2, \ldots, K$, $\beta_k =< \mathbf{X},\, \mathbf{v}_k >$, $k = (K+1),\, (K+2),\, \ldots$, while the convergence is understood the L_2 norm sense.

Collections of coefficients a_k's and β_k's are both random, but they play different roles in our derivations. Namely, a_k's are regarded as descriptors that are informative for curves classification, while β_k's are interpreted as coefficients of non-informative error $\mathbf{R}_K$.

Denote by $\mathbb{E}$ the expectation with respect to a_k's and β_k's. Although their distributions are not known, the following assumptions are made:

$$\mathbb{E}[\beta_k] = 0, \quad \mathbb{E}[\beta_k\, \beta_j] = 0, \quad k \neq j\; k, j = (K+1),\, (K+2),\, \ldots, \tag{7}$$

$$\gamma(K) \stackrel{def}{=} \mathbb{E}||\mathbf{R}_K||^2 \to 0, \text{ as } K \to \infty, \tag{8}$$

$$\mathbb{E}(a_k^2) < \infty,\; \mathbb{E}(a_k\, \beta_l) = 0,\; k = 1, 2 \ldots, K,\; l = (K+1), (K+2), \ldots \tag{9}$$

Assumption (8) implicitly imposes constraints on the variability of the residual curve $\mathbf{R}_K(t)$, $t \in [0,\, T]$ for large K.

For simplicity, $1 \leq K \leq \infty$ is assumed to be fixed and known. In practice, one should select K so as to minimize an estimate of the classification error plus a penalty term for too complicated model, such as in the AIC, BIC etc. criterions.

2.2 The Relationship Between Descriptors of a Curve and Its Derivative

In the next section, we provide details of learning moments of $\mathbf{X}'$ from equidistant observations of $\mathbf{X}$ only. Here, we outline a general idea. If $\mathbf{v}_k$'s are differentiable and the series (5) and (6) is term-by-term differentiable, then,

$$\mathbf{X}'(t) = \sum_{k=1}^{K} a_k\, \mathbf{v}_k'(t) + \mathbf{R}_K'(t), \quad t \in [0,\, T], \tag{10}$$

On the other hand, for $\mathbf{w}_k$, $k = 1, 2, \ldots$ being an orthonormal and complete sequence in $L_2(0,\, T)$, $\mathbf{X}' \in L_2(0,\, T)$ has the representation

$$\mathbf{X}'(t) = \sum_{k=1}^{K} b_k\, \mathbf{w}_k(t) + \mathbf{r}_K(t), \quad b_k = <\mathbf{X}',\, \mathbf{w}_k> \tag{11}$$

$$\mathbf{r}_K(t) = \sum_{k=(K+1}^{\infty} \eta_k\, \mathbf{w}_k(t), \quad \eta_k = <\mathbf{X}',\, \mathbf{w}_k>,\ k = (K+1).\, \ldots \tag{12}$$

For sufficiently large K, according to (8), we approximate $\mathbf{X}'$ in (10) by the first summand, which yields, after substituting it into (11),

$$b_k = \sum_{j=1}^{K} a_j <\mathbf{w}_k,\, \mathbf{v}_j'>, \quad k = 1,\, 2,\, \ldots,\, K. \tag{13}$$

Observe that these formulas are exact, if $<\mathbf{w}_k,\, \mathbf{R}_K' >= 0$, $k = 1,\, 2, \ldots$, but this is not postulated here.

Summarizing, moments b_k's of $\mathbf{X}'$ with respect to basis $\mathbf{w}_k$'s can be expressed as linear combinations of moments a_k's that are estimable from observations of $\mathbf{X}$ itself, assuming that for each $k = 1,\, 2,\, \ldots,\, K$

$$\text{at least one } <\mathbf{w}_k,\, \mathbf{v}_j'> \neq 0, \quad j = 1,\, 2,\, \ldots,\, K. \tag{14}$$

Additionally, elements $<\mathbf{w}_k,\, \mathbf{v}_j'>$ of $K \times K$ transformation matrix, say B, are either known or they can be approximated to any desired accuracy by quadrature formulas.

3 Learning Classifiers Based on Curves' Derivative Descriptors

Suppose, for simplicity of formulas only, that random curves like $\mathbf{X}$ are drawn from two classes, labelled by I and II, that are formed as follows: firstly, vector $\bar{a} \stackrel{def}{=} [a_1, a_2, \ldots, a_K]^{tr}$ is drawn from a cumulative distribution function (c.d.f.), which is either F_I or F_{II}. These c.d.f.'s are not known. We do not impose any special restrictions on them, except for the existence of the second moments of a_k's and (9). In this way, a large class of classification problems for informative part of $\mathbf{X}$ can be stated. The second step in modeling $\mathbf{X}$ is to draw β_k's. Their distributions are also unknown and only conditions (7), (8) and (9) are assumed to hold. Finally, $\mathbf{X}$ is formed according to (5) and (6). Thus, $\mathbf{X}$ may come from class I or II, depending whether $\bar{a}$ was according to c.d.f. F_I or F_{II}. The existence of a priori probabilities $0 < p_I < 1$, $0 < p_{II} < 1$, $p_I + p_{II} = 1$ that $\mathbf{X}$ is from class I or II is postulated, but they are unknown. Their estimation by fractions in the learning sequence is a simple task, unless an essential class imbalance does not appear, which is excluded in this paper.

3.1 Learning Sequence

A learning sequence that we have at our disposal is of the form:

$$\mathcal{L}_N \stackrel{def}{=} \{(\bar{x}^{(1)}, j_1), (\bar{x}^{(2)}, j_2), \ldots, (\bar{x}^{(N)}, j_N)\}, \tag{15}$$

where $j_n \in \{I, II\}$ are correct class labels (provided by an expert), while $\bar{x}^{(n)}$ are equidistant, in $[0, T]$, samples from curves $\mathbf{X}^{(n)}$, taken at time instants t_i, $i = 1, 2, \ldots, m$, $n = 1, 2, \ldots, N$. Samples forming $\bar{x}^{(n)}$'s have the following form:

$$x_i^{(n)} = \mathbf{X}^{(n)}(t_i) = \bar{\mathbf{v}}^{tr}(t_i)\, \bar{a}^{(n)} + \mathbf{R}_K(t_i), \quad i = 1, 2, \ldots, m, \tag{16}$$

where $\bar{a}^{(n)}$ are drawn either according to F_J or F_{II}, while

$$\bar{\mathbf{v}}^{tr}(t) \stackrel{def}{=} [\mathbf{v}_1(t), \mathbf{v}_2(t), \ldots \mathbf{v}_K(t)]. \tag{17}$$

Analogously, new $\mathbf{X}$ to be classified is represented only by $\bar{x}$ with elements

$$x_i = \mathbf{X}(t_i) = \bar{\mathbf{v}}^{tr}(t_i)\, \bar{a} + \mathbf{R}_K(t_i), \quad i = 1, 2, \ldots, m, \tag{18}$$

Problem Formulation. Using learning sequence $\mathcal{L}_N$, derive a classifier that classifies $\mathbf{X}$, represented only by $\bar{x}$, to class I or II. This classifier should be shape sensitive in the sense that, for a preselected orthonormal and complete sequence $\mathbf{w}_k$'s, the classifier decision is based on learning descriptors $b_k = < \mathbf{X}', \mathbf{w}_k >$, $k = 1, 2, \ldots, K$, which are directly not available.

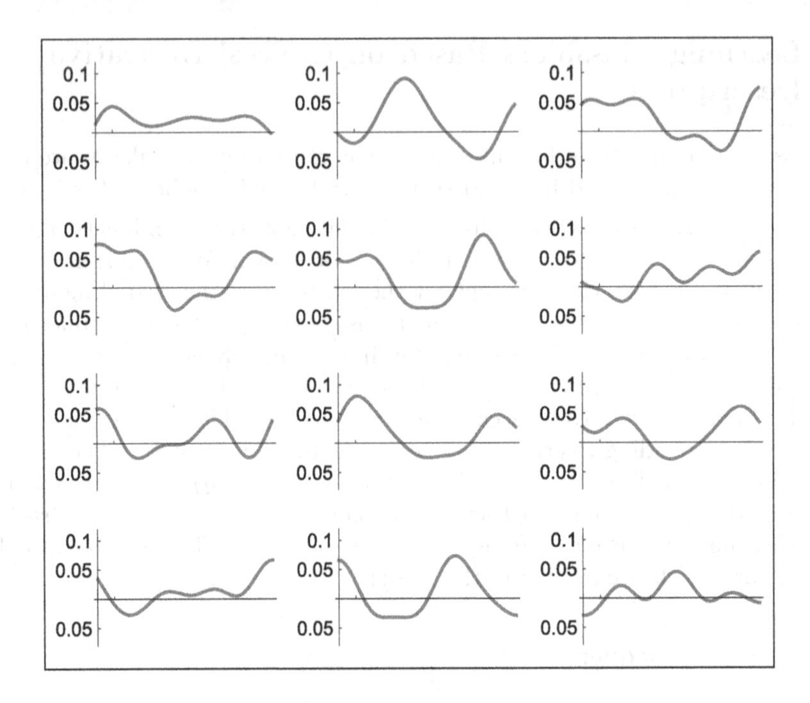

Fig. 1. Examples of curves to be classified

Fig. 2. Descriptors of curves to be classified, stacked together, and displayed as images. Upper panel – classic DCT descriptors, lower panel – descriptors based on learning derivatives.

3.2 Learning Descriptors

Model of observations (18) and (16) suggest that for estimating primary descriptors $\bar{a}$ and $\bar{a}^{(n)}$'s one may use the method of minimizing the least squares error (LSE) in the nonparametric setting with deterministic regressors t_i's (see [12]).

However, in this case the ordinary (unweighted) LSE approach is not recommended, since $\mathbf{R}_K(t_i)$'s are correlated for moderate K.

Thus, $\bar{a}$ is estimated in more classic way as

$$\hat{\bar{a}} = \Delta_m \sum_{i=1}^{m} x_i \, \bar{\mathbf{v}}(t_i) = \Delta_m \, \bar{\mathbf{V}} \, \bar{x}, \quad \Delta_m \stackrel{def}{=} T/m, \tag{19}$$

where $\bar{\mathbf{V}}$ is $K \times m$ matrix composed of the columns: $\bar{\mathbf{v}}(t_i)$'s. It is not difficult to show that $\hat{\bar{a}}$ is asymptotically (as $m \to \infty$) unbised for $\bar{a}$. It is more tedious to bound the variances of $\hat{\bar{a}}_k$'s by $\zeta(K)/m^2$, where $\zeta(K) > 0$ depends on K in a polynomial way.

Transforming samples of the learning curves in the same way as in (19), we obtain the learning sequence, denoted as $\mathcal{A}_N$, composed of the classic descriptors:

$$\mathcal{A}_N = \{(\hat{\bar{a}}^{(n)}, \, j_n), \; n = 1, 2, \ldots, N\}, \; \hat{\bar{a}}^{(n)} = \Delta_m \, \bar{\mathbf{V}} \, \bar{x}^{(n)}, \tag{20}$$

where j_n's are joined at the original order.

Using the plug-in idea and (13), we can learn derivative-sensitive descriptors as follows:

$$\hat{b}_k = \sum_{j=1}^{K} \hat{a}_j <\mathbf{w}_k, \, \mathbf{v}_j'>, \quad k = 1, 2, \ldots, K \tag{21}$$

and they are also asymptotically unbised and with finite variances that can be reduced faster sampling (m larger).

Transforming the Learning Sequence. Elements of $\mathcal{L}_N$ are transformed into descriptors in the same way as in (21), providing learning sequence $\mathcal{B}_N$, say, of the form:

$$\mathcal{B}_N = \{(\hat{\bar{b}}^{(n)}, \, j_n), \; n = 1, 2, \ldots, N\}, \; \hat{\bar{b}}^{(n)} = \Delta_m \, B \, \bar{\mathbf{V}} \, \bar{x}^{(n)}, \tag{22}$$

while labels j_n's are rewritten from $\mathcal{L}_N$, accordingly.

Summarizing, original learning sequence $\mathcal{L}_N$, with usually long sequences of samples $\bar{x}^{(n)}$, was transformed into learning sequence $\mathcal{B}_N$ with descriptors for derivatives. Furthermore, this transformation is linear in $\bar{x}^{(n)}$, which allows for speeding up computations.

At this stage, it suffices to select a proper classifier, to learn and test it using $\mathcal{B}_N$ and to apply it for newly coming sample $\bar{x}$, after transforming it to $\hat{\bar{b}} = \Delta_m \, B \, \bar{\mathbf{V}} \, \bar{x}$. For brevity, the obtained classifier will be denoted as $CLname[\mathcal{B}_N; \, \bar{x}]$ or $CLname[\mathcal{A}_N; \, \bar{x}]$, when the learning is based on standard descriptors, for the sake of comparisons. For example, the support vector machine (SVM) classifier that was trained on $\mathcal{B}_N$ is denoted as $SVM[\mathcal{B}_N; \, \bar{x}]$ and its output is I or II class label.

As we shall see in the next section, this obvious route of building a classifier may lead to moderate or essential improvements of the classification accuracy, depending on the choice of a classifier.

4 Testing and Comparisons on Augmented Acceleration Data

Operators' cabins of large working machines are frequently subject to relatively high, repetitive accelerations. Benchmark data of this kind are freely available from [17], while in [18] their detailed description is provided.

The benchmark consists from $N = 43$ learning curves, each containing $m = 1K$ samples (see Fig. 1, where examples of curves are shown, after a low-pass filtering). Labels, either I or II, were attached to each curve, corresponding to lighter or heavier working conditions. Notice that curves in Fig. 1 differ mainly by shape rather than in amplitude.

As orthogonal systems in $L_2(0, T)$, we have selected the cosine series as $\mathbf{v}_k$'s and the sine series as $\mathbf{w}_k$'s. Descriptors $\hat{\bar{a}}^{(n)}$'s were computed according to (20) for $K = 16$. For illustration purposes, these $N = 43$ vectors were stacked into 43×16 matrix that is displayed in Fig. 2 – upper panel (dark places correspond to lower values of the descriptors).

Descriptors of derivatives $\hat{\bar{b}}^{(n)}$'s were computed according to (22). They are analogously visualized in Fig. 2 – lower panel. By a visual inspection of these two panels, we conclude that the variability of $\hat{\bar{b}}^{(n)}$'s is much larger than $\hat{\bar{a}}^{(n)}$'s. Thus, one may hope that the classification accuracy will also be larger.

Data Augmentation. Unfortunately, the learning sequence of the length $N = 43$ is far too short for learning and comparisons. Therefore, we augmented the original data as follows: each estimated $\hat{\bar{a}}^{(n)}$'s was repeated 1000 by adding to it the Gaussian perturbations with zero mean and dispersions 0.02 and keeping the same label. This augmentation corresponds to about 11 % perturbations amplitudes. Notice that it suffices to add perturbations to $\hat{\bar{a}}^{(n)}$'s, instead adding them to original samples, since $\hat{\bar{a}}^{(n)}$'s depend on them linearly. In this way, we have obtained the augmented learning sequence $\mathcal{A}_{N_e}$ of the length $N_e = 43000$. Each descriptor from this sequence was transformed by (22), which led to the augmented learning sequence of the derivatives descriptors $\mathcal{B}_{N_e}$.

The next step was to learn the logistic regression classifier twice in order to obtain: $LogR(\mathcal{A}_{N_e};\, \bar{x})$ and $LogR(\mathcal{B}_{N_e};\, \bar{x})$ and their characteristics, such as accuracy, precision, recall etc. These characteristics are collected as pairs, separated by |, in the *LogR* column in Table 1.

In the same way, the following classifiers were learned and validated:

LogR – the logistic regression classifier,
SVM – the support vector machine,
DecT – the decision tree classifier,
gbTr – the gradient boosted trees,
RFor – the random forests classifier,
5NN – the 5 nearest neighbors classifier.

The results are summarized in Table 1. Notice that the result on the left hand side in each cell of this table is intentionally the same as in [15], for the sake of comparisons.

Table 1. An account on testing the popular classifiers when the cosine moments (the left result) and the shape sensitive descriptors (the right result) are used for learning them. Abbreviations: Cohen – the Cohen κ coefficient, MCC – the Matthews Correlation Coefficient. For the abbreviations of the classifiers' names – see the text.

Classifier	LogR	SVM	DecT	gbTr	RFor	5 NN
Accuracy	0.91|0.97	0.94|0.96	0.84|0.86	0.92|0.90	0.91|0.92	0.90|0.92
Cohen	0.76|0.91	0.82|0.88	0.60|0.65	0.77|0.72	0.72|0.77	0.70|0.73
MCC	0.76|0.91	0.82|0.88	0.61|0.66	0.77|0.72	0.72|0.78	0.71|0.75
Precision	0.96|0.99	0.94|0.95	0.93|0.95	0.94|0.93	0.92|0.93	0.90|0.90
Recall	0.92|0.97	0.98|0.99	0.86|0.87	0.96|0.94	0.96|0.98	0.98|0.99
Specificity	0.88|0.95	0.80|0.83	0.60|0.85	0.79|0.77	0.72|0.75	0.65|0.65
F1 score	0.94|0.98	0.96|0.97	0.89|0.91	0.95|0.94	0.94|0.95	0.94|0.95

The analysis of this table leads to the following conclusions:

- when the *LogR* classifier is used together with descriptors based on derivatives b_k's, it provides a noticeable increase of the accuracy and other indicators (the Cohen and MCC) in comparison to applying the *LogR* classifier to classic descriptors a_k's,
- also the *SVM* classifier performs better on b_k's than on a_k's, but the improvements are less spectacular,
- only slight improvements, but pertaining all of the indicators, are visible when the decision trees, random forests and 5 NN classifiers are applied,
- somewhat unexpectedly to the authors, the gradient boosted trees classifier provided a slightly worse results when applied to b_k's descriptors, in other words, the *gbTr* was not able to take advantages from the derivative based descriptors.

5 Conclusions

The new way of learning descriptors of functional data is proposed and investigated from the view-point of the classification accuracy. Its essence is in learning descriptors of a curve derivative, without estimating it directly. Extensive simulations indicate that using these descriptors one may expect a better classification accuracy, but the improvement is essential when simultaneously an appropriate classifier of these descriptors is used. In the case study of accelerometer data, the proper choice was the logistic regression classifier, followed by the SVM.

The results are promising, but further efforts are necessary to reveal an influence of a kind of functional data on the choice of the classifier.

One of possible directions of generalizations of the proposed approach is to allow curves having derivatives with a finite number of jumps. Before learning their descriptors, it would be necessary to smooth samples in a jump-preserving way, as it was proposed in [11].

Acknowledgments. The authors express their thanks to Professor P. Moczko and to Dr. J. Więckowski from the Faculty of Mechanical Engineering, Wroclaw University of Science and Technology for the permission to use acceleration signals in our case studies.

References

1. Abdulla, L., Al-Ani, M.: A review study for electrocardiogram signal classification. UHD J. Sci. Technol. **4**(1), 103–117 (2020). https://doi.org/10.21928/uhdjst.v4n1y2020.pp103-117. http://journals.uhd.edu.iq/index.php/uhdjst/article/view/711
2. Ahsan, M.R., Ibrahimy, M.I., Khalifa, O.O., et al.: EMG signal classification for human computer interaction: a review. Eur. J. Sci. Res. **33**(3), 480–501 (2009)
3. Azlan, W.A., Low, Y.F.: Feature extraction of electroencephalogram (EEG) signal - a review. In: 2014 IEEE Conference on Biomedical Engineering and Sciences (IECBES), pp. 801–806 (2014). https://doi.org/10.1109/IECBES.2014.7047620
4. Gandhi, T., Panigrahi, B.K., Anand, S.: A comparative study of wavelet families for EEG signal classification. Neurocomputing **74**(17), 3051–3057 (2011)
5. Garrett, D., Peterson, D.A., Anderson, C.W., Thaut, M.H.: Comparison of linear, nonlinear, and feature selection methods for EEG signal classification. IEEE Trans. Neural Syst. Rehabil. Eng. **11**(2), 141–144 (2003). https://doi.org/10.1109/TNSRE.2003.814441
6. Harris, T., Tucker, J.D., Li, B., Shand, L.: Elastic depths for detecting shape anomalies in functional data. Technometrics 1–11 (2020)
7. Lotte, F., Congedo, M., Lécuyer, A., Lamarche, F., Arnaldi, B.: A review of classification algorithms for EEG-based brain-computer interfaces. J. Neural Eng. **4**(2), R1–R13 (2007). https://doi.org/10.1088/1741-2560/4/2/r01
8. Marron, J.S., Ramsay, J.O., Sangalli, L.M., Srivastava, A.: Functional data analysis of amplitude and phase variation. Stat. Sci. **30**(4), 468–484 (2015). https://doi.org/10.1214/15-STS524
9. Marron, J.S., Ramsay, J.O., Sangalli, L.M., Srivastava, A.: Functional data analysis of amplitude and phase variation. Stat. Sci. **30**, 468–484 (2015)
10. Mironovova, M., Bíla, J.: Fast Fourier transform for feature extraction and neural network for classification of electrocardiogram signals. In: 2015 Fourth International Conference on Future Generation Communication Technology (FGCT), pp. 1–6 (2015). https://doi.org/10.1109/FGCT.2015.7300244
11. Pawlak, M., Rafajłowicz, E.: Jump preserving signal reconstruction using vertical weighting. Nonlinear Anal. Theory, Methods Appl. **47**(1), 327–338 (2001)
12. Rafajlowicz, E.: Nonparametric least squares estimation of a regression function. Statistics **19**(3), 349–358 (1988)
13. Rutkowski, L.: A general approach for nonparametric fitting of functions and their derivatives with applications to linear circuits identification. IEEE Trans. Circu. Syst. **33**(8), 812–818 (1986). https://doi.org/10.1109/TCS.1986.1086001
14. Rutkowski, L., Rafajłowicz, E.: On optimal global rate of convergence of some nonparametric identification procedures. IEEE Trans. Automatic Control **AC-34**, 1089–1091 (1989)
15. Skubalska-Rafajłowicz, E., Rafajłowicz, E.: Classifying functional data from orthogonal projections - model, properties and fast implementation. In: International Conference on Computational Science, Krakow, Poland (2021, accepted)

16. Srivastava, A., Klassen, E., Joshi, S.H., Jermyn, I.H.: Shape analysis of elastic curves in Euclidean spaces. IEEE J. Sel. Areas Commun. **10**(2), 391–400 (1992). https://doi.org/10.1109/49.126990
17. Więckowski, J.: Data from vibration in SchRs1200, Mendeley Data, V1. http://dx.doi.org/10.17632/htddgv2p3b.1. Accessed Jan 2021
18. Więckowski, J., Rafajlowicz, W., Moczko, P., Rafajlowicz, E.: Data from vibration measurement in a bucket wheel excavator operator's cabin with the aim of vibrations damping. Data Brief, 106836 (2021)
19. Xie, W., Chkrebtii, O., Kurtek, S.: Visualization and outlier detection for multivariate elastic curve data. IEEE Trans. Visual. Comput. Graph. **26**(11), 3353–3364 (2020). https://doi.org/10.1109/TVCG.2019.2921541

An Approach to Contextual Time Series Analysis

Anton Romanov(✉), Aleksey Filippov, and Nadezhda Yarushkina

Ulyanovsk State Technical University, Street Severny Venets 32, 432027 Ulyanovsk, Russian Federation
{al.filippov,jng}@ulstu.ru
http://www.ulstu.ru/

Abstract. This article presents and formally describes an ontology-based approach to domain context formation for time series analysis. Considered the logical representation of the ontology using the descriptive logic ALCHI(D). Also described the experimental results that confirm the correctness and effectiveness of the proposed approach.

Keywords: Predictive analytics · Ontology · Time series · Context

1 Introduction

The activities of any modern organization often require urgent management decisions. The decision-maker must know the domain and have the skills in various decision support systems and tools for working with knowledge. Data mining techniques are required to automate and speed up the state evaluation of some complex system. These techniques are focused on working with dynamic data of a concrete system considering the domain context. The context allows considering the specific and limitations of the domain in analyzing data represented by time series.

The context allows using of additional domain-dependent knowledge in the process of describing the system behavior. Usually, the description of the system behavior is represented as qualitative assessments of its state. The same dataset in different domains (in different contexts) will have different models and analysis results.

2 The State of the Art

Forecasting methods, with all conventions and limitations, are the development of descriptive analytics mechanisms. Any model of real-world objects works only under conditions of restrictions and agreements. Also, the following conclusion can be drawn for a forecast – the future state of the control objects cannot be predicted 100%. However, there is a task to build the most accurate forecast in the given conditions.

L. Rutkowski et al. (Eds.): ICAISC 2021, LNAI 12854, pp. 496–505, 2021.
https://doi.org/10.1007/978-3-030-87986-0_44

The development of methods for intellectual analysis of enterprise processes in integration with knowledge engineering methods is dictated by various needs. Data mining methods are not able to works with such components of processes than not implemented in models.

The works [1–3] describe that the hybridization of analyzing methods and the usage of context can improve the quality and efficiency of the analysis. Ontologies are used to formed the context in the presented works. The ontologies are used for: improving the accuracy of search engines requests in the predictive analytic [1]; creation of relations [2] between time series of social networks messages and features of a domain: company structure, names of employees, etc.; detecting changes in a group of related time series [3].

The works [4,5] presented the approach to the process mining based on the analysis of a set of time series. However, these works do not consider the features of the domain. This fact reduces the accuracy of process discovery. The analyzed process can be more complex than the vision through a time series model.

Thus, methods for modeling time series need a component that stores additional knowledge about the modeled process. Especially in tasks of analytic and forecasting of time series. The proposed approach is based on the hypothesis about the possibility of applying modeling methods and intelligent analysis of time series extracted from various information systems to detect hidden information processes specific to a specific domain. Knowledge engineering methods are used for the detection and validation of information processes with the domain features.

3 Contextual Time Series Forecasting

The scheme of analysis of dynamic data with the context considering presents in Fig. 1.

The framework of the forecasting system uses the entire set of modeled objects O. The decision-maker (DM) initiates the entire procedure: forms a request for managing of interest objects, sets goals.

Two functions formulated to study the dynamics of the behavior of domain objects:

1. Function for converting object data to numeric representation $num(O) \rightarrow O^{num}$. The implementations of this function can be following: obtaining the values of the numerical attributes of an object, obtaining the number of objects of the same type included as a property in the parent object. Another implementation option is periodically obtaining the indicator values.
2. The function of converting data about an object into a representation of the object behavior $rule(O) \rightarrow O^{rule}$. For simplicity, we can represent patterns by a list of "IF-THEN" rules. $o_i^{rule} = \{A, C\}$, where A is a set of antecedents, C is a set of consequencies.

The information aggregation block is used for the comparison of the numerical and meaningful indicators:

$$aggr(O^{num}, O^{rule}) \rightarrow \{o_i^{num}, \hat{O}^{rule}\}, \quad (1)$$

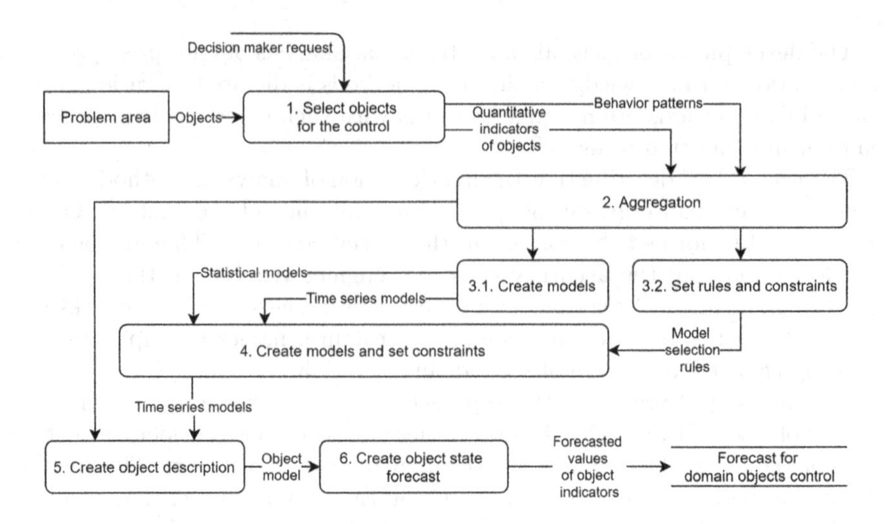

Fig. 1. Analyzing dynamic data with the context considering

where $\hat{O}^{rule}$ is a subset of the behavior rules for a numeric attribute o_i^{num}.

Time series models for predictive analytics can be built using different methods: statistical, neural network, fuzzy. A large number of existing time series models suggest that there is no best and universal way. In studies [6–8] it is noted that a separate time series model may not give a correct forecast, and a combination and teams of models can improve the quality of the forecast. The proposed scheme defines choosing the model. Extracted from the domain information helps to select the model. As selection criteria we can use: presence of a tendency; presence of a periodic component; length of time series; presence of outliers and the need for preliminary smoothing; the degree of uncertainty about the representation of levels, trends, and lengths of time series intervals.

Then selection function can be defined as follows:

$$SelectModel(o_i^{num}, \hat{O}^{rule}, TS) \rightarrow \{o_i^{num}, \hat{O}^{rule}, \hat{TS}\}, \tag{2}$$

where TS is a set of available time series models; $\hat{TS}$ is a set of selected models.

The time series model works like a converting function of the values attribute o_i^{num} into a set of time series points:

$$TS_i(o_i^{num}) \rightarrow \{y_1, y_2, ..., y_n\}, \tag{3}$$

where TS_i is a time series model represents an attribute value into a time-dependent value – ts_i.

The time series model in the classical sense does not use additional information. They use only the history of the time series. The main task is to create a time series model that considers the constraints and conventions of the time series behavior. The behavior extracted from the context of the object's functioning. Let's describe the information extracted from the context that affects the modeling:

1. Time series baseline y_{base}.
2. The main tendency of time series $tend_{base}$.
3. Max-min value limitations of the levels of the time series $bound = (y_{min}, y_{max})$.
4. The rate of change of trends (inertia) of the time series $tend_{\Delta}$.
5. Range of acceptable values that are not anomalous $accept = (y_{min}^{accept}, y_{max}^{accept})$.

The specified information categorized by the following types. The first class defines time series behavior. This class includes the following items: time series baseline, the primary tendency. You can create a crude time series model that used information only from the context, not analyzing the time series. The second class that describes the modeling result and is used to evaluate predicted values: min-max values restrictions, the rate of tendency change, the interval of acceptable values. Also, defining information can be used to validate predicted values.

The formal definition of both classes is:

$$\begin{aligned} Def &= \{y_{base}, tend_{base}\}, \\ Desc &= \{bound, tend_{\Delta}, accept, y_{base}, tend_{base}\} \end{aligned} \tag{4}$$

The time series model o_i^{num} can define as a combination of the following components:

$$\{o_i^{num}, \hat{O}^{rule}, \hat{TS}, \hat{Def}, \hat{Desc}\}, \tag{5}$$

where $\hat{TS}$ is a set of selected time series models. $\hat{TS}$ is used for modeling and forecasting of numerical attribute representation O^{num} of control object with using additional components $\hat{Def}$, $\hat{Desc}$, formed as $Extract(\hat{O}^{rule}) \rightarrow (\hat{Def}, \hat{Desc})$.

New components $\hat{Def}$, $\hat{Desc}$ of the time series model work with the following expression:

$$y_{Def} = \alpha * y_{base} + \beta * y_{tend_{base}}, \tag{6}$$

where α, β are the weight coefficients of the model components. The context $\hat{O}^{rule}$ determines these values.

The expression shows that the model built only from information extracted from the context does not use the values of the original numeric representation of the time series. The proposed model of time series contain a representation with weights:

$$y_{model} = w_{Def} * y_{Def} + (1 - w_{Def}) * y, \tag{7}$$

where w_{Def} is a weight of model values of the time series. In this case, the result of another model can be used as y. Forecast values are weighted in the same way.

The *Decs* model component is used for the validation modeling and forecasting results. It helps evaluate the difference between the context-oriented model and the model obtained from the numerical representation of the time series in modeling and forecasting:

$$Err_{valid} = \frac{\sum_{i=1}^{n} isValid(desc_i)}{n * |Desc|}, \tag{8}$$

where the function $isValid(desc_i)$ has a range $[0, 1]$ and helps to check the constraints.

4 Forming a Context for Time Series Analysis and Forecasting

The context for analysis and forecasting of time series can be represented as an ontology. This ontology contains a description of domain objects, their properties, and restrictions on values of properties. The ontology can also contain logical rules and additional entities for prescriptive analytics. Also, the ontology contains a description of the methods of analysis and forecasting of time series. This description is used to select methods depending on properties: tendency, seasonality, frequency, length of the time series, etc.

The ontology for context formation for the analysis and forecasting of time series is formally presented as the following expression:

$$Domain = \langle O, M, A, R \rangle, \tag{9}$$

where O is a component for describing the domain specifications; M is a component for describing the characteristics of methods of analysis and forecasting of time series; A is a component for prescriptive analytics; R is a set of relations between ontology components.

The O and M components are described in this article. The A component is described in the work [9].

The $\mathcal{ALCHI}(\mathcal{D})$ extension of the descriptive logic [10] is used for the logical representation of the *Domain* ontology (Eq. 9):

$$Domain = \{TBox, ABox\},$$

where $TBox$ is set of terminological axioms; $ABox$ is a set of assertional axioms.

Let's consider the TBox of the *Domain* ontology in more detail.

$$M \sqsubseteq \top \quad O^m \sqsubseteq \top \quad O^{num} \sqsubseteq \top \quad TS \sqsubseteq \top \quad Def \sqsubseteq \top \quad Desc \sqsubseteq \top \quad Interval \sqsubseteq \top$$

$$Interval \equiv \top \sqcap \exists hasName.String \sqcap \forall hasName.String \sqcap \exists hasMinValue.Integer \sqcap$$
$$\sqcap \forall hasMinValue.Integer \sqcap \exists hasMaxValue.Integer \sqcap \forall hasMaxValue.Integer$$

$$M \equiv \top \sqcap \exists hasName.String \sqcap \forall hasName.String \sqcap \exists hasTendency.Boolean \sqcap$$
$$\sqcap \forall hasTendency.Boolean \sqcap \exists hasPeriod.Boolean \sqcap \forall hasPeriod.Boolean \sqcap$$
$$\sqcap \exists hasSeason.Boolean \sqcap \forall hasSeason.Boolean \sqcap \exists hasSmooth.Boolean \sqcap$$
$$\sqcap \forall hasSmooth.Boolean \sqcap \exists hasFuzzy.Boolean \sqcap \forall hasFuzzy.Boolean \sqcap$$
$$\sqcap \exists length.Interval \sqcap \forall length.Interval$$

$$O^m \equiv \top \sqcap \exists hasName.String \sqcap \forall hasName.String \sqcap \forall hasProperty.O^{num}$$

$$Def \equiv \top \sqcap \exists hasName.String \sqcap \forall hasName.String \sqcap \exists hasBase.Integer \sqcap$$
$$\sqcap \forall hasBase.Integer \sqcap \exists hasTendBase.Integer \sqcap \forall hasTendBase.Integer$$

$$O^{num} \equiv \top \sqcap \exists hasName.String \sqcap \forall hasName.String \sqcap \exists hasDef.Def \sqcap$$
$$\sqcap \forall hasDef.Def \sqcap \forall hasTS.TS$$
$$Desc \equiv \top \sqcap \exists hasName.String \sqcap \forall hasName.String \sqcap \exists hasBase.Integer \sqcap$$
$$\sqcap \forall hasBase.Integer \sqcap \exists hasTendBase.Integer \sqcap \forall hasTendBase.Integer \sqcap$$
$$\sqcap \exists hasTendDelta.Integer \sqcap \forall hasTendDelta.Integer \sqcap \exists hasBound.Interval \sqcap$$
$$\sqcap \forall hasBound.Interval \sqcap \exists hasAccept.Interval \sqcap \forall hasAccept.Interval$$
$$TS \equiv \top \sqcap \exists hasName.String \sqcap \forall hasName.String \sqcap \exists hasDef.Def \sqcap$$
$$\sqcap \forall hasDef.Def,$$

where $Interval$ is a concept representing an integer interval; $hasName$ is a functional role for "has a name" axiom; $hasMinValue$ and $hasMaxValue$ are functional roles for "has a minimal value" and "has a maximal value" axioms; $String$ is a string data type; $Integer$ is an integer data type; M is a concept representing some method for analyzing or forecasting a time series; $hasTendency$ is a functional role for "has the ability to work with tendencies" axiom; $hasPeriod$ is a functional role for "has the ability to work with periodicity" axiom; $hasSeason$ is a functional role for "has the ability to work with seasonality" axiom; $hasSmooth$ is a functional role for "has the ability to use smoothing" axiom; $hasFuzzy$ is a functional role for "has the ability to use fuzzy values" axiom; $length$ is a functional role for "has an acceptable interval of time series length" axiom; $Boolean$ is a boolean data type; O^m is a concept representing some control object; $hasDef$ is a functional role for "has a time series behavior" axiom; $hasProperty$ is a functional role for "has a property" axiom; Def is a concept representing a time series behavior; $hasBase$ is a functional role for "has a time series baseline" axiom; $hasTendBase$ is a functional role for "has a main tendency" axiom; O^{num} is a concept representing a control object property; $hasDesc$ is a functional role for "has a modeling result description" axiom; $hasTS$ is a functional role for "has a time series" axiom; $Desc$ is a concept representing a modeling result description; $hasTendDelta$ is a functional role for "has a rate of change of trends" axiom; $hasBound$ is a functional role for "has a range of acceptable values" axiom; TS is a concept representing a time series of values of control object property.

5 An Experiment on Contextual Time Series Forecasting

This experiment demonstrates the principle of the proposed methods. The problem area is the planning of the production capacity of an aircraft enterprise. The problems of extraction and forecasting of time series of production processes were described earlier in work [11]. The production process is represented by the dynamics of indicators of enterprise resources and production load. It is significant how much work a particular department can perform for the production planning process. This information is represented by the data about downtime of equipment stored in information systems of enterprise. For example, such a task can be determined as the need for additional capacity, considering the current production plan. We analyze in detail how the methods work according to the scheme shown in Fig. 1 and describe the performed actions and the results.

Let's build the description of the context in the form of the following fragment of the ontology: object *department*, object property *operationalTime*, the time series of property values *otTS*:

$$bound\colon Interval \quad accept\colon Interval \quad otDesc\colon Desc \quad otDef\colon Def$$
$$otTS\colon TS \quad operationalTime\colon O^{num} \quad department\colon O^{m}$$
$$(otTS, otDesc) : hasDesc \quad (operationalTime, otTS) : hasTS$$
$$(operationalTime, otDef) : hasDef \quad (department, operationalTime) : hasProperty$$

The enterprise equipment was connected to the monitoring system. The operational time of each machine can be observed. The experiment shows analyze the quantitative indicator represented by the time series of the useful operating time of the department equipment. The time series is formed as one of the output datasets of block 1.

Block 3.1 of the scheme requires the construction of time series models. A set of models to be able to choose the best one for a particular time series. The following methods used as an example for forecasting: Holt time series model, Holt-Winters time series model, F-Transform method with the "If-Then" rule base.

In the ontology, the Holt time series model are declared as follows:

$$tsLength\colon Interval \quad holt\colon M$$
$$(tsLength, 5) : hasMinValue \quad (tsLength, MAX) : hasMaxValue$$
$$(holt, true) : hasTendency \quad (holt, false) : hasPeriod \quad (holt, false) : hasSeason$$
$$(holt, true) : hasSmooth \quad (holt, false) : hasFuzzy \quad (holt, tsLength) : length$$

The forecast contains 3 points. The small length of the series requires a small forecast horizon. Forecast tested on known values of the series.

Figure 2 shows forecasts on test interval by described methods without using the context. Also, the figure shows a forecast estimate on the SMAPE criterion (the lower, the better) [12].

According to the expression (7) we can build the following forecasts (Fig. 3). Forecasts apply context information by block 4. Let's represent the components of the model (4) used for the forecast:

$$(bound, 0) : hasMinValue \quad (bound, 100) : hasMaxValue \quad (accept, 50) : hasMinValue$$
$$(accept, 60) : hasMaxValue \quad (otDesc, 50) : hasBase \quad (otDesc, 0) : hasTendBase$$
$$(otDesc, 5) : hasTendDelta \quad (otDesc, bound) : hasBound \quad (otDesc, accept) : hasAccept$$
$$(otDef, 50) : hasBase \quad (otDef, 0) : hasTendBase$$

The value of the model parameter w_{Def} is set to 0.3. Then the weight of the model forecast in contrast with the forecast by context is greater.

Based on the values of the given context model components, the following actions are performed:

- select the models by restrictions on seasonality, frequency. All proposed models satisfy this requirement.
- then select the models that satisfy the expression (8). The following models were chosen to build the forecast: Holt, Holt-Winters. The F-transform method is not suitable because it does not satisfy the requirement: the forecast values are within acceptable boundaries (50, 60) with the obtained value of the last point at 47.81.

The control object analytical description based on the modeling result (block 5): the operational time of the equipment in the department for the nearest predicted points will decrease, but it will retain acceptable values, as indicated in the context. This prediction will be true if the operational time of the equipment keeps the stability tendency.

All previously selected models that combined with information from the context (by the expression 7) are suitable for predicting future values. The leading model will be the Holt-Winters model. For example, if the context will change and production volumes increase or equipment is modified to eliminate emergency downtime, then the model components can take the following form:

$$(bound, 0) : hasMinValue \quad (bound, 100) : hasMaxValue \quad (accept, 50) : hasMinValue$$
$$(accept, 60) : hasMaxValue \quad (otDesc, 50) : hasBase \quad (otDesc, 1) : hasTendBase$$
$$(otDesc, 5) : hasTendDelta \quad (otDesc, bound) : hasBound \quad (otDesc, accept) : hasAccept$$
$$(otDef, 50) : hasBase \quad (otDef, 1) : hasTendBase$$

The forecast for these conditions show on the Fig. 4.

This forecast is given as an example to demonstrate the change in forecast results. The models must be re-selected due to the context change.

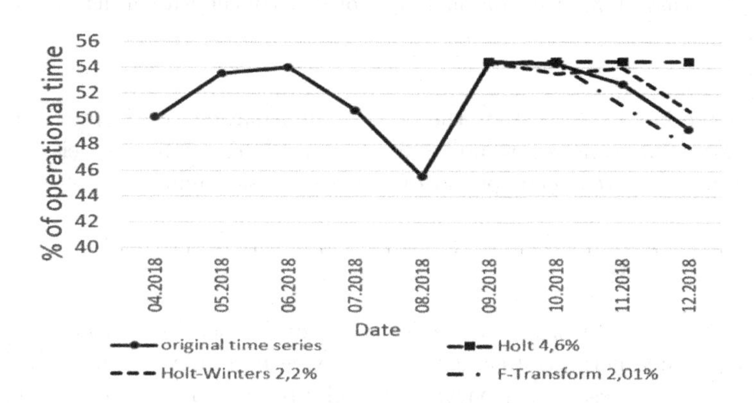

Fig. 2. Forecast of % of operational time of department

The Holt-Winters model does not pass the test there is a downward trend (Eq. 8) although it shows the best estimate for SMAPE. Therefore, it is excluded from the final forecasting in such a context.

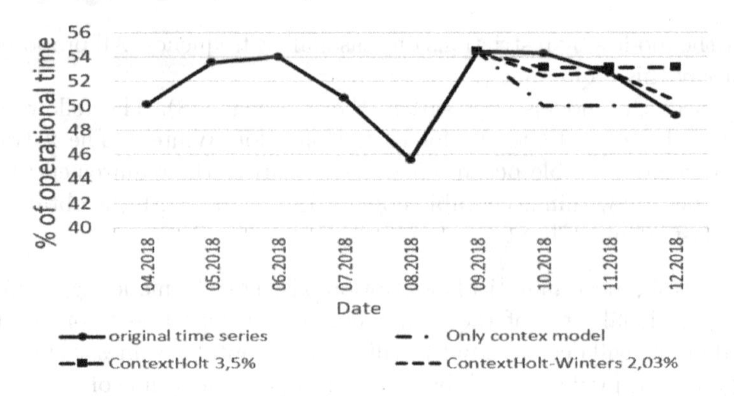

Fig. 3. Forecast of % of operational time of department with using of contextual information

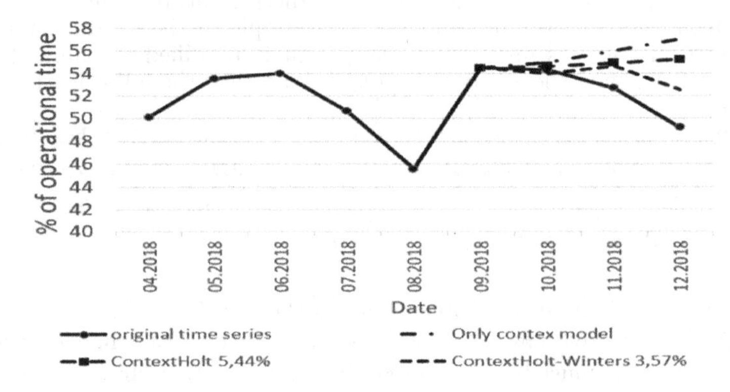

Fig. 4. Forecast of % of operational time of department with using of contextual information

The experiments show that the context information can improve the forecast quality while time series modeling. Also, allow excluding models with high quality but not match the expectations of the decision-maker.

6 Conclusion

The main aim of the proposed models is to exclusion the limitations of time series forecasting methods. In case when those methods analyzed only the history of changes in the researched parameters. The presented experiments show the possibility of context usage in the time series modeling processes. The advantage of the proposed approach is the ability to use existing methods for time series modeling. The set of model parameters extracted from context can be expanded to improve the forecasting accuracy for a more detailed selection of models, setting their limitations, and adjusting the forecast.

Acknowledgments. The authors acknowledge that the work was supported by the framework of the state task of the Ministry of Science and Higher Education of the Russian Federation No. 075-00233-20-05 "Research of intelligent predictive multimodal analysis of big data, and the extraction of knowledge from different sources". The reported study was funded by RFBR and the government of Ulyanovsk region according to the research projects: 18-47-732016, 18-47-730022, and 19-47-730005.

References

1. Li, Z., Xu, W., Zhang, L., Lau, R.Y.: An ontology-based Web mining method for unemployment rate prediction. Decis. Support Syst. **66**, 114–122 (2014)
2. Qu, H., Sardelich Nascimento, M., Qomariyah, N.N., Kazakov, D.L.: Integrating time series with social media data in an ontology for the modelling of extreme financial events. In: Proceedings of the European Language Resources Association (ELRA), LREC 2016, pp. 57–63 (2016)
3. Chen, X.C., Steinhaeuser, K., Boriah, S., Chatterjee, S., Kumar, V.: Contextual time series change detection. In: Proceedings of the 2013 SIAM International Conference on Data Mining, pp. 503–511. Society for Industrial and Applied Mathematics (2013)
4. Dunkl, R., Rinderle-Ma, S., Grossmann, W., Froschl, K.A.: A method for analyzing time series data in process mining: application and extension of decision point analysis. In: International Conference on Advanced Information Systems Engineering, pp. 68–84 (2014)
5. Dunkl, R., Rinderle-Ma, S., Grossmann, W., Froschl, K.A.: Decision point analysis of time series data in process-aware information systems. In: Pre-proceedings of CAISE 2014 Forum (2014). https://eprints.cs.univie.ac.at/4083/2/PaperVision05.pdf. Accessed 21 Nov 2020
6. Fildes, R., Hibon, M., Makridakis, S., Meade, N.: Generalising about univariate forecasting methods: further empirical evidence. Int. J. Forecast. **14**(3), 339–358 (1998)
7. Makridakis, S., Hibon, M.: The M3-competition: results, conclusions and implications. Int. J. Forecast. **16**(4), 451–476 (2000)
8. Small, G., Wong, R.: The validity of forecasting. A Paper for Presentation at the Pacific Rim Real Estate Society International Conferencem, Christchurch, New Zealand, pp. 1–14 (2002)
9. Romanov, A., Yarushkina, N., Filippov, A.: Application of time series analysis and forecasting methods for enterprise decision-management. In: Rutkowski, L., Scherer, R., Korytkowski, M., Pedrycz, W., Tadeusiewicz, R., Zurada, J.M. (eds.) ICAISC 2020. LNCS (LNAI), vol. 12415, pp. 326–337. Springer, Cham (2020). https://doi.org/10.1007/978-3-030-61401-0_31
10. Baader, F., Calvanese, D., McGuinness, D., Nardi, D., Patel-Schneider, P.F.: The Description Logic Handbook: Theory, Implementation, and Applications. Cambridge University Press (2003)
11. Romanov, A., Filippov, A., Yarushkina, N.: Extraction and forecasting time series of production processes. In: Dolinina, O., Brovko, A., Pechenkin, V., Lvov, A., Zhmud, V., Kreinovich, V. (eds.) ICIT 2019. SSDC, vol. 199, pp. 173–184. Springer, Cham (2019). https://doi.org/10.1007/978-3-030-12072-6_16
12. SMAPE criterion by Computational Intelligence in Forecasting (CIF). http://irafm.osu.cz/cif/main.php. Accessed 21 Nov 2020

Simulation Analysis of Tunnel Vision Effect in Crowd Evacuation

Akira Tsurushima(✉)

Intelligent Systems Laboratory, SECOM CO., LTD., Mitaka, Tokyo, Japan
a-tsurushima@secom.co.jp

Abstract. Excessive cognitive demands, fear, or stress narrow evacuees' functional fields of view (FFV) in disaster evacuation situations. This tunnel vision hypothesis leads to a new model of evacuee behavior deviating significantly from the previously accepted understanding, and possibly altering conventional evacuation protocol designs. In this study, we analyze the impacts of narrowed vision of evacuees on crowd evacuation efficiency through simulated evacuations. The simulated room to be evacuated included multiple exits, of which only one was correct, as well as a single visual sign designating the correct exit, and an agent found the correct exit via this sign if it was within their FFV. We designed an evacuation decision model for the simulated agents based on herd behavior, including cognitive biases frequently observed during evacuations, to which evacuees were assumed to be subject.

Keywords: Tunnel vision · Herd behavior · Evacuation decision model

1 Introduction

Mackworth (1965) introduced the concept of tunnel vision, a narrowing of the human FFV owing to excessive cognitive demand, stress, or fear [7]; numerous studies on the subject have since been conducted [1–3,6]. In tunnel vision, human cognitive resources focus on the center of the visual field, resulting in a loss object perception outside of this focus, especially in the peripheral vision. Furthermore, tunnel vision has been shown to occur during violent crimes and emergency situations, and to affect operators of complex systems and vehicle drivers; its effects are known to lead to loss of life or damage to property in some cases.

Although several conditions common in disaster evacuation situations are understood to cause tunnel vision, prior research on crowd evacuation has rarely focused on the human FFV. Most crowd evacuation studies using experiments or simulations have assumed the FFV of an agent arbitrary or implicitly, or simply considered the range at which an agent could perceive objects as their FFV. As yet, few studies have explored the human FFV based on physiological or psychological factors.

The FFV of an evacuee is crucial in crowd evacuations for two reasons. (1) The FFV of any agent bounds other agents considered in models of herd behavior among evacuees, affecting the behavior of the crowd as a whole. (2) Visual

L. Rutkowski et al. (Eds.): ICAISC 2021, LNAI 12854, pp. 506–518, 2021.
https://doi.org/10.1007/978-3-030-87986-0_45

information such as exit signs or evacuation route signs are commonly used to efficiently guide crowd evacuations, and agents are able to recognize such signs only within their FFV.

Presenting a simulation analysis of evacuation behaviors among people appearing in a video clip captured during the Great East Japan Earthquake, Tsurushima (2020) introduced the *tunnel vision hypothesis* that the human FFV during emergency evacuation situations narrows to an angle of 20° toward the heading direction, with a relatively long range [14]. While numerous studies on crowd evacuation have adopted a wide visual field, such as 120° or 360°, the tunnel vision hypothesis introduced a much narrower visual field, restricted to 20°. The significant difference in visual range and scope between the predictions of the tunnel vision hypothesis and the conventional assumptions may alter previously accepted research results on crowd evacuation. The impact of the tunnel vision hypothesis may be crucial, especially for research on emergency evacuation protocol design using visual communication for critical information.

This study analyzes the impact of tunnel vision effects on crowd evacuation using two simple simulation problems. Both problem scenarios concern an environment with two exits, one being designated by a visual sign. The objective of both problems was to maximize the number of agents reaching a correct exit designated by a visual sign, by varying the position of the sign within the simulated environment. Exit choice decisions in crowd evacuations are highly complex; aside from visual signs, herd behaviors among evacuees also affect individuals' decisions by propagating both correct and incorrect information within a crowd. Thus, directing a crowd to a correct exit can prove challenging. We investigated the impact of tunnel vision on crowd exit choice decisions by comparing simulated agents with tunnel vision and with normal (wider) vision.

2 Related Works

Several studies of psychological and cognitive factors on evacuation behavior have been conducted using virtual reality (VR) devices. Tucker et al. (2018) investigated the impact of hazard levels on evacuees' anxiety and exit route decisions using VR, and showed that evacuees tended significantly to select a major exit rather than peripheral exits under highly hazardous conditions [15]. Meng and Zhang (2014) conducted evacuation experiments simulating a hotel fire incident using VR, analyzing simulations with and without conditions of virtual fire, and found that evacuees required more time to find evacuation signs and exits in simulated fire emergency conditions [9]. New approaches using shared virtual spaces with multiple subjects have recently emerged. Mousaid et al. (2018) conducted VR evacuation experiments in settings with four exits, of which one was randomly selected as correct, in which a subset of the subjects knew the correct exit in advance. They analyzed evacuation behaviors under stressful emergency conditions and under non-stressful conditions, and showed that a higher number of collisions occurred under high-stress conditions, and a majority of participants moved as a herd in the same direction [10]. VR evacuation experiments

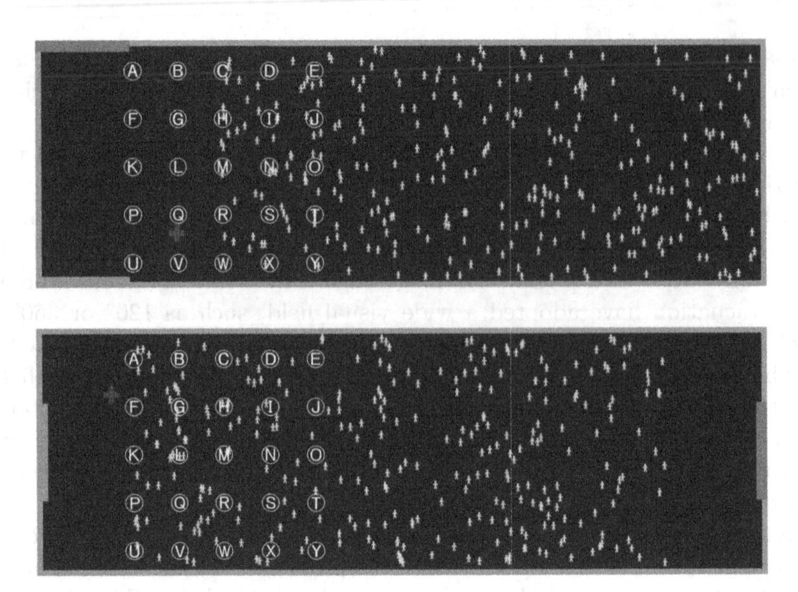

Fig. 1. Room1 (top) and Room 2 (bottom) (Color figure online)

have shown promise in investigations of cognitive factors in evacuations; however, to the best of our knowledge, as yet no prior studies have been conducted using VR concerning the visual fields of evacuees.

Li et al. (2019) introduced a visibility function and investigated the effects of several factors on evacuation efficiency, such as psychological tension, vision radius, and pedestrian density. They found that vision radius and initial density affected evacuation time, and that vision radius decreased with increasing evacuation time [5]. The authors did not restrict the angular width of the evacuees' visual fields, assuming instead that they could acquire information within 360°. Similar assumptions have been adopted in numerous studies [16–18]. Some studies have been conducted employing visual field widths other than 360°, such as 60° [8] and 90° [11]; however, their reasons for choosing these assumptions were not clearly stated.

3 Problem

Three hundred simulated agents ($a_i \in A$) were randomly distributed in a square room with XY coordinates $x \in [-65, 65]$ and $y \in [-20, 20]$, and agents were required to evacuate through either of two exits $B = \{b^+, b^-\}$. The coordinates of an exit b are denoted by (b_x, b_y). We investigated two layouts according to the location of the exits. Room 1 had two exits on the north (blue) and south (green) sides of the western edge of the room (Fig. 1 – top). Agents were initially allocated to $x_i \in [-32, 65]$ and faced west. Room 2 had two exits on the east (blue) and west (green) edges of the room (Fig. 1 – bottom). Agents were initially

Algorithm 1. Agent i's action ($X_i = 0$)

1: $M \leftarrow \{a_j \in V_i | \eta_j = moving\}$; $N \leftarrow \{a_j \in V_i | \eta_j = not\_moving\}$
2: **if** $|M| > |N|$ **then**
3:     **if** $\exists m \forall n |\{a_j \in V_i | \pi_j = \beta_m\}| \geq |\{a_k \in V_i | \pi_k = \beta_n\}|$ **then** $\pi_i \leftarrow \beta_m$ **end if**
4:     **if** $\pi_i = undecided$ **then**
5:         $\Delta x(t) \leftarrow \Delta x(t-1)$; $\Delta y(t) \leftarrow \Delta y(t-1)$
6:     **else**
7:         $b \leftarrow \pi_i$; $G_x \leftarrow b_x$; $G_y \leftarrow b_y$
8:         Solve $\mathcal{P}$ with respect to $\Delta x(t)$ and $\Delta y(t)$ for given G_x and G_y
9:     **end if**
10:     $\eta_i \leftarrow moving$; Give $(\Delta x(t), \Delta y(t))$ to the SFM to obtain the new position
11: **else**
12:     $\eta_i \leftarrow not\_moving$
13: **end if**

allocated to $x_i \in [-48, 48]$ and randomly faced either east or west. Symbols Ⓐ to Ⓨ in Fig. 1 indicate candidate positions for a visual sign employed in visual sign simulations (VSS), as discussed in Sect. 5.2.

An agent was required to choose one of two exits to evacuate the room. One was the correct exit (b^+), leading to safe evacuation, and the other (b^-) lead to an improper route, considered an evacuation failure. Thus, the agents had to select the correct exit to evacuate successfully. In our examples, the correct exits were the north and east exit for Rooms 1 and 2, respectively (both exits are indicated in blue in Fig. 1).

A visual sign κ was present (a small blue cross in Fig. 1), designating the correct exit from the room. An agent could identify the correct exit via this visual sign if it was within their visual field ($V_i \subset \{A \cup \{\kappa\}\}$). However, it was uncertain whether an agent with the correct information could always choose the correct exit, because they were also subject to herd behavior biases that could lead them to the incorrect exit. Two additional variables specified the current statuses of each agent, including a movement status $\eta_i(t) \in \{moving, not\_moving\}$, and a current decision $\pi_i(t) \in \{B \cup \{undecided\}\}$.

Using these problems, we investigated two crucial factors affecting the efficiency of crowd evacuations, including varying positions of a visual sign and varying widths of the FFV of agents. The impacts of these factors were evaluated in terms of the number of agents that chose the correct exit.

4 Agent Model

The agents in our simulations incorporated an evacuation decision model (EDM) [12,13] and a social force model (SFM) [4]. The EDM represented the herd behavior of agents during evacuations, and SFM represented physical factors affecting evacuations.

In the EDM, an agent a_i has a mental state $X_i \in \{0, 1\}$, making decisions intentionally when $X_i = 1$ and unintentionally by following the behaviors of

Algorithm 2. Agent i's action ($X_i = 1$)

1: **if** $\kappa \in V_i$ **then**
2: $\quad \pi_i \leftarrow b^+$
3: **else if** $\pi_i = undecided$ **then**
4: $\quad$ Randomly select $b \in B$ and $\pi_i \leftarrow b$
5: **end if**
6: **if** $\pi_i = undecided$ **then**
7: $\quad \Delta x(t) \leftarrow \Delta x(t-1)$; $\Delta y(t) \leftarrow \Delta y(t-1)$
8: **else**
9: $\quad b \leftarrow \pi_i$; $G_x \leftarrow b_x$; $G_y \leftarrow b_y$
10: $\quad$ Solve $\mathcal{P}$ with respect to $\Delta x(t)$ and $\Delta y(t)$ for given G_x and G_y
11: **end if**
12: $\eta_i \leftarrow moving$; Give $(\Delta x(t), \Delta y(t))$ to the SFM to obtain the new position

others in the agent's vicinity when $X_i = 0$. The transition probabilities between $X_i = 0$ and $X_i = 1$ are $P(X_i = 0 \rightarrow X_i = 1) = s_i^2(s_i^2 + \theta_i^2)^{-1}$ and $P(X_i = 1 \rightarrow X_i = 0) = \epsilon$, where s_i denotes a local estimation of the stimulus in the environment associated with a_i, θ_i denotes a response threshold, and ϵ denotes a constant probability common to all agents. The local estimation of stimulus in the environment is $s_i(t+1) = max\{s_i(t) + \delta - \alpha(1-R)F,\ 0\}$, where δ denotes an increase in the stimulus per unit time and α a scale factor of the stimulus. The variable R denotes the risk-perception function $R(r)$ of the objective risk r in the environment. We let $R(r) = (1 + e^{-g(r-\mu_i)})^{-1}$, where g denotes the activation gain affecting the shape of a sigmoid function, and μ_i denotes agent i's risk perception, representing individuals' varying sensitivities to risk. The term F denotes an evacuation progress function indicating a local observation of the total evacuation progress from the viewpoint of agent i. We let $F(n) = 1 - n/N_{max}$ (if $n < N_{max}$) or 0 (otherwise), where n denotes the number of agents in a vicinity V_i and N_{max} the maximum possible number of agents in a given vicinity. The agents estimate the total progress of the evacuation using n/N_{max}, which is the population density of their vicinities.

In this simulation, we assumed the FFV of an agent as their vicinity. Thus, the range associated with the FFV of a_i (V_i) could affect (1) the estimation of the evacuation progress F, (2) agents moving in herd behavior patterns, and (3) the visual field within which an agent can recognize the visual sign κ.

Agent i executes Algorithm 1 if their mental state is $X_i = 0$ and Algorithm 2 if $X_i = 1$. Algorithm 1 represents herd behavior such that an agent selects the exit chosen by the greatest number of agents in their vicinity. Algorithm 2 represents intentional behavior pattern in which an agent chooses the designated exit if the sign is within their visual field or randomly selects an exit otherwise. A vector $(\Delta x(t), \Delta y(t))$, representing the difference between the current location and the location of an agent's next step is calculated by solving Problem $\mathcal{P}$: $min\ (x(t) + \Delta x(t) - G_x)^2 + (y(t) + \Delta y(t) - G_y)^2$, subject to $\Delta x(t)^2 + \Delta y(t)^2 = 1.0$.

To calculate the new coordinates of an agent, we considered physical factors in the environment. Based on the desired vector $(\Delta x(t), \Delta y(t))$ (line 10 in

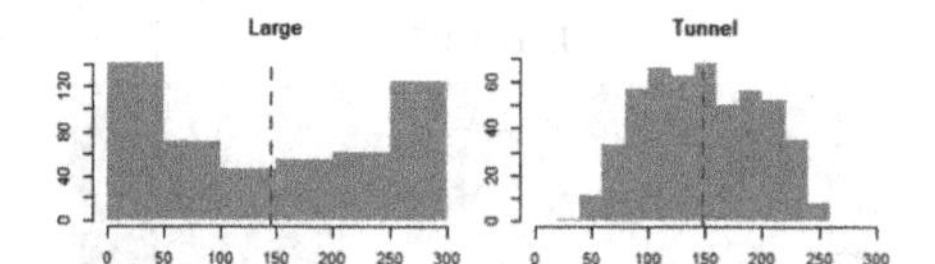

Fig. 2. Baseline histogram for Room1 (Color figure online)

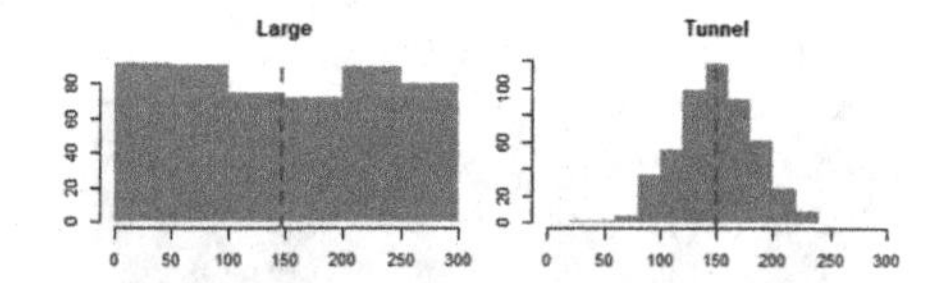

Fig. 3. Baseline histogram for Room2 (Color figure online)

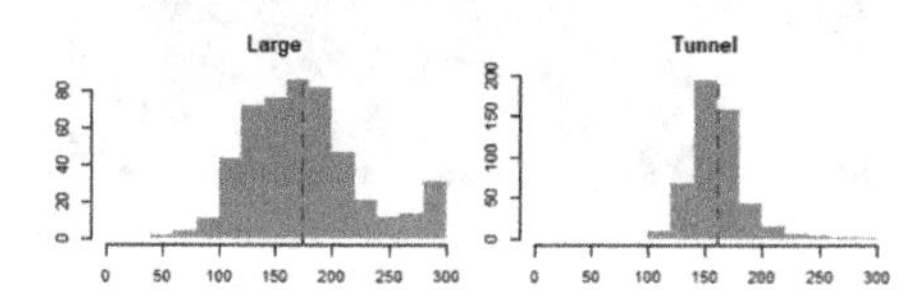

Fig. 4. VSS histogram for Room1 (Color figure online)

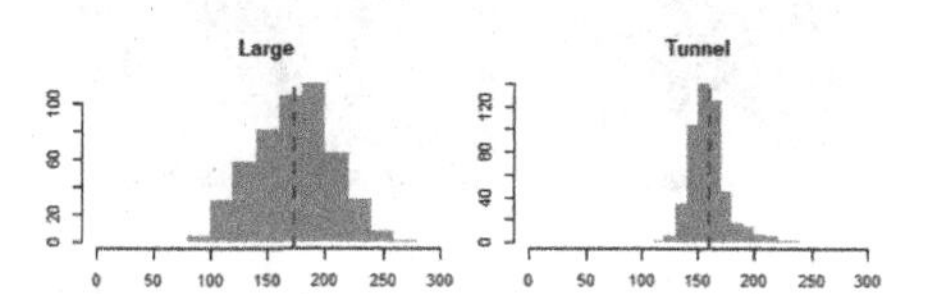

Fig. 5. VSS histogram for Room2 (Color figure online)

Algorithm 1, and Line 12 in Algorithm 2), the SFM [4] was used to process the calculation. The parameter values used in the experiments were $\epsilon = 0.5$, $\delta = 1.0$, $\alpha = 0.5$, $N_{max} = 10$, $g = 1.0$, $d = 5$, $\Delta r = 0.5$, and $\theta_i, \mu_i \sim U(0, 100)$.

5 Experiment

We conducted experiments to explore efficient evacuation protocols through the simulation problem presented in Sect. 3. The problem aimed to maximize the number of agents selecting the correct exit (O) by varying the position of the visual sign (κ). The mean of the objective values $\bar{O}$ was adopted as the evaluation value for each sample position.

The FFV (V_i) of a_i was assumed to be a fan shape with a radius of d and an angle of ω toward each agent's heading direction. In this experiment, we investigated two types of V_i, including Large (V^L) and Tunnel (V^T). The values of d and ω for V^L and V^T were 10 120° and 10 20°, respectively.

5.1 Baseline

Prior to the experiment, we conducted baseline simulations (BS) without using a visual sign. Figure 2 depicts the histograms of 500 BSs with V^L and V^T for Room1 and Fig. 3 for Room2. The red dashed lines show the mean number of agents selecting the north exit ($\bar{O}$). Agents simply selected one of the two exits randomly in the BS; thus, the theoretical value of $\bar{O}$ was 150. Samples were distributed around $\bar{O}$; however, variances differed in each histogram. Samples far from $\bar{O}$ indicate an occurrence of exit choice pattern symmetry breaking caused by herd behaviors.

In comparison with Fig. 2 and 3, we may observe from the data shown that Room1 had larger variances than Room2. At the beginning of the simulations, all the agents flowed together in the same direction in Room1, and the flow

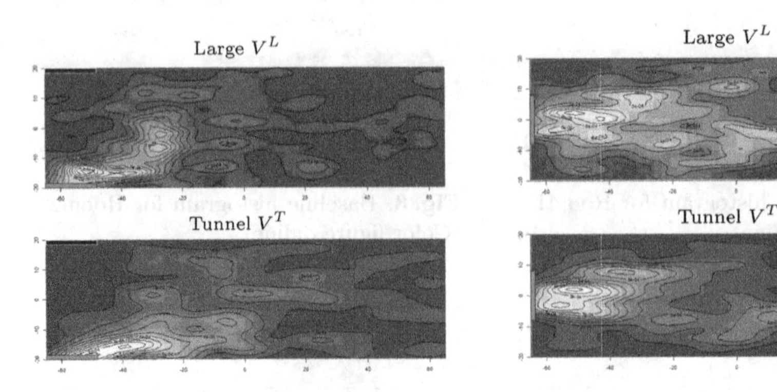

Fig. 6. Kernel density of Room1 for V^L (top) and V^T (bottom). (Color figure online)

Fig. 7. Kernel density of Room2 for V^L (top) and V^T (bottom). (Color figure online)

subsequently split into two directions, whereas, in Room2, agents formed two flows in opposite directions. The results of BS are summarized in the row labeled *Base* in Table 2 in the appendix.

We also investigated the impact of evacuation information by providing information about the correct exit to a portion of the agents. In this setting, informed agents always selected the correct exit whenever $X = 1$. The ratio of informed agents and the resulting $\bar{O}$ values are summarized in Table 1 and presented in Fig. 12 and 13 for Room1 and Room2.

Table 1. Ratio of informed agents and $\bar{O}$

Ratio (%)		10	20	30	40	50	60	70	80	90
Room1	V^L	197.86	239.12	264.56	282.12	288.66	294.23	296.40	298.26	299.24
	V^T	199.56	230.16	250.91	265.69	274.70	282.75	288.27	292.56	296.35
Room2	V^L	201.32	243.30	268.41	282.33	287.20	291.58	294.38	297.04	298.62
	V^T	190.72	217.92	240.91	259.28	270.18	279.57	286.15	291.40	295.84

5.2 Visual Sign Simulation

We conducted VSS using one visual sign κ for Room1 and Room2. Five hundred positions of κ were randomly generated over $x \in [-65, 65]$ and $y \in [-25, 25]$. To estimate the value of $\bar{O}$, 10 simulations for each position were conducted.

Figure 4 shows histograms of VSS results for Room1 with V^L and V^T; the red dashed lines show the values of $\bar{O}$. Figure 5 shows the same information for Room2. Comparing the histograms from VSS and BS, we note that the presence of a single visual sign reduced variances of the results significantly. In contrast, as the mean increased in VSS, the differences were negligible, implying that the

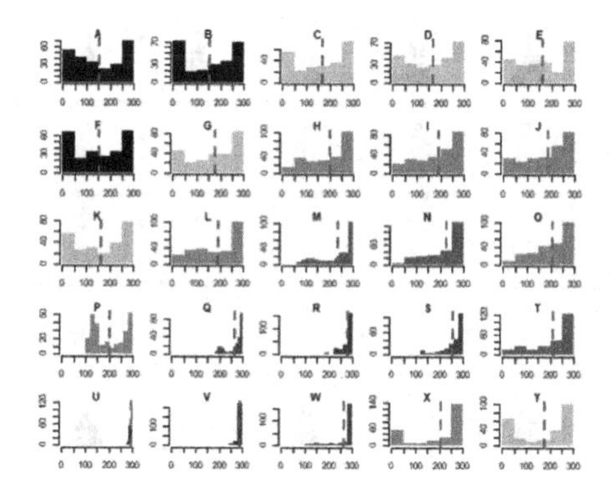

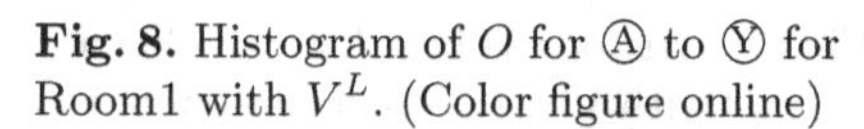

Fig. 8. Histogram of O for Ⓐ to Ⓨ for Room1 with V^L. (Color figure online)

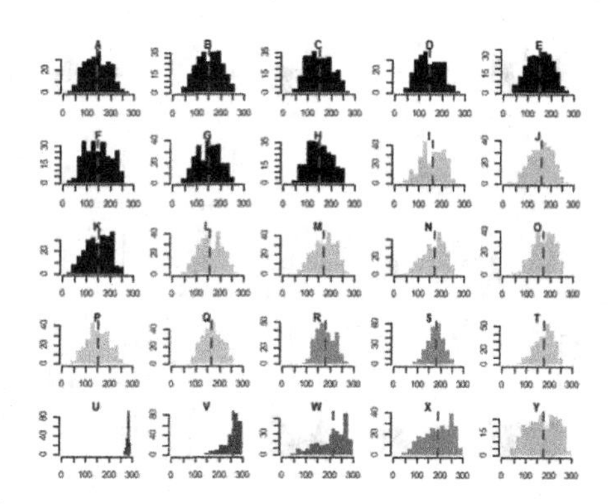

Fig. 9. Histogram of O for Ⓐ to Ⓨ for Room1 with V^T. (Color figure online)

effects were sensitive to the position of κ. The results of these simulations are summarized in the row labeled *Rand* in Table 2 in the appendix.

Figure 6 depicts heatmaps of the Kernel density of O in Room1 for V^L and V^T, from top to bottom, respectively. The Kernel density was calculated as $\hat{O} = max\{0,\ \bar{O} - 150\}$. The north exit, denoted by κ, is presented in blue, and the south exit in green. Light-colored regions represent positions of κ with high $\bar{O}$ values, indicating efficient evacuation. The same applies to Fig. 7, which belongs to Room2.

Heatmaps in Fig. 6 and 7 illustrate that the value of O changed significantly depending on location; the results were sensitive to changes in the positions of κ. If the position of κ was carefully adjusted, κ was able to guide numerous evacuees to the correct exit. However, the effect of κ could be negligible if it was improperly positioned in ineffective locations. In all the cases, the most effective position of κ was in front of the exit not designated by κ (green exits in Fig. 6 and 7). The efficient range of κ positions was not broad, especially in V^T.

We also conducted VSS with κ positions at Ⓐ to Ⓨ, as shown in in Fig. 1. Simulations were conducted 250 times for each position with V^L and V^T. Histograms of the results are presented in Fig. 8 and 9 for Room1 and in Fig. 10 and 11 for Room2; the means and standard deviations of the results are summarized in the columns labeled μ and δ in Table 2 in the Appendix. We analyzed significant differences in $\bar{O}$ between the results of VSS and BS using Welch's T-test. The P values of the t-test are also summarized in the columns labeled p in Table 2. P values with significant differences ($P < 0.05$) are indicated in italic letters in Table 2, and in blue color in Fig. 8, 9, 10 and 11. The darker the blue, the larger the mean value $\bar{O}$. Note that histograms in black show insignificant results in Fig. 8, 9, 10 and 11.

We also evaluated the results of VSS in comparison with simulations with informed agents (Table 1). Results are presented in Fig. 12 and 13 for Room1 and Room2, respectively. The x-axis shows the ratio (%) of informed agents, and the y-axis shows $\bar{O}$ values. The results with V^L are shown in blue, and

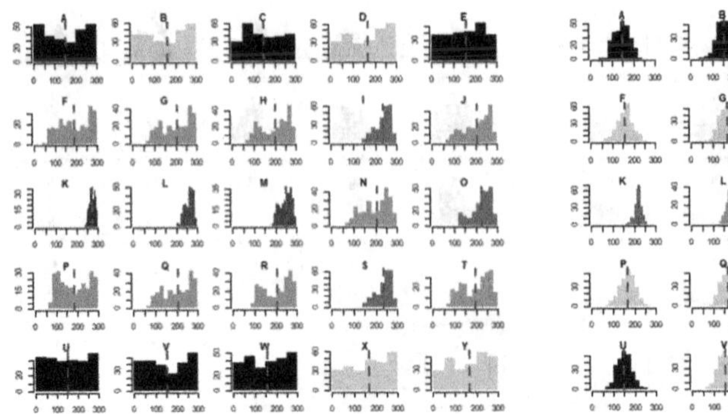

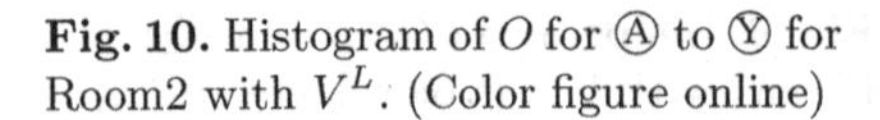

Fig. 10. Histogram of O for Ⓐ to Ⓨ for Room2 with V^L. (Color figure online)

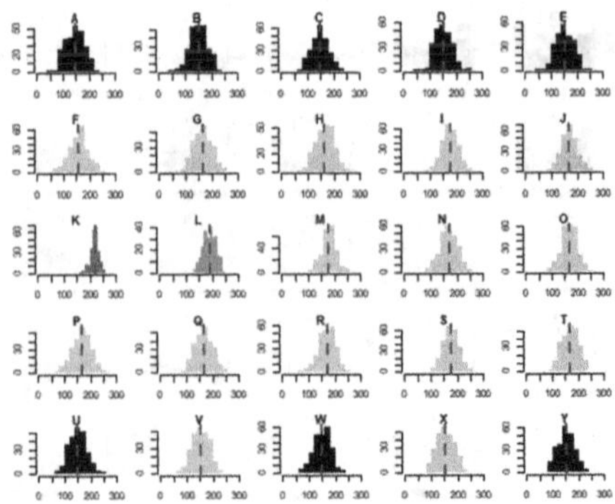

Fig. 11. Histogram of O for Ⓐ to Ⓨ for Room2 with V^T. (Color figure online)

V^T in red. Positions with $\bar{O} > 200$ in VSS are indicated on the left (V^T) and right (V^L) by corresponding letters. For example, the maximum value of $\bar{O}$ with V^L in Room1 was 293.6 (at Ⓤ), which was estimated to be equivalent to the result of conditions where 58.9 % of agents were informed (the column labeled % in Table 2). The position Ⓤ also reached a maximum $\bar{O}$ value (283.8) in the simulation with V^T, equivalent to the results with 61.8 % of agents informed. This value is almost equal to that with V^L (58.9 %). However, except for Ⓤ , these values in V^T (the column labeled % in Table 2) are significantly smaller than those in V^L. This implies that the impacts of κ were almost equivalent in V^L and V^T if the position was carefully selected. Otherwise, the effect of κ significantly dropped in V^T. The same applies to Room2; however, the values in the column labeled % were smaller than those in Room1.

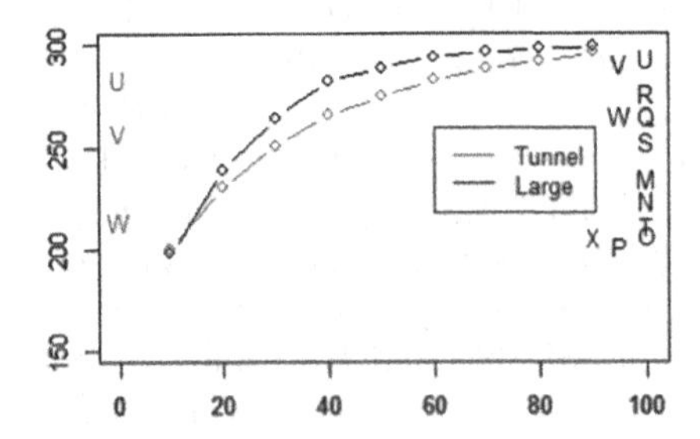

Fig. 12. $\bar{O}$ values for ratio of informed agents and κ positions for Room1. (Color figure online)

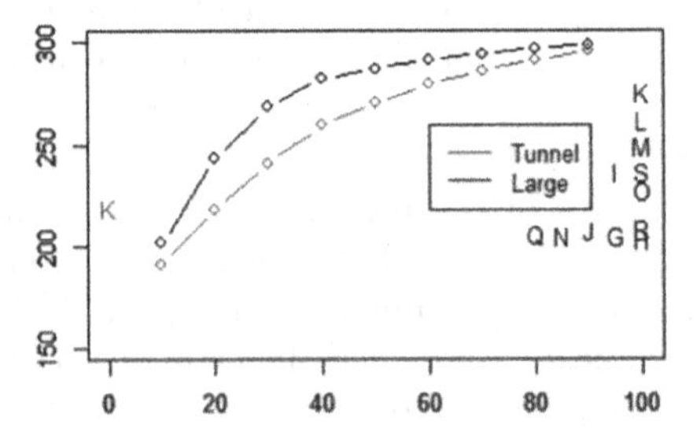

Fig. 13. $\bar{O}$ values for ratio of informed agents and κ positions for Room2. (Color figure online)

6 Discussion and Conclusion

In this study, we have assumed that excessive mental stresses or cognitive demands in evacuation situations limit human cognitive resources, leading to a narrowed visual field, which can be considered a reasonable assumption regarding tunnel vision under evacuation situations [6,7]. V^L had an area six times larger than V^T, implying significant impacts of both herd behaviors and visual sign recognition. While the broad area of V^L was efficient for visual sign recognition, it also strengthened the effects of herd behavior, decreasing the entropy in exit choice decisions (large standard deviations of V^L in Table 2), leading to an increase in incorrect exit selections. Compared with V^L, tunnel vision V^T decreased the mean number of evacuees who chose the correct exit in most positions from Ⓐ to Ⓨ ; the standard deviations also decreased in most positions. There were some cases where all evacuees chose the incorrect exit in V^L; these were rare in V^T. Because a large variance renders control difficult in general, this variance reduction in tunnel vision may easily control the evacuee behavior distribution. For example, in exploring an optimal position of visual signs through black-box optimization, large variances in V^L make estimations of mean values in each candidate point difficult, requiring many attempts in a trial-and-error method. Reduction in variances in V^T may lead to a decrease in the number of iterations required for optimization. However, this does not immediately mean that efficient positions for visual evacuation signs may be easily obtained.

VSS experiments in Room1 and Room2 showed that the results of visual sign positionings depended on both the FFV of evacuees and the exit layout in the room to be evacuated. Our experience with VSS implies that controlling the flow of evacuees is a key factor, but we expect it to be challenging in general. Assigning a visual sign at Ⓤ in Room1 produced surprisingly good results, such as 293.6 and 283.8 for V^L and V^T, respectively. This is a result of using just one visual sign; increasing these values by introducing more signs may be difficult. One factor accounting for this good result is that one of the evacuees' movement directions within the crowd flow tend to connect on this point, which could affect many other evacuees. However, this value dropped significantly with slight changes in the position of the sign, especially in the case of V^T. Moreover, some positions had insignificant effects ($p \geq 0.05$), meaning they showed no difference to the baseline. Our experiments in this study reveal the extremely high sensitivity of the tunnel vision effect to visual sign location optimization. This finding implies that tunnel vision effects may be expected to present severe difficulties and significant design challenges in terms of optimizing visual evacuation sign positions within buildings.

In this study, we concentrated on maximizing the number of agents choosing the correct exit, ignoring another crucial objective, that of minimizing evacuation time. The evacuation time in VSS, baseline, and simulations with informed agents is summarized in Table 3 The optimum positions in VSS (Ⓤ and ⓚ for Room1 and Room2) resulted in longer evacuation times because agents moved to the incorrect exit first and changed their headings to the correct exit later. This pattern of movement may be considered unnecessary and unpreferable in evacuations; however, these issues are left for future work.

Appendix

Table 2. Columns labeled μ and δ denote mean and standard deviations of $\bar{O}$. Columns labeled p denote p values of T test with BS. Columns labeled % denote the ratio of informed agents with equivalent impact to the visual sign positioning at the position corresponding to the row name.

	Room1								Room2							
	$Large(V^L)$				$Tunnel(V^T)$				$Large(V^L)$				$Tunnel(V^T)$			
	μ	δ	p	%	μ	δ	p	%	μ	δ	p	%	μ	δ	p	%
Base	148.0	50.4	–	–	145.3	103.4	–	–	149.6	33.5	–	–	146.5	86.2	–	–
Rand	161.1	24.1	–	–	175.1	50.2	–	–	160.0	16.8	–	–	173.0	33.6	–	–
A	152.7	102.1	0.45	0.6	143.9	50.1	0.30	0.0	152.5	91.6	0.52	0.5	146.3	35.1	0.36	0.0
B	154.8	106.7	0.31	1.0	152.5	49.7	0.18	0.5	161.0	90.0	*0.04*	2.2	153.1	32.4	0.07	0.8
C	166.3	99.7	*0.00*	3.4	151.3	51.3	0.33	0.3	149.8	84.0	0.83	0.0	149.2	33.4	0.86	0.0
D	167.6	98.6	*0.00*	3.7	146.9	49.5	0.84	0.0	168.1	87.3	*0.00*	3.5	152.1	32.5	0.16	0.5
E	164.7	98.3	*0.01*	3.1	154.0	51.2	0.08	0.8	156.3	81.9	0.15	1.2	153.4	33.8	0.06	0.8
F	154.3	102.3	0.32	0.9	149.2	53.7	0.69	0.0	192.8	71.9	*0.00*	8.3	160.5	36.0	*0.00*	2.6
G	177.8	95.7	*0.00*	5.8	151.9	48.2	0.23	0.4	205.1	66.6	*0.00*	10.9	167.1	34.1	*0.00*	4.2
H	199.4	85.6	*0.00*	10.4	154.5	49.2	0.05	0.9	202.7	65.6	*0.00*	10.3	165.2	34.7	*0.00*	3.7
I	194.0	84.4	*0.00*	9.2	160.9	48.4	*0.00*	2.2	236.6	35.5	*0.00*	18.4	176.8	29.6	*0.00*	6.6
J	187.8	88.8	*0.00*	7.9	160.6	47.1	*0.00*	2.2	207.0	60.4	*0.00*	11.4	169.9	29.6	*0.00*	4.9
K	164.5	100.6	*0.01*	3.0	151.8	50.5	0.25	0.4	275.7	13.6	*0.00*	35.3	219.4	16.9	*0.00*	20.6
L	193.0	91.7	*0.00*	9.0	160.6	48.4	*0.00*	2.1	262.0	21.3	*0.00*	27.5	191.6	24.0	*0.00*	10.3
M	234.6	70.2	*0.00*	18.9	174.3	46.3	*0.00*	4.9	248.8	28.6	*0.00*	22.2	179.6	27.4	*0.00*	7.3
N	223.5	72.7	*0.00*	16.2	170.5	46.5	*0.00*	4.1	204.3	60.1	*0.00*	10.7	170.5	32.4	*0.00*	5.1
O	206.3	75.2	*0.00*	12.1	172.1	45.8	*0.00*	4.5	205.1	43.6	*0.00*	16.2	167.9	30.4	*0.00*	4.4
P	201.2	68.1	*0.00*	10.8	154.6	47.1	*0.04*	0.9	186.1	70.6	*0.00*	7.0	167.3	35.7	*0.00*	4.3
Q	265.0	37.8	*0.00*	30.2	170.2	41.3	*0.00*	4.1	205.1	66.6	*0.00*	10.9	167.1	34.1	*0.00*	4.2
R	276.5	29.2	*0.00*	36.8	182.4	39.9	*0.00*	6.5	208.1	62.7	*0.00*	11.6	171.0	33.8	*0.00*	5.2
S	252.4	52.4	*0.00*	25.2	180.0	34.6	*0.00*	6.1	236.6	35.5	*0.00*	18.4	176.8	29.6	*0.00*	6.6
T	211.4	84.9	*0.00*	13.3	178.8	40.7	*0.00*	5.8	198.6	63.7	*0.00*	9.5	169.3	32.6	*0.00*	4.7
U	293.6	5.2	*0.00*	58.9	283.8	6.1	*0.00*	61.8	151.4	88.3	0.64	0.3	150.7	33.7	0.43	0.2
V	290.6	17.3	*0.00*	53.5	257.2	32.5	*0.00*	34.2	154.8	89.7	0.29	0.9	153.8	33.7	0.05	0.9
W	265.7	55.1	*0.00*	30.6	214.2	61.2	*0.00*	14.8	156.9	87.5	0.15	1.3	150.8	33.0	0.39	0.2
X	205.6	106.5	*0.00*	11.9	192.4	60.3	*0.00*	8.6	166.5	86.0	*0.00*	3.2	154.5	33.1	*0.02*	1.1
Y	176.0	112.9	*0.00*	5.4	175.5	63.2	*0.00*	5.1	166.6	83.2	*0.00*	3.2	152.1	32.5	0.17	0.5

Table 3. Evacuation time of VSS (Ⓤ and Ⓚ for Room1 and Room2), baseline, and 100% informed agent simulation. Means and standard deviations are provided in the columns labeled μ and δ.

	(V^T)						(V^L)					
	VSS		Baseline		100%		VSS		Baseline		100%	
	μ	δ	μ	δ	μ	δ	μ	δ	μ	δ	μ	δ
Room1	193.14	10.48	168.93	9.88	170.56	10.20	148.73	8.68	143.27	4.88	146.13	3.29
Room2	232.80	19.95	169.23	12.49	176.10	8.21	196.94	32.91	147.14	19.04	167.91	9.80

References

1. Burke, A., Heuer, F., Reisberg, D.: Remembering emotional events. Memory Cogn. **20**, 277–290 (1992)
2. Christianson, S.A.: Emotional stress and eyewitness memory: a critical review. Psyhol. Bull. **112**(2), 284–309 (1992)
3. Christianson, S.A., Loftus, E.F.: Memory for traumatic events. Appl. Cogn. Psychol. **1**(4), 225–239 (1987)
4. Helbing, D., Farkas, I., Vicsek, T.: Simulating dynamical features of escape panic. Nature **407**(28), 487–490 (2000)
5. Li, X., Guo, F., Kuang, H., Geng, Z., Fan, Y.: An extended cost potential field cellular automaton model for pedestrian evacuation considering the restriction of visual field. Physica A Stat. Mech. Appl. **515**, 47–56 (2019)
6. Loftus, E., Loftus, G., Messo, J.: Some facts about "weapon focus". Law Hum. Behav. **11**, 55 (1987)
7. Mackworth, N.H.: Visual noise causes tunnel vision. Psychon. Sci. 67–68 (1965). https://doi.org/10.3758/BF03343023
8. Mas, E., Suppasri, A., Imamura, F., Koshimura, S.: Agent-based simulation of the 2011 great East Japan earthquake/tsunami evacuation: an integrated model of tsunami inundation and evacuation. J. Nat. Dis. Sci. **34**(1), 41–57 (2012)
9. Meng, F., Zhang, W.: Way-finding during a fire emergency: an experimental study in a virtual environment. Ergonomics **57**, 816–827 (2014)
10. Moussad, M., Schinazi, V.R., Kapadia, M., Thrash, T.: Virtual sensing and virtual reality: how new technologies can boost research on crowd dynamics. Front. Robot. AI **5**, 1–14 (2018)
11. Moussaïd, M., Helbing, D., Theraulaz, G.: How simple rules determine pedestrian behavior and crowd disasters. Proc. Natl. Acad. Sci. **108**(17), 6884–6888 (2011)
12. Tsurushima, A.: Modeling herd behavior caused by evacuation decision making using response threshold. In: Davidsson, P., Verhagen, H. (eds.) MABS 2018. LNCS (LNAI), vol. 11463, pp. 138–152. Springer, Cham (2019). https://doi.org/10.1007/978-3-030-22270-3_11
13. Tsurushima, A.: Symmetry breaking in evacuation exit choice: impacts of cognitive bias and physical factor on evacuation decision. In: van den Herik, J., Rocha, A.P., Steels, L. (eds.) ICAART 2019. LNCS (LNAI), vol. 11978, pp. 293–316. Springer, Cham (2019). https://doi.org/10.1007/978-3-030-37494-5_15
14. Tsurushima., A.: Reproducing evacuation behaviors of evacuees during the great East Japan earthquake using the evacuation decision model with realistic settings. In: Proceedings of the 13th International Conference on Agents and Artificial Intelligence - Volume 1: ICAART, pp. 17–27. INSTICC, SciTePress (2021). https://doi.org/10.5220/0010167700170027
15. Tucker, A., March, K.L., Gifford, T.: The effects of information and hazard on evacuee behavior in virtual reality. Fire Saf. J. **99**, 1–11 (2018)
16. Wijerathne, M., Melgar, L., Hori, M., Ichimura, T., Tanaka, S.: HPC enhanced large urban area evacuation simulations with vision based autonomously navigating multi agents. Procedia Comput. Sci. **18**, 1515–1524 (2013). 2013 International Conference on Computational Science

17. Xu, Y., Huang, H.J.: Simulation of exit choosing in pedestrian evacuation with consideration of the direction visual field. Physica A Stat. Mech. Appl. **391**(4), 991–1000 (2012)
18. Yue, H., Guan, H., Shao, C., Liu, Y.: Simulation of pedestrian evacuation with affected visual field and absence of evacuation signs. In: 2010 Sixth International Conference on Natural Computation, vol. 8, pp. 4286–4290 (2010)

Author Index